Sustainable Resource Recovery and Zero Waste Approaches

Sustainable Resource Recovery and Zero Waste Approaches

Edited by

MOHAMMAD J. TAHERZADEH, PhD
Swedish Centre for Resource Recovery
University of Borås
Borås, Sweden

KIM BOLTON, PhD
Swedish Centre for Resource Recovery
University of Borås
Borås, Sweden

JONATHAN WONG, PhD
Institute of Bioresource and Agriculture
Hong Kong Baptist University
Kowloon Tong, Hong Kong, China

ASHOK PANDEY, PhD
Centre for Innovation and Translational Research
CSIR-Indian Institute of Toxicology Research
Lucknow, India

ELSEVIER

Sustainable Resource Recovery and Zero Waste Approaches ISBN: 978-0-444-64200-4

Publisher: Susan Dennis
Acquisition Editor: Marinakis, Kostas KI
Editorial Project Manager: Michael Lutz
Production Project Manager: Sreejith Viswanathan
Cover Designer: Miles Hitchen

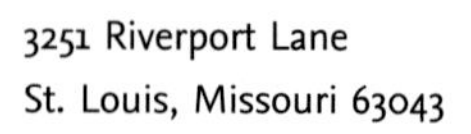
3251 Riverport Lane
St. Louis, Missouri 63043

List of Contributors

Haris Nalakath Abubackar, PhD
Chemical Engineering Laboratory
Faculty of Sciences and Center for Advanced Scientific Research (CICA)
University of La Coruña
La Coruña, Spain

Swarnima Agnihotri, PhD
Swedish Centre for Resource Recovery
University of Borås
Borås, Sweden

Teguh Ariyanto, PhD
Waste Refinery Center
Faculty of Engineering
Universitas Gadjah Mada
Yogyakarta, Indonesia

Mukesh Kumar Awasthi, PhD
College of Natural Resources and Environment
Northwest A&F University
Yangling, Shaanxi, China

Sanjeev Kumar Awasthi, PhD
College of Natural Resources and Environment
Northwest A&F University
Yangling, Shaanxi, China

Istna Nafi Azzahrani, STP
Department of Food and Agricultural Product Technology
Faculty of Agricultural Technology
Universitas Gadjah Mada
Yogyakarta, Indonesia

Kim Bolton, PhD
Swedish Centre for Resource Recovery
University of Borås
Borås, Sweden

Pedro Brancoli, MSc
Swedish Centre for Resource Recovery
University of Borås
Borås, Sweden

Rochim Bakti Cahyono, PhD
Waste Refinery Center
Faculty of Engineering
Universitas Gadjah Mada
Yogyakarta, Indonesia

Hongyu Chen, Master
College of Natural Resources and Environment
Northwest A&F University
Yangling, Shaanxi, China

Paul Chen, PhD
Center for Biorefining
Department of Bioproducts and Biosystems Engineering
University of Minnesota
St Paul, MN, United States

Yanling Cheng, PhD
Center for Biorefining
Department of Bioproducts and Biosystems Engineering
University of Minnesota
St Paul, MN, United States

Yunlei Cui, MSc
School of Energy Science and Engineering
Harbin Institute of Technology
Harbin, Heilongjiang, China

Kuan Ding, PhD
Center for Biorefining
Department of Bioproducts and Biosystems Engineering
University of Minnesota
St Paul, MN, United States

Yumin Duan, MSc
College of Natural Resources and Environment
Northwest A&F University
Yangling, Shaanxi, China

Panagiotis Evangelopoulos, PhD
Department of Material Science and Engineering
Royal Institute of Technology (KTH)
Stockholm, Sweden

Liangliang Fan, PhD
Center for Biorefining
Department of Bioproducts and Biosystems Engineering
University of Minnesota
St Paul, MN, United States

State Key Laboratory of Food Science and Technology
Nanchang University
Nanchang, Jiangxi, China

Jorge A. Ferreira, PhD
Swedish Centre for Resource Recovery
University of Borås
Borås, Sweden

V.K. Garg, PhD
Department of Environmental Science and Engineering
Guru Jambheshwar University of Science and Technology
Hisar, Haryana, India

Centre for Environmental Sciences and Technology
Central University of Punjab
Bathinda, Punjab, India

Efthymios Kantarelis, PhD
Assistant Professor
Department of Chemical Engineering
Royal Institute of Technology (KTH)
Stockholm, Sweden

Christian Kennes, PhD
Chemical Engineering Laboratory
Faculty of Sciences and Center for Advanced Scientific Research (CICA)
University of La Coruña
La Coruña, Spain

Samir Kumar Khanal, PhD
Department of Molecular Biosciences and Bioengineering
University of Hawaii at Manoa
Honolulu, HI, United States

Joakim Krook, PhD
Department of Management and Engineering
Environmental Technology and Management
Linköping University
Linköping, Sweden

Sumit Kumar, PhD
College of Natural Resources and Environment
Northwest A&F University
Yangling, Shaanxi, China

Bingxi Li, PhD
School of Energy Science and Engineering
Harbin Institute of Technology
Harbin, Heilongjiang, China

Richen Lin, PhD
Environmental Research Institute
School of Engineering
University College Cork
Cork, Ireland

Shiyu Liu, MSc
Center for Biorefining
Department of Bioproducts and Biosystems Engineering
University of Minnesota
St Paul, MN, United States

Tao Liu, PhD
College of Natural Resources and Environment
Northwest A&F University
Yangling, Shaanxi, China

Yuhuan Liu, PhD
State Key Laboratory of Food Science and Technology
Nanchang University
Nanchang, Jiangxi, China

Maria Cecília Loschiavo dos Santos, PhD
School of Architecture and Urbanism
University of São Paulo
São Paulo, Brazil

Stefan Luidold, PhD
Christian Doppler Laboratory for Extractive Metallurgy of Technological Metals/Nonferrous Metallurgy
Montanuniversitaet Leoben
Leoben, Austria

Ria Millati, PhD
Department of Food and Agricultural Product Technology
Faculty of Agricultural Technology
Universitas Gadjah Mada
Yogyakarta, Indonesia

Min Min, PhD
Center for Biorefining
Department of Bioproducts and Biosystems Engineering
University of Minnesota
St Paul, MN, United States

Jerry D. Murphy, PhD
Civil, Structural and Environmental Engineering
University College Cork
Cork, Ireland

Saoharit Nitayavardhana, PhD
Department of Environmental Engineering
Chiang Mai University
Chiang Mai, Thailand

Ashok Pandey, PhD
Centre for Innovation and Translational Research
CSIR-Indian Institute of Toxicology Research
Lucknow, India

Regina J. Patinvoh, PhD
Department of Chemical and Polymer Engineering
Faculty of Engineering
Lagos State University
Lagos, Nigeria

Peng Peng, PhD
Center for Biorefining
Department of Bioproducts and Biosystems Engineering
University of Minnesota
St Paul, MN, United States

Rininta Utami Putri, STP
Department of Food and Agricultural Product Technology
Faculty of Agricultural Technology
Universitas Gadjah Mada
Yogyakarta, Indonesia

Karthik Rajendran, PhD
Department of Environmental Science
SRM University – AP
Amaravati, Andhra Pradesh, India

Tobias Richards, PhD
Swedish Centre for Resource Recovery
University of Borås
Borås, Sweden

Kamran Rousta, PhD
Swedish Centre for Resource Recovery
University of Borås
Borås, Sweden

Roger Ruan, PhD
Nanchang University
Nanchang, Jiangxi, China

Center for Biorefining, and Department of Bioproducts and Biosystems Engineering
University of Minnesota
St. Paul, MN, United States

Kavita Sharma, PhD
Department of Environmental Science and Engineering
Guru Jambheshwar University of Science and Technology
Hisar, Haryana, India

Parimala Gnana Soundari, PhD
College of Natural Resources and Environment
Northwest A&F University
Yangling, Shaanxi, China

Niclas Svensson, PhD
Department of Management and Engineering
Environmental Technology and Management
Linköping University
Linköping, Sweden

Mohammad J. Taherzadeh, PhD
Swedish Centre for Resource Recovery
University of Borås
Borås, Sweden

Tjokorda Gde Tirta Nindhia, PhD
Department of Mechanical Engineering
Engineering Faculty
Udayana University
Bali, Indonesia

Karel Van Acker, PhD
Department of Materials Engineering
KU Leuven
Leuven, Belgium

Steven Van Passel, PhD
Department of Engineering Management
Faculty of Business and Economics
University of Antwerp
Antwerp, Belgium

Centre for Environmental Sciences
Hasselt University
Hasselt, Belgium

María C. Veiga, PhD
Chemical Engineering Laboratory
Faculty of Sciences and Center for Advanced Scientific Research (CICA)
University of La Coruña
La Coruña, Spain

Teresa Villac, PhD
Office of the Attorney General (AGU)
University of São Paulo
São Paulo, Brazil

David M. Wall, PhD
School of Engineering
University College Cork
Cork, Ireland

Yiqin Wan, PhD
State Key Laboratory of Food Science and Technology
Nanchang University
Nanchang, Jiangxi, China

Quan Wang, PhD
Institute of Bioresource and Agriculture and Department of Biology
Hong Kong Baptist University
Kowloon Tong, Hong Kong

College of Natural Resources and Environment
Northwest A&F University
Yangling, Shaanxi, China

Yunpu Wang, PhD
State Key Laboratory of Food Science and Technology
Nanchang University
Nanchang, Jiangxi, China

Jonathan W.C. Wong, PhD
Institute of Bioresource and Agriculture
Hong Kong Baptist University
Kowloon Tong, Hong Kong

Weihong Yang, PhD
Department of Material Science and Engineering
Royal Institute of Technology (KTH)
Stockholm, Sweden

Yaning Zhang, PhD
School of Energy Science and Engineering
Harbin Institute of Technology
Harbin, Heilongjiang, China

Zengqiang Zhang, PhD
College of Natural Resources and Environment
Northwest A&F University
Xianyang, Shaanxi, China

Junchao Zhao, Master
College of Natural Resources and Environment
Northwest A&F University
Yangling, Shaanxi, China

Nan Zhou, MSc
Center for Biorefining
Department of Bioproducts and Biosystems Engineering
University of Minnesota
St Paul, MN, United States

Preface

Since industrial revolution, there has been constant improvement in the lifestyle of humans with improved hygienic conditions and better healthcare. The world population has increased to more than 7 billion from 1 billion in just two centuries. The global economy has also developed linearly, where the resources consumption has been increased dramatically, due to which greenhouse gases, solid wastes, and wastewater are also generated in huge quantities. One example is the CO_2 level in the atmosphere that fluctuated between c. 200 and 300 ppm in million years, but just in last 50 years, passed over 400 ppm. This situation is not sustainable and needs urgent attention to create balance in nature, where we should replace faster natural process to treat the wastes (solid, liquid, and gases) and develop new resources for human need. These considerations led to the establishment of 17 Sustainable Development Goals by the United Nations, which have also been linked to circular economy as a popular concept. This clearly has led to the concept of resource recovery from the waste for a sustainable society. In other words, there is no waste but just resources, unless our knowledge would not be enough to utilize it.

This book is dedicated to the latest development of resource recovery of solid wastes and residuals. Treatment and management of solid wastes, especially municipal solid waste, is one of the most difficult and challenging tasks. While some countries still dump them in open space, or adopt landfilling for their disposal, some others have developed methods to treat such wastes to produce new materials and energy. This book makes an overview of the amount of wastes and residuals produced in the world, gives an example of how zero wastes has been achieved in Sweden compared to other waste management practices, and lists which factors affect sustainable waste management. The book then considers several aspects of these developments, including soft aspects such as laws and regulations and life-cycle assessments. Physical sorting of wastes is the first step in the treatment that is done at source, e.g., by people at household or by machines. The treatment of organic wastes is done by composting, vermicomposting, or anaerobic digestion, including the recent methods of dry digestion. The recycling of wastes, metals, and electronic wastes has particularly been considered. Combustion or incineration is common to burn the wastes and extract energy, but while other processes could be of interest, pyrolysis and gasification are often employed. Thermal processes can be used to produce syngas and then use fermentation to produce a variety of biochemical products from syngas. All these aspects are discussed in various chapters. The book has a final chapter on how to extract materials from older landfills using landfill mining.

Editors greatly appreciate the efforts made by the authors in compiling the relevant information on different aspects of solid waste treatment and management for resource recovery and circular economy, which we believe will be very useful to the scientific community. We gratefully acknowledge the reviewers for their valuable comments, which substantially improved the scientific content of this volume. We thank Elsevier team comprising Dr. Kostas Marinakis, Senior Book Acquisition Editor, and Michael Lutz, Editorial Project Manager, and the entire Elsevier production team for their consistent hard work in the publication of this book.

Editors
Mohammad Taherzadeh
Kim Bolton; Jonathan Wong
Ashok Pandey

Contents

CHAPTER 1

Agricultural, Industrial, Municipal, and Forest Wastes: An Overview

RIA MILLATI, PHD • ROCHIM BAKTI CAHYONO, PHD • TEGUH ARIYANTO, PHD • ISTNA NAFI AZZAHRANI, STP • RININTA UTAMI PUTRI, STP • MOHAMMAD J. TAHERZADEH, PHD

INTRODUCTION

According to the EU Waste Directive [1], waste is any substance or object that the holder discards or intends to discard or is required to discard. However, in the concept of sustainable development, waste is considered as a resource that is useful in the production of various valuable products. By doing so, the use of raw materials in manufacturing processes is reduced. The environmental impact caused by waste accumulation would likewise be minimized. In this chapter, different groups of wastes from the agricultural, forest, municipal, and industrial sectors are discussed. The discussion covers the quantity and quality of both waste characteristics and waste treatment and the potential utilization of wastes for producing, e.g., bioenergy, biomaterials, and biochemicals.

AGRICULTURAL WASTE

Sugarcane, corn, rice, and wheat are the top four of the world's largest produced crops by quantity [2]. Together, rice, corn, and wheat supply more than 42% of all calories consumed by the whole human population [3], whereas palm oil produced from the oil palm tree is one of the most consumed and produced oils in the world. Cultivation and manufacturing processes, which use these crops as raw material, generate solid wastes such as rice straw, rice husk, corncob, wheat straw, and oil palm empty fruit bunch (OPEFB). As the crops are considered to be the world's most produced and consumed agricultural products, accordingly, the associated solid wastes are available in plentiful amount. It is not only the amount of these solid wastes that creates significant concern but also their characteristic as a lignocellulose, which poses a challenge in treating or utilizing them. Hence, it would be beneficial to be aware of the developing technologies, as their progress creates a niche in science related to this specific material as well as lignocellulosic materials in general. Based on these facts, the following six types of wastes were chosen to represent solid wastes in the agricultural sector.

Corncob

Corncob is a byproduct of corn, which is the central core of an ear of maize (*Zea mays* sp.), where the kernels grow. Corncob can be separated from corn kernels manually or by using a machine during the corn-based manufacturing process. A total of 1 kg of ear corn yields 0.22 kg of corncob [4]. Pursuant to the United States Department of Agriculture (USDA) report, the global production of corn in 2017 was 1061 million tons and, accordingly, the total corncob generation is estimated to be around 230 million tons [5]. The top three producers of corn in the world are the United States, China, and Brazil. Corn is a major crop in the United States, where it is mostly utilized as feedstock for ethanol, distillers' grain, and livestock feed, e.g., beef cattle, dairy, hogs, and poultry [5]. According to the same report [5], the total corn production in the United States was 370 million tons. Based on this amount and using the share ratio of corncob [4], the amount of corncob generated in the United States would be around 81 million tons (Table 1.1). China and Brazil produced 225 and 94.5 million tons of corn, respectively [5], making their respective corncob production about 49 and 20 million tons (Table 1.1).

Corncob is a lignocellulosic material that is high in cellulose content, with up to 69.2%. Other lignocellulosic components, i.e., hemicellulose and lignin, comprise up to 22.8% and 8%, respectively [6] (Table 1.2). Corncob is rich in carbon compounds [6]

Sustainable Resource Recovery and Zero Waste Approaches. https://doi.org/10.1016/B978-0-444-64200-4.00001-3

TABLE 1.1
Estimated Production of Agricultural Waste.

Types	Production (Million Tons)	Country of Origin	References
Corncob[a]	81	USA	[5]
	49	China	[4]
	20	Brazil	
	13	EU	
	9	Argentina	
Oil palm empty fruit bunch[a]	37	Indonesia	[121]
	19	Malaysia	[21]
	2	Thailand	
	1	Colombia	
	0.9	Nigeria	
Rice husk[a]	39	China	[33]
	30	India	[32]
	10	Indonesia	
	9	Bangladesh	
	8	Vietnam	
Rice straw[a]	149	China	[33]
	114	India	[32]
	39	Indonesia	
	36	Bangladesh	
	30	Vietnam	
Sugarcane bagasse[a]	94	Brazil	[57]
	93	India	[56]
	55	EU	
	38	Thailand	
	29	China	
Wheat straw[b]	128	EU	[68]
	110	China	[43]
	40	USA	
	25	Canada	
	22	Pakistan	

[a] In the year 2017.
[b] In the crop year 2017/2018.

as well as in silica content (79.95%) [7] (Table 1.2). It is frequently milled into small sizes and further used as animal feed. In the past, corncob had been used in direct combustion as a fuel for cooking and heating [8]. Studies have been conducted to investigate if corncob has more value, with results showing that corncob can be used as feedstock for the production of bioethanol [9,10], xylooligosaccharides [11], biomass-degrading enzymes, antioxidants, and fermentable sugars [12], xylitol [13], adsorbents [14], and lactic acid [15]. In an effort toward commercialization, pilot plants of corncob-based feedstock have been set up. An example is the pilot plant production of xylooligosaccharides in China, with a final product quantity of 10 L from every 40 kg of corncob [16]; xylose in China with a yield of 3.575 kg xylose from every 22 kg of corncob [17]; and dimethyl ether in China with a capacity of 45–50 kg corncob/h [18]. In China, corncob is used as feedstock for furfural production by Hebei Furan International Co., Ltd. [19]. Corncob is also used as one of the raw materials by Praj Industries in India to produce ethanol, with a capacity of 1 million L/year [20].

Oil Palm Empty Fruit Bunch

OPEFB is a biomass generated in the palm oil industry. In the crude palm oil (CPO) mill, the oil is extracted from the fruit pulp, leaving OPEFB, fibers, and shell from the kernel as solid wastes. A total of 1 kg of fresh fruit bunch produces 0.234 kg CPO, 0.123 kg fibers, 0.071 kg shell, and 0.217 kg OPEFB [21]. In some cases, fibers and shell are utilized as fuels for mill boilers to produce steam and electricity, whereas OPEFB is used

TABLE 1.2
Chemical Composition of Agricultural Waste.

Types	Cellulose (%)	Hemicellulose (%)	Lignin (%)	Organic Compound (%)	Inorganic Compound (%)	References
Corncob	69.2	22.8	8.0	C 42.0 H 6.7 O 48.1 N 1.5	SiO_2 79.9 CaO 1.2 K_2O 1.5 Fe_2O_3 3.9 Al_2O_3 5.2	[6] [7]
Oil palm empty fruit bunch	39.1	23.0	34.4	C 42.8 N 1.5 P 0.5 K 7.3 Mg 0.9	Silica 1.8	[23] [122] [24]
Rice husk	43.3	28.6	22.0	C 39.8 H 5.7 O 0.5 N 37.4	SiO_2 99.50 MgO 0.02 Al_2O_3 0.17 P_2O_5 0.11 SO_3 0.02	[34,35] [36]
Rice straw	36.0	24.0	15.6	C 48.7 H 5.9 O 44.2 N 1.1 S 0.1 Cl 0.5	SiO_2 69.0 K_2O 8.3 Na_2O 3.4 CaO 1.8 MgO 1.7	[43] [44]
Sugarcane bagasse	46.4	23.9	28.1	C 44.1 H 5.7 O 47.7 N 0.2 S 2.3	SiO_2 3.0	[58,123]
Wheat straw	37.8	26.5	17.5	C 49.0 H 5.9 O 43.7 N 0.8 S 0.2 Cl 0.5	SiO_2 46.32 K_2O 25.12 Na_2O 3.12 MgO 2.56 CaO 1.96	[43] [44]

as fertilizer in plantation. Being the highest share among the other solid wastes, it would be more beneficial if OPEFB were optimally utilized in order to increase the economic benefit and the sustainability of the palm oil industry itself. Indonesia and Malaysia dominate the world palm oil production with 56% and 29%, respectively, of the 73 million tons of annual global production in 2017 [22]. This means that Indonesia and Malaysia produced 40.5 and 21 million tons of CPO, respectively. Using the mass balance of the CPO mills [21], the global OPEFB accumulation would be approximately 67 million tons. Indonesia and Malaysia generated around 37 and 19 million tons of OPEFB, respectively (Table 1.1).

OPEFB is a lignocellulosic material that is composed of 39.13% cellulose, 23.04% hemicellulose, and 34.37% lignin [23]. In addition to cellulose, hemicellulose, and lignin, OPEFB is rich in inorganic elements such as silica and metal ions, e.g., copper, calcium, manganese, iron, and sodium [24]. Previously, the practice of treating OPEFB was mainly incineration. However, this causes air pollution around the palm oil mill areas. Other alternatives to process OPEFB are to turn it into mulch at the oil plantation or process it into compost.

Being a natural fiber and plentiful, OPEFB has attracted interest in being used to make a composite [25]. Other attempts for utilizing OPEFB include production of linerboard coating [26], fermentable sugars [27], polyhydroxybutyrate [28], biogas [29], and ethanol [23,30]. In Indonesia, an ethanol pilot plant is run to produce 144.4 kg anhydrous ethanol from 1000 kg OPEFB [31].

Rice Husk

There are two major types of residues from rice cultivation, i.e., rice straw and rice husk. Rice husk is also commonly called rice hulls. Rice husk is the coating on a seed or a grain of rice. It is formed from hard materials, including silica and lignin, to protect the seed during the growing season. Each kilogram of milled white rice results in roughly 0.28 kg of rice husk as a byproduct of rice production during milling [32]. The global production of milled rice was 488 million tons in 2017 [33]. Using the ratio aforementioned, the total global rice husk generated was approximately 136 million tons. China, India, and Indonesia are the three largest producers of rice, with 142.20,109, and 37.30 million tons, respectively [33]. Therefore the corresponding amounts of rice husk produced by these countries were approximately 39,30, and 10 million tons (Table 1.1).

Rice husk contains 43.30% cellulose, 28.6% hemicellulose, and 22% lignin [34,35]. Also, it is known to have a very high silica content, which is about 99.5% over other inorganic compounds [36]. Dry rice husk in its loose form is traditionally used as an energy source for households in rural areas or home industries. Direct burning of rice husk in a furnace for drying paddy is also a common practice for farmers. There are more innovative ways to use rice husk as a solid fuel, e.g., by using it in the form of husk charcoal briquettes or husk charcoal, which is used in industrial boilers to replace fossil fuel. Husk charcoal is produced by thermal decomposition of the rice husk under a limited supply of oxygen (O_2) and at a high temperature. Other than being used as an energy source, husk charcoal can also be used as an activated carbon. The ash of rice husk is also useful as a fertilizer. The ash has various types of chemical elements that are good for soil fertilization (Table 1.2). As the content of silica in rice husk is quite high, rice husk is a good raw material to produce mesoporous silica [37], anticoagulants [38], and nanocrystalline materials [39]. Furthermore, several pilot plants have tested using rice husk as their feedstock, i.e., ceiling board production in Nigeria [40]; bio-oil production in China, with a yield reaching 53.2 wt% [41]; and production of producer gas, as a renewable energy carrier, in India, with a gas yield of 2.7 m^3/kg [42].

Rice Straw

The other major residue from paddy cultivation is rice straw. Rice straw is a byproduct of paddy cultivation, produced during the harvesting. Rice straw is separated from the grain after the plants are threshed, either manually or by using a machine. It includes stem, leaves, and spikelets. Each kilogram of milled rice gives approximately 1.05 kg of rice straw [32]. With the total global milled rice production in 2017, i.e., 488 million tons [33], the total rice straw accumulated would be about 510 million tons. As China, India, and Indonesia are the top three rice producers in the world, the three countries also generated the highest amount of rice straw. It can be seen in Table 1.1 that the amount of rice straw in China, India, and Indonesia are calculated to be around 149, 114, and 39 million tons, respectively [32,33].

Rice straw is a lignocellulosic material, mainly composed of cellulose (36%), hemicellulose (24%), and lignin (15.6%) [43]. Rice straw is rich in carbon content but poor in nitrogen source (Table 1.2). Furthermore, it has high ash content. The ash is high in silica content (SiO_2 is 69.02%) [44] and low in alkali content [45]. Burning rice straw in the field is still practiced in different parts of the world. Traditionally, rice straw has also been used as animal feed [45]. Other than traditional usages, some studies showed that rice straw can be utilized as feedstock for the production of some value-added products such as anticoagulants [38], food-grade glucose [46], laccase enzyme [47], biogasification [48], biogas [49,50], and ethanol [51]. On a pilot scale, a rice straw biogas plant in China is operated with a digester capacity of 300 m^3 [52]. There is also a pilot plant of ethanol production in Taiwan with a fermenter capacity of 100 L [53]. Moreover, a pilot scale to recover sugar from rice straw is also tested in India, with a capacity of 250 kg biomass/day [54]. In commercial production, an Italian company (Beta Renewables) declared using rice straw as one of its raw materials to produce ethanol [55].

Sugarcane Bagasse

Sugarcane bagasse is the fibrous residue remaining after the sugarcane stalk has been crushed and the juice removed. Every 100 tons of sugarcane results in 10 tons of cane sugar and 25–30 tons bagasse [56]. Based on the USDA report [57], the total global production of cane sugar in 2017 was 1882 million tons. Using this proportion of sugarcane bagasse, the global amount of sugarcane bagasse produced is calculated to be about 510 million tons. Brazil is the first largest producer of cane sugar, with a total production of 342 million tons in 2018 [57]. India is the second largest cane sugar producer, with a total production of 338.30 million tons in the same year. Accordingly, the amount of sugarcane bagasse generated in the two countries is estimated to be around 94 and 93 million tons, respectively (Table 1.1).

Sugarcane bagasse is often collected after the milling process to be fed into a boiler for electricity generation. The electricity is used as energy supply in the mills. Because of its fibrous nature, sugarcane bagasse has been most widely used as a fuel, in paper and pulp industries, in structural material manufacture, and in agriculture. Analysis of sugarcane bagasse indicates that its main constituents are cellulose, 46.42%; hemicellulose, 23.97%; and lignin, 28.09% [58]. Sugarcane bagasse is also high in carbon content (Table 1.2). The composition of sugarcane bagasse makes it an ideal ingredient to be applied and utilized as a reinforcement fiber in composite materials [59]. Research shows that sugarcane bagasse can be used as a substrate for the production of biodiesel [60], cellulose acetates [61], cement composites [62], and ethanol [63]. In Brazil, a pilot-scale production of ethanol from sugarcane bagasse is available, with a capacity of 83.03 m^3 ethanol/h [64]. A pilot-scale production of hemicellulosic sugars is also studied to increase the production of the sugar yield from sugarcane bagasse using a 65-L steam gun reactor [65]. In commercial production, sugarcane bagasse, together with sugarcane straw, is used by GranBio [66] and Raízen [67] in Brazil for ethanol production, with a capacity of 82 and 42 million L/year, respectively.

Wheat Straw

The main agricultural residue associated with wheat is wheat straw. Similar to rice straw, wheat straw is collected after wheat grain harvesting. It includes major parts of the stem, leaves, and spikelets. The weight ratio of wheat straw over wheat is 0.85 kg/kg [43]. The total global production of wheat grain in the crop year 2017/2018 was 758.0 million tons. Taking into account the proportion of wheat straw, the amount of wheat straw accumulated globally was about 640 million tons. The largest wheat grain producers are the European Union, China, and the United States, with their production in the crop year 2017/2018 reaching 151.7, 129.8, and 47.4 million tons, respectively [68]. Their respective wheat straw accumulation was approximately 128, 110, and 40 million tons (Table 1.1).

Wheat straw can be ploughed into the field or used as mulch covering the topsoil. Wheat straw can also be collected in bales using baling machines for off-field utilization. Nevertheless, the traditional practice, i.e., open field burning, is still done in some regions [69]. Wheat straw represents a valuable source of cellulose, hemicellulose, and lignin. Wheat straw contains 37.8% cellulose, 26.5% hemicellulose, and 17.5% lignin [43] (Table 1.2). Wheat straw is rich in carbon components and other organic and inorganic compounds [44]. Studies show that wheat straw can be used for the production of ethanol [70,71] and fermentable sugars [72]. In the United States, an ethanol pilot plant for wheat straw with a reactor capacity of 20 L was set up to study the feasibility of the process before commercialization [73]. Other examples of pilot-scale production plants using wheat straw as the raw material are the coproduction of bioethanol (from sugars) and electricity (from lignin) in Denmark with a capacity of 120–150 kg straw/h [74] and bio-oil production with a capacity of 25 kg straw/h in China [75]. Together with corncob, wheat straw is used as one of the raw materials by Praj Industries to produce ethanol [20].

FOREST WASTE

Hardwoods and softwoods are the two major wood types. Hardwood belongs to deciduous trees, a tree that loses its leaves during the autumn season. Hardwood includes oak, maple, hickory, and birch. Softwood belongs to coniferous trees, an evergreen tree. Softwood includes pine, spruce, fir, and juniper. Hardwood lumbers are mostly produced in regions such as East Asia, Oceania, America, South and Central Asia, and Europe (Table 1.3). In 2013, the global hardwood lumber production reached 117.5 million m^3 [76]. East Asia and Oceania countries are the first largest hardwood producers, with the global share production of 48%, or approximately 56.8 million m^3. American countries, i.e., North America and Latin America, are the second largest producers, with the global share of 17% (20 million m^3) and 10% (11.6 million m^3), respectively [76]. In 2016 the global softwood lumber production was 333.4 million m^3 [77] (Table 1.3). Softwood lumbers are mostly produced in Europe, in an amount of 142 million m^3. The second largest softwood producers are North American countries, with a total quantity of 103.8 million m^3. The other producers of softwoods are countries in Asia (53.4 million m^3), South America (20 million m^3), Oceania (8.7 million m^3), Africa (2.9 million m^3), and Central America (2.6 million m^3).

TABLE 1.3
Estimated Quantity of Hardwood and Softwood Residues.

Type	Quantity Global Statistic (Million m^3)	Waste Quantity (Approximately) (Million m^3)	Country of Origin	References
Hardwoods[a]	56.8	28	East Asia and Oceania	[76]
	20.0	10	North America	
	11.6	6	Latin America	
	11.5	6	Russia, South and Central Asia, and Middle East	
	10.9	5	Europe	
	6.7	3	Africa	
Total	117.5	58		
Softwoods[b]	142.0	71	Europe	[77]
	103.8	52	North America	
	53.4	27	Asia	
	20.0	10	South America	
	8.7	4	Oceania	
	2.9	1	Africa	
	2.6	1	Central America	
Total	333.4	166		

[a] In the year 2013.
[b] In the year 2016.

Processing wood into timber or other valuable wood products results in wood residues or wood wastes as byproducts. Approximately 50% of wood is turned into valuable products, and the rest becomes waste [78]. Considering this ratio and the global wood production, the forest waste from hardwood and softwood would be approximately 58 and 166 million m^3, respectively (Table 1.3). Wood wastes generated from primary manufacturing processes include bark, slabs, sawdust, chips, coarse residues, planer shavings, peeler log cores, and end trimmings. Wood wastes generated from secondary manufacturing processes include chips, sawdust, sander dust, end trimmings, used or scrapped pallets, coarse residues, and planer shavings [78]. Bark comprises 8%–12% of the total percentage in woods; sawdust, 11%–15%; and chippable wastes (slabs or edgings), 30%–40% [78].

The chemical composition of wood consists of structural and nonstructural substances. The structural components are cellulose, hemicellulose, and lignin, while the nonstructural components are extractives, water-soluble organics, and inorganics [79]. Structural substances constitute the major part of the chemical composition of wood. Generally, wood contains cellulose (40%–45%), hemicellulose (20%–30%), and lignin (20%–32%) [80]. Cellulose is composed of β-D-glucose units, which are linked together with 1,4-glycosidic bonds to form long linear chains. Hemicellulose is composed of short, highly branched copolymers of glucose, mannose, galactose, xylose, and arabinose. Lignin is an aromatic polymer synthesized from phenylpropane units [80].

Each part of the tree has its own lignocellulosic composition (Table 1.4). For example, the cellulose content in bark varies in the range of 10%–25%. The cellulose content in the bark of birch, aspen, pine, and spruce is 10.7%, 25.4%, 25.4%, and 19%, respectively [81–84]. Specifically in its bark, birch has a hemicellulose content of 11.2%, aspen has 23.4%, pine has 14.7%, and spruce has 11% [81,82,85,86]. The bark of birch has a lignin content of 27.9%, aspen has 22.6%, pine has 31.15%, and spruce has 22.6% [82,87,88]. The carbohydrate composition in hardwoods and softwoods differs in every species and wood parts. For example, hardwoods such as birch and aspen mainly consist of glucan and xylan [88,89], whereas softwoods such as pine and spruce have more glucan composition and relatively less xylan than hardwoods [88]. Table 1.5 shows the details of carbohydrate composition in wood residues. The organic elements found in woods are mostly C, H, O, N, and P. Additionally, hardwoods and softwoods also consist of inorganic materials such as Ca, Mg, Na, K, P, Al, Si, Zn, Cu, and others.

Conventionally, wood wastes were mainly used for combustion for cooking or left in the forest to maintain

TABLE 1.4
Chemical Composition of Hardwoods and Softwoods.

Commodity	Cellulose (%)	Hemicellulose (%)	Lignin (%)	Ash (%)	References
HARDWOODS					
Birch (bark)	10.7	11.2	27.9	2.9	[81] [87]
Aspen (bark)	25.4	23.4	22.6	0.4	[82] [124]
SOFTWOODS					
Pine (bark)	25.4	14.7	27.6	3.3	[83] [88]
Pine (chips)	32.1	14.2	31.2	2.4	[85]
Spruce (bark)	19.0	11.0	22.6	3.9	[84] [86] [88]

TABLE 1.5
Carbohydrate Composition in Hardwood and Softwood Residues.

Commodity	Glucan (%)	Mannan (%)	Galactan (%)	Xylan (%)	Ara (%)	References
HARDWOODS						
Birch (bark)	13.3	0.5	0.9	7.5	10.3% (in total of monosaccharides)	[88] [87]
Aspen	44.0	2.4	0.9	19.0	0.5	[89]
SOFTWOODS						
Pine (bark)	19.8	3.1	1.8	3.4	1.5	[88] [125]
Pine (chips)	33.8	10.9	5.6	8.5	2.6	[126]
Spruce (bark)	16.8	1.5	2.1	2.1	1.2	[88] [125]

a nutritional balance in the soil. As wood waste or forest waste is an abundant source of cellulose, hemicellulose, and lignin, forest waste is a potential source that can be utilized for many beneficial products. In the biorefinery concept, biomass can be transformed into a sustainable feedstock for fuels, chemicals, and materials that are currently produced from petroleum [90]. For example, lignocellulosic waste biomass can be an inexpensive alternative substrate for fuel ethanol production [91]. The isolated cellulose in nanocrystal forms, including those in the forest residues, can be utilized into reinforcing agents in polymer matrices [92], barrier films [93], flexible displays [94], drug delivery excipients [95], security paper [96], and templates for electronic components [97]. Hemicellulose can be produced into plant gum for thickeners, adhesives, protective colloids, emulsifiers, and stabilizers [98]. Additionally, hemicellulose can be utilized as a biodegradable oxygen barrier film [99,100]. High-quality lignin can be utilized as a substitute for polymeric materials such as phenolic powder resins, which can be used as a binder in friction products, automotive brake pads, molding, polyurethane and polyisocyanurate foams, and epoxy resins, which can be used as printed circuit board resins [101]. Lignin also has the potential of being converted into carbon fiber and being used as a precursor for dimethyl sulfoxide, vanilla, phenol, and ethylene [101–103]. Lignin is applicable in the agricultural sector, e.g., as a biodegradable ultraviolet-light antioxidant absorbent, slow-release fertilizer, and soil conditioner [104]. Lignin is sulfonated to water-soluble lignosulfonates in sulfite pulping processes and can be used in active packaging to protect against oxidative damage [105]. Pilot-scale production using forest residues as substrate has been studied. In Sweden, a pilot plant for ethanol production from softwood residues was inaugurated in 2004. The capacity is 1 ton dry biomass/24 h, if it is run in continuous operation [106]. In the United States, a pilot plant producing ethanol from the Douglas-fir forest harvest residues has been set up, with a capacity of 50 kg/batch [107]. A pilot-scale production of cellulose nanocrystal (CNC) (integrated with bioethanol pilot plant in Sweden) with a yield of 600 g/day CNC has also been studied [108]. In 2013, the Indian River BioEnergy Center in the United States began producing cellulosic ethanol at commercial volumes. Together with other cellulosic plants, which were under construction at that time, the plants are designed to produce approximately 80 million gallons annually [109]. Borregaard, a company in Norway, is producing value-added products from different components in wood, for example, ethanol, lignin products, and vanillin, in order to replace oil-based products [110].

MUNICIPAL SOLID WASTE

Waste Generation

Municipal solid waste (MSW) includes waste from households, commercial and trade activities, institutional/office activities, construction and demolition, medical waste, and municipal waste. Table 1.6 shows the waste generator and the type of solid waste from each source in detail.

Generally, MSW is collected and treated by city governments as well as private companies by using some technologies that depend on the onsite conditions. Based on Table 1.6, many types of materials are available within the MSW that require different treatment methods and give different products during final processing. The final types of MSW could also be classified as recyclables (paper, plastic, glass, metals, etc.), toxic substances (paints, pesticides, used batteries, medicines), compostable organic matter (fruit and vegetable peels, food waste), and soiled waste (blood stained cotton, sanitary napkins, disposable syringes). Many factors play important roles in MSW generation, such as population, level of income, consumption rate, and location. Among these, population and level of income are the two most significant factors contributing to the quantity of MSW.

The current world MSW generation is approximately 1.7 billion tons per year and will continue to increase following the world population growth [111]. Table 1.7

TABLE 1.6
Waste Generator and Type of Solid Waste for Each Source of Municipal Solid Waste.

Source	Waste Generator	Type of Solid Waste
Residential/household	Single and multifamily lodging	Food waste, paper, cardboard, plastics, textiles, leather, yard waste, wood, glass, metals, electronic waste, etc.
Commercial and trade	Stores, hotels, restaurants, markets, office buildings	Paper, cardboard, plastics, wood, food waste, glass, metals, electronic waste, etc.
Institutional/office	Schools, hospitals, prisons, government buildings, airports, train station, etc.	Paper, cardboard, plastics, wood, food waste, glass, metals, electronic waste, etc.
Construction and demolition	New construction sites, road repair, renovation sites, demolition of buildings	Wood, steel, concrete, dirt, bricks, tiles
Medical waste	Hospitals, nursing homes, clinics	Infectious wastes, hazardous wastes radioactive waste from cancer therapies, pharmaceutical waste
Municipal services	Street cleaning, landscaping, parks, recreational areas, water and wastewater treatment plants	Street sweepings, landscape and tree trimmings, general wastes from parks, beaches, and other recreational areas

TABLE 1.7
Estimated Total World MSW Generation in Each Region [111].

Region	Countries Included	Urban Population (Million)	MSW Generation (kg/Capita/day)	Total MSW Generation (Million Tons/year)	Global Contribution (%)
Asia	39	1475	1.4	743	44
America	35	663	2.3	544	32
Africa	41	332	0.8	95	6
Europe	35	493	1.7	298	18
Australia and Oceania	4	20	2.8	21	1
Total world MSW				1701	

MSW, municipal solid waste.

TABLE 1.8
Estimated MSW Generation in the Top 10 Most Populous Countries.

Country	Income Level	Urban Population (Million)	MSW Generation (kg/Capita/day)	Total MSW Generation (Million Tons/year)	References
China	Upper middle	511	1.0	190	[127]
India	Lower middle	321	0.3	40	[128]
USA	High	242	2.6	228	[129]
Brazil	Upper middle	144	1.0	54	[111]
Indonesia	Lower middle	117	0.5	22	[130]
Russia	Upper middle	107	0.9	36	[111]
Japan	High	84	1.7	53	[129]
Mexico	Upper middle	80	1.2	36	[129]
Nigeria	Low	73	0.6	15	[131]
Germany	High	61	2.1	47	[129]

MSW, municipal solid waste.

represents the estimation of the world MSW generation for each region, and Asia contributes 44% to the global MSW. The MSW generation per capita ranges from 0.78 to 2.8 kg/capita/day, which depends on the economic development, degree of industrialization, public habits, and local climate. Urbanization is a common problem in most countries; people want to move from rural areas to the city to find jobs and for lifestyle. As waste generation is much higher in cities/urban areas than that in rural areas, urbanization would lead to higher waste volumes. Compared with rural areas, urban residents also produce a higher fraction of inorganic wastes (e.g., plastics and aluminum) than organic wastes (food waste). Owing to the high work demands, buying food in stores instead of cooking at home becomes a habit, thus decreasing organic waste.

In addition to population, the main factor for MSW generation is income level/gross domestic product (GDP) in each country. Table 1.8 represents the estimation of MSW generation in the top 10 most populous countries. A high income level would promote better prosperity and wealth, which transfers into larger waste generation per capita. The United States and Germany, as two well-developed countries, have the highest MSW generation per capita, above 2 kg/capita/day. In addition to lifestyle, the societies fulfill not only the basic needs but also additional needs such as cars, housing facilities, clothes, various foods, and entertainment activities.

As low-income countries, India, Nigeria, and Indonesia own lower MSW generation per capita, which is below 1 kg/capita/day. Compared with well-developed countries, these societies only have the ability to meet their basic needs such as housing, limited food, and clothing so the MSW generation is quite limited. In order to have a solid perspective on this issue, an analysis was conducted on similar societies with increasing GDP level so other factors would remain constant, i.e., culture, location, and climate. In the period from 1960 until 1980, when the per capita GDP for the United States rose from US $3000 to US $23,000, the MSW generation per capita increased from 1.3 to 1.8 kg/capita/day [112].

Fig. 1.1 presents the MSW generation of the top 10 countries in the world. These countries contribute to almost a half of the global MSW generation, with varying income levels, cultural values, climate, and lifestyle. Thus data analysis from these countries could represent global MSW generation more precisely. Therefore China as the first most populous country contributes to around 11% of the global MSW generation, which is less than that of the United States that contributes 14%. India, as the second most populous country, also contributes less than Germany.

Current Disposal Treatment

There are several steps in the hierarchy of solid waste management, including reduce, reuse, recycle, recover, and disposal. It is quite challenging to get global data related to reduce and reuse of MSW because of the close system of its utilization. Therefore only recycle, recovery for energy and material, and the disposal method will be discussed in this section. Open dumping means to put MSW in a specific area without any treatment. In most countries, this method is an illegal waste disposal practice and should not be confused with a permitted municipal solid waste landfill or a recycling facility. Landfilling is the disposal method of MSW, with various different layers and finally with earth covering the specific location, which is designed for safe disposal but limited benefits. Landfills produce landfill gas containing CH_4 and liquid leachate, which contain water-soluble materials. The bottom-liners and top earth cover are considered as the most critical components to prevent the negative impacts of liquid leachate and landfill gas. When the landfill system fails, liquid leachate causes land and water pollution, while landfill gas leads to high quantity of greenhouse gasses and climate change. Recycling is one of the most sustainable ways to handle waste by converting waste into valuable products without changing the composition. Typical wastes that are suitable for this method are metal, glass, paper, and plastics. When facing difficulties during recycling, MSW could be recovered as compost material and energy sources by both bioprocess and thermochemical process.

Fig. 1.2 shows the treatment and disposal processes of MSW in the world and selected countries with different levels of income/GDP. Landfilling of waste is still the most common method of MSW disposal in the world, i.e., around 43%, whereas recycling and compositing are the second and third options, respectively, for MSW treatment. Open dumping still exists

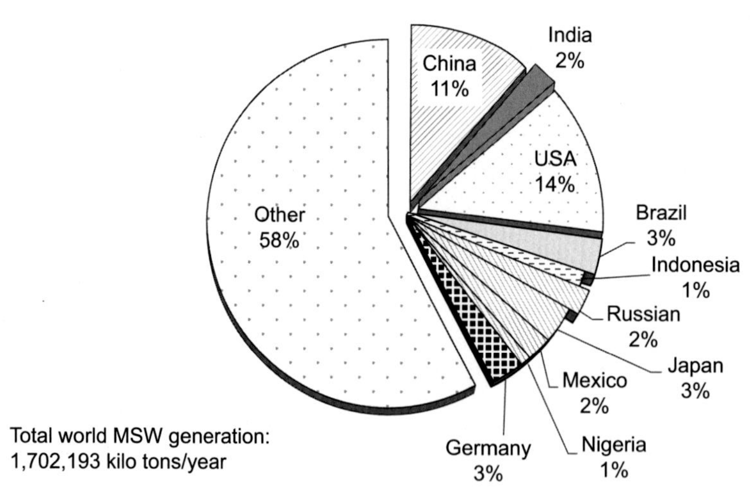

FIG. 1.1 Distribution of municipal solid waste (MSW) generation in the world.

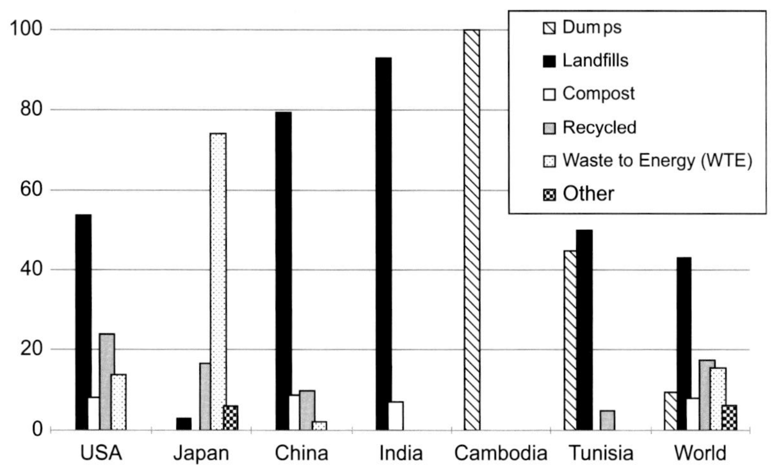

FIG. 1.2 Disposal method for municipal solid waste in selected countries with different income levels [111].

and is predominant until now, especially in low-income/GDP countries such as Cambodia and Tunisia because of the lack of financial, social, and management consideration. Based on this data, high-income countries already have proper MSW treatments by implementing the waste-to-energy concept as well as recycling. Converting MSW into compost is one of the appropriate options in countries that have a large agricultural sector, such as China, the United States, and Brazil. The compost material can be used directly in agricultural lands and as a chemical fertilizer substitute. Nevertheless, around the world, there is still a huge amount of MSW that is treated by open dumping and landfilling, waiting for suitable options of utilization. The available options depend on several considerations such as economic factors, technology, and local situations.

Physical Characteristics of Municipal Solid Waste

In the MSW stream, waste is broadly categorized into organic and inorganic materials. As many technologies are available for treating the MSW, knowing its detailed composition, including organic, paper, plastic, glass, metals, etc., is essential to determine the best and suitable method. The organic fraction consists of food scraps, yard waste, wood, and process residues, while metals include cans, foil, tins, railing, and bicycles. Paper scraps, cardboard, newspapers, magazines, bags, and boxes are classified as paper waste. Paper waste classification is driven by a higher economic value than the organic fraction. The remaining other fraction is composed of textiles, leather, rubber, e-waste, appliances, and inert materials. Table 1.9 shows the physical composition of MSW in the top 10 most populous countries, which is also a representative of the different levels of income/GDP.

Physical waste composition is influenced by various factors such as income level/GDP, cultural value, energy sources, and climate. In countries with high income level/GDP, the usage of inorganic materials such as plastic, paper, and metals is higher and the organic fraction is used relatively lower when compared with low-income countries. This statement is expressed by comparing the United States, Germany, and Japan with Indonesia, Nigeria, and Pakistan. Cultural values also influence waste composition, e.g., building material choice (e.g., wood vs. concrete). Furthermore, the waste composition is slightly different due to energy sources used for cooking and heating. Communities that use coal and hardwoods as an energy source for cooking generate more ash waste than those that use natural gas and electricity. Based on the climate factor, waste composition would be different between countries with subtropical (four seasons) and countries with tropical (two seasons) climates because of the consumption of heating energy and organic decomposition rates.

The most influencing factor that determines the physical composition of MSW is income level/GDP, presented in Fig. 1.3. For a more clear perspective, the top 10 most populous countries are categorized into four different levels of income, i.e., low income (GDP $<$ \$975), lower-middle income (\$976 $\leq$ GDP $\leq$ \$3855), upper-middle income (\$3856 $\leq$ GDP $\leq$ \$11,905), and high income (GDP $\geq$ \$11,906) [111]. Low-income countries have the highest

TABLE 1.9
Physical Composition of MSW in the Top 10 Most Populous Countries.

Country	Year	MSW COMPOSITION (%MASS)						References
		Organic	Paper	Plastics	Glass/ Ceramic	Metals	Textile and Other	
China	2002	59.0	8.0	10.0	3.0	1.0	19.0	[132]
India	2008	40.0	10.0	2.0	0.2	0.0	47.8	[133]
USA	2009	12.7	31.0	12.0	4.9	8.4	31.0	[112]
Brazil	2007	36.1	17.1	23.3	3.5	2.4	17.6	[134]
Indonesia	2007	74.0	10.0	8.0	2.0	2.0	4.0	[130]
Russia	1996	31.5	28.0	4.0	6.0	2.5	28.0	[135]
Japan	2003	42.6	22.3	11.4	1.6	9.0	13.0	[129]
Mexico	2000	52.4	14.1	4.4	5.9	2.9	20.3	[136]
Nigeria	2000	68.0	10.0	7.0	4.0	3.0	8.0	[137]
Germany	2005	30.0	24.0	13.0	10.0	1.0	22.0	[138]
Pakistan	2009	67.0	5.0	18.0	2.0	0.0	7.0	[139]

MSW, municipal solid waste.

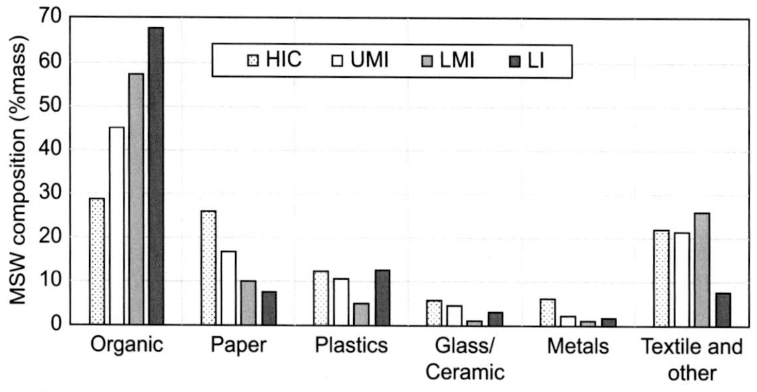

FIG. 1.3 Physical composition of municipal solid waste (MSW) in countries with different income levels [111]. *HI*, high income; *LI*, low income; *LMI*, lower-middle income; *UMI*, upper-middle income.

proportion of organic waste (more than 60%). In order to minimize the daily expenses, people in low-income countries prefer to cook by themselves, leading to the remaining foods being an organic fraction in their MSW collection. Paper, metals, and plastics make up a predominant proportion in high-income countries. High utilization of paper in education and also large application of metal in high-upper-middle-income countries leads to the presence of huge amounts of those materials in the MSW. Textile fraction is also quite interesting to discuss due to a large difference between low- and high-income countries. People in high-income countries make fashion as lifestyle, thus changing and updating of textile materials drives more MSW generation from this fraction. As large differences in MSW composition are found in different countries, the technology applicable for MSW treatment in each country would also be different.

Fig. 1.4 shows the effect of income level/GDP on changing the physical MSW composition in China.

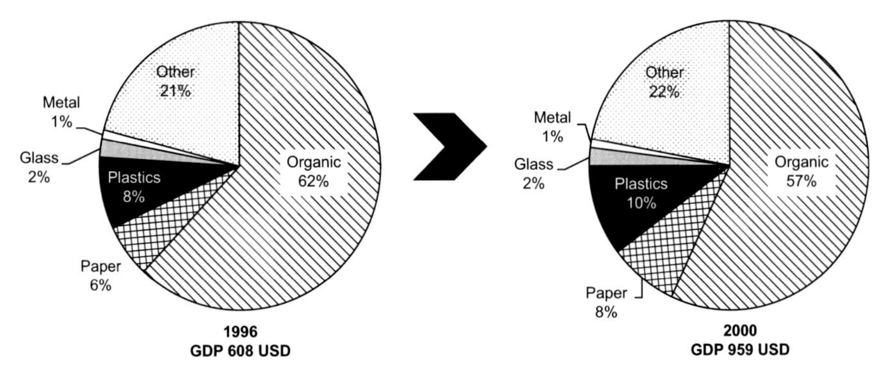

FIG. 1.4 Changing physical composition of municipal solid waste due to lifestyle and prosperity [120]. *GDP*, gross domestic product.

TABLE 1.10
Chemical Composition of Municipal Solid Waste in Selected Countries.

	ELEMENTAL ANALYSIS (%)								
Country	C	H	O	N	S	Moisture (%)	Ash (%)	Lower Heating Value (kJ/Kg)	References
Brazil	42.0	5.3	26.5	1.1	0.2	36.0	16.0	7376.5	[140]
China	29.4	3.9	28.7	1.6	0.5	55.4	18.9	4695.0	[141,142]
India	27.4	4.8	33.1	1.1	0.8	29.1	28.8	4270.0	[143–145]

Clearly, the growing GDP would accelerate high standards of living by decreasing the organic fraction that is replaced by plastic and paper. A high-standard community would prefer to use plastic and paper as the packaging material rather than organic material. The lifestyle would also shift, in relation to both housing facilities and food services. Modern people would like to choose inorganic materials such as plastic and steel as housing facilities rather than organic materials (hardwood). In addition, people would avoid cooking food by themselves due to the high work demands; thus, the amount of organic food waste would decrease in their MSW stream. This fact reconfirms that the physical composition of MSW is highly affected by the income level. It is crucial to determine a suitable technology for processing MSW, especially in developing counties where the income level/GDP is always rising.

Chemical Characteristics of Municipal Solid Waste

MSW is a series of heterogeneous materials, whose chemical characteristics are closely related with the chemical properties of various constituent components. In addition to physical components, the chemical characteristics are also a determining factor in establishing the final treatment of MSW. Table 1.10 shows the chemical characteristics of MSW in several countries around the world.

Each country has its own composition because of its special characteristics. The data are quite diverse among those countries due to several factors such as income level, climate, cultural values, and current available treatment. When the waste-to-energy concept is one of the promising options, data related to the heating value and moisture content would offer precise estimation on how much energy could be generated. The data of ash content would provide information related to the amount of solid waste (ash) after the final treatment of MSW. Hence, chemical characteristics would complete the information in determining the right technology for final treatment of MSW.

INDUSTRIAL WASTE

Types, Amount, and Origin

Industrial waste, generally, can be categorized into two types, i.e., nonhazardous and hazardous. Nonhazardous industrial waste is the waste from industrial

TABLE 1.11
Characteristic Wastes Produced From Various Industrial Activities.

Industrial Sector	Description	Typical Waste
Mining and quarrying	Extraction, beneficiation, and processing of minerals	Solid rock, slag, phosphogypsum, muds, tailings
Energy	Electricity, gas, steam, and air-conditioning supply	Fly ash, bottom ash, boiler slag, particulates, used oils, sludge
Manufacturing	Chemical	Spent catalyst, chemical solvents, reactive waste, acid, alkali, used oils, particulate waste, ash, sludge
	Food	Plastic, packaging, carton
	Textile	Textile waste, pigments, peroxide, organic stabilizer, alkali, chemical solvents, sludge, heavy metals
	Paper	Wood waste, alkali, chemical solvents, sludge
Construction	Construction, demolition activity	Concrete, cinder blocks, gypsum, masonry, asphalt, wood shingles, slate, metals, glass, and plaster
Waste/water services	Water collection, treatment, and supply	Spent adsorbent, sludge

activity, which does not pose a threat to public health or environment, e.g., carton, plastic, metals, glass, rock, and organic waste. In contrast, hazardous waste is a residue from industrial activity that can harm public health or environment, e.g., flammable, corrosive, active, and toxic materials. The characteristics of waste produced by different sectors of industry is shown in Table 1.11. Regarding the amount, typically nonhazardous waste is tremendously higher than the hazardous one, despite there being a distinct classification of hazardous waste for different countries [113]. It was reported in Europe (EU-28) that only 3.8% of the total industrial waste was classified as hazardous waste [114]. This is also consistent with the observation in other big countries such as the United States (less than 10%) [113], China (1.1%) [115], and India (1.5%) [116].

As a representative, waste composition in industrial sectors in Europe (in 2014) is depicted in Fig. 1.5A. In Fig. 1.5A, waste from households (MSW) featuring only 8.3% of the total waste is also presented to highlight the dominant part of the industrial waste. Indeed, the ratio between industrial and household wastes may differ from one country to another, mostly depending on the economic activities. However, it is widely known that more industrialized countries will produce a higher amount of industrial waste. Fig. 1.5A indicates that most of the waste stems from construction (34.7%) and mining (28.1%). These two sectors produce a high amount of solid wastes, e.g., for construction, concrete, cinder blocks, gypsum, masonry, asphalt and wood shingles, slate, and plaster and for mining, solid rock and tailings.

The production of industrial waste changes from time to time. Fig. 1.5B shows the evolution of industrial waste from 2004 to 2014 in EU-28. As mineral waste contributes to more than 50% of the total waste generated, mineral waste was excluded to magnify the production of other waste categories. Different sectors in industries show different trends. Manufacturing (including chemical production) displays a linear decrease, whereas construction exhibits the opposite. For energy and mining, typically a more or less constant generation of waste is observed. Holistically, the production of industrial waste in Europe has decreased by approximately 5.3% from 2004 to 2014.

Fig. 1.6A,B show the change in industrial waste generation in Japan [117] and China [118]. To highlight the alteration from one time to another, the delta value of waste production is given. The baseline is production in 2003, when Japan and China produced 412 and 1004 million tons of industrial waste, respectively. It can be observed that in Japan, waste generation changed dynamically from 2003 to 2010, but it led to a slight decrease. In China, however, which is a more developing country, the production of waste rose rapidly, generating more than double the amount in 2010. Observing the fact of waste production in Japan

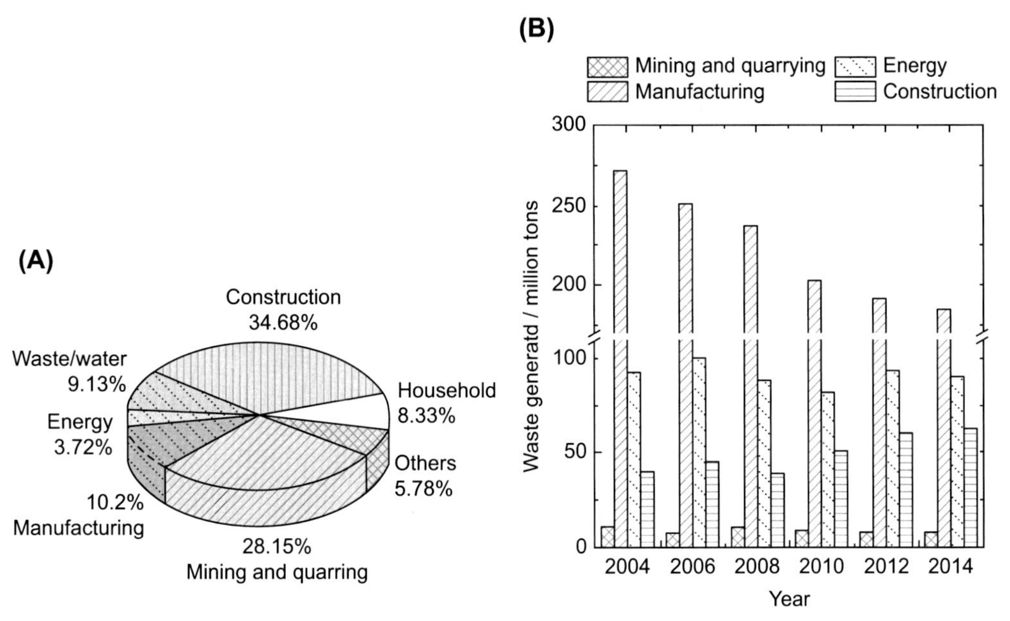

FIG. 1.5 **(A)** Waste composition generated in Europe in 2014. **(B)** Major industrial activity year to year in Europe, excluding mineral waste [114].

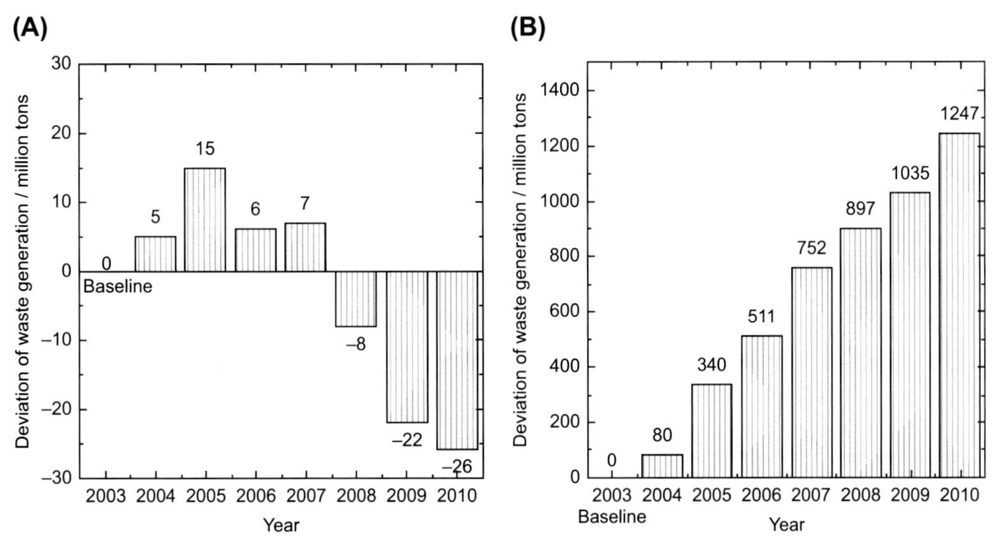

FIG. 1.6 Alteration of industrial waste generation in **(A)** Japan and **(B)** China. (**(A)** Adapted from [117] and **(B)** From [118].)

TABLE 1.12
Amount of Industrial Waste Generated in Different Regions (in 2011) in Million Tons.

Items	Europe	Americas	Africa and Middle East	Asia Pacific	Global
Industrial waste generated	1933	915	921	5357	9177
Key countries	Germany, United Kingdom, France, Russia, Bulgaria	United States, Brazil, Canada, Chile, and Columbia	South Africa, Saudi Arabia, United Arab Emirates, Egypt, and Tunisia	China, Japan, India, South Korea, Australia	

Adapted from Frost & Sullivan. The Global industrial waste recycling & services markets. 2012.

and EU-28, it may be generally accepted that in developed countries the production of industrial waste displays a reduction, whereas in developing countries, it shows the opposite trend.

Frost and Sullivan [119] reported the production of industrial waste in different regions, i.e., Europe, the United States, Africa, Middle East, and Asia Pacific (Table 1.12). The top producer of industrial waste is Asia Pacific, with 5357 million tons (58.4% of the global production). This was contributed by key countries such as China, Japan, India, South Korea, and Australia. Surprisingly, Americas generated the least amount, with 10% of the total share.

Physical and Chemical Characteristics

The characteristics of nonhazardous industrial waste are similar to those of household waste, by its nature and composition. This waste is not toxic and can thus be recycled or disposed of safely. In contrast, hazardous wastes may cause danger to health or environment, either alone or when in contact with other wastes. Therefore a special treatment must be applied. The properties of the materials are flammable, corrosive, active, radioactive, and toxic. Some hazardous wastes may be recycled because they contain important components, e.g., silica, alumina, iron, and precious metals. Among these, the majority of wastes are from metallurgical activities and a minor amount from other processes such as spent catalyst. Table 1.13 shows the chemical composition of some hazardous wastes and the efforts to utilize them.

CONCLUSIONS AND PERSPECTIVES

The accumulation of agricultural waste, i.e., corncob, rice husk, rice straw, OPEFB, sugarcane bagasse, and wheat straw, is approximately 2 billion tons worldwide. Forest waste accounts for approximately 0.2 billion m^3; MSW, 1.7 billion tons; and industrial waste, 9.1 billion tons. Altogether, it makes more than 10 billion tons of wastes and residuals, which is a considerably huge amount; the number tends to increase over time. Without proper treatment, these solid wastes would create many challenges and negatively affect the sustainability of the related industries and societies. Information about the physical and chemical composition of the waste is necessary to find a suitable treatment and technology, which likewise depends on several considerations such as economics and local situations. Due to its vast amount, the wastes, especially from agricultural and forest sectors, have the potential of being used as feedstock for industries, which is discussed in the rest of this book. However, being able to collect them from where they are generated is a challenging issue. Studies are being conducted to increase their added value. There are developed technologies that work, as some of them have gone into commercialization. The products are more environmentally friendly because they are not oil-based products, and at the same time, by converting waste into valuable products, the problems associated with waste accumulation can be alleviated. More efforts in science are expected to reduce the cost for industries that use these wastes as feedstock. Meanwhile, landfilling is still the traditional practice of treating MSW and nonhazardous industrial wastes. Proper treatments such as recycling as well as material and energy recovery need to be introduced and practiced even more because they can result in high economic benefits. In the end, successful utilization and proper treatment of wastes would contribute to realizing the concept of sustainable development for the sake of a better nation and global society.

TABLE 1.13
Chemical Composition and Utilization of Some Hazardous Wastes Generated From Industrial Activities.

	CHEMICAL COMPOSITION/%										
Material	**SiO_2**	**Al_2O_3**	**Fe_2O_3**	**CaO**	**MgO**	**SO_3**	**Na_2O**	**K_2O**	**Others**	**Application**	**References**
Spent catalyst	34.9	59.2	1.3	0.4	0.3	2.4	NA	NA	Precious metals	Blended cement	[146]
Copper slag	24.0–41	2.0–16.0	39.0–45.0	0.7–10.0	NA	NA	NA	NA	0.5–2.1 (Cu)	Tiles, mine backfill materials	[147]
Blast furnace slag	33.1	21.6	0.9	33.0	8.9	NA	NA	NA		Blended cement	[148]
Electric furnace slag	20.3	7.3	42.4	22.8	8.0	NA	0.6	1.5	0.3 (TiO_2)	Vitreous ceramic tiles	[149]
Jarosite	1.2	16.9	39.7	NA	0.5	18.9	NA	NA	Toxic substances	Mortar restoration	[150]

NA, data not available.

REFERENCES

[1] Environment Agency. Guidance on the legal definition of waste and its application. 2012. p. 9. August.

[2] FAO. FAO statistical pocketbook 2015. 2015.

[3] The global staple - Ricepedia. 2018 [Online]. Available: http://ricepedia.org/rice-as-food/the-global-staple-rice-consumers.

[4] Basalan M, Bayhan R, Secrist D, Hill J, Owens F, Witt M, Kreikemeier K. "Corn maturation: changes in the grain and cob," animal science research report. 1995. p. 92–8.

[5] USDA. Grain: world market and trade coarse grains COARSE GRAINS CORN PRICES corn daily FOB export bids Argentina up river black sea Brazil paranagua US Gulf. 2018.

[6] Lu JJ, Chen WH. Product yields and characteristics of corncob waste under various torrefaction atmospheres. Energies 2014;7(1):13–27.

[7] Wardhani GAPK, Nurlela N, Azizah M. Silica content and structure from corncob ash with various acid treatment (HCl, HBr, and citric acid). Molekul 2017;12(2):174.

[8] Zych D. The viability of corn cobs as a bioenergy feedstock. West Cent. Res. Outreach Cent.; 2008. p. 1–25.

[9] Velmurugan R, Muthukumar K. Utilization of sugarcane bagasse for bioethanol production: sono-assisted acid hydrolysis approach. Bioresource Technology 2011; 102(14):7119–23.

[10] Sewsynker-Sukai Y, Gueguim Kana EB. Simultaneous saccharification and bioethanol production from corn cobs: process optimization and kinetic studies. Bioresource Technology 2018;262:32–41.

[11] Gowdhaman D, Ponnusami V. Production and optimization of xylooligosaccharides from corncob by Bacillus aerophilus KGJ2 xylanase and its antioxidant potential. International Journal of Biological Macromolecules 2015;79:595–600.

[12] Michelin M, Ruiz HA, Polizeli Mde LTM, Teixeira JA. Multi-step approach to add value to corncob: production of biomass-degrading enzymes, lignin and fermentable sugars. Bioresource Technology 2018; 247:582–90.

[13] El-Batal AI, Khalaf SA. Xylitol production from corn cobs hemicellulosic hydrolysate by *Candida tropicalis* immobilized cells in hydrogel copolymer carrier. International Journal of Agriculture and Biology 2004;6(6):1073. 1066.

[14] Bhatti HN, Sadaf S, Aleem A. Treatment of textile effluents by low cost agricultural wastes: batch biosorption study. Journal of Animal and Plant Sciences 2015; 25(1):284–9.

[15] Miura S, Arimura T, Itoda N, Dwiarti L, Feng JB, Bin CH, Okabe M. Production of l-lactic acid from corncob. Journal of Bioscience and Bioengineering 2004;97(3): 153–7.

[16] Yuan QP, Zhang H, Qian ZM, Yang XJ. Pilot-plant production of xylo-oligosaccharides from corncob by steaming, enzymatic hydrolysis and nanofiltration. Journal of Chemical Technology and Biotechnology 2004; 79(10):1073–9.

[17] Zhang H-J, Fan X-G, Qiu X-L, Zhang Q-X, Wang W-Y, Li S-X, Deng L-H, Koffas MAG, Wei D-S, Yuan Q-P. A novel cleaning process for industrial production of xylose in pilot scale from corncob by using screw-steam-explosive extruder. Bioprocess and Biosystems Engineering 2014;37(12):2425–36.

[18] Li Y, Wang T, Yin X, Wu C, Ma L, Li H, Sun L. Design and operation of integrated pilot-scale dimethyl ether synthesis system via pyrolysis/gasification of corncob. Fuel 2009;88(11):2181–7.

[19] CCM International. Corn products. 2010.

[20] Home | Praj Industries, 2018. [Online]. Available: https://www.praj.net/. Accessed 05.09.2018.

[21] Hayashi K. Schematic flow of palm oil industry, vol. 07; 2007. p. 646–51.

[22] USDA. Soybean and oilseed meal import prospects higher as severe weather damages EU crops. 2018.

[23] Ishola MM, Isroi, Taherzadeh MJ. Effect of fungal and phosphoric acid pretreatment on ethanol production from oil palm empty fruit bunches (OPEFB). Bioresource Technology 2014;165:9–12.

[24] Law KN, Daud WRW, Ghazali A. Morphological and chemical nature of fiber strands of oil palm empty-fruit-bunch (OPEFB). BioResources 2007;2(3):351–62.

[25] Razak NWA, Kalam A. Effect of OPEFB size on the mechanical properties and water absorption behaviour of OPEFB/PPnanoclay/PP hybrid composites. Procedia Engineering 2012;41:1593–9.

[26] Narapakdeesakul D, Sridach W, Wittaya T. Novel use of oil palm empty fruit bunch's lignin derivatives for production of linerboard coating. Progress in Organic Coatings 2013;76(7–8):999–1005.

[27] Zanirun Z, Bahrin EK, Lai-Yee P, Hassan MA, Abd-Aziz S. Enhancement of fermentable sugars production from oil palm empty fruit bunch by ligninolytic enzymes mediator system. International Biodeterioration & Biodegradation 2015;105:13–20.

[28] Zhang Y, Sun W, Wang H, Geng A. Polyhydroxybutyrate production from oil palm empty fruit bunch using Bacillus megaterium R11. Bioresource Technology 2013;147: 307–14.

[29] Nieves DC, Karimi K, Horváth IS. Improvement of biogas production from oil palm empty fruit bunches (OPEFB). Industrial Crops and Products 2011;34(1): 1097–101.

[30] Duangwang S, Sangwichien C. Utilization of oil palm empty fruit bunch hydrolysate for ethanol production by baker's yeast and loog-pang. Energy Procedia 2015; 79:157–62.

[31] Jeon H, Kang K-E, Jeong J-S, Gong G, Choi J-W, Abimanyu H, Ahn BS, Suh D-J, Choi G-W. Production of anhydrous ethanol using oil palm empty fruit bunch in a pilot plant. Biomass and Bioenergy 2014;67: 99–107.

[32] Rice Knowledge Bank, 2016. [Online]. Available: http://www.knowledgebank.irri.org/step-by-step-production/postharvest/rice-by-products/rice-straw. Accessed 10.08.2018.

[33] USDA. Grain: world market and trade rice. 2018.

[34] Chandrasekhar S, Satyanarayana KG, Pramada PN, Raghavan P, Gupta TN. Review Processing, properties and applications of reactive silica from rice husk—an overview. Journal of Materials Science 2003;3(3):3159−68.

[35] Wannapeera J, Worasuwannarak N, Pipatmanomai S. Product yields and characteristics of rice husk , rice straw and corncob during fast pyrolysis in a drop-tube/fixed-bed reactor. Songklanakarin Journal of Science and Technology 2007;30(3):393−404.

[36] Bakar RA, Yahya R, Gan SN. Production of high purity amorphous silica from rice husk. Procedia Chemistry 2016;19:189−95.

[37] Bhagiyalakshmi M, Yun LJ, Anuradha R, Jang HT. Utilization of rice husk ash as silica source for the synthesis of mesoporous silicas and their application to CO2 adsorption through TREN/TEPA grafting. Journal of Hazardous Materials 2010;175(1−3):928−38.

[38] Ragab TIM, Amer H, Mossa AT, Emam M, Hasaballah AA, Helmy WA. Anticoagulation, fibrinolytic and the cytotoxic activities of sulfated hemicellulose extracted from rice straw and husk. Biocatalysis and Agricultural Biotechnology 2018;15:86−91.

[39] Islam MS, Kao N, Bhattacharya SN, Gupta R, Choi HJ. Potential aspect of rice husk biomass in Australia for nanocrystalline cellulose production. Chinese Journal of Chemical Engineering 2018;26(3):465−76.

[40] Ajiwe VIE, Okeke CA, Ekwuozor SC, Uba IC. A pilot plant for production of ceiling boards from rice husks. Bioresource Technology 1998;66(1):41−3.

[41] Cai W, Liu R, He Y, Chai M, Cai J. Bio-oil production from fast pyrolysis of rice husk in a commercial-scale plant with a downdraft circulating fluidized bed reactor. Fuel Process Technology 2018;171:308−17.

[42] Makwana JP, Pandey J, Mishra G. Improving the properties of producer gas using high temperature gasification of rice husk in a pilot scale fluidized bed gasifier (FBG). Renewable Energy 2019;130:943−51.

[43] Bakker R, Elbersen W, Poppens R, Lesschen JP. Rice straw and wheat straw. Potential feedstocks for the biobased economy potential feedstocks for the biobased economy. The Netherlands: NL Agency, NL Energy and Climate Change; 2013. p. 8.

[44] Thy P, Yu C, Jenkins BM, Lesher CE. Inorganic composition and environmental impact of biomass feedstock. Energy & Fuels 2013;27(7):3969−87.

[45] Abraham A, Mathew AK, Sindhu R, Pandey A, Binod P. Potential of rice straw for bio-refining: an overview. Bioresource Technology 2016;215:29−36.

[46] Wang C-H, Chen W-H, Liu H-S, Lai J-T, Hsu C-C, Wan B-Z. Process development for producing a food-grade glucose solution from rice straws. Chinese Journal of Chemical Engineering 2018;26(2):386−92.

[47] Niladevi KN, Sukumaran RK, Prema P. Utilization of rice straw for laccase production by Streptomyces psammoticus in solid-state fermentation. Journal of Industrial Microbiology and Biotechnology 2007;34(10):665−74.

[48] Zhang R, Zhang Z. Biogasification of rice straw with an anaerobic-phased solids digester system. Bioresource Technology 1999;68(3):235−45.

[49] Kim M, Kim B-C, Choi Y, Nam K. Minimizing mixing intensity to improve the performance of rice straw anaerobic digestion via enhanced development of microbe-substrate aggregates. Bioresource Technology 2017;245:590−7.

[50] Guan R, Li X, Wachemo AC, Yuan H, Liu Y, Zou D, Zuo X, Gu J. Enhancing anaerobic digestion performance and degradation of lignocellulosic components of rice straw by combined biological and chemical pretreatment. The Science of the Total Environment 2018;637−638:9−17.

[51] Phi Trinh LT, Lee J-W, Lee H-J. Acidified glycerol pretreatment for enhanced ethanol production from rice straw. Biomass and Bioenergy 2016;94:39−45.

[52] Zhou J, Yang J, Yu Q, Yong X, Xie X, Zhang L, Wei P, Jia H. Different organic loading rates on the biogas production during the anaerobic digestion of rice straw: a pilot study. Bioresource Technology 2017;244(30):865−71.

[53] Lin TH, Guo GL, Hwang WS, Huang SL. The addition of hydrolyzed rice straw in xylose fermentation by Pichia stipitis to increase bioethanol production at the pilot-scale. Biomass and Bioenergy 2016;91:204−9.

[54] Kapoor M, Soam S, Agrawal R, Gupta RP, Tuli DK, Kumar R. Pilot scale dilute acid pretreatment of rice straw and fermentable sugar recovery at high solid loadings. Bioresource Technology 2017;224:688−93.

[55] Chemicals Technology, 2018. [Online]. Available: https://www.chemicals-technology.com/projects/mg-ethanol/. Accessed 05.09.2018.

[56] Birru E. Sugar cane industry overview and energy efficiency considerations. KTH School of Industrial Engineering and Management 2016;(01):1−61.

[57] USDA. Sugar: world markets and trade. 2018.

[58] Candido RG, Godoy GG, Gonçalves AR. Characterization and application of cellulose acetate synthesized from sugarcane bagasse. Carbohydrate Polymers 2017;167:280−9.

[59] Loh YR, Sujan D, Rahman ME, Das CA. Sugarcane bagasse—the future composite material: a literature review. Resources, Conservation and Recycling 2013;75:14−22.

[60] Alves MJ, Cavalcanti ÍV, de Resende MM, Cardoso VL, Reis MH. Biodiesel dry purification with sugarcane bagasse. Industrial Crops and Products 2016;89:119−27.

[61] Shaikh HM, Pandare KV, Nair G, Varma AJ. Utilization of sugarcane bagasse cellulose for producing cellulose acetates: novel use of residual hemicellulose as plasticizer. Carbohydrate Polymers 2009;76(1):23−9.

[62] Cabral MR, Nakanishi EY, dos Santos V, Palacios JH, Godbout S, Savastano Junior H, Fiorelli J. Evaluation of pre-treatment efficiency on sugarcane bagasse fibers for the production of cement composites. Archives of Civil and Mechanical Engineering 2018;18(4):1092–102.
[63] Hofsetz K, Silva MA. Brazilian sugarcane bagasse: energy and non-energy consumption. Biomass and Bioenergy 2012;46:564–73.
[64] Nakanishi SC, Nascimento VM, Rabelo SC, Sampaio ILM, Junqueira TL, Rocha GJM. Comparative material balances and preliminary technical analysis of the pilot scale sugarcane bagasse alkaline pretreatment to 2G ethanol production. Industrial Crops and Products 2018;120:187–97.
[65] Silveira MHL, Chandel AK, Vanelli BA, Sacilotto KS, Cardoso EB. Production of hemicellulosic sugars from sugarcane bagasse via steam explosion employing industrially feasible conditions: pilot scale study. Bioresource Technology Reports 2018;3:138–46.
[66] Granbio, 2018. [Online]. Available: http://www.granbio.com.br/en/. Accessed 05.09.2018.
[67] Raízen [Online]. Available: https://www.raizen.com.br/en/home; 2018.
[68] USDA. World agricultural supply and demand estimates. 2018.
[69] Zhong C, Wang C, Huang F, Wang F, Jia H, Zhou H, Wei P. Selective hydrolysis of hemicellulose from wheat straw by a nanoscale solid acid catalyst. Carbohydrate Polymers 2015;131:384–91.
[70] Yuan Z, Wen Y, Li G. Production of bioethanol and value added compounds from wheat straw through combined alkaline/alkaline-peroxide pretreatment. Bioresource Technology 2018;259:228–36.
[71] Ren J, Liu L, Zhou J, Li X, Ouyang J. Co-production of ethanol, xylo -oligosaccharides and magnesium lignosulfonate from wheat straw by a controlled magnesium bisulfite pretreatment (MBSP). Industrial Crops and Products 2018;113:128–34.
[72] Xie X, Feng X, Chi S, Zhang Y, Yu G, Liu C, Li Z, Li B, Peng H. A sustainable and effective potassium hydroxide pretreatment of wheat straw for the production of fermentable sugars. Bioresource Technology Reports 2018;3:169–76.
[73] Saha BC, Nichols NN, Qureshi N, Kennedy GJ, Iten LB, Cotta MA. Pilot scale conversion of wheat straw to ethanol via simultaneous saccharification and fermentation. Bioresource Technology 2015;175:17–22.
[74] Thomsen MH, Thygesen A, Thomsen AB. Hydrothermal treatment of wheat straw at pilot plant scale using a three-step reactor system aiming at high hemicellulose recovery, high cellulose digestibility and low lignin hydrolysis. Bioresource Technology 2008;99(10):4221–8.
[75] Fu P, Bai X, Yi W, Li Z, Li Y. Fast pyrolysis of wheat straw in a dual concentric rotary cylinder reactor with ceramic balls as recirculated heat carrier. Energy Conversion and Management 2018;171:855–62.
[76] FAOSTAT [Online]. Available: http://www.fao.org/faostat/en/#data/FO; 2014.
[77] Pöschel F. Global softwood lumber production. 2017 [Online]. Available: https://www.timber-online.net/schnittholz/2017/10/nadelschnittholzproduktion-weltweit.html.
[78] Alderman DR, Smith RL, Reddy VS. Assessing the availability of wood residues and residue markets in Virginia. Forest Products Laboratory 1998;49(4):1–188.
[79] Yang G, Jaakkola P. Wood chemistry and isolation of extractives from wood. Literature study for BIOTULI project 2011:47.
[80] Sjöström. Wood chemistry: fundamentals and applications. 1993.
[81] Räisänen T, Athanassiadis D. Basic chemical composition of the biomass components of pine, spruce and birch. For. Refine; 2013. p. 4.
[82] Yemele Ngueho MC, Blanchet P, Koubaa A, Cloutier A. Effects of bark content and particle geometry on the physical and mechanical properties of particleboard made from black spruce and trembling aspen bark. Forest Products Laboratory 2008;58(11):48–56.
[83] Valentín L, Kluczek-Turpeinen B, Willför S, Hemming J, Hatakka A, Steffen K, Tuomela M. Scots pine (Pinus sylvestris) bark composition and degradation by fungi: potential substrate for bioremediation. Bioresource Technology 2010;101(7):2203–9.
[84] Laks PE. Chemistry of bark. In: Hon DN-S, Shiraishi N, editors. Wood and cellulosic chemistry. New York: Marcel Dekker Inc.; 1991.
[85] Cotana F, Cavalaglio G, Gelosia M, Nicolini A, Coccia V, Petrozzi A. Production of bioethanol in a second generation prototype from pine wood chips. Energy Procedia 2014;45:42–51.
[86] Krogell J. Intensification of hemicellulose extraction from spruce extraction from spruce wood by parameter tuning parameter tuning. 2015.
[87] Miranda I, Gominho J, Mirra I, Pereira H. Fractioning and chemical characterization of barks of Betula pendula and Eucalyptus globulus. Industrial Crops and Products 2013;41(1):299–305.
[88] Taherzadeh MJ, Eklund R, Gustafsson L, Niklasson C, Lidén G. Characterization and fermentation of dilute-acid hydrolyzates from wood. Industrial & Engineering Chemistry Research 1997;36(11):4659–65.
[89] Wang Z, Winestrand S, Gillgren T, Jönsson LJ. Chemical and structural factors influencing enzymatic saccharification of wood from aspen, birch and spruce. Biomass and Bioenergy 2018;109:125–34.
[90] Vermerris W, Abril A. Enhancing cellulose utilization for fuels and chemicals by genetic modification of plant cell wall architecture. Current Opinion in Biotechnology 2015;32:104–12.
[91] Przybysz Buzała K, Kalinowska H, Małachowska E, Przybysz P. The utility of selected kraft hardwood and softwood pulps for fuel ethanol production. Industrial Crops and Products 2017;108:824–30.

[92] Dufresne A. Processing of polymer nanocomposites reinforced with polysaccharide nanocrystals. Molecules 2010;15(6):4111−28.
[93] Belbekhouche S, Bras J, Siqueira G, Chappey C, Lebrun L, Khelifi B, Marais S, Dufresne. Water sorption behavior and gas barrier properties of cellulose whiskers and microfibrils films. Carbohydrate Polymers 2011; 83(4):1740−8.
[94] Nakagaito AN, Nogi M, Yano. Displays from transparent films of natural nanofibers. MRS Bulletin 2010;35(3): 214−8.
[95] Jackson JK, Letchford K, Wasserman BZ, Ye L, Hamad WY, Burt. The use of nanocrystalline cellulose for the binding and controlled release of drugs. Nanomedicine 2011;6:321−30.
[96] Revol JF, Godbout L, Gray. Solid self-assembled films of cellulose with chiral nematic order and optically variable properties. Journal of Pulp and Paper Science 1998;24(5):146−9.
[97] Azizi Samir MAS, Alloin F, Sanchez J-Y, Dufresne. Cross-linked nanocomposite polymer electrolytes reinforced with cellulose whiskers. Macromolecules 2004;37(13): 4839−44.
[98] Kamm B, Kamm M. Principles of biorefineries. Applied Microbiology and Biotechnology 2004;64:137−45.
[99] Grondahl M, Eriksson L, Gatenholm P. Material properties of plasticized hardwood xylans for potential application as oxygen barrier films. Biomacromolecules 2004;5: 1528−35.
[100] Hartman J, Albertsson A, Lindblad M, Sjöberg J. Oxygen barrier materials from renewable sources: material properties of softwood hemicellulose-based films. Journal of Applied Polymer Science 2006;100: 2985−91.
[101] Lora J, Glasser W. Recent industrial applications of lignin: a sustainable alternative to nonrenewable materials. Journal of Polymers and the Environment 2002;10:39−48.
[102] Reddy N, Yang Y. Biofibers from agricultural byproducts for industrial application. Trends in Biotechnology 2005;23:22−7.
[103] Eckert C, Liotta C, Ragauskas A, Hallettac J, Kitchens C, Hill E, Draucker L. Tunable solvents for fine chemicals from the biorefinery. Green Chemistry 2007;9:545−8.
[104] Faix O. New aspects of lignin utilization in large amounts. Papier 1992;12:733−40.
[105] Ten E, Vermerris W. Functionalized polymers from lignocellulosic biomass: state of the art. Polymers 2013;5(2):600−42.
[106] Collaboration 2gen ethanol − SEKAB. [Online]. Available: http://www.sekab.com/about-us/cooperation-partners/collaboration-2gen-ethanol/. Accessed 05.09.2018.
[107] Zhu JY, Chandra MS, Gu F, Gleisner R, Reiner R, Sessions J, Marrs G, Gao J, Anderson D. Using sulfite chemistry for robust bioconversion of Douglas-fir forest residue to bioethanol at high titer and lignosulfonate: a pilot-scale evaluation. Bioresource Technology 2015; 179:390−7.
[108] Mathew AP, Oksman K, Karim Z, Liu P, Khan SA, Naseri N. Process scale up and characterization of wood cellulose nanocrystals hydrolysed using bioethanol pilot plant. Industrial Crops and Products 2014;58:212−9.
[109] U.S. Department of Energy's National Renewable Energy Laboratory. 2013 renewable energy data book. 2014. p. 134.
[110] Borregaard [Online]. Available: https://borregaard.com/; 2018.
[111] Hoornweg D, Bhada-Tata P. What a waste: a global review of solid waste management. 2012.
[112] Environmental Protection Agency (EPA). 2009 facts and figures municipal solid waste in the United States. 2009.
[113] Allen DT, Behmanesh N. Non-hazardous waste generation. In: Hazardous waste and hazardous materials, vol. 9. Mary Ann Liebert, Inc.; 1992.
[114] Union E. Waste statistics. 2017.
[115] Duan H, Huang Q, Wang Q, Zhou B, Li J. Hazardous waste generation and management in China: a review. Journal of Hazardous Materials 2008;158(2−3):221−7.
[116] Pappu A, Saxena M, Asolekar SR. Solid wastes generation in India and their recycling potential in building materials. Building and Environment 2007;42(6): 2311−20.
[117] UNEP. The Japanese industrial waste experience: lessons for rapidly industrializing countries. 2013.
[118] Song Q, Li J, Zeng X. Minimizing the increasing solid waste through zero waste strategy. Journal of Cleaner Production 2015;104:199−210.
[119] Frost & Sullivan. The Global industrial waste recycling & services markets. 2012.
[120] Chen X, Geng Y, Fujita T. An overview of municipal solid waste management in China. Wastes Management 2010;30(4):716−24.
[121] USDA. Table 11: palm oil: world supply and distribution. 2018.
[122] Darmosarkoro W, Rahutomo S. Tandan kosong kelapa sawit sebagai bahan pembenah tanah. J. Lahan dan Pemupukan Kelapa Sawit Ed. I 2007;C3:167−80.
[123] A. Anukam, E. Meyer, O. Okoh, and S. Mamphweli, "Gasification characteristics of sugarcane bagasse."
[124] Misra MK, Ragland KW, Baker AJ. Wood ash composition as a function of furnace temperature. Biomass and Bioenergy 1993;4(2):103−16.
[125] Sixta H. Handbook of pulp, vol. 1. Strauss GmbH, Mörlenbach: The Federal Republic of Germany; 2006.
[126] Karunanithy C, Muthukumarappan K, Gibbons WR. Extrusion pretreatment of pine wood chips. Applied Biochemistry and Biotechnology 2012;167(1):81−99.
[127] Hoornweg D, Lam P, Chaudhry M. Waste management in China: issues and recommendations. 2005.
[128] Hanrahan D, S S. Improving management of municipal solid waste in India. Environmental Sociology Development Unit 2006:1−72.
[129] Organization for Economic Cooperation and Development (OECD). Environmental data compendium. 2009. Paris.

[130] Shekdar AV. Sustainable solid waste management: an integrated approach for Asian countries. Wastes Management 2009;29(4):1438–48.

[131] Solomon UU. The state of solid waste management in Nigeria. Wastes Management 2009;29(10):2787–8.

[132] Huang Q, Wang Q, Dong L, Xi B, Zhou B. The current situation of solid waste management in China. J Mater Cycles Waste Manag 2006;8(1):63–9.

[133] Unnikrishnan S, Singh A. Energy recovery in solid waste management through CDM in India and other countries. Resources, Conservation and Recycling 2010;54(10):630–40.

[134] Machado SL, Carvalho MF, Gourc J-P, Vilar OM, do Nascimento JCF. Methane generation in tropical landfills: simplified methods and field results. Wastes Management 2009;29(1):153–61.

[135] Rodionov M, Nakata T. Design of an optimal waste utilization system: a case study in St. Petersburg, Russia. Sustainability 2011;3(9):1486–509.

[136] Fehr M. The prospect of municipal waste landfill diversion depends on geographical location. Environmentalist 2002;22(4):319–24.

[137] Kofoworola OF. Recovery and recycling practices in municipal solid waste management in Lagos, Nigeria. Wastes Management 2007;27(9):1139–43.

[138] Mühle S, Balsam I, Cheeseman CR. Comparison of carbon emissions associated with municipal solid waste management in Germany and the UK. Resources, Conservation and Recycling 2010;54(11):793–801.

[139] Batool SA, Ch MN. Municipal solid waste management in lahore city District, Pakistan. Wastes Management 2009;29(6):1971–81.

[140] Leme MMV, Rocha MH, Lora EES, Venturini OJ, Lopes BM, Ferreira CH. Techno-economic analysis and environmental impact assessment of energy recovery from Municipal Solid Waste (MSW) in Brazil. Resources, Conservation and Recycling 2014;87:8–20.

[141] Zhou H, Meng A, Long Y, Li Q, Zhang Y. An overview of characteristics of municipal solid waste fuel in China: physical, chemical composition and heating value. Renewable & Sustainable Energy Reviews 2014;36:107–22.

[142] Wang H, Nie Y. Municipal solid waste characteristics and management in China. Journal of the Air and Waste Management Association 2001;51(2):250–63.

[143] Gupta N, Yadav KK, Kumar V. A review on current status of municipal solid waste management in India. Journal of Environmental Sciences 2015;37:206–17.

[144] Sethi S, Kothiyal NC, Nema AK, Kaushik MK. Characterization of municipal solid waste in Jalandhar city, Punjab, India. Journal of Hazardous, Toxic, and Radioactive Waste 2013;17(2):97–106.

[145] Katiyar RB, Suresh S, Sharma AK. Characterisation of municipal solid waste generated by the city of bhopal, India. International Journal of ChemTech Research CODEN 2013;5(2):623–8.

[146] Lin KL, Lo KW, Hung MJ, Cheng TW, Chang YM. Recycling of spent catalyst and waste sludge from industry to substitute raw materials in the preparation of Portland cement clinker. Sustainable Environment Research 2017;27(5):251–7.

[147] Gorai B, Jana RK, Premchand. Characteristics and utilisation of copper slag – a review. Resources, Conservation and Recycling 2003;39(4):299–313.

[148] Kumar S, Kumar R, Bandopadhyay A, Alex TC, Ravi Kumar B, Das SK, Mehrotra SP. Mechanical activation of granulated blast furnace slag and its effect on the properties and structure of portland slag cement. Cement and Concrete Composites 2008;30(8):679–85.

[149] Sarkar R, Singh N, Das SK. Utilization of steel melting electric arc furnace slag for development of vitreous ceramic tiles. Bulletin of Materials Science 2010;33(3): 293–8.

[150] Katsioti M, Mauridou O, Moropoulou A, Aggelakopoulou E, Tsakiridis PE, Agatzini-Leonardou S, Oustadakis P. Utilization of jarosite/alunite residue for mortars restoration production. Materials and Structures 2010;43(1–2):167–77.

CHAPTER 2

Life Cycle Assessment of Waste Management Systems

PEDRO BRANCOLI, MSC • KIM BOLTON, PHD

INTRODUCTION

Waste management systems have become increasingly complex over time in order to be able to comply with stricter legal frameworks and environmental protection targets defined by governments [1]. The EU Waste Framework Directive [2] has established, for example, the waste hierarchy that defines a priority order for waste management. According to this hierarchy, prevention of waste is the preferred option, followed by reuse, material recycling, energy recovery, and disposal as the least preferred alternative. Yet there are situations when the hierarchy does not represent the best treatment pathway and deviations from it are allowed if supported in a scientifically robust manner. Often this support is obtained from the life cycle assessment (LCA) methodology, which systematically evaluates the environmental impacts of waste management alternatives. Divergences from the waste hierarchy can occur for distinct reasons. For example, the environmental impacts of different waste management systems are significantly influenced by local conditions, such as the composition of the waste, the energy supply mix, and the level of technology used in the system.

The holistic approach of LCA, which considers the whole life cycle of the system, has the advantage that it reduces the risk of shifting environmental burdens from one part of the waste management system to another, and hence reduces the risk for suboptimization of the system [1,3].

LCA quantifies the environmental impacts by assessing the resources used and the emissions related to the production, use and end of life of a product, and the consequent impact on human health and ecosystems. The European Union considers LCA as the best framework currently available for quantifying the environmental impacts of products [4]. In the context of waste management, LCAs are often used by policy makers and in businesses to compare specific technologies in a particular context and region. The information provided by an LCA is used together with other information, such as economic and social aspects, to support decision-making processes.

The LCA of waste management systems is often comparative and can be used to answer questions such as the following:

- Is composting the organic fraction of municipal solid waste (OFMSW) better than anaerobic digestion (AD) in the city of Stockholm?
- What parameters (e.g., recycling rates, sorting efficiency, emissions in waste treatment facilities) have the largest contribution to the environmental impacts in the waste management system in Sweden?

This chapter aims to introduce the LCA methodology and to describe the steps in LCA studies that focus on waste management. It describes the principles and methodology and highlights the modeling aspects for LCAs of specific waste management systems.

OVERVIEW OF THE STEPS IN LCA IN THE CONTEXT OF WASTE MANAGEMENT

This section presents the different parts of the LCA methodology within the waste management perspective. LCA consists of four iterative steps, namely, goal and scope definition, inventory analysis, impact assessment, and interpretation (Fig. 2.1).

Goal and Scope Definition

Goal

Defining the goal is usually the first step in an LCA and this describes the purpose of the study. In the waste management context, often the objective of the study is to compare the environmental impacts of different waste treatment pathways. The management of waste is a complex and multidisciplinary task. Therefore the

Sustainable Resource Recovery and Zero Waste Approaches. https://doi.org/10.1016/B978-0-444-64200-4.00002-5

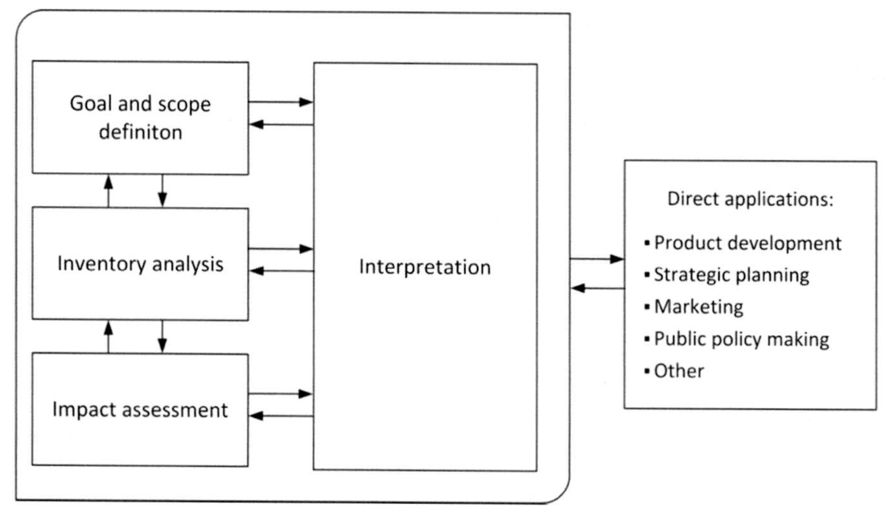

FIG. 2.1 Life cycle assessment framework. (Adapted from ISO (2006) [31])

environmental profile provided by the LCA is only part of the information needed in a decision-making process. Technical, economic, and social aspects must also be considered and a broad understanding of the relationships between them is crucial in the development of sustainable waste management practices.

The goal also defines the audience intended for the study and the questions that it proposes to answer. This is particularly important because the results of a study might vary significantly due to the differences in geographic boundaries, legislation, technologic aspects, waste composition, etc. Thus, the results should not be generalized or extrapolated outside the context of the study without careful analysis.

The goal of a waste management LCA could be, for example, to compare the environmental impacts of recycling of plastic bottles and their incineration with energy recovery in Sweden. In this case, the availability of the necessary infrastructure, the waste sorting scheme and its efficiency, and the energy supply mix of the country will influence the results significantly. If the same study would be conducted in another region under different conditions, it could lead to different conclusions.

LCA considers the upstream and downstream processes in a system. This feature is particularly important to avoid suboptimization of a system. By considering the entire waste management system and the full range of environmental impacts, LCA helps alleviate the problem of shifting the burden from one part of the waste management system to another. For example, the apparent benefits of a certain waste treatment might be offset by the increase in transportation distances.

Scope

The scope of a study defines the methodology to be used, the assumptions, the functional unit, the requirements for data quality, and the method for impact assessment.

Functional unit

The functional unit is the basis for comparison that is used in an LCA and serves as a reference unit for the inventory analysis. It aims to quantify the function of the product or service under study. In the context of waste management, the functional unit is often the management of certain amount of waste with a particular composition in a specified region. For example, if the goal of a study is to evaluate different alternatives for managing waste in a specific region, the functional unit could be defined as the amount of municipal solid waste (MSW) that is treated during 1 year in that region. Alternatively, other functional units could be used depending on the objectives of the study. For example, if the LCA focuses on AD, the functional unit could be 1000 tons of waste that enters the AD plant or it can be based on the output, such as 10 m^3 of biogas produced in the AD plant.

System boundaries

Typically, the system boundaries in waste management include all the activities from the collection, treatment, and through the final disposal of the residue. The system boundaries include the necessary infrastructure, such as material recovery facilities, biogas plants, waste-to-energy plants, collection vehicles, the energy used in the processes and facilities, and accessory equipment such as bailing wires.

Frequently, upstream processes before waste collection are not included. For example, in an LCA of the waste management of a plastic bottle, the impacts associated with the production and use of the bottle are often not included, i.e., it is assumed that the waste that enters the system has a zero burden associated with it [5]. Nevertheless, studies that consider waste prevention or minimization must take into account the impacts avoided during the production and use phases.

The geographic boundaries of waste management LCAs are often related to the city, region, or country where the waste is generated and treated. The geographic boundaries determine several important aspects in the LCA, such as the composition of the waste to be treated, the technologic level of facilities, and the type of electricity that is used. Most importantly, the system boundaries must be clearly defined in the LCA report. This step is crucial for the planning and collection of data in the inventory analysis.

Fig. 2.2 shows the system boundaries when performing an LCA of the management of some waste fractions, namely, food, residual, and plastic packaging wastes from households, in the city of Borås, Sweden. In the city's waste management system, the inhabitants are responsible for sorting the food and residual waste in bags of different colors, which are collected together and later separated via optical sorting in the city's materials recovery facility (MRF). The food waste is treated via AD, whereas the residual waste is incinerated with energy recovery. Another important stream in the system comprises packaging materials, here exemplified by plastic waste. In reality, the packaging stream comprises plastic and other fractions such as glass, newspaper, cardboard, and metals, which are collected separately. The system boundaries in Fig. 2.2 also include the utilization of the coproducts generated in the waste treatment facilities, such as electricity, heat, biogas, fertilizers, and recyclables (shown in the dashed boxes).

Multifunctionality and allocation

One of the challenges when modeling waste treatment systems is the fact that most of the processes are multifunctional, i.e., they provide the service of managing the waste as well as producing products such as heat, electricity, biogas, and fertilizer (Fig. 2.2). Thus the resources consumed and emissions from the facility must be divided among these services. The challenges associated with including multifunctional systems in LCAs have been debated since the LCA methodology was first developed. There is still no consensus on how to deal with this problem, and different approaches have been developed over time.

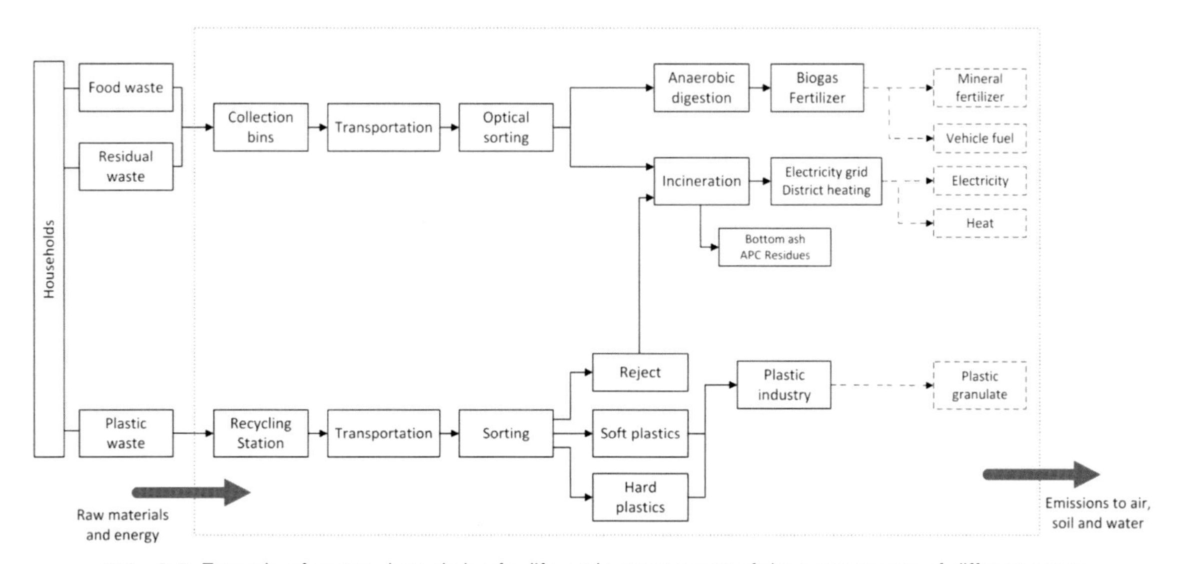

FIG. 2.2 Example of system boundaries for life cycle assessment of the management of different waste fractions in the city of Borås, Sweden. *APC*, air pollution control.

In a multifunctional process, the burdens must be shared between the different services provided by the waste treatment facility. For example, the environmental impacts of an incineration plant must be shared with the heat and electricity production, whereas the environmental burdens of a recycling plant must be shared with the secondary materials produced, e.g., paper, plastic, glass, and metals. Fig. 2.2 shows the additional functionality of the system as dashed boxes.

Consider that a waste manager wants to know if is better to incinerate or anaerobically digest the OFMSW in his/her municipality. Besides having different energetic and material demands—and consequently different emissions into air, water, and land—the two waste treatment systems also generate different products, namely, heat and electricity from the former and biogas and biofertilizer from the latter. One way to deal with this issue is to use system expansion, in which the system boundaries are expanded to include the additional services. Fig. 2.3 shows a hypothetical example in which two systems for waste management are being compared. System 1 provides the service of waste treatment and additionally produces some electricity, whereas system 2 only provides the service of treating the waste. In order to make the systems comparable, both must have the same outputs, i.e., both systems must provide comparable services. Therefore the environmental impacts of the 500 kWh of electricity that would have been produced by some other means are subtracted from the system 1 in order to make the systems comparable. The type of electricity substituted could vary according to the type of study. Typically, a consequential approach would use the marginal electricity market, whereas an attributional approach would use the average market for electricity production in the region.

Allocation is an alternative method to the partition of environmental burdens in a multifunctional system. Similar to system expansion, allocation distributes the environmental loads of the inputs and outputs of a system over its coproducts [6]. Allocation can be based on different criteria, with the two most common criteria being physical (mass, volume, energy, etc.) and economic. For example, when considering the transportation of a specific waste flow, e.g., plastic collected together with other waste materials, the burdens of the transportation should be divided relative to the mass or the volume of each product in the truck. Likewise, an incineration plant producing combined heat and power (CHP) might partition the burdens between the electricity and heat by using energy or exergy of the products as the criterion [7].

Economic allocation uses the revenue of the coproducts as the allocation criterion. Considering the same example as abovementioned for the CHP, if heat generates 30% of the plant's revenue and electricity generates 70%, the environmental burdens would be allocated using these percentages.

Inventory Analysis

This step is usually the most time-consuming step in an LCA. It includes the construction of a flowchart according to the boundaries defined in the goal and scope, the collection of data related to the activities within the system boundaries, and the calculation of the inputs and outputs in relation to the functional unit [8]. The inventory should include energy, raw materials and ancillary inputs, coproducts, waste, and emissions into air, water, and land.

The inventory of an MSW collection scheme should include, for example, the containers used for temporarily storing the disposed waste, the fuel used by the

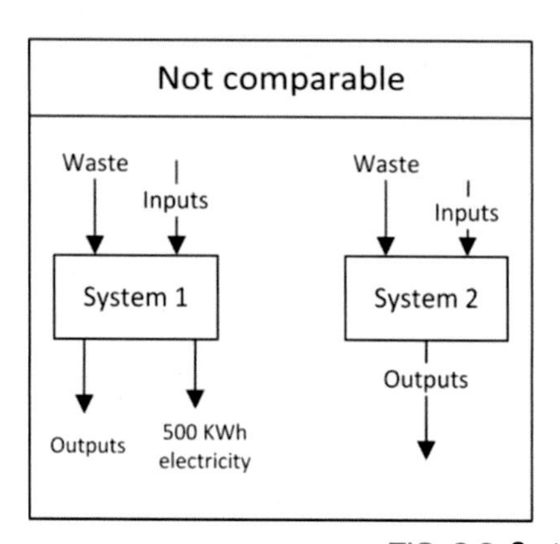

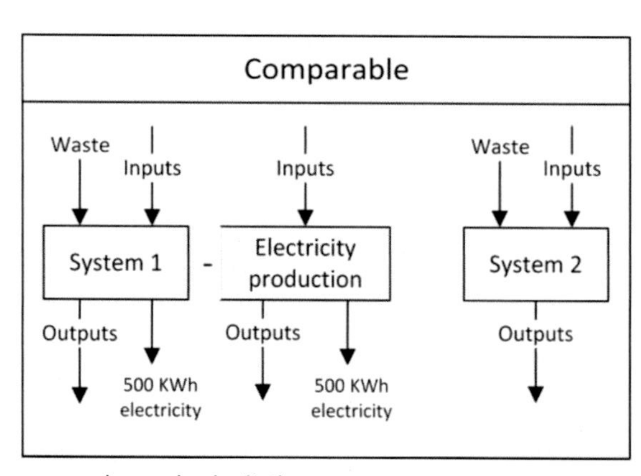

FIG. 2.3 System expansion and substitution.

collection vehicles, the production of the collection vehicles, and the emissions related to the process, such as the exhaust gases from the vehicles. It is clear that the amount of information necessary for the study can be large because, for example, steel is necessary to produce the collection vehicles, so steel production should also be included in the study. Most of the time, this background data can be found in the literature and in databases such as ecoinvent [9].

The focus during data collection is usually on the foreground data, i.e., data that refers to a particular process in the modeling. In the context of a waste management LCA, the foreground could be the waste composition in the area of the study, the sorting efficiency in a particular MRF, the energy consumption, type of trucks used for collection, or the emissions from a specific plant. The Overview of Modeling Aspects for LCAs of Waste Management section gives more examples regarding the data collection and other modeling aspects for different waste management systems. It is important to collect data over a long period in order to ensure that the data represents the average operation of the plant and that it accounts for typical variations in the data [3].

Impact Assessment

In this step, the environmental impacts are calculated based on the inputs and outputs gathered during the inventory analysis. The inventory analysis provides an extensive list of the different resources used and the emissions in to different compartments (air, water, and land) that must be translated into units that reveal the environmental impacts, such as global warming, acidification, eutrophication, and impacts on human health. There are numerous impact assessment methodologies that consider impacts in more or less detail, and most of these calculate the environmental impacts at the midpoint or end point of the impact chain. An end point indicator represents the environmental impact in the end of the cause-effect chain, whereas a midpoint indicator evaluates the impacts earlier in the chain [7]. For example, the emission of nitrogen compounds and their deposits in lakes increases the nutrient concentration, intensifies algae growth, and consequently increases the oxygen demand (midpoint), causing damage and increasing the risk of extinction of freshwater species (end point).

The Joint Research Centre of the European Union recommends midpoint characterization methods for several impact categories in the International Reference Life Cycle Data System (ILCD) Recommendations for Life Cycle Impact Assessment methods handbook [10]. Table 2.1 lists the recommended impact categories in the ILCD Life Cycle Impact Assessment method.

TABLE 2.1
Overview of the ILCD Recommend Impact Categories.

ILCD Midpoint Categories
Climate change
Ozone depletion
Human toxicity, cancer effects
Human toxicity, noncancer effects
Particulate matter
Ionizing radiation (human health)
Ionizing radiation (ecosystems)
Photochemical ozone formation
Acidification
Terrestrial eutrophication
Freshwater eutrophication
Marine eutrophication
Freshwater ecotoxicity
Land use
Water resource depletion
Mineral, fossil, and renewable resource depletion

ILCD, International Reference Life Cycle Data System.

Interpretation

The interpretation step includes the identification of the significant results based on the inventory and impact assessment analysis, the evaluation of the data quality, limitations, conclusions, and recommendations [11].

Hotspot analysis can be used to evaluate and identify central life cycle stages, processes, or elementary flows that have large contributions to environmental impacts. Moreover, the results, and their presentation, must consider the goal and scope of the study. LCA is an iterative process and preliminary results might lead to a revision of other parts of the LCA, such as the definition of the goal and scope.

Information provided by the LCA is often used together with other types of information, such as economic feasibility or social aspects, to support a decision-making process in a policy or business context. Thus it is crucial to understand that the results from an LCA are influenced by methodological choices. Assumptions, limitations, and uncertainties must be clearly defined in the LCA report and uncertain data should be tested through a sensitivity analysis, which investigates the influence of variations in these aspects on the results. For example, in recycling processes the recovery efficiency and market ratio can play a major role in the results, and if these parameters are not precisely known, it is crucial to perform a sensitivity analysis to understand the extent to which the results are influenced by them [12]. In a study on the performance of waste management systems for household waste in seven countries, Andreasi Bassi et al. [13] noted that the substitution rates for paper and metals, emissions from waste-to-energy plants, and household sorting efficiencies are among the most sensitive parameters in the modeling.

It is also important to check the quality of the data in relation to aspects such as time, geographic and technologic representativeness, completeness, and reliability [14]. The data quality in LCA can be assessed with the aid of a pedigree matrix [15]. The temporal representativeness indicates the correlation between the time when the data was collected and the time when the data was used in the study. Similarly, the geographic representativeness indicates if the data is from the area under the study, from another area with similar conditions, or from an area with different characteristics. Technologic representativeness indicates if the data represents similar processes and materials as the ones that are being studied. It is important to be careful when using data that is old, from other regions, or from systems that have different technologic levels, as this might introduce large uncertainties into the study [12]. The completeness criterion focuses on the statistical properties of the data, such as size of the sample and if the period of data collection is large enough to represent normal fluctuations in the operation, whereas reliability relates to the data sources and collection methods, i.e., if the data is based on measurements and assumptions or if it is estimated [16].

OVERVIEW OF MODELING ASPECTS FOR LCAS OF WASTE MANAGEMENT

This section describes some of the most important methodological issues and modeling aspects that must be taken into consideration when performing LCAs of various waste management systems.

Prevention

Waste prevention is the top priority in the waste hierarchy defined by the European Union [2]. Nevertheless, relative to other waste management pathways, the potential impacts and benefits of prevention measures are seldom quantified in LCA. Some research has studied the benefits of the prevention of unsolicited mail, packaging, and food waste [17,18]. Other examples of prevention measures are increasing the life span of products and promoting reusable instead of disposable packaging.

One way to assess the potential benefits of prevention measures is to expand the system boundaries to include the affected upstream processes. For example, the prevention of food waste in households would result in avoided burdens in agricultural production, transportation, processing, use, and waste management.

Indirect effects are often not included in the analysis of prevention measures, but they can have significant influence on the results. For example, when households decrease food waste, it results in less food being purchased that, in turn, results in economic savings, also known as the rebound effect. These savings can be used to purchase other products or services instead of food. Martinez-Sanchez and Tonini [18] concluded that depending on the type of products or services assumed to be purchased, the associated environmental burdens could offset the savings by preventing food waste.

Collection

Waste collection is the first step in the technical part of a waste management system. The collection scheme can gather mixed waste, waste that has been segregated into different fractions, or a combination of mixed and segregated wastes. The efficiency of the waste management system is directly influenced by the collection structure. For instance, the level of contamination of the materials that are sent for further treatment is determined by the collection scheme and by other factors such as sorting efficiencies at the households. In turn, the quality of the material is crucial when determining what type of treatment is feasible for each waste fraction, e.g., collection of pure plastic waste fractions allows higher rates of recycling, whereas highly contaminated mixed materials might be better suited for other treatment pathways, such as incineration [19].

The fuel consumption of a collection vehicle and consequently its emissions are often different from other types of transportation, as it operates in a stop-and-go mode and has additional equipment, such as hydraulic compactors, that increases the fuel consumption. Moreover, the fuel consumption is also affected by the amount that is loaded into the truck (load factor), which depends on factors such as the type and density of the waste. This determines if the transport is limited by the weight or volume of the waste. For instance, lower load factors are observed for biodegradable waste, which needs to be collected more frequently than other waste fractions to avoid inconveniences such as degradation of the waste and odor generation [1].

Recycling

Recycling processes can be categorized according to the quality of the material that is obtained from the recycling process. For example, when an aluminum can is recycled, the metal is melted and reprocessed into another aluminum can or another aluminum product, i.e., the quality of the material remains roughly the same as the primary material. This type of process is known as closed-loop recycling. In open-loop recycling the quality of the material changes, often with a loss in quality, to the extent that it cannot be used to produce products with the same quality (or value) as the original product. It is therefore used as a secondary material for another product. For example, plastic packaging can be recycled into fibers to be used in clothes. In this scenario, after the life cycle of the garment, the material will most likely not be recycled into fibers again.

It can be a difficult task to determine what products are substituted in an open-loop recycling. For instance, one might follow what happens to a recycled plastic until the plastic granulate is produced. These plastic granulates can be used not only to replace virgin plastic but also to substitute other materials, such as wood in the production of furniture.

A substitution rate is often used to describe how much production of the virgin material is avoided by the recycling process. The substitution ratio is calculated as the product of two variables, namely, the recovery efficiency and the market ratio [13,20]. The recovery efficiency refers to the efficiency of the process, i.e., the quantity of the material recovered from the recycling process in relation to the quantity of the material entering the process. The market ratio refers to the quality of the product after the recycling process, e.g., during recycling, paper loses some of its esthetic and mechanical qualities and the addition of new fibers is required to achieve a similar quality as the original paper. Table 2.2 lists the recovery efficiency and market ratio for different materials. Other indicators are also used to describe the efficiency of the system, such as the collection rate, which is the amount of material collected divided by the amount of material produced [21].

TABLE 2.2
Recovery Efficiency and Market Ratio for Different Materials.

Material	Recovery Efficiency	Market Ratio
Aluminum	0.93	1
Paper	1	0.83
PET	0.755	0.81
Glass	1	1

Adapted from Andreasi Bassi S, Christensen TH, Damgaard A. Environmental performance of household waste management in Europe – an example of 7 countries. Waste Management, 2017; 69: 545–557.

Anaerobic Digestion

AD is the process where microorganisms convert organic material in the absence of oxygen into mainly carbon dioxide and methane. The material remaining after AD is called the digestate. When modeling AD, some aspects must be taken into consideration, such as the operational parameters, e.g., temperature (thermophilic or mesophilic), hydraulic retention time, the degradation rate of the waste fraction, and leakage of biogas. Other factors, such as the electricity and heat efficiency generated during the combustion of the biogas, are also important.

TABLE 2.3
Potential for Methane Production From Different Substrates.

Substrate	Methane Yield (m^3 CH_4 $kg^{-1}VS$)	References
Expired food	0.47–1.10	[32]
Garden waste	0.1–0.2	[33]
Municipal solid waste	0.2–0.22	[34]
Household waste	0.4–0.50	[33]
Cow manure	0.15–0.30	[33]
Molasses	0.31	[33]

VS, volatile solids.
Adapted from Kabir MM. et al., Biogas from wastes: processes and applications, in Resource recovery to approach zero municipal waste. Taherzadeh MJ, Richards T, editors. USA:CRC Press; 2016.

The type and composition of the biogas that is generated depend on the substrate used. Table 2.3 exemplifies the potential for methane production in relation to the amount of volatile solids entering the system as different substrates.

Treatment of organic waste via AD generates two products: biogas and digestate. Biogas consists of carbon dioxide, methane, and traces of other gases such as hydrogen sulfide and ammonia. It can be combusted to generate electricity and heat, upgraded and injected into the grid for natural gas, or used as vehicle fuel. Depending on the intended use of biogas, different upgrading processes are necessary, and the environmental burdens associated with the downstream processes are therefore different. Similarly, the different uses of biogas substitute different products, which have different environmental burdens.

One important aspect that should be taken into account during the modeling of AD processes is the presence of rejects from the pretreatment steps. AD plants that use the OFMSW as substrate often receive wrongly sorted materials, such as packaging waste, mixed with the OFMSW. The inorganic material is not digested and often accumulates in the reactor, decreasing the available reaction volume. To prevent this, the waste fraction undergoes pretreatment in many plants. This can consist of optical sorting, a drum sieve, magnetic sorting of metals, and a mechanical filter in which most of the food waste is washed from the packaging material using water. The subsequent rejected material can contain large amounts of organic material and is hence not suitable for material recycling. It is therefore usually incinerated. For example, a composition analysis of the rejected material in a biogas plant in Borås, Sweden, showed that 44% of the organic mass is lost via the rejected material [22], leading to lower biogas and fertilizer production.

Composting

Composting is a biological aerobic process in which organic substrates decompose and stabilize, yielding a product that can be used for land applications [23]. There are different composting technologies, such as composting at homes, on fields, and at centralized plants. The environmental performance of a certain composting technology will primarily depend on the technology itself, the composition of the waste, and the operation of the process. Some important inventory aspects of the LCA are the waste composition and decomposition and the stabilization rates of the material that enters the system. Moreover, waste collection and distribution and spreading of the compost should be included in the LCA [24]. According to Saer et al. [25], the hotspot in the LCAs of composting is the emissions into air from the facility, such as methane, carbon dioxide, and ammonia.

If a composting process is properly operated, i.e., if aerobic conditions are maintained, methane emission is negligible. However, methane and hydrogen sulfide emissions can be produced if anaerobic conditions are formed during composting. Other typical emissions into air from composting are nitrous oxide (N_2O) and ammonia [1].

The material generated from the composting process has beneficial properties for application in soil, if it has adequate quality. It can provide nutrients, improve the structure of the soil, and increase microbial activity in the soil [24]. One benefit of using compost is the reduced amount of mineral fertilizer that is needed, and in this case the substitution rate should be based on the amount of nutrients available in the compost, which are often nitrogen, phosphorus, and potassium. Some research has also considered the fact that compost can substitute peat for soil conditioning [25].

Combustion

Combustion of waste has traditionally been performed primarily to reduce its volume, but over time, technology has developed and new systems have been introduced to reduce the environmental impact of incineration plants. This has been done by reducing harmful emissions into air, water, and soil and by recovering and using the energy that is produced during combustion. Current waste management technologies, especially incineration, not only reduce the volume of the waste but also provide a large part of the energy that is used in many countries.

Certain aspects must be taken into consideration when performing the LCAs of thermal technologies because they influence the emissions from, and the efficiency of, the plants. The composition of the waste is very important, as it affects the calorific content, humidity, and carbon and other elemental content. Ideally, any LCA should report the waste type (e.g., source segregated, mixed waste, industrial waste) and the chemical composition of the waste. The type of plant is also critical because of the different efficiencies in electricity and heat generation (Table 2.4) as are the technologies used for the treatment of flue gases and NO_x and dioxin and dust removal [1,26]. The type of management of the residues generated in the facilities can also affect the LCA results. For instance, bottom ash, after the removal of metals, can be used as a substitute for gravel in road construction, landfilled, or stored in underground deposits.

The heat and electricity produced in waste-to-energy facilities are of major importance for the environmental performance of a plant. The most common approach is to use system expansion (see Section 2.1.5) to consider the avoided burdens from energy production [26]. The type of energy that is substituted varies according to the scope of the study. First, a facility can produce both heat and electricity, or only one of them. Furthermore, the

TABLE 2.4
Range of Energy Recovery Efficiencies for Different Waste Incineration Technologies.

	Gross Electricity Efficiency (%)	Net Electricity Efficiency (%)	Net Heat Efficiency (%)
Incineration	0–34	−2 to 30	0–87.7
Co-combustion in cement kilns	4.38		
Co-combustion in power plants	34–40	34	26–40

Adapted from Astrup TF. et al. Life cycle assessment of thermal waste-to-energy technologies: review and recommendations. Waste Management 2015;37:104–115.

energy substitution in the system expansion can be based on marginal sources (consequential approach) or the average mix from the energy supply mix of the region (attributional approach), or the energy produced can substitute specific fuels, such as coal in the case of combustion in cement plants.

Some studies might be interested in the combustion of a specific waste fraction in order to identify more sustainable waste management practices for this fraction. For instance, one could compare the incineration and recycling of plastic bottles. In this case, part of the energy produced in the incineration plant must be allocated specifically to plastic bottles. In this case, it is recommended to allocate the electricity and/or heat production based on the calorific value of the plastic bottles relative to the other materials in the waste. Similarly, emissions into air, water, and land should also be allocated based on some criteria. Carbon dioxide emissions, for example, are proportional to the carbon content of the waste fraction and should thus be allocated using the carbon content of the material as the criterion. Other emissions are also specific to the type of waste fraction, but they are also influenced by the technology employed. In this case, one might use the calorific content of the residue as the allocation criterion [1].

Landfill

Modeling the environmental impacts of landfills is complex because, in contrast to other types of waste treatments, it is difficult to make direct measurements in landfills. Thus it is often necessary to model and estimate landfill emissions [27].

There are several important aspects to consider when modeling the environmental impacts of landfills. The emissions and performance depend on the technology employed, which varies from open dumps that do not have any measures to control the emissions of leachates and gas to landfills with gas collection and energy recovery [28]. Leachate and gas collection, with subsequent treatment, is critical to reduce the environmental impacts of landfills.

Emissions from landfill can occur over a long time. Consequently, definition of the time horizon of the LCA is key when considering, for instance, the amount of pollutants emitted and the quantity of sequestrated carbon, i.e., carbon that is left in the landfill at the end of the stipulated time. Carbon sequestration may be considered as a positive aspect of landfills because it decreases, for example, the impact to climate change [28]. Hence, the time horizon of the study should be long enough so that most of the material has decomposed. However, in practice, there are limited data available that consider long periods [1].

The location of the landfill or dump determines the precipitation, temperature, and soil characteristics, which also influence the emissions, e.g., leachate generation [1]. The type of waste, i.e., mixed waste or inert, will define the type, quantity, and rate of production of gas and leachate. The amount of gas generated and its distribution over time depend on the degradability of the waste, which is often modeled using a decomposition rate. The decomposition rate varies according to the type of waste, which can be slow, e.g., paper and wood; moderate, e.g., garden waste; or rapid, e.g., sewage sludge and food waste [29]. The gas generated from the landfills, if collected, can be used to produce energy, namely, heat and electricity, and can thus avoid environmental impacts when substituted for other sources of energy, e.g., coal.

CONCLUSION AND PERSPECTIVES

LCA is an important tool to support decision-making processes and it is used in the waste management context to identify alternatives with low environmental impacts. Its holistic approach, which covers the life cycle of the service or product, decreases the risk of shifting activities that have large environmental impacts between different stages in the waste management system.

Nevertheless, sustainable solutions should combine the environmental information given by an LCA with technologic, social, and economic data.

This chapter provides guidance in some of the modeling aspects that should be considered when assessing the environmental impacts of waste management systems. LCA modeling of waste management systems is influenced by the conditions in and the specificity of the area under study, because of variables such as waste composition, sorting efficiencies, and technologies employed in the system. Hence, extrapolation of results to other regions should be done with caution. Yet in a review of 222 LCA studies of solid waste management systems, Laurent et al. [30] noted some trends in the results. For example, the majority of the studies for paper and plastic waste show that recycling and thermal treatment are not as harmful for the environment as landfilling. However, the results also show that there is no consensus over which types of technology have the lowest environmental impacts for each type of waste fraction, with the exception of landfilling, which is ranked as having the worst performance for all materials in the majority of studies. This not only reinforces the argument that LCA is an important tool that can guide authorities to make better environmental decisions but also shows that regional characteristics play an important role and extrapolations must be done with caution.

REFERENCES

[1] Manfredi S, et al. Supporting environmentally sound decisions for waste management with LCT and LCA. International Journal of Life Cycle Assessment 2011; 16(9):937–9.

[2] European Commission. Directive 2008/98/EC of the european parliament and of the council of 19 November 2008 on waste. Official Journal of the European Union L 2008:22. 11.

[3] Hauschild M, Barlaz MA. LCA in waste management: Introduction to principle and method. Solid Waste Technology & Management 2010;1:111–36.

[4] European Commission, COM. In: C.o.t.E. Communities, editor. 302 – integrated product policy building on environmental life-cycle thinking; 2003. 2003: Brussels.

[5] Ekvall T, et al. What life-cycle assessment does and does not do in assessments of waste management. Waste Management 2007;27(8):989–96.

[6] Weidema B. Avoiding co-product allocation in life-cycle assessment. Journal of Industrial Ecology 2000;4(3): 11–33.

[7] Goedkoop M, et al. Introduction to LCA with SimaPro 7. The Netherlands: PRé Consultants; 2008.

[8] Baumann H, Tillman A-M. The Hitch Hiker's guide to LCA. Studentlitteratur; 2004.

[9] Frischknecht R, et al. Implementation of life cycle impact assessment methods. Data v2. 0. Ecoinvent Centre; 2007. Ecoinvent report No. 3. 2007.

[10] EC-JRC. International Reference Life Cycle Data System (ILCD) handbook recommendations for life cycle impact assessment in the European context. Luxembourg: European Commission Joint Research Centre, Institute for Environment and Sustainability; 2011.

[11] ISO. ISO 14044:2006 environmental management: life cycle assessments: requirements and guidelines. I.S.O. Geneva: International Standardization Organization; 2006.

[12] Laurent A, et al. Review of LCA studies of solid waste management systems – Part II: methodological guidance for a better practice. Waste Management 2014;34(3):589–606.

[13] Andreasi Bassi S, Christensen TH, Damgaard A. Environmental performance of household waste management in Europe – an example of 7 countries. Waste Management 2017;69:545–57.

[14] Zampori L, et al. Guide for interpreting life cycle assessment result. EUR; 2016.

[15] Weidema BP, Wesnæs MS. Data quality management for life cycle inventories—an example of using data quality indicators. Journal of Cleaner Production 1996;4(3): 167–74.

[16] Weidema BP, et al. Overview and methodology: data quality guideline for the ecoinvent database version 3. 2013.

[17] Gentil EC, Gallo D, Christensen TH. Environmental evaluation of municipal waste prevention. Waste Management 2011;31(12):2371–9.

[18] Martinez-Sanchez V, et al. Life-cycle costing of food waste management in Denmark: importance of indirect effects. Environmental Science and Technology 2016;50(8): 4513–23.

[19] De Feo G, Malvano C. The use of LCA in selecting the best MSW management system. Waste Management 2009; 29(6):1901–15.

[20] Rigamonti L, Grosso M, Giugliano M. Life cycle assessment of sub-units composing a MSW management system. Journal of Cleaner Production 2010;18(16):1652–62.

[21] Ligthart TN. Modelling of recycling in LCA. In: Post-consumer waste recycling and optimal production. InTech; 2012.

[22] Brancoli P, Rousta K, Bolton K. Life cycle assessment of supermarket food waste. Resources, Conservation and Recycling 2017;118:39–46.

[23] Haug R. The practical handbook of compost engineering. Routledge; 1993.

[24] Sánchez A, et al. Composting of wastes. In: Taherzadeh MJ, Richards T, editors. Resource recovery to approach zero municipal waste. CRC Press; 2015.

[25] Saer A, et al. Life cycle assessment of a food waste composting system: environmental impact hotspots. Journal of Cleaner Production 2013;52:234–44.

[26] Astrup TF, et al. Life cycle assessment of thermal Waste-to-Energy technologies: review and recommendations. Waste Management 2015;37:104–15.

[27] Obersteiner G, et al. Landfill modelling in LCA – a contribution based on empirical data. Waste Management 2007;27(8):S58–74.
[28] Damgaard A, et al. LCA and economic evaluation of landfill leachate and gas technologies. Waste Management 2011;31(7):1532–41.
[29] Pipatti R, et al. Solid waste disposal. In: Eggleston S, et al., editors. 2006 IPCC guidelines for national greenhouse gas inventories. Japan: Institute for Global Environmental Strategies (IGES); 2006.
[30] Laurent A, et al. Review of LCA studies of solid waste management systems – Part I: lessons learned and perspectives. Waste Management 2014;34(3):573–88.
[31] ISO. ISO 14040:2006 Environmental management–life cycle assessment–principles and framework. I.S.O. London: International Standardization Organization; 2006.
[32] Braun R, Brachtl E, Grasmug M. Codigestion of proteinaceous industrial waste. Applied Biochemistry and Biotechnology 2003;109(1–3):139–53.
[33] Angelidaki I, Ellegaard L. Codigestion of manure and organic wastes in centralized biogas plants. Applied Biochemistry and Biotechnology 2003;109(1–3):95–105.
[34] Chynoweth DP, Owens JM, Legrand R. Renewable methane from anaerobic digestion of biomass. Renewable Energy 2001;22(1–3):1–8.
[35] Kabir MM, et al. Biogas from wastes: processes and applications. In: Taherzadeh MJ, Richards T, editors. Resource recovery to approach zero municipal waste. USA: CRC Press; 2016.

CHAPTER 3

Waste Biorefinery

JORGE A. FERREIRA, PHD • SWARNIMA AGNIHOTRI, PHD •
MOHAMMAD J. TAHERZADEH, PHD

INTRODUCTION

Fossil fuels have long been recognized as nonrenewable and related to atmospheric accumulation of greenhouse gases. This led to the development of renewable processes using both nonbiomass sources (e.g., wind, solar, geothermal, and hydroelectric) and biomass sources (for direct combustion or production of value-added products via microbial conversion). By combining microorganisms with biomass, an avenue was established for the commercial production of biofuels, biochemicals, and miscellaneous materials leading to the genesis of a bioeconomy.

Industrialization, population growth, and improvement of quality of life exacerbated the pollution problem related to the Earth source's exploration and the steadily increasing amount and diversity of wastes. Due to the increasing public awareness, the society has assisted to a paradigm shift from a linear (make, use, dispose) to a circular economy based on a reuse and recycle/recover philosophy.

Microorganisms have been playing an increasing role in the establishment of a circular bioeconomy (Fig. 3.1), through valorization of wastes or residuals and industrial side streams to various value-added products within low environmental footprint closed-loop processes [1]. This approach has received increasing attention over the years by addressing not only environmental and energy security concerns but also ethical aspects related to the diversion of human food substrates for microbial bioconversion. A good example is the production of the so-called first-generation bioethanol, which relies to a major extent on agricultural crops (corn, wheat, sugarcane, etc.) [2].

The use of lignocellulosic materials as a platform for the production of various products is one excellent example of waste or residual management that has received a great deal of attention. A significant part of the research carried out has been toward the production of bioethanol; however, intrinsic process constraints have rendered their commercial establishment inexistent at present despite decades of research [3]. A more recent approach includes their integration into existent ethanol facilities to alleviate capital costs [4]. This can also be widespread to all kinds of wastes together with valorization of intrinsic industrial side streams. This approach can fasten the commercialization of processes centered on waste management via upgrading existing processes. This would, in turn, lead to an optimized use of resources and a wider range of products produced [3]. This represents the concept of biorefinery referred in the International Energy Agency (IEA) Bioenergy Task 42 [5]: "Biorefining is the sustainable processing of biomass into a spectrum of marketable products and energy." Thus the overall goal is similar to petroleum refineries, producing multiple fuels and chemicals [6]. The global market for biorefinery technologies is projected to have a compound annual growth rate (CAGR) of 8.9% within the period 2016–21, reaching $714.6 billion [7].

The supply of wastes to biorefineries requires a good characterization of their effect on established processes. At first glance, separation at source should be needed for removal of impurities such as glass and metal from the biomass-derived wastes because they are not degraded during biological conversion. Moreover, mixed lignocellulose-containing substrates can represent a challenge because, as will be discussed further, a substrate composition–specific pretreatment step is needed to have access to sugar polymers.

This chapter aims to provide an overview of the available industrial microbial processes and the research avenues studied for their upgrade into waste biorefineries.

Sustainable Resource Recovery and Zero Waste Approaches. https://doi.org/10.1016/B978-0-444-64200-4.00003-7

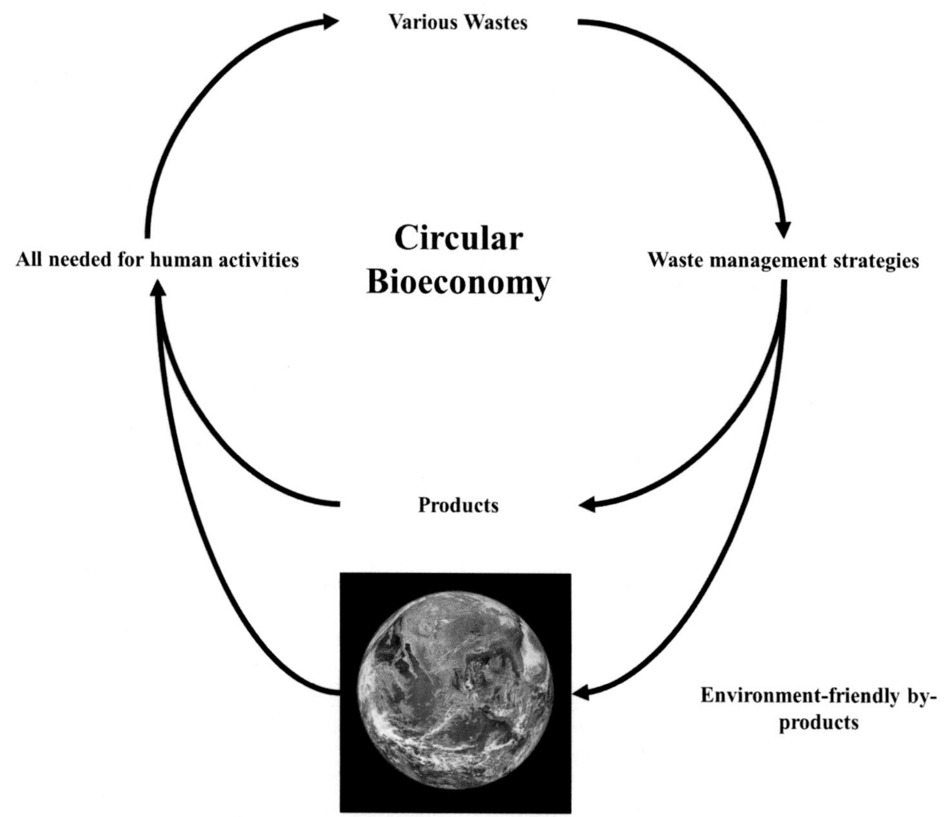

FIG. 3.1 Overall reasoning behind a circular bioeconomy concept. In case there are by-products originating from waste valorization strategies, they are of low-harm level to the environment (e.g., well-treated process waste streams).

VALORIZATION OF WASTE FOR FUELS AND CHEMICALS BY MICROORGANISMS

The diversity of microorganisms existing on Earth with their related diversified enzymatic machineries opens up the possibility to valorize a wide range of substrates that are available as wastes (Fig. 3.2). Substrates for microbial valorization range from simple sugar-containing substrates, such as waste streams from paper and pulp industries rich in hexose and pentose sugars, to more complex substrates containing dimers (e.g., sucrose in vinasse from sugarcane-based ethanol industries) and polymers such as starch-containing streams and all kinds of lignocellulosic materials. Overall, municipal solid waste (MSW) and streams from paper and pulp, food, ethanol (both as fuel and spirit), agricultural, and forest industries represent the dominant waste sources that require management routes [2,8].

The use of lignocellulosic materials gives rise to the most complex processes due to the material's recalcitrance. Lignocellulosic materials are mostly composed of cellulose, hemicellulose, and lignin organized in a compact and recalcitrant structure, so a simple enzymatic step is not enough to gain access to the structural polymers and depolymerize them into monomeric sugars [9]. A pretreatment step, being physical, chemical, biological, or their combination, is needed to open up the structure so that enzymes can catalyze the hydrolysis of cellulose and hemicellulose into single sugars [10]. This adds considerable costs to the overall production of value-added products (e.g., biofuels) and, moreover, increases complexity, as a single pretreatment step cannot be used for a wide range of substrates [11]. Lignocelluloses vary based on the polymer composition, which determines which pretreatment strategy will be the most suitable. A wide variety of pretreatment methods have been developed leading to different outcomes [11]. For instance, pretreatment by acid bisulfite, the method used by paper and pulp industries, dissolves both lignin and hemicellulose leaving cellulose intact [12], whereas the main effect

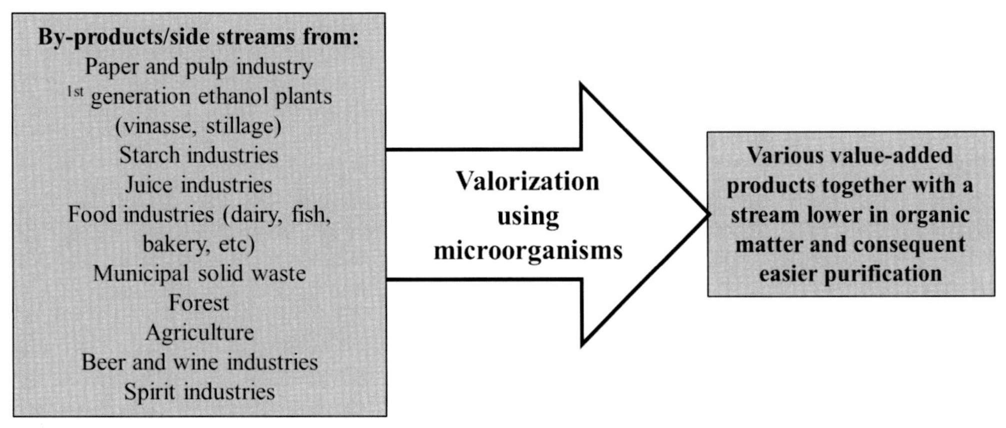

FIG. 3.2 Examples of available wastes that can be converted to various value-added products by microorganisms.

of hydrothermal pretreatment is the dissolution of hemicellulose [13]. Moreover, if lignin is of interest, organosolv pretreatment is used for easier polymer separation [14]. Therefore the choice of pretreatment will be based on the type of substrate and which fraction is of interest for further processing. In addition to pretreatment costs and the related solvent recycling, the efficient processing of lignocelluloses is further hampered by the inhibitors formed during pretreatment that can inhibit microbial growth and by the lack of cost-effective enzymes in the market [15].

It should be noted that a pretreatment step can also include the separation of impurities, such as plastics, glass, metals, or removal of any potential growth inhibitors before the substrate undergoes the biological conversion step.

Nowadays, the scientific community has accepted the slogan "waste as a value" and research has been developed toward valorization of, in principle, all kinds of wastes available. Nonetheless, due to their structural and compositional differences, the establishment of industrial processes including efficient waste management routes has proven troublesome. This is related to several issues including separation and purification challenges due to low yields, lack of a robust microorganism, and intrinsic process requirements (nutrient supplementation, proper substrate, and oxygen transfer rates).

Evidently, specific management routes have to be constructed that meet the specific needs of each substrate or of a closely related group of substrates (e.g., having similar composition and/or can be pretreated by using the same strategy). The following sections provide examples of the success of industrial processes using waste materials for production of biogas, compost/vermicompost, volatile fatty acids (VFAs), and biodiesel. A section is also dedicated to bioethanol because of the existing research on the integration of waste materials into established processes.

CURRENTLY ESTABLISHED PRODUCTS: PROCESSES AND APPLICATIONS

Biogas From Organic Municipal Wastes

Anaerobic digestion (AD) for biogas production, together with composting (see the section Compost and Vermicompost From Organic Municipal Wastes), plays a crucial role in the management of food waste (FW) considering that it accounts for 35%–45% of the total MSW with an annual food loss of 1.3 billion tonnes with exponential increasing trends [16]. Other substrates within the organic fraction of municipal solid waste (OFMSW), including paper, cardboard, and garden waste, are also potential substrates for AD [17]. FWs are recycled mainly as animal feed and compost. Considering their composition that includes carbohydrates, protein, and lipids, they are also potential substrates for AD for biogas production or other processes producing other value-added products [18]. This composition actually dictates if a given substrate is of high or low potential for AD for biogas production. AD has become a very relevant technology where it is applied in developing countries using small-scale, homemade bioreactors for cooking, whereas industrial processes have been established in developed countries for heat and electricity supply or replacement of natural gas if biogas is upgraded to biomethane [19].

Biogas is actually a mixture of components in which methane (45%–75%) and carbon dioxide (25%–50%) dominate, together with water vapor (1%–5%), nitrogen (0%–5%), and lower amounts of ammonia and hydrogen sulfide, as well as trace amounts of hydrogen and carbon monoxide [18]. Differentiated impurity removal is carried out according to the final application [20]. Biogas is a result of a complex degradation process composed of four interdependent, sequential, parallel pathways (Fig. 3.3). The organic matter is first decomposed into its building blocks through a hydrolysis step, and these are converted to carboxylates via acidogenesis or to acetate and hydrogen and carbon dioxide via acetogenesis where such conversion is also carried out from the originated carboxylates. Lastly, the products from acetogenesis are converted to methane and CO_2 through methanogenesis [21]. A more extensive description of each of the degradation steps in AD can be found in the work by Kabir and Forgács [19]. Different consortia of microorganisms, belonging to both bacteria and archaea domains, with different growth requirements, are involved in each degradation stage [22]. Reasonably, despite established AD processes, knowledge gaps exist on the biochemistry and microbiology of such complex systems [19].

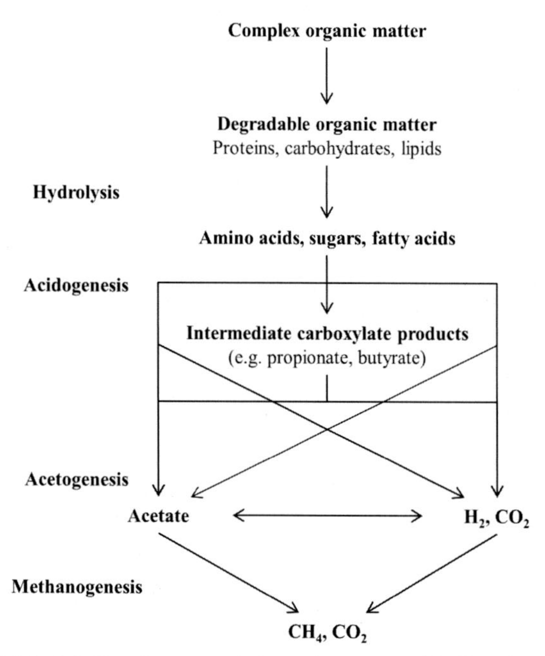

FIG. 3.3 Reaction cascade throughout anaerobic digestion. (Adapted from [19].)

The performance of AD can be influenced by operational conditions, including reactor design (most used designs are continuous stirred tank and plug flow bioreactors), substrate characteristics (e.g., composition and particle size and shape), substrate mixing (codigestion), and by environmental factors including temperature, nutrients, C/N ratio, pH and alkalinity, hydraulic retention time, and organic loading rate [23]. A major fraction of anaerobic digesters operate under mesophilic (around 35°C) and thermophilic (around 55°C) conditions [24]. The hydraulic retention time becomes a more influencing factor when complex and difficult-to-digest substrates are used [25]. Substrate impurities such as inert materials (e.g., sand, clay, and glass) or floating materials (e.g., plastic) need to be removed before AD, with concomitant increase in the process costs [19]. Interesting aspects are related to codigestion because of AD stabilization by complementing C/N ratio and nutrients and by tackling ammonium inhibitions, among others, where a centralized AD concept was created via agricultural waste digestion leading to waste contribution by several farms to a single, large-scale digestion plant [26]. The importance of codigestion, leading to a more stable process and higher biogas yield, has been proven by mixing slaughterhouse waste with animal manure and agricultural and FWs [19]. Research and development has led to the establishment of AD processes working under batch or continuous, wet or dry (total solid content higher than 15%), and single-, two-, or multistage systems [19]. Two- or multistage systems have been developed based on the variety of microbial consortia involved in AD processes, where more favorable or optimal conditions can be more specifically provided at each stage [27].

Global biogas production technologies are expected to grow at a CAGR of 10.6% and reach $10.1 billion by 2022. Around 60% of the global biogas production comes from landfill gas, food and municipal wastes, and agricultural segment, while the remaining originates from the wastewater/sludge and industrial segment [28].

Volatile Fatty Acids From Organic Wastes

VFAs are produced as important intermediate products during the acidogenic and acetogenic processes of AD of organic materials, usually FW. At present, 90% of the commercially available VFAs rely on petro-based methods. The total global market demand for VFAs (acetic, butyric, and propionic acid) is estimated to be

18,500 kilotonnes by 2020 [29], as they have several applications as potentially renewable carbon sources [30–32]. An added value at \$50–\$130/tonne [33] compared with that of methane at \$0.72/m^3 is also projected because VFAs are preferred over methane and biohydrogen to be recovered from anaerobic fermentation of FW [34]. Generally, the degradation of complex polymers such as lignocellulosic materials, lipids, and proteins in FW to smaller molecules is the rate-limiting step during AD process [35]. Hence, hydrolysis is regarded as the rate-limiting step in the acidogenic fermentation of organic wastes. Thus the pretreatment of FW is considered as a favorable way to enhance VFA production from acidogenic fermentation of FW. Chemical (acid and alkaline), physical (thermal, microwave, and ultrasound), and biological (enzymes) methods can be used in the pretreatment of organic waste such as FW, and these methods were shown by Kim et al. [36] to have enhanced the soluble chemical oxygen demand (COD) generation and VFA production. The rate-limiting hydrolysis step can be enhanced by optimizing the key operating factors, which would in turn increase the readily available carbon substrate for subsequent conversion to VFAs. Promoting the acidogenic process to maximize the conversion of hydrolyzed soluble substrates to VFAs, while removing the inhibiting factors and preventing the methanogens, has been shown to positively influence VFA production from FW [37,38]. Critical factors that impact the quality of VFAs and their production during the acidogenic process include substrate, inoculum, pH, temperature, hydraulic retention time, organic loading rate, and headspace gas [39]. More importantly, an optimum pH (5.2–6.4) obtained by maintaining a good buffering capacity of the fermentation system was shown to improve the hydrolysis and acidogenesis of substrates, which leads to an increased production of VFAs [40].

Compost and Vermicompost From Organic Municipal Wastes

Through composting, biological decomposition and stabilization of organic matter takes place at conditions allowing the development of biological heat-originated thermophilic conditions leading to a stable compost free of pathogens and plant seeds for land application [41]. Therefore the efficiency of composting depends on factors related to biological processes including those that favor microbial growth (C/N ratio, pH, moisture, etc.) together with aeration and related porosity [42]. In order to reunite optimal conditions for biological conversion, combining waste substrates, within the concept of co-composting, is common practice [43]. Several studies, reunited in the reviews made by Godlewska and Schmidt [42] and Wu and Lai [44], have been conducted on the effect of adding biochar to the organic matter for co-composting. Biochar influences the physicochemical parameters of composting, including compost stability and pH, lower loss of nutrients, increased nitrification, reduced bioavailability of heavy metals, and reduced greenhouse gas emissions. It should be highlighted that although the compost is applied to land, it does not function as a land fertilizer, but as a complement of some nutrients and mainly an organic amendment [43].

Waste biodegradability is also an important composting factor because it will dictate the compost's quality and the environmental impacts of composting. Accordingly, having a good knowledge of the waste that is going to be used for composting is of utmost importance, as it will point toward the best composting strategy and use of the originated compost. Normally, a dynamic respirometric method is used to evaluate waste biodegradability [45]. A discussion of the methods used for biodegradability evaluation and their limitations has been reviewed elsewhere [43].

By considering respiratory indices, wastes have been classified into three categories [46]:

1. Highly degradable wastes, with respiratory activity higher than 5 g O_2 kg^{-1} DM h^{-1} (where DM stands for dry matter) (includes nondigested municipal wastewater sludge and animal by-products).
2. Moderately biodegradable wastes, with respiratory activity within 2–5 g O_2 kg^{-1} DM h^{-1} (includes mixed MSW, digested municipal wastewater sludge, and manure).
3. Wastes of low biodegradability, with respiratory activity lower than 2 g O_2 kg^{-1} DM h^{-1} (includes digested sewage sludge).

Respiratory indices have been playing a role in the formulation of waste mixtures for optimal composting conditions [43].

Composting has its greatest influence on the reduction of one of the mainstreams of MSW, that is, FW [43]. Having a circular economy concept in mind, the so-called 2V "vegetal to vegetal" model has been developed (Fig. 3.4). Considering the model, home composting can have important influences including the commitment and awareness of the society, concomitantly avoiding the transport of waste, organic fertilizers, and vegetables to retailers [43].

The composting technology is more than 100 years old with established home and industrial processes [41]. The limiting step of composting is not achieving

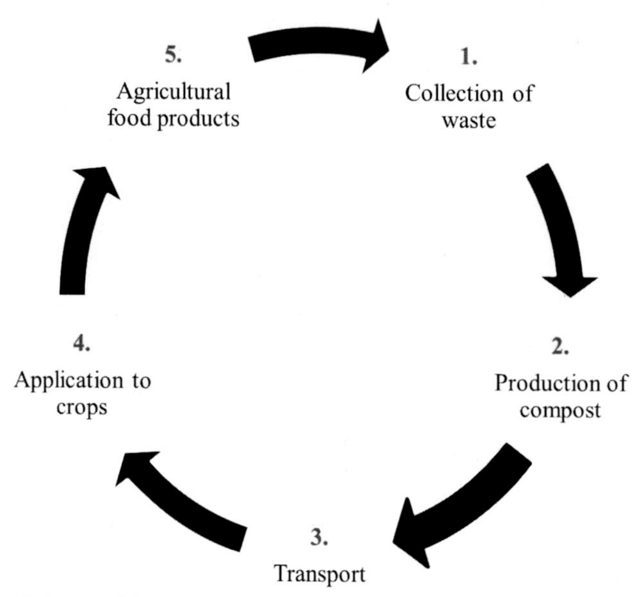

FIG. 3.4 2V "vegetable to vegetable" model. (Adapted from [43].)

high biodegradation rates but choosing the best technology regarding cost efficiency and associated environmental impacts [43].

In addition to home and industrial composting, the latter being divided into simple and in-vessel systems, vermicomposting has been implemented at industrial scale, but to a lower extent. In vermicomposting, microorganisms work together with the main drivers, earthworms, to attain stabilization of organic solid wastes through earthworm consumption converting waste to earthworm castings (vermicast) rich in plant nutrients and free of pathogens [47]. The earthworm species *Eisenia andrei* and *Eisenia fetida* have been found to be the most suitable for waste treatment. Vermicomposting has been shown to be a suitable technique for biodegradation of wastes such as sewage sludge, agro-industrial wastes and sludge, cattle manure, and urban solid wastes [48]. A description and discussion of home composting as well as of the different industrial processes can be found elsewhere [41,43].

Although composting and AD lead to important results in the management of MSW, OFMSW, waste sludge, and manures, their output is distinct [43]. On one hand, it has been shown that the stability of compost is higher than that of the digestate obtained after AD, which can lead to phytotoxicity [49] and malodor [50] when the latter is applied to land. On the other hand, AD is an energy-producing technology (100–150 kWh per tonne of OFMSW) [51], whereas composting is an energy-consuming technology (65.5–242 kWh per tonne of OFMSW) [43]. Accordingly, composting or AD followed by composting has been proposed, with the former being considered as the suitable strategy to treat OFMSW containing low to medium level of impurities in rural areas of low- to medium-density population, whereas the combined anaerobic and aerobic plants with mechanical pretreatment (the so-called mechanical biological treatment [MBT] plants) are considered as the suitable strategy for areas of high-density population where OFMSW might have a higher fraction of impurities [43]. On the negative side, the compost obtained from MBT plants ends mostly in landfills owing to its unsuitability for land application [43].

According to the 2018 report from Eurostat, in 2016, EU-28 has composted 16.9% of the municipal waste, whereas the United States and Canada have composted 8.9% and 7.9%, respectively. Composting is absent in countries such as Mexico, Turkey, China, Australia, South Korea, and Japan [52].

Bioethanol From Lignocellulosic Waste or Residuals

Bioethanol production, together with biogas and biodiesel production, although at a comparatively higher extent, represents the most successfully green process in the production of biofuels as an alternative to fossil fuels. The primordial end use for this alcohol is in the transport sector replacing oil-based gasoline but its use/presence in chemicals and beverages (e.g., produced during wine and beer production) is also accountable.

Bioethanol is classified as first, second, and third generations when produced from food-competing crops, lignocellulosic materials, and algal biomass, respectively [4,53]. Nowadays, commercial-scale production of bioethanol is carried out using food-competing crops such as corn in the United States, sugarcane in Brazil, and wheat and sugar beets in Europe (c.f. Renewable Fuels Association). Other substrates, normally used for human consumption, including cassava, sweet potato, oats, rice, and fruits, are also used for ethanol production [54–56].

Irrespective of whether the debate "food versus fuel" still holds true presently, the research community has accepted that the full replacement of fossil fuels cannot rely solely on the bioethanol produced from food-competing crops. Consequently, lignocellulosic materials became the *"in focus"* substrate for bioethanol production. However, using lignocellulosic materials adds complexity to the process of bioethanol production for reasons aforementioned. Production of bioethanol can be carried out directly after sugarcane crushing and bagasse separation because the baker's yeast *Saccharomyces cerevisiae* can consumed directly sucrose, and when producing bioethanol from starchy crops, starch dissolution and hydrolysis by the enzymes α-amylase and glucoamylase, respectively, to glucose monomers is needed before yeast fermentation. The use of lignocellulosic materials for bioethanol production adds another step to the overall process, namely, the pretreatment step to open up the recalcitrant structure (Fig. 3.5).

A great deal of research has taken place on the production of bioethanol from lignocelluloses, and facilities have been built in the United States, Brazil, Italy, and Slovakia using crop residues (rice, rapeseed, and wheat straw; corn stover; sugarcane bagasse and straw; corn kernel), wood chips, dedicated energy crops such as switchgrass, and MSW, producing 14–114 million L/year [57,58]. There is also production of cellulosic ethanol in Sweden and Norway from spent sulfite liquor, a waste stream from the production of pulp and lignin processing [58]. Interestingly, the ethanol produced by Sekab in Sweden is used for chemical production [58]. However, only 1.1% of the overall ethanol production in the United States in 2017 was of cellulosic origin (c.f. Renewable Fuels Association). Plans exist for construction of plants in the United States using *Miscanthus* and switchgrass, in Norway from sawdust, and in Finland using sawdust and recycled wood [58]. The Indian company Praj also plans to build several facilities through collaboration with Indian oil marketing companies. According to the IEA forecasts, a sevenfold increase in the production of biofuels such as cellulosic ethanol is expected to take place by 2022 [59]. The year 2017 was considered a good year for cellulosic ethanol production, and further development forecasts are supported by the need to meet the goals of the Paris agreement [59]. It should be highlighted, however, that the successful establishment of cellulosic ethanol is dependent on external economic supports, as a stand-alone second-generation facility is not feasible [4].

Another strategy has been developed that can increase the rate at which cellulosic ethanol reaches large-scale production, namely, the integration of first- and second-generation ethanol plants [4]. This strategy takes advantage of all previous investments on equipment and, therefore, lowers capital investment risks. Additionally, all investments done for the first-generation ethanol plants could be used for upgrading the plants into biorefineries with related social and environmental impacts via intake of waste materials [4] (Fig. 3.6). The analysis of such an integration strategy has been more dominant within the Brazilian sugarcane-based ethanol production industry due to the tradition of using bagasse for energy production [60]. Altogether, the substrates and process strategies that will dominate large-scale cellulosic ethanol production remain to be seen, as the technology is still under development.

Feed Products From Side Streams

As a result of increase in human population, pressure on the production of protein sources, mainly meat and fish, has intensified. Accordingly, an increased need for feed products has been seen throughout the years to increase the production of poultry, beef cattle, dairy cattle, and swine, as well as fish produced within the steadily growing aquaculture sector. Naturally, food-competing sources have also been used for animal feed, including soya and fish oil. However, a number of animal feed products have been created through valorization of waste streams of various processes. One example includes the waste streams from ethanol production, for both fuel and spirit applications. After distillation of ethanol for fuel applications, all the remaining components in the medium leave the distillation column and undergo centrifugation, evaporation, and drying for animal feed production

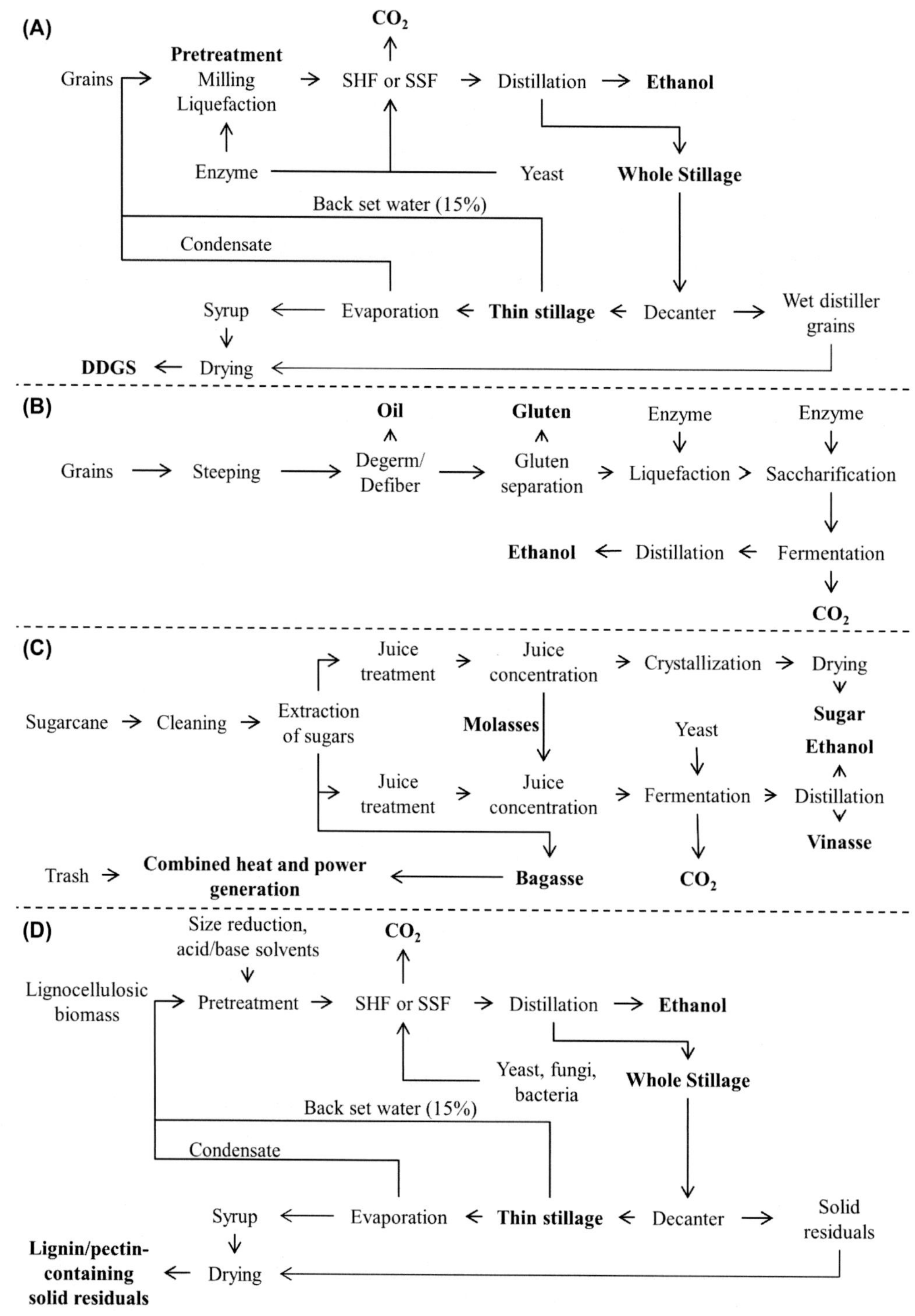

FIG. 3.5 Overview of the process steps in the production of **(A–C)** first-generation ethanol from starch by **(A)** dry and **(B)** wet processes and from **(C)** sugarcane and the production of **(D)** second-generation ethanol from lignocelluloses. The dry and wet mill processes differ on the number of final value-added products [3]. *SHF*, separate hydrolysis and fermentation; *SSF*, simultaneous saccharification and fermentation.

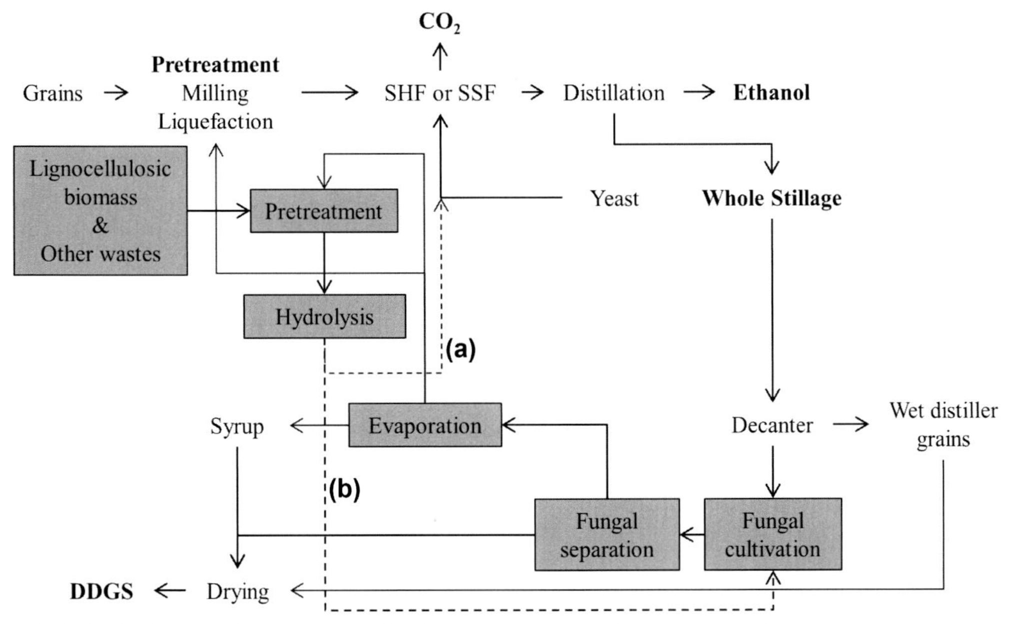

FIG. 3.6 Strategies for the integration of first- and second-generation ethanol production processes, namely, at yeast fermentation level or at the cultivation step with filamentous fungi where the side stream thin stillage is the cultivation medium [3]. *DDGS*, dried distillers' grains with solubles; *SHF*, separate hydrolysis and fermentation; *SSF*, simultaneous saccharification and fermentation.

(Figs. 3.5 and 3.6). Such a stream is known as stillage in starch-based ethanol facilities and vinasse in sugarcane-based processes. Gluten feed and feed oil can also be produced by ethanol facilities in addition to the by-products from beer and wine production industries. Only the US ethanol industry produced 45 million metric tonnes of feed products in 2017 [61]. Animal feed products have also been produced from dairy side streams, for example, protein concentrates from cheese whey [62].

Another source of protein for feed and human consumption, ancient to the feed products produced by first-generation ethanol plants, is single-cell protein (SCP), and the term was coined in 1966 by Carl L. Wilson, referring to the biomass obtained after cultivation of various groups of microorganisms (microbial protein) in different substrates [63]. Its potential for feed or human food applications arises from its high-protein and high-fat fractions composed of essential amino acids and polyunsaturated fatty acids, respectively. A limited number of microbial species (exacerbated if human consumption is considered) within the group of algae, fungi (filamentous fungi and yeast), and bacteria are used for SCP production [64]. Food-grade substrates are commonly used for production of SCPs for human consumption, mostly from fungal and algal origin [64]. Algae are used as, ingredients in pastas, baked products, snacks, etc. [64] or as a source of omega fatty acids and carotenoids in aquaculture [65]. Products from *Saccharomyces* (yeast extracts), *Fusarium* (Quorn products characterized by a meat texture and perhaps the only one product used for human consumption), and *Torulopsis* (rich in glutamic acid and therefore used as a replacement for monosodium glutamate as a flavor enhancer) are commercially available [64]. In addition to starch that has been used for production of Quorn and yeast extracts, lignocellulosic sugars present in sulfite liquor from paper mills, methanol, alkanes, and whey have also been used for SCP production [64].

A great deal of research attention has been paid to the production of SCP from various wastes of diversified composition, particularly using edible filamentous fungi [2,8]. In addition to the common high protein and fat levels, the production of pigmented fungal biomass for feed or for pigment extraction, as an alternative to synthetic dyes with clear applications in, e.g., food, feed, and textile industries, has also been a matter of research [66,67]. Bacterial SCP production from methane has also been proposed by applying a

U-loop fermenter and a production facility for the product FeedKind has been opened in the United Kingdom in 2016.

The inclusion of filamentous fungi in waste biorefineries has several implications:

- biomass is always a granted product even if biomass production is not the ultimate goal;
- it opens the possibility to produce protein sources from several wastes because of its versatile enzymatic machinery;
- by growing on wastes, fungi produce enzymes with applications in several industrial sectors, such as in feed industries, creating a closed loop;
- the fungal cell wall is composed of chitin/chitosan that has increasing applications in various sectors such as biomedicine.

Biodiesel From Waste Vegetable Oil

Waste vegetable oil is rich in free fatty acids and is no longer suitable for consumption. Additionally, with the global consumption of cooking oil being at 17 million tonnes with an annual raise of approximately 2% [68], the amount of waste vegetable oil is increasing rapidly with serious disposal problems. This highly viscous waste oil can instead be converted to a low-viscosity, renewable, and carbon-neutral fuel, namely, biodiesel. This is accomplished by the transesterification (Fig. 3.7) of lipids present in waste oil to fatty acid methyl esters (biodiesel), giving it an economic as well as waste management solution [69]. The main by-product obtained in the process is glycerol, and the transesterification process takes place in the presence of an alcohol, usually methanol for its low cost, and a catalyst to obtain higher biodiesel yields [70]. Different catalysts have been studied to increase the yield of biodiesel during the transesterification process [70,71], while the reaction temperature and the molar ratio of alcohol to vegetable oil are the most significant variables affecting the biodiesel yield [72].

Hydrotreating the vegetable oil is an alternative process to esterification for biodiesel production. Hydrotreated vegetable oil (HVO), which is marketed by Neste Oil as "the highest quality diesel in the world," is a renewable diesel produced by the process of hydrogenation, i.e., treatment with hydrogen. As compared with regular biodiesel production by esterification, the production of HVO uses hydrogen and not methanol as the catalyst. As a result, propane, not glycerol, is the by-product of hydrogenation. Additionally, HVO does not have the detrimental effects of ester-type biodiesel fuels, such as increased NO_x emission, deposit formation, storage stability problems, more rapid aging of engine oil, and poor cold properties [73].

The total global biodiesel production was projected to be around 25 billion liters [75], with the United States and Brazil being the largest producers in the world totaling 6 and 4.3 billion liters, respectively, in 2017. Germany, Argentina, Indonesia, and France were also accounted as big producers with a yearly total of 11.6 billion liters [76]. In addition, glycerol, the by-product of biodiesel industry, is a value-added product itself. It is obtained in a crude state after the transesterification reaction with impurities including traces of glycerides, alcohol, and inorganic salts and needs to be further purified to be used as a commodity in chemical industries. Development of purification techniques such as microfiltration, simple and vacuum distillation, and ion exchange techniques would help obtain highly purified glycerol for industrial applications, e.g., the development of personal care products, food, alkyd resins, and polyether [77]. On the downside, a continuous increase of glycerol availability can lead to selling

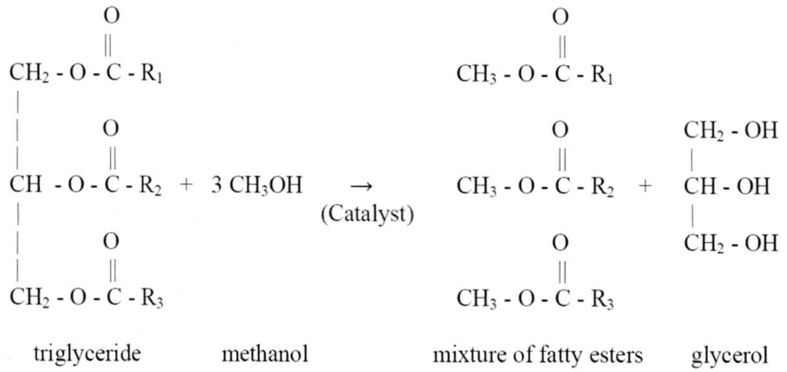

$$
\begin{array}{lcccl}
CH_2-O-\overset{\displaystyle O}{\overset{\|}{C}}-R_1 & & & CH_3-O-\overset{\displaystyle O}{\overset{\|}{C}}-R_1 & \\
\vert & & & & CH_2-OH \\
CH-O-\overset{\displaystyle O}{\overset{\|}{C}}-R_2 & + \; 3\,CH_3OH & \underset{\text{(Catalyst)}}{\longrightarrow} & CH_3-O-\overset{\displaystyle O}{\overset{\|}{C}}-R_2 \; + & \vert \\
\vert & & & & CH-OH \\
CH_2-O-\overset{\displaystyle O}{\overset{\|}{C}}-R_3 & & & CH_3-O-\overset{\displaystyle O}{\overset{\|}{C}}-R_3 & \vert \\
 & & & & CH_2-OH \\
\text{triglyceride} & \text{methanol} & & \text{mixture of fatty esters} & \text{glycerol}
\end{array}
$$

FIG. 3.7 The transesterification reaction. (Adapted from [74].)

price reduction. Therefore other valorization strategies should be considered, such as those that have included the use of glycerol for the production of fungal biomass, single-cell oil, and organic acids [78].

FUTURE PRODUCTS: ANAEROBIC DIGESTION SIDE PRODUCTS, ALCOHOLS, BIOPLASTICS, AND LIGNIN

While bringing together different waste substrates and microorganisms, a continuous discovery/development of new processes and value-added products takes place. In addition to the successful commercial green products supplied nowadays, there are "rounds" of value-added products waiting to reach the same development level. Naturally, this will be influenced by cost-efficiency parameters.

A better characterization of underdevelopment processes is the main driver to continuously unveil new potential products, for example, the biogas process. New products including biohydrogen, biohythane, and a range of carboxylates and VFAs can be produced from a better understanding of the biogas processes or by identifying the potential of those products as alternatives to fossil fuels or even the existing green biofuels and biochemicals. As presented in Fig. 3.3 (see the section Biogas From Organic Municipal Wastes), hydrogen is obtained as a side product from AD, even if methane, the main component of biogas, is sought. Attention has also been given to the optimization of hydrogen production that is considered as another potential green fuel combustion from which water originates [18]. Similar to biomethane production, FW has also been the first-choice substrate for production of biohydrogen via dark fermentation [16]. Being an intermediate by-product, efforts have been focused on codigestion strategies and on specific enrichment of microbial communities for enhancing biohydrogen production. An example is the codigestion of 80%–90% FW and the remaining of sewage sludge giving rise to a carbohydrate-to-protein ratio of 1.66 g COD/gCOD from which 2 L of cumulative biohydrogen was produced [18].

Research emphasis has also been given to the production of biobutanol through the acetone-butanol-acetone pathway in which *Clostridium acetobutylicum* is, at present, the first-choice biocatalyst [79]. During processing of lignocellulosic materials, lignin represents a fraction that is not consumed by microorganisms and research efforts have been carried out to successfully separate this polymer and extend its potential applications. Moreover, alcohols have also been considered for applications other than as a biofuel. For instance, there are some facilities, helped by subsidies, that convert first-generation ethanol to bioethylene with applications as building blocks in several areas [80].

Lignin Valorization

Lignin is present in a variety of lignocellulose-rich waste streams arising majorly from pulp and paper industries after pulping and bleaching processes, from agricultural residues generated during or after processing of agricultural crops (e.g., sugarcane bagasse, wheat straw, rice straw, oat hulls, coconut shells), from ethanol biorefineries, and in small quantities from MSW. Out of these four waste streams, the pulp and paper industry is the biggest producer of lignin waste and had produced 50 million tonnes of low-quality lignin in 2010. Only 2% of this lignin was used commercially as dispersants or binding source, while the remaining was incinerated for energy production [81]. The amount of lignin produced every year from pulp and paper industry and agricultural residues is increasing significantly with lignocellulosic biorefineries finally coming into existence. As a cheap, natural, and renewable material having the highly functionalized and aromatic nature, researchers have explored its potential to be converted into a variety of value-added products such as fuel, aromatics, chemicals, polymers, and high-performance materials instead of simply burning it as a waste for energy production. However, these conversion processes depend on improvements and innovations in the field of catalysis and product separation [82] and all these developments are often taken into account in the context of integrated biorefineries.

As all native lignins are heterogeneous biopolymers linked to polysaccharides, a chemical pretreatment process is necessary to separate lignin from carbohydrate fibers. These lignins isolated from biomass by, e.g., wood pulping or lignocellulosic pretreatment processes are termed as technical lignins (Table 3.1). Further lignin processing and its valorization should be accompanied by studying the composition and structure of technical lignins and the specific features of its production [83]. Kraft process is the dominant pulping technology worldwide. It produces 130 million tonnes of kraft pulp annually [84], 50 million tonnes of which refers to low-purity kraft lignin originating as a waste stream. Around 1 million tonne of higher quality commercial kraft lignin is retrieved by the LignoBoost process (Table 3.2). This is a unique technology for extracting high-quality lignin from kraft spent liquor, developed by the Swedish company RISE Innventia AB [85]. For short- to medium-term opportunities,

TABLE 3.1
Technical Lignins Isolated From Biomass by Four Chemical Pulping Processes.

Pulping Process	Type of Lignin Produced	Solvent	Purity	Products	Scale
Sulfite	Lignosulfonates	SO_2 aqueous solution	Low-medium/ S rich	Low-quality products: binders/vanillin	Commercial
Soda	Soda lignins	NaOH-Na_2S solution	Impure/alkaline/ S free	Energy by incineration	Commercial
Kraft	Kraft lignins	NaOH	Low/S free	Aromatics/carbon fibers with commercial kraft	Laboratory/pilot
Organosolv	Organosolv lignins	Organic alcohols/acids	Highest purity/ S free	High end use/carbon fibers, biocomposites, antioxidants	Laboratory

TABLE 3.2
Lignin Types and Production [94–96].

Lignin Type	World Annual Production, Tonnes/year
Kraft	50,000,000
Commercial kraft[a]	114,000
Lignosulfonates	1,000,000
Organosolv[b]	1000
Ethanol biorefineries[c]	180,000
Total	51,295,000

[a] Commercial kraft lignin (high purity) is produced by MeadWestvaco and Domtar in the United States (about 60,000 tonnes per year), Stora Enso Sunila Mill in Finland (about 50,000 tonnes), and Metso Corporation at a demonstration plant in Sweden (about 4000 tonnes per year) [95–97].
[b] Organosolv lignin is mostly produced in pilot plants by CIMV (France) and Lignol (Canada).
[c] About 0.3% of the total potential is produced today [95].

thermochemical process routes are interesting to obtain, e.g., low-purity lignin. However, most of the available kraft lignin can be either utilized to produce aromatics via catalytic oxidation or pyrolyzed (200–760°C) and further liquefied to bio-oil. Lignin of any type, including very-low-quality soda lignin, can be directly gasified (480–1650°C) to syngas, which is a valuable fuel [86]. Additionally, another popular pulping process at pulp and paper mills is sulfite pulping that produces lignosulfonates (Table 3.1), which are sulfur-rich low-purity lignin and can be converted to low-quality products, e.g., binders. These lignosulfonates are converted to vanillin, an unique value-added product, at the Borregaard biorefinery in Norway, which is one of the world's most advanced and sustainable biorefineries. Here, 90% of the woody biomass input exits as marketable products [87]. Developments in lignin valorization include the production of low-cost carbon fibers and biocomposites, but these applications need higher quality lignin such as that obtained through organosolv or commercial kraft (Table 3.1). Organosolv pretreatment is being studied for conversion of lignocellulosic feedstock to achieve a relatively pure bio-lignin fraction that is sulfur free, low in polydispersity, and rich in functionality with limited carbohydrate contamination [88–92]. It typically results in more than 50% lignin removal through cleavage of lignin carbohydrate bonds and β-O-4 interunit linkages and subsequent solubilization in an organic solvent [93].

The high-purity lignin can be converted to a high-value product, carbon fibers, which are lightweight, strong, and flexible and can be used in both structural (load-bearing) and nonstructural applications (e.g., thermal insulation). They are currently produced via an expensive conventional fossil precursor polyacrylonitrile and pitches. The manufacture of low-cost carbon fibers with high added value from a renewable, abundant, and cheap precursor such as lignin (Tables 3.3 and 3.4) is one of its most interesting and promising valorization routes, from both the economic and environmental point of view [99,100]. These fibers have been manufactured by utilizing different methods of spinning, including wet and dry spinning, melt spinning, and electrospinning [101–104]. Additionally, lignin has shown the capacity to be developed as a sustainable component for various polymer composite

TABLE 3.3
A Comparison of Fossil (Polyacrylonitrile) and Renewable Precursor (Lignin) of Carbon Fibers [98].

Precursor	Scale	Carbon Fiber Production Cost	Market Price ($/kg)
Polyacrylonitrile	Commercial	High/fossil derived (90% of total production)	17–26
Lignin	Bench	Low/renewable	6.5–11

TABLE 3.4
Lignin-Based CF Cost Savings Estimates Compared With Conventional PAN-Based CF [113].

Process Cost Category	PAN-Based CF Cost Estimate ($21.78/kg)	Lignin-Based CF Cost Estimate ($8.17/kg)
Precursors	11.11	1.10
Stabilization and oxidation	3.40	2.18
Carbonization and graphitization	5.11	3.26
Surface treatment	0.82	0.73
Spooling and packaging	1.34	0.90

CF, carbon fiber; *PAN*, polyacrylonitrile.

applications such as surfactants, plasticizers, superabsorbent gels, coating, stabilizing agents, and superabsorbent hydrogels and can replace their commercial chemical precursors in the future [105–107]. A wide range of technical lignins from the industry can be used to develop uniform lignin nanoparticles. This could be a propitious multifaceted platform to ameliorate the nanocomposite field [108–111], given that the global biocomposite market is projected to reach $36.76 billion by 2022 from $16.46 billion in 2016, with a CAGR of around 14% from 2017 to 2022 [112].

Owing to the tremendous industrial potential of lignin, the valorization of industrial waste streams and agricultural wastes containing lignin seems promising toward contributing to a circular bioeconomy in the near future (Fig. 3.1).

INTEGRATION OF WASTES IN ESTABLISHED INDUSTRIAL PROCESSES

The research reasoning, to a great extent, has been toward the establishment of commercial processes from scratch. Nonetheless, this strategy can lead to the slow development of renewable processes mostly because of the investment risks associated with poor support from governmental entities and the complex and immature nature of the developed processes. Therefore integration of substrates in wastes in established commercial processes is believed to be a strategy that can lead to faster establishment of renewable products. In addition to enhancing sustainability, integration of wastes could result in beneficial synergies, e.g., nutrient complementation or equilibrated energy balances. Brazilian first-generation ethanol plants using sugarcane compensate outweigh the high energy costs for ethanol recovery by integrating energy-generating systems using substrate-processing residuals (e.g., bagasse) [3]. Potential candidates include those that are used to handle sugar-containing substrates and that employ biological conversions. Examples include paper and pulp industries handling lignocellulosic materials and biogas and first-generation ethanol plants. Industries producing FW, such as dairy or juice companies, can have biogas facilities installed, whereas if the goal is, e.g., bioethanol production the development would be faster if integration takes place in first-generation ethanol plants. This is due to lower investment costs because the distillation equipment, normally expensive, is already available.

Some examples of integration, presented in Fig. 3.8, are as follows:

- Lignocellulosic (e.g., forest and agricultural residues and wooden MSW) and citrus wastes can be integrated into biogas or first-generation ethanol plants. If a suitable pretreatment step is employed, lignin and pectin are produced as a value-added product of the updated biorefinery together with ethanol and fungal

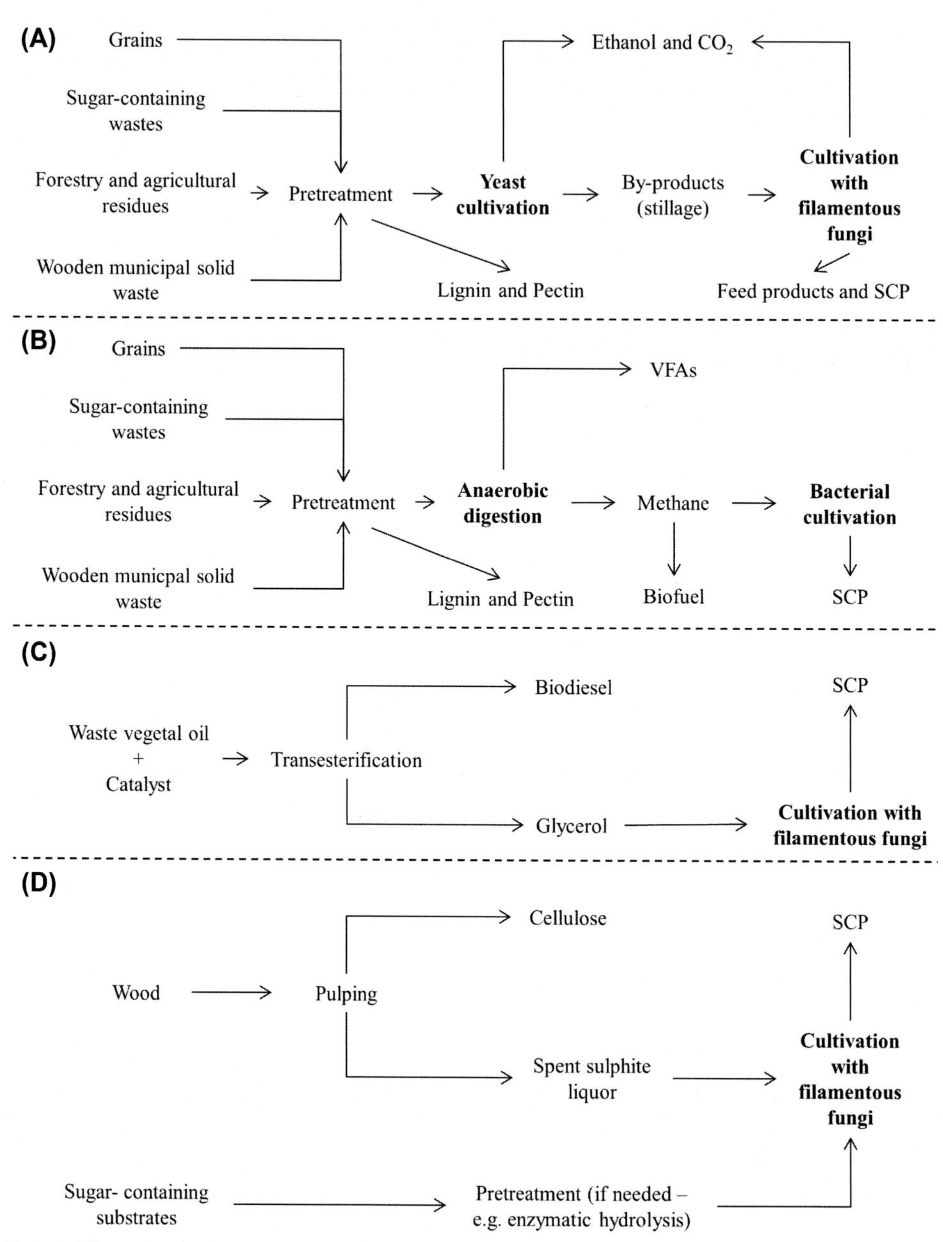

FIG. 3.8 Waste integration strategies in **(A)** first-generation ethanol, **(B)** biogas, **(C)** biodiesel, and **(D)** paper and pulp established plants. "Pretreatment" can include milling, cooking, enzymatic hydrolysis, or energy-intensive methods. *SCP*, single-cell protein.

biomass (via conversion of by-products). These join the group of products already produced by the plant (biogas, digestate, ethanol, animal feed, and CO_2).

- Fungal cultivation can be followed at paper and pulp industries for conversion of both hexoses and pentoses, and this opens the possibility for integration of streams that do not need an energy-intensive pretreatment step but perhaps only enzymatic hydrolysis (e.g., streams containing starch, sucrose, cellulose).
- Bacterial cultivation can be included in biogas plants for the production of SCPs from methane.

- Fungal cultivation can be integrated at biodiesel plants for production of SCPs from glycerol.

By having a good characterization of the wastes available as well as the potential products that can be produced, multiple integration strategies can be developed.

CONCLUSIONS AND PERSPECTIVES

Efficient management routes are needed for the increasing amount and heterogeneity of wastes generated by human activities. Biological conversion has a high potential for the development of diversified processes producing a panoply of fuels and chemicals. This can represent an important step toward waste management, replacement of fossil fuels, circular bioeconomy, and consequently overall sustainability. Integration of wastes in established commercial plants such as first-generation ethanol plants can potentially lead to faster development of waste management routes, thus concomitantly enhancing their biorefinery ability. Such development requires committed incentives by policy makers for research and application and for social awareness programs. The establishment of direct communication bridges between academia and industry is, naturally, of utmost importance. In line with these barriers, the chapter describes the successful processes of waste management and the strategies of integration of other wastes for beneficial synergies.

ACKNOWLEDGMENTS

The authors are grateful to the Swedish Agency for Economic and Regional Growth for its financial support through the European Regional Development Fund.

REFERENCES

[1] Venkata Mohan S, et al. Waste biorefinery models towards sustainable circular bioeconomy: critical review and future perspectives. Bioresource Technology 2016; 215:2–12.

[2] Ferreira JA, et al. Waste biorefineries using filamentous ascomycetes fungi: present status and future prospects. Bioresource Technology 2016;215:334–45.

[3] Ferreira JA, et al. A review of integration strategies of lignocelluloses and other wastes in 1st generation bioethanol processes. Process Biochemistry 2018;75.

[4] Lennartsson PR, Erlandsson P, Taherzadeh MJ. Integration of the first and second generation bioethanol processes and the importance of by-products. Bioresource Technology 2014;165:3–8. CESE 2013 & Special Issue: ICABB 2013.

[5] IEA. IEA bioenergy Task 42 on biorefineries: co-production of fuels, chemicals, power and materials from biomass. Copenhagen, Denmark: International Energy Agency; 2007.

[6] Cherubini F. Life cycle assessment of biorefinery systems based on lignocellulosic raw materials: concept development, classification and environmental evaluation. 2009.

[7] Gobina E. Biorefinery technologies: global markets. bccResearch; 2016.

[8] Ferreira JA, et al. Zygomycetes-based biorefinery: present status and future prospects. Bioresource Technology 2013;135(0):523–32.

[9] Diaz AB, Blandino A, Caro I. Value added products from fermentation of sugars derived from agro-food residues. Trends in Food Science & Technology 2018;71:52–64.

[10] Chen H, et al. A review on the pretreatment of lignocellulose for high-value chemicals. Fuel Processing Technology 2017;160:196–206.

[11] Kumar P, et al. Methods for pretreatment of lignocellulosic biomass for efficient hydrolysis and biofuel production. Industrial & Engineering Chemistry Research 2009;48(8):3713–29.

[12] Tan L, et al. Pretreatment of empty fruit bunch from oil palm for fuel ethanol production and proposed biorefinery process. Bioresource Technology 2013;135: 275–82.

[13] Petersen MØ, Larsen J, Thomsen MH. Optimization of hydrothermal pretreatment of wheat straw for production of bioethanol at low water consumption without addition of chemicals. Biomass and Bioenergy 2009; 33(5):834–40.

[14] Wildschut J, et al. Ethanol-based organosolv fractionation of wheat straw for the production of lignin and enzymatically digestible cellulose. Bioresource Technology 2013;135:58–66.

[15] Jönsson LJ, Martín C. Pretreatment of lignocellulose: formation of inhibitory by-products and strategies for minimizing their effects. Bioresource Technology 2016; 199:103–12.

[16] Sen B, et al. State of the art and future concept of food waste fermentation to bioenergy. Renewable and Sustainable Energy Reviews 2016;53:547–57.

[17] Hoornweg D, Bhada-Tata P. What a waste: a global review of solid waste management. Washington DC, USA: World Bank; 2012.

[18] Sarkar O, Butti SK, Mohan SV. Acidogenic biorefinery: food waste valorization to biogas and platform chemicals. In: Bhaskar T, et al., editors. Waste biorefinery: potential and Perspectives. Amsterdam, The Netherlands: Elsevier B. V.; 2018.

[19] Kabir MM, Forgács G, Taherzadeh MJ. Biogas from wastes: processes and applications. In: Taherzadeh MJ, Richards T, editors. Resource recovery to approach zero municipal waste. Boca Raton FL, USA: CRC Press, Taylor & Francis Group; 2016.

[20] Bioenergy I. Biogas upgrading and utilisation, Task 24. 1999.

[21] Ziemiński K, Frąc M. Methane fermentation process as anaerobic digestion of biomass: transformations, stages and microorganisms. African Journal of Biotechnology 2012;11(18):4127–39.
[22] Deublein D, Steinhauster A. Biogas from waste and renewable resources: an introduction. Weinheim, Germany: Wiley; 2008.
[23] Ward AJ, et al. Optimisation of the anaerobic digestion of agricultural resources. Bioresource Technology 2008; 99(17):7928–40.
[24] Astals S, Romero-Güiza M, Mata-Alvarez J. Municipal solid waste. In: Ferreira G, editor. Alternative energies, advanced structured materials. Berlin, Germany: Springer; 2013.
[25] Forgács G. Biogas production from citrus wastes and chicken feather: pretreatment and Co-digestion. In: Chemical and biological engineering. Gothenburg, Sweden: Chalmers University of Technology; 2012.
[26] Mata-Alvarez J, Mace S, Llabres P. Anaerobic digestion of organic solid wastes. An overview of research achievements and perspectives. Bioresource Technology 2000; 74(1):3–16.
[27] Griffin LP. Anaerobic digestion of organic wastes: the impact of operating conditions on hydrolysis efficiency and microbial community composition. Fort Collins, FO: Colorado State University; 2012.
[28] Eckard R. Waste-derived biogas: global markets for anaerobic digestion equipment. bccResearch; 2018.
[29] Mordor Intelligence. Acetic acid market - segmented by end-user industry, application, and geography – growth, trends, and forecast (2018–2023). Hyderabad (India). 2018.
[30] Chen Y, et al. Enhanced production of short-chain fatty acid by co-fermentation of waste activated sludge and kitchen waste under alkaline conditions and its application to microbial fuel cells. Applied Energy 2013;102: 1197–204.
[31] Fei Q, et al. The effect of volatile fatty acids as a sole carbon source on lipid accumulation by Cryptococcus albidus for biodiesel production. Bioresource Technology 2011;102(3):2695–701.
[32] den Boer E, et al. Volatile fatty acids as an added value from biowaste. Waste Management 2016;58:62–9.
[33] Fei Q, et al. Lipid production by microalgae Chlorella protothecoides with volatile fatty acids (VFAs) as carbon sources in heterotrophic cultivation and its economic assessment. Bioprocess and Biosystems Engineering 2015;38(4):691–700.
[34] Oleskowicz-Popiel P, et al. Co-production of ethanol, biogas, protein fodder and natural fertilizer in organic farming – evaluation of a concept for a farm-scale biorefinery. Bioresource Technology 2012;104:440–6.
[35] Marin J, Kennedy KJ, Eskicioglu C. Effect of microwave irradiation on anaerobic degradability of model kitchen waste. Waste Management 2010;30(10):1772–9.
[36] Kim HJ, et al. Effect of pretreatment on acid fermentation of organic solid waste. Water Science and Technology 2005;52(1–2):153–60.
[37] Wang K, et al. Anaerobic digestion of food waste for volatile fatty acids (VFAs) production with different types of inoculum: effect of pH. Bioresource Technology 2014; 161:395–401.
[38] Zhou M, et al. Enhanced volatile fatty acids production from anaerobic fermentation of food waste: a mini-review focusing on acidogenic metabolic pathways. Bioresource Technology 2018;248:68–78.
[39] Lee WS, et al. A review of the production and applications of waste-derived volatile fatty acids. Chemical Engineering Journal 2014;235:83–99.
[40] Wu Q, et al. Enhancement of volatile fatty acid production using semi-continuous anaerobic food waste fermentation without pH control. RSC Advances 2015; 5(126):103876–83.
[41] Haug RT. The practical handbook of compost engineering. Boca Raton, Florida, USA: Lewis Publishers; 1993.
[42] Godlewska P, et al. Biochar for composting improvement and contaminants reduction. A review. Bioresource Technology 2017;246.
[43] Sánchez A, et al. Composting of wastes. In: Taherzadeh MJ, Richards T, editors. Resource recovery to approach zero municipal waste. Boca Raton, Florida, USA: CRC Press, Taylor & Francis Group; 2016.
[44] Wu H, et al. The interactions of composting and biochar and their implications for soil amendment and pollution remediation: a review. Critical Reviews in Biotechnology 2017;37(6):754–64.
[45] de Guardia A, et al. Comparison of five organic wastes regarding their behaviour during composting: Part 1, biodegradability, stabilization kinetics and temperature rise. Waste Management 2010;30(3):402–14.
[46] Barrena R, et al. Categorizing raw organic material biodegradability via respiration activity measurement: a review. Compost Science & Utilization 2011;19(2):105–13.
[47] Singh RP, et al. Management of urban solid waste: vermicomposting a sustainable option. Resources, Conservation and Recycling 2011;55(7):719–29.
[48] Lleó T, et al. Home and vermicomposting as sustainable options for biowaste management. Journal of Cleaner Production 2013;47:70–6.
[49] Andruschkewitsch M, Wachendorf C, Wachendorf M. Effects of digestates from different biogas production systems on above and belowground grass growth and the nitrogen status of the plant-soil-system. Grassland Science 2013;59(4):183–95.
[50] Tchobanoglous G, Kreith F. Handbook of solid waste management. New York, USA: McGraw Hill Professional; 2002.
[51] Hartmann H, Ahring BK. Strategies for the anaerobic digestion of the organic fraction of municipal solid waste: an overview. Water Science and Technology 2006;53(8):7–22.
[52] Eurostat. The EU in the world. Luxembourg: European Union; 2018.
[53] Goh CS, Lee KT. A visionary and conceptual macroalgae-based third-generation bioethanol (TGB) biorefinery in

Sabah, Malaysia as an underlay for renewable and sustainable development. Renewable and Sustainable Energy Reviews 2010;14(2):842–8.
[54] Chu-Ky S, et al. Simultaneous liquefaction, saccharification and fermentation at very high gravity of rice at pilot scale for potable ethanol production and distillers dried grains composition. Food and Bioproducts Processing 2016;98:79–85.
[55] Nair RB. Integration of first and second generation bioethanol processes. In: Swedish centre for resource recovery. Sweden: University of Borås: Borås; 2017.
[56] Lin Y, Tanaka S. Ethanol fermentation from biomass resources: current state and prospects. Applied Microbiology and Biotechnology 2006;69(6):627–42.
[57] Energy USDo. 2015 bioenergy market report. USA: Colorado; 2017.
[58] E4tech. Ramp up of lignocellulosic ethanol in Europe to 2030. 2017 [London, UK].
[59] IEA. World energy outlook. Paris, France: International Energy Agency; 2017.
[60] García CA, et al. Life-cycle greenhouse gas emissions and energy balances of sugarcane ethanol production in Mexico. Applied Energy 2011;88(6):2088–97.
[61] RFA. Animal feed production - protein power. 2017.
[62] Yogarathinam LT, et al. Concentration of whey protein from cheese whey effluent using ultrafiltration by combination of hydrophilic metal oxides and hydrophobic polymer. Journal of Chemical Technology and Biotechnology 2018;93(9):2576–91.
[63] Ali S, et al. Production and processing of single cell protein (SCP) – a review. European Journal of Pharmaceutical and Medical Research 2017;4(7):86–94.
[64] Ritala A, et al. Single cell protein—state-of-the-art, industrial landscape and patents 2001–2016. Frontiers in Microbiology 2017;8:2009.
[65] Muller-Feuga A. The role of microalgae in aquaculture: situation and trends. Journal of Applied Phycology 2000;12(3–5):527–34.
[66] Gmoser R, et al. Filamentous ascomycetes fungi as a source of natural pigments. Fungal biology and biotechnology 2017;4(1):4.
[67] Gmoser R, et al. Pigment production by the edible filamentous fungus Neurospora intermedia. Fermentation 2018;4(1):11.
[68] Authority AaFD. Waste oils and fats as biodiesel feedstocks: an assessment of their potential in the EU. In: ALTENER program NTB-NETT phase IV, Task 4, final report; March 2000.
[69] Math MC, Kumar SP, Chetty SV. Technologies for biodiesel production from used cooking oil — a review. Energy for Sustainable Development 2010;14(4):339–45.
[70] Elkady MF, Zaatout A, Balbaa O. Production of biodiesel from waste vegetable oil via KM micromixer. Journal of Chemistry 2015;2015:9.
[71] Modiba E, Enweremadu C, Rutto H. Production of biodiesel from waste vegetable oil using impregnated diatomite as heterogeneous catalyst. Chinese Journal of Chemical Engineering 2015;23(1):281–9.
[72] Demirbas A. Progress and recent trends in biodiesel fuels. Energy Conversion and Management 2009; 50(1):14–34.
[73] Aatola H, et al. Hydrotreated vegetable oil (HVO) as a renewable diesel fuel: trade-off between NOx, particulate emission, and fuel consumption of a heavy duty engine. SAE International Journal of Engines 2009; 1(1):1251–62.
[74] Dixit S, Dixit S. Optimization and fuel properties of water degummed linseed biodiesel from transesterification process. Chemical Science Journal 2016;7.
[75] Estelvina Rg, et al. Analysis of the effect of biodiesel energy policy on markets, trade and food safety in the international context for sustainable development. 2011.
[76] Statista. Leading biodiesel producers worldwide in 2017, by country (in billion liters). August 23, 2018. Available from: https://www.statista.com/statistics/271472/biodiesel-production-in-selected-countries/.
[77] Tan HW, Abdul Aziz AR, Aroua MK. Glycerol production and its applications as a raw material: a review. Renewable and Sustainable Energy Reviews 2013;27:118–27.
[78] André A, et al. Biotechnological conversions of bio-diesel derived waste glycerol into added-value compounds by higher fungi: production of biomass, single cell oil and oxalic acid. Industrial Crops and Products 2010;31(2):407–16.
[79] Stoeberl M, et al. Biobutanol from food wastes – fermentative production, use as biofuel an the influence on the emissions. Procedia Food Science 2011;1:1867–74.
[80] Mohsenzadeh A, Zamani A, Taherzadeh MJ. Bioethylene production from ethanol: a review and techno-economical evaluation. ChemBioEng Reviews 2017;4(2): 75–91.
[81] Pan X, Saddler JN. Effect of replacing polyol by organosolv and kraft lignin on the property and structure of rigid polyurethane foam. Biotechnology for Biofuels 2013;6(1):12.
[82] Strassberger Z, Tanase S, Rothenberg G. The pros and cons of lignin valorisation in an integrated biorefinery. RSC Advances 2014;4(48):25310–8.
[83] Krutov SM, et al. Lignin wastes: past, present, and future. Russian Journal of General Chemistry 2014;84(13): 2632–42.
[84] Ferreira JA, et al. Spent sulphite liquor for cultivation of an edible Rhizopus sp. Bioresources 2012;7(1):173–88.
[85] Tomani P. The lignoboost process, 44; 2010.
[86] Demirbas A. Biorefineries: current activities and future developments. Energy Conversion and Management 2009;50:2782–801.
[87] Borregaard. January 5, 2014. Available from: http://www.borregaard.com/content/view/full/10231.
[88] Benjelloun, B., The CIMV organosolv Process in "Tomorrow's biorefineries in Europe". 11–12 February 2014: Brussels, Belgium.
[89] Berlin A, et al. Inhibition of cellulase, xylanase and β-glucosidase activities by softwood lignin preparations. Journal of Biotechnology 2006;125(2):198–209.

[90] Zhao X, et al. Organosolv fractionating pre-treatment of lignocellulosic biomass for efficient enzymatic saccharification: chemistry, kinetics, and substrate structures. Biofuels, Bioproducts and Biorefining 2017;11(3): 567–90.
[91] Nitsos C, Rova U, Christakopoulos P. Organosolv fractionation of softwood biomass for biofuel and biorefinery applications. Energies 2017;11(1):50.
[92] Agnihotri S, et al. Ethanol organosolv pretreatment of softwood (Picea abies) and sugarcane bagasse for biofuel and biorefinery applications. Wood Science and Technology 2015;49(5):881–96.
[93] Zhao X, Cheng K, Liu D. Organosolv pretreatment of lignocellulosic biomass for enzymatic hydrolysis. Applied Microbiology and Biotechnology 2009;82(5): 815.
[94] Smolarski N. High-value opportunities for lignin: unlocking its potential. Frost & Sullivan; November 7, 2012. Available from: http://www.frost.com/sublib/display-market-insight-top.do?id=269017995.
[95] lake M, Scouten C. What are we going to do with all this lignin? St. Simons Island, Georgia: Presented to Frontiers in BioRefining; December 2015. October 24.
[96] Wertz J, Bédué O. Lignocellulosic biorefineries. Lausanne: Switzerland EPFL Press; 2013.
[97] Valmet-supplied LignoBoost plant now handed over to Stora Enso's Sunila mill in Finland. Finland: Valmet Corporation; 2015. Press release.
[98] LIGNIMATCH. Future use of lignin in value added products – a roadmap for possible Nordic/Baltic innovation. 2010.
[99] Ragauskas AJ, et al. Lignin valorization: improving lignin processing in the biorefinery. Science 2014; 344(6185).
[100] Frank E, et al. Carbon fibers: precursor systems, processing, structure, and properties. Angewandte Chemie International Edition 2014;53(21):5262–98.
[101] Otani S FY, Igarashi B, Sasaki K. Method for producing carbonized lignin fiber. 1969. U.S. Patent.
[102] Baker DA, H O. High glass transition lignins and lignin derivatives for the manufacture of carbon and graphite fibers. 2014 [U.S].
[103] Hosseinaei O, et al. Improving processing and performance of pure lignin carbon fibers through hardwood and herbaceous lignin blends. International Journal of Molecular Sciences 2017;18(7):1410.
[104] Darren A, Baker MS, Landmer A, Friman L, Echardt L. Structural carbon fibre from kraft lignin. Stockholm, Sweden: NWBC; 2017. 2017.
[105] Haihua W, et al. Preparation and colloidal properties of an aqueous acetic acid lignin containing polyurethane surfactant. Journal of Applied Polymer Science 2013; 130(3):1855–62.
[106] AC FF, et al. Lignin-based polyurethane doped with carbon nanotubes for sensor applications. Polymer International 2012;61(5):788–94.
[107] Cerrutti BM, et al. Carboxymethyl lignin as stabilizing agent in aqueous ceramic suspensions. Industrial Crops and Products 2012;36(1):108–15.
[108] Qian Y, et al. Formation of uniform colloidal spheres from lignin, a renewable resource recovered from pulping spent liquor. Green Chemistry 2014;16(4): 2156–63.
[109] Lievonen M, et al. A simple process for lignin nanoparticle preparation. Green Chemistry 2016;18(5): 1416–22.
[110] Zhao W, et al. From lignin association to nano-/micro-particle preparation: extracting higher value of lignin. Green Chemistry 2016;18(21):5693–700.
[111] Tian D, et al. Lignin valorization: lignin nanoparticles as high-value bio-additive for multifunctional nanocomposites. Biotechnology for Biofuels 2017; 10(1):192.
[112] Biocomposites market by fiber (wood fiber and non-wood fiber), polymer (synthetic and natural), product (hybrid and green), end-use industry (transportation, building & construction and consumer goods), and region – global forecast to 2022. 2017. Available from: https://www.marketsandmarkets.com/Market-Reports/biocomposite-market-258097936.html?gclid=Cj0KCQjwl7nYBRCwARIsAL7O7dHEoiQOt923eVvzh2nH-gPiA6HfHJAtcX0bdB2HLwGNHy4TeRJEhOUaAr0IEALw_wcB.
[113] Baker F. Utilization of sustainable resources for production of carbon fiber materials for structural and energy efficiency applications.". In: Frontiers in Biorefining: Biobased products from renewable carbon products; 2010. St. Simons Island, Georgia.

CHAPTER 4

Solid Waste Management Toward Zero Landfill: A Swedish Model

KIM BOLTON, PHD • KAMRAN ROUSTA, PHD

INTRODUCTION

The Waste Problem

Municipal solid waste (MSW) has several sources, including households and small businesses [1]. It is complex in that it contains many fractions with diverse chemical compositions, including garden, food, packaging (metal, paper, glass, plastic), bulky, and hazardous wastes. It is this complexity that, unless handled correctly, makes it difficult to recycle these fractions into high-value products.

The quantity of MSW is expected to increase due to the global population growth combined with the increased economic power of the developing countries. The World Bank estimates that the quantity of global MSW produced will almost double from 3.5 million tonnes per day in 2012 to more than 6.0 million tonnes per day in 2025 [1]. Most of the present waste is from the countries that belong to the Organisation for Economic Co-operation and Development (OECD), who have 1.6 million tonnes per day but, because the total urban population is only expected to grow from 729 million to 842 million by 2025, the waste quantity is predicted to rise by only 6%, i.e., 1.7 million tonnes per day, by 2025. In contrast, the developing countries typically produce lower amounts of waste, but the percentage increase by 2025 is expected to be larger than that for the OECD countries. For example, countries from the East Asian and Pacific regions, including China, are projected to almost treble from 740,000 tonnes per day in 2012 to 1.9 million tonnes per day in 2025.

It may be noted that MSW is only a small fraction of the total global waste. According to the Swedish Environmental Protection Agency [2], household waste was only 2.5% of the total waste generated in Sweden in 2014. About 83% of the total waste is from mining industries. However, a lot of the nonhousehold waste is generated to make products to be used by households. For example, manufacture of a cellular phone that weighs 169 g generates 85 kg of waste material in upstream processes [3]. In this perspective, much of the larger waste fractions, such as mining waste, can be considered as waste because of household consumption. In addition, recycling of MSW will reduce the need for virgin resources and will thereby reduce the amount of these other waste fractions.

Lack of proper waste management not only has environmental impacts but also leads to health and social challenges [1,4]. If the different fractions of MSW are mixed, then it can lead to landfill or dumping of this waste. As mentioned earlier, virgin materials must then be used to make products to replace this waste. In addition, dumping and landfilling that are not properly regulated release emissions into air, water, and soil. For example, methane, which has a much larger global warming potential than carbon dioxide [5], is emitted into air from dumping and many landfill sites. Similarly, leachates can be emitted into water, which can contaminate a municipality's potable water leading to sickness.

Some citizens, often known as scavengers, make their living from sorting and collecting material fractions from landfills or dumps. Hence, there is an economic incentive to scavenge. This is testament to the fact that these materials have value, provided they are managed correctly. However, as illustrated in Fig. 4.1, scavengers often live under undesirable circumstances. More controlled conditions are possible if the sorting is done at sorting facilities, which can offer the possibility of job creation. In principle, this is, from a social perspective, preferred to uncontrolled scavenging.

This discussion may indicate that we cannot approach zero MSW waste, as the amount of waste generated is predicted to increase. This is because our waste management models have been based on a linear process in which everything that is discarded by actors in the municipality is traditionally called waste. This

Sustainable Resource Recovery and Zero Waste Approaches. https://doi.org/10.1016/B978-0-444-64200-4.00004-9

FIG. 4.1 An example of improper waste management that leads to uncontrolled landfill or dumping. Scavengers can forge a livelihood from these sites, but usually under undesirable conditions.

is no longer correct in a circular economy perspective, where something discarded by one actor is considered as a resource for other actors. Hence, by increasing the percentage of discarded material that is considered as a resource, zero municipal waste can be attained, with only a small fraction of the waste going to landfills.

Sustainable Development and the Waste Hierarchy

As discussed in the previous section, all three aspects of sustainable development, namely, economic, environmental, and social aspects, are important when discussing the proper management of MSW. In fact, several of the United Nations' 17 sustainable development goals are directly or indirectly linked to proper waste management. Table 4.1 shows the three goals that are probably the most directly coupled to MSW management.

TABLE 4.1
Three United Nations' Sustainable Development Goals That are Directly Coupled to MSW Management.

	Correct management of food, including edible food that is discarded by actors along the supply chain, is important not only to alleviate environmental burdens [28] but also to provide food security [29]. According to the Food and Agricultural Organization of the United Nations [30], the global food waste in 2007 was estimated to be 1.6 gigatonnes of primary product equivalents and the corresponding edible part was estimated as 1.3 gigatonnes. In fact, one-third of all the food produced for human consumption is wasted [31], which raises ethical concerns because 795 million people suffer from undernourishment [32]. The proper management of uneaten, yet edible, food will become even more important, given that the global food demand is projected to increase by 70% by 2050 [33].
	In 2012, 3 billion of the planet's citizens lived in cities [1]. This is an increase of 3.5% from 2002 and is expected rise by 43% by 2025 [1]. Hence, it is predicted that by 2025, 4.3 billion people will be urban residents. As a result, there will be an increased amount of commodities that will be discarded within cities and municipalities. These must be seen as resources to be reused and recycled, instead of wastes that must be deposited within or outside the city limits.
	The Earth has limited resources that should be shared in a fair and just way by the ever-increasing population. Overconsumption must be combatted and the subsequent amount of discarded commodities must be minimized. In addition, these discarded commodities must not be perceived as waste, but as resources that are needed in a circular economy. This forms the basis for sustainable management of MSW and will reduce production from virgin materials and increase production from discarded materials.

MSW, municipal solid waste.

In addition to the goals shown in Table 4.1, the following goals are also related to waste management:

- Good health and well-being (Goal 3): Uncontrolled landfills and dumps can create conditions for the spreading of diseases and the emissions of leachates into drinking water. Proper waste management is also needed for sanitary living conditions.
- Quality education (Goal 4): Citizens will need to be educated on the importance of proper waste management and their role in the waste management system.
- Clean water and sanitation (Goal 6): See discussion related to Goal 3.
- Affordable and clean energy (Goal 7): Emphasis is placed on energy from renewable sources, which includes energy from combustion and biotreatment of waste.
- Climate actions (Goal 13): Greenhouse gases, such as methane from uncontrolled landfills, cause global warming, which drives climate change. Proper handling of waste is needed to combat these changes.
- Life below water (Goal 14): Discarded materials, such as microplastics, must be properly managed to prevent them from entering the oceans and water reservoirs.
- Partnerships for the goals (Goal 17): In order to address challenges such as global warming, proper waste management must be implemented in all countries. This worldwide proper management of waste can be achieved through global partnerships.

Fig. 4.2 shows the waste hierarchy, which is a strategy developed by the European Union to manage waste [6]. In general, the waste treatment method that is at the top of the hierarchy is preferred. Reduction in the amount of waste is at the top of the hierarchy, so reduction in the amount of the discarded commodities is preferred. Hence, even if a commodity can be reused (and is therefore not waste), it is preferable not to consume the commodity in the first place.

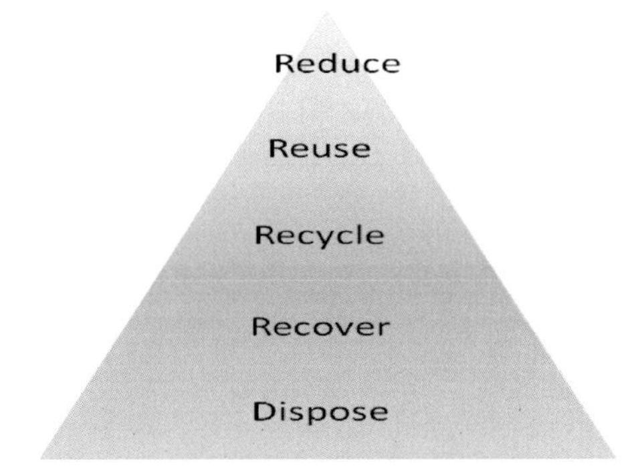

FIG. 4.2 The waste hierarchy.

If one discards a product then, according to the waste hierarchy, it is best to reuse it. Hence, the reused product has the same function as the virgin product and, in this sense, there is no loss in value. If the product cannot be reused then it should be recycled. Although not specified, the recycled product should have as high a value, or quality, as possible. Recycling that leads to a large decrease in value should be avoided. If the quality of the discarded product, or the quality of the material of which the product is made, is too low to warrant recycling, then it should be used for energy recovery. The least preferred option is disposal, with controlled landfilling being preferred to uncontrolled landfilling or dumping.

Other terms have been used to describe the waste hierarchy, or parts of this hierarchy. For example, the three *R*'s—reduce, reuse, recycle—have been used to describe this strategy. As with the waste hierarchy, reduction is preferred to reuse that, in turn, is preferred to recycling. These three activities are preferred to disposal. Another term is the circular economy, for which the European Union has developed an action plan [7]. This terminology underscores the vision to continually reuse or recycle discarded products without a loss in quality; that is, the recycled products should have the same value as the original product. This is not possible with the present waste management systems but the hierarchy shows the path that is envisaged for the future development of sustainable waste management.

THE SWEDISH MODEL

Sweden had very little, if any, waste sorting before 1986. Now Sweden is recognized as one of the world leaders in sustainable waste management. Waste is presently a valuable commodity in Sweden, and private businesses profit from sorting waste, reusing or recycling the waste, or converting waste into electricity and heat. Hence, the system in Sweden not only takes care of the problems associated with waste generation but also, by implementing programs such as producer responsibility, ensures that waste is used to produce useful products and energy. In fact, due to the competitiveness in this area, some actors import waste from other countries to increase their profit margins.

Although all Swedish municipalities are required to sort their wastes into three main waste streams (described in Fig. 4.5), they are responsible for developing strategies and procedures for sorting and the subsequent waste treatment. Most procedures for

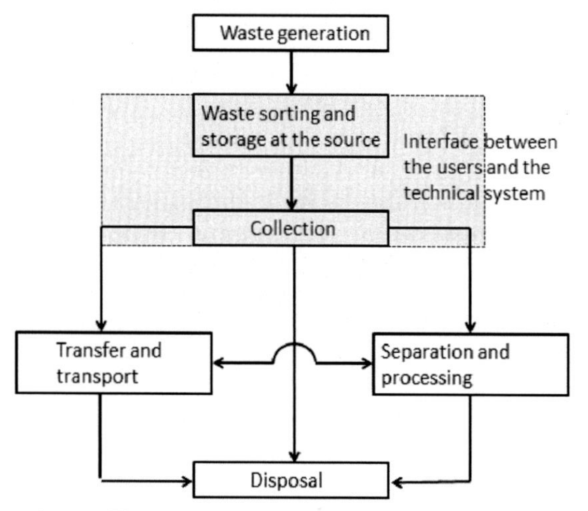

FIG. 4.3 The six activities in the model by Tchobanoglous et al. [8] for managing municipal solid waste.

the treatment of solid waste are similar to the model introduced by Tchobanoglous et al. in 1993 [8]. As shown in Fig. 4.3, this model has six activities in the waste management system. The first activity is waste generation. This is done by various actors along the value chain, which, in municipalities, include restaurants, supermarkets, schools, and households. If one considers the preferred option in the waste hierarchy, then it is these actors that should consume less and reduce the amount of their wastes. However, a certain amount of waste is inevitable, and this should be treated in line with the waste hierarchy. Therefore investigating the amount and types of wastes is essential and is the first functional element for the subsequent processes in this model. Tchobanoglous et al. [8] suggest that sorting of the waste into certain fractions (different types of wastes) should be done at the source, i.e., at the location where the waste is generated. These fractions, which are discussed in more detail later, can then be collected (by the municipalities or other companies) and transferred either to actors that further manage the waste (e.g., sorting into finer fractions, reusing, recycling, and recovery) or to disposal. The waste that goes directly to disposal includes a large fraction that could have gone to a higher level in the waste hierarchy, but that was not correctly sorted at the source.

In order to put this model into the perspective of approaching zero waste, the same terminology used earlier in the chapter can be used for this model. The first activity, called waste generation, could also be called "generation of unwanted commodities." These commodities are only waste if they are not sorted correctly, and therefore cannot be placed in the preferred level in the waste hierarchy. The commodities that are sorted correctly are valuable resources that can be reused or recycled. In fact, in the Swedish MSW management system, many of the commodities that are not correctly sorted are used for energy recovery. Thus as very little MSW goes to disposal, we can approach zero waste. However, it must be reiterated that correct sorting is required for optimal waste management according to the waste hierarchy.

The model shown in Fig. 4.3 places a large burden on the actors, or users, in the municipalities, including households, who are required to sort at the source. In principle, it is possible for these actors to mix the different waste fractions and for the technical system to sort them at a later stage. This can be done, for example, at the material recovery facilities. However, mixing food waste, for example, with packaging waste would require subsequent cleaning of the packaging that results in economic costs and environmental impacts. In addition, Xevgenos et al. [9] have observed that cities that introduced waste sorting at the source have also obtained a dramatic decrease in landfilling and increase in recycling rates. In practice, a combination of sorting at the source and subsequent sorting into finer fractions by the technical system may be a preferred option. For example, plastic packaging may be a fraction that it sorted at the source, and this fraction can be subsequently sorted into different types of plastics (polyethylene, polypropylene, etc.) at a later stage. However, it is not reasonable to expect actors such as households to separate these different types of plastics because this takes time, requires expert competence, and requires space for storing the sorted fractions.

Fig. 4.3 also shows that the actors that generate the waste need to interact with the technical waste management system. A review by Rousta et al. [10] revealed that almost all the studies that are done by researchers from the social science field focus on the actors, whereas almost all studies that are done by people from the engineering sciences focus on the technical system. The review identified a need for multidisciplinary or cross-disciplinary research that focuses on the interface between the actors and the technical system.

The need for a functioning interface between the users and the technical part of the system increases the complexity of the MSW management system. Many actors in the society, from politicians through waste management companies to citizens, need to be motivated to develop and use the system. The key concepts that were used when debating and developing the envisaged waste management system in Sweden were *modernization of society, health, economy* (especially due to the dramatic increase in oil prices in 1973–74), *ecology*, and *climate change*. These terms were used by politicians, media, and the citizens. However, these are only guiding concepts and concrete and effective interventions that motivate the actors that generate and manage the waste need to be identified and implemented. Some interventions that were used in Sweden are listed in the following:

- Formal education: Much of the initial focus was placed on educating school children, who are believed to be receptive to new ideas. In addition, it was seen that these children motivated their parents to sort correctly at home.
- A reasonable number of waste fractions: Sorting must be simple and convenient. To enable this, the number of waste fractions for household sorting must be sufficiently small. At the same time, they must be sufficiently large to enable the technical treatment of the different fractions.
- Policies, laws, and strategies: Environmental strategies and debates have been on the Swedish political agenda for at least four decades [11]. Methods to manage hazardous materials, including those at waste treatment facilities, were described in the Environmental Protection Act of 1969 [12]. This is complemented by overarching laws regarding environmental protection and waste handling in both Sweden [13] and the European Union [6]. These initiatives are continually being updated, as exemplified by the United Nations sustainable development goals discussed earlier.
- Development of a Swedish Waste Management Association: A Waste Management Association was founded in Sweden in 1947. It consists of all Swedish municipalities, as well as municipality and private companies involved in waste management. Its main goals are to provide education and information, to share experiences, and to monitor and develop the waste management system [4].
- Producer responsibility: Sweden introduced the Ordinance on Producer Responsibility [14,15] according to which the producers of packaging, as well as actors that import packaging, are responsible for the proper collection of the packaging and its transport for further treatment. This fee, which is ultimately borne by the consumers, enables municipalities or private companies to perform these tasks.
- Landfill taxes and bans: As discussed by Milios [16], landfilling of sorted combustible waste was banned in Sweden in 2002 and landfilling of organic waste was banned in 2005. Tax for landfilling was imposed in 2000 and has been regularly increased to favor waste management methods that are preferred according to the waste hierarchy [16].
- Proper delegation of waste handling: Municipalities are responsible for developing and implementing their own waste management systems. This can be done by themselves or in collaboration with other actors, such as private companies. This is paid for by municipal taxes and waste tariffs (e.g., based on the weight or volume of the waste), and the users are obliged to follow the waste management system implemented in their municipality. The municipalities are mainly responsible for handling the waste that is not covered by the producer responsibility discussed earlier, and hence, these activities need to be coordinated. However, this format is currently under development in Sweden and the municipalities may, in the future, be responsible for handling all the waste fractions.
- Economic incentives: Economic incentives can be used especially when introducing new waste management systems or when changing existing waste management systems. Examples are refunds on PET or glass bottles. Here consumers pay an extra fee when purchasing the bottled drink, and they obtain a refund when returning the empty bottle. This can lead to a large number of returns [17].

It is evident from this discussion that in order for the waste management system to function, the actors in the municipality that generate waste must interact properly with the technical part of the waste system. Improving this interaction will lead to a more efficient waste management system. A procedure that can be used for achieving this improvement is the recycling behavior transition procedure [18], which is discussed in detail in Chapter 8.

The Swedish waste management system, described earlier, has led to improved sorting of municipal waste. This, in turn, has enabled the "waste" to be used as a valuable resource for material recycling, biological treatment to yield biogas, and for energy recovery from incineration. Fig. 4.4 shows the amount of municipal waste that was treated using different methods in 1975 and 2016 [19]. In 1975, 62% of all municipal waste was landfilled. Due to, among other things, the landfill taxes and bans, discussed earlier, this amount became less than 1% in 2016. This waste has, instead, been treated using methods that are preferred according to the waste hierarchy, discussed earlier with reference to Fig. 4.2. The amount of waste that is recycled has increased from 6% to 35% between 1975 and 2016, the amount treated biologically has increased from

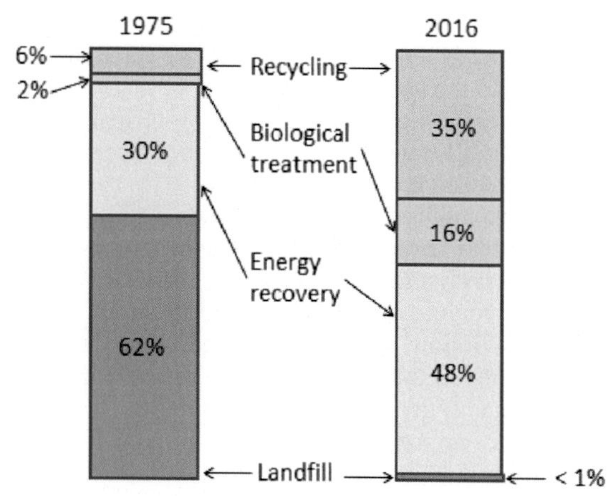

FIG. 4.4 Amount of municipal waste that was treated using different methods in 1975 and 2016.

2% to 16%, and the amount incinerated has increased from 30% to 48%.

Implementing the Swedish Model in the City of Borås

The city of Borås, which is a medium-sized Swedish city with approximately 110,000 citizens, was one of the pioneering cities to implement a sustainable waste management system. Before 1980, more than 90% of the city's waste was landfilled [20]. This was identified as a challenge in 1986, and an investigation to implement an integrated waste management system was initiated in the same year. The city's first waste management plan, which had reduction of landfill as its main goal, was established in 1987. The plan included waste separation at the source and introducing or upgrading biological and thermal waste treatment facilities.

The city's second waste management plan was established for the period 1991–2000 [4]. As discussed earlier, one needs to have separate waste fractions in order to be able to treat the waste using biological or thermal methods, and this involves households, among other actors in the city. A pilot project involving 3000 households was begun in 1988, and during the subsequent 3 years, face-to-face communication was used to describe the waste sorting model in these households [4].

At the same time the technical system was developed, and in 1991, the city's combined transfer facility and material recycling (TF/MR) plant was inaugurated. As some landfill was deemed to be inevitable, a sanitary landfill was opened in 1992. Composting and production of biogas began at the TF/MR plant in 1995, and temporary storage of hazardous waste began in 1998 [4].

The city's third waste management plan covered the period 2001–11 [4]. This plan emphasized the need to reduce waste as well as reduce the amount sent to landfill. In 2002 a new anaerobic digester was inaugurated. This new plant enabled upgrading of biogas to vehicle fuel. In addition, a new combustion plant was opened in 2005. It was a combined heat and power (CHP) plant that provided heat and electricity for the city.

The city's fourth waste management plan is for 2012–20. The aim is to further improve the city's waste management system. The motto of the city is "A city free from fossil fuels" and the municipal buses are, at present, all powered by biogas produced from waste. One of the next steps in the plan is locate the city's facilities for material recovery, biogas treatment, incineration, and waste water treatment at a single site. The aim is to improve the efficiency of the integrated waste management system.

Fig. 4.5 shows the three main municipal waste streams, each divided into substreams, which are collected in Borås. The first stream is food waste and residual waste sorted by different actors in black bags and white bags, respectively. Residual waste is the material that cannot be collected in any of the other waste fractions. Hence, food waste, newsprint, packaging material, bulky commodities, hazardous waste, etc. should not be sorted in the white bags. Used diapers and envelopes that are composed of both plastic and paper are examples of residual waste. The black bags and white bags are provided free of charge by the city. Most actors, including households, usually have these bags at the place where the waste is generated, e.g., in the kitchen. The full bags are subsequently taken by the various actors to the collection bins. The black bags and white bags are usually collected in the same containers, and then optically sorted at the TF/MR plant. Collection of the bags and further treatment is the responsibility of the municipality. The food waste, collected in the black bags, is treated biologically and the white bags, including their contents, are shredded and sent to the city's CHP plant.

The second waste stream is newsprint and packaging. The packaging fractions are metal, plastic, colored glass, transparent glass, and paper packaging. The fractions of this waste stream are collected at the recycling stations. Actors in the municipality, including the households, are expected to take these waste fractions to the recycling stations and sort them correctly. There are more than 80 recycling stations in Borås [20],which are distributed in accordance with the population

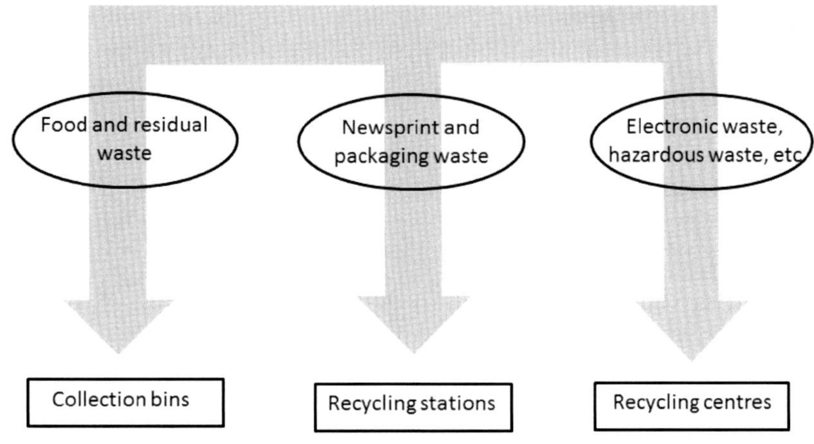

FIG. 4.5 The three main municipal waste streams in Borås.

distribution in the city. Some recycling stations only offer sorting for newsprint and colored and transparent glass bottles. These were the fractions that were collected when the first waste management system was implemented in Borås. According to the producer responsibility discussed earlier, producers and importers of the packaging fractions are responsible for providing the recycling stations and collecting and recycling the discarded packaging. A service organization with representatives from the recycling industries coordinates this service.

The third waste stream shown in Fig. 4.5 consists of fractions such as bulky waste, hazardous waste, electronic waste, and garden waste. Bulky waste includes large commodities such as fridges and garden furniture; hazardous waste includes paints and solvents; electronic waste includes discarded smart phones, TVs, etc.; and garden waste is organic waste from the maintenance of gardens and parks. Separate containers are provided for these fractions at recycling centers. In addition, the recycling centers provide containers for metal, glass, cardboard, and plastic commodities that are not used for packaging. The municipality is responsible for the management of the third waste stream, and there are five recycling centers in Borås [20]. There is no direct charge for using these facilities.

As mentioned earlier, some waste fractions such as PET bottles and aluminum cans are not included in the waste streams in Fig. 4.5. These are returned to supermarkets to obtain a refund. Also, unused medicine is returned to pharmacies for proper waste treatment.

Implementation of the waste treatment strategies and management system in Borås has reduced the MSW that is landfilled from 100,000 tons in 1990 to less than 200 tonnes in 2010 [21]. This was achieved even though the total waste increased during this period. Fig. 4.6 shows the relative amounts of the waste fractions that were treated using the different treatment methods in 2009 [21].

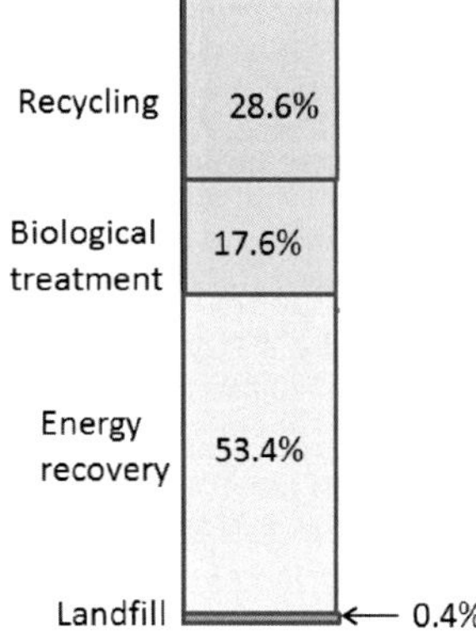

FIG. 4.6 Relative amounts of municipal solid waste fractions that were treated using different treatment methods in Borås in 2009.

As discussed earlier, all cities in Sweden must sort their waste into the three waste streams shown in Fig. 4.5. However, the infrastructure for collecting the waste fractions within each of the three waste streams is the responsibility of each municipality and can therefore differ between municipalities. For example, sorting of the food waste fraction is compulsory in Borås, and this is done by sorting this waste into a black plastic bag. This is different in the Gothenburg municipality, which lies just 60 km from Borås. Separate sorting of the food waste fraction is encouraged, but not obligatory, in Gothenburg, and when it is sorted, it is collected in paper bags. When the food waste is not sorted separately, it is collected together with the residual waste fraction.

Improving the Waste Management System

The waste management systems in Sweden and Borås are continuously being improved. This is a result of the country's strategy to be world-leading in waste management, as well as the municipality's aim to be a national leader. Based on the United Nations sustainable development goals discussed earlier, Sweden has developed an action plan. As an example, under Goal 12, sustainable consumption and production, Sweden has identified, among others, the following actions [22]:

- Implement the 10-year framework for sustainable consumption and production patterns. All countries will take actions, with developed countries taking a leading role and taking into consideration the conditions and need for development in developing countries.
- Implement sustainable management and effective utilization of natural resources by 2030 at the latest.
- By 2030, halve the global food waste per person at the retail and consumer levels and reduce food waste along the entire supply chain, including losses after harvesting.
- By 2020, implement an environment-friendly management system for chemicals and all types of wastes during their life cycles, in accordance with international agreements. Also, by 2020, significantly reduce their emissions into air, water, and soil to minimize their negative consequences on human health and the environment.
- By 2030, significantly reduce the amount of waste by implementing actions to prevent, reduce, reuse, and recycle waste.
- By 2030 ensure that citizens have the necessary information and knowledge that is required for sustainable development and lifestyles that are in harmony with nature.

These actions require development of both the technical system and the interaction between the technical system and the actors that use the system. The heart of the technical waste management system in Borås is the city's TF/MR plant [4], which was established in 1991. The main activities of this plant are listed in the following:

1. To separate the black bags and white bags collected in the first waste stream shown in Fig. 4.5. This is done using optical sorting.
2. To produce biogas from food waste from actors such as households, restaurants, schools, hospitals, supermarkets, and butchers. This biogas is upgraded and used as vehicle fuel. This fuel is used, for example, by the municipal busses.
3. To temporarily store most of the waste fractions shown in the third stream in Fig. 4.5, such as electronic and hazardous wastes. This waste is subsequently transported to the companies that manage these fractions.
4. To prepare combustible waste, e.g., from the white bags and recycling centers, for feedstock to the municipal incineration plants. These plants provide heating and electricity to the city.
5. To landfill waste such as asbestos and some other fractions from construction and demolition activities.

The TF/MR plant is presently being extended to increase its capacity and to colocate the five abovementioned activities with a new CHP incineration plant and a facility for sludge and water treatment. Increased capacity is needed for the increasing population in Borås. Colocation of the different waste management facilities is expected to create synergies and make the waste management more efficient.

WASTE MANAGEMENT—AN INTERNATIONAL CHALLENGE

To reach zero landfill is a global mission. It is important that Sweden and other countries that are leaders in waste management continue to set higher goals and improve their waste management systems, but it is of equal importance that this knowledge and technology is adapted by, and implemented in, all countries. Many of the environmental impacts that arise due to improper handling of waste are global. For example, emissions of methane from dumps and uncontrolled landfills lead to *global* warming, irrespective of where the dumps or landfills are located. The fact that proper waste management is a global challenge is also made clear by the sustainable development goals discussed earlier, which are *global* goals. All countries that have signed the Agenda 2030 are responsible for these goals, including improved health, access to drinking water, etc., which can be negatively affected by improper handling of waste. The 17th goal, "partnerships for the goals," advises all countries to work in partnership to realize the goals.

The advantages of improving waste management systems are enormous. One country that has made progress in its waste management is Nigeria. For example, significant progress has been made in Lagos State with the introduction of, among other things, transfer loading stations and sorting facilities. One such sorting facility is the privately owned Solous materials recovery facility near Ojo, shown in Fig. 4.7. This facility is based on a previous landfill and receives 1000–2000 tonnes of (mixed) waste per day. Approximately 300 workers, many of them having been scavengers on the original landfill, are employed at the facility. They manually sort the different fractions of the waste (e.g., plastic, glass, metal), which are then sold by the company for further processing. The company sorts 20%–25% of waste for selling and the remainder goes to landfill. Power plants, such as Akute and the Island Power Plant, have also been installed to assist with Lagos' need for electricity.

However, most of the waste generated in Lagos is landfilled with minimal benefit gained from recycling and/or conversion to energy. In addition, the demographics of Lagos with its approximately 20 million inhabitants make the challenge even more pressing. In fact, there is an enormous potential for the reuse, recycling, and conversion of waste to energy in Lagos. This is evidenced by the presence of informal and formal sectors that, even today, make livelihoods by sorting and selling valuable fractions of waste. Also, previous studies [23–25] have revealed the potential of using waste in Nigeria to generate energy and electricity. For example, Abila [23] estimated that the municipal waste in Nigeria in 2006 was approximately 75,000 tonnes per day. Abila estimated that 2 million cubic meters per day of biogas can be produced from this waste. Abila makes the point that this biogas can be used for cooking, which would alleviate the health problems associated with cooking using firewood, which is the main fuel used for cooking in Nigeria. In addition, proper waste management will prevent this gas from entering the environment.

In order to implement and improve waste management systems, one needs to include all the relevant stakeholders. The number of stakeholders, and who they are, depends on the aim and size of the project as well as the country of implementation. For example, a project that was done in a pilot area in the city of Borås and that resulted in the building of a recycling station [26] required a university (researcher), the citizens (who used the management system), the property owner (who decided to build a recycling station on the property), the municipality, and the company that collects the packaging and newsprint (both of these actors needed to change their logistics for collection).

FIG. 4.7 The Solous materials recovery facility located in Ojo, Lagos, Nigeria.

An Example of an International Partnership

The Gemah Ripah fruit market in Yogyakarta, Indonesia, discarded 4–10 tons of rotten fruit per day, which was transported and dumped 40 km from the marketplace. In an attempt to alleviate this problem, a collaboration began between the Swedish Centre for Resource Recovery at the University of Borås and Gadjah Mada University in Yogyakarta. The collaboration included several student and researcher exchanges, during which plans were initiated to convert this waste into biofuel. Preliminary dimensions of the bioreactor and its working conditions were calculated. However, implementation of the bioreactor at the market required the involvement of all relevant actors, both in Sweden and in Indonesia.

Thanks to funding from the Swedish International Development Cooperation Agency and the Gadjah Mada University, a Swedish-Indonesian cluster was established. The Swedish team included the University of Borås, the Borås municipality (including the municipality-owned company that manages waste in Borås), and RISE (Research Institutes of Sweden). The Indonesian team consisted of Gadjah Mada University, local governments, embassies, municipalities, companies, nongovernmental organizations, and workers at the fruit market [27]. It was important that the responsibilities and roles of each partner were made clear before, during, and after implementation of the project. It was also important to include local expertise in the project from an early stage. One cannot always take technology that works in one country and implement it in another country. For example, the bioreactor requires a specific temperature to operate (so that the enzymes or bacteria are biologically active) and the outdoor temperature in Borås is very different from that in Yogyakarta.

The bioreactor was commissioned in 2011. About 4 tons of fruit waste per day is used as feed for the bioreactor. This significantly reduces the waste that needs to be transported to the dump, thus reducing the environmental impact of the transport and of the dumped fruit. Also, the electricity that is generated is used to light the marketplace, instead of lighting based on fossil fuel. This has environmental, health, and economic advantages.

CONCLUSIONS AND PERSPECTIVES

This chapter describes the importance of proper waste management in order to approach zero waste. Proper waste management will pave the way toward a circular economy, with different types of wastes being treated at the proper level in the waste hierarchy. The Swedish waste management system for MSW is used as an example.

MSW is very complex because it contains many fractions such as food, metal, paper, plastics, hazardous, and bulky waste. These fractions need to be sorted in order to make them accessible for efficient reuse, recycling, or energy recovery. In Sweden, this sorting is done at the source. Actors, such as households, are required to sort their waste at the place where the waste is generated, e.g., at the homes. If this is not done properly then the management of the waste in the downstream technical system will not be optimal. Proper implementation of the waste management system in Sweden, including education, laws, and taxes, has led to a large reduction in landfilling of waste and a consequent increase in material recycling and energy recovery. The changing habits and culture of the Swedish citizens require that the waste management system must be continually modified and developed to suit the needs of the households.

Waste management is central to global sustainable development and impacts many of the United Nations' development goals. In addition, the ever-increasing global population, urbanization, and standards of living lead to larger quantities of MSW. It is therefore vital to share knowledge and experiences gained in developing sustainable type of management systems on a global level.

REFERENCES

[1] Hoornweg D, Bhada-Tata P. What a waste: a global review of solid waste management. Urban development series knowledge papers, vol. 15. Washington DC: W. Bank; 2012. p. 1–98.

[2] SMED. Avfall i sverige 2014 [waste in Sweden 2014]. Stockholm, Sweden (in Swedish): Swedish Environmental Protection Agency; 2016.

[3] Laurenti R, Stenmark Å. Produkters totala avfall – studie om avfallets fotavtryck och klimatkostnader [Products total waste – study of the waste's footprint and climate costs]. Malmö (in Swedish): Swedish Waste Management Association; 2015.

[4] Rousta K, Richards T, Taherzadeh MJ. An overview of solid waste management toward zero landfill: a Swedish model. In: Taherzadeh MJ, Richards T, editors. Resource recovery to approach zero municipal waste. Boca Raton, USA: CRC Press; 2016. p. 1–22.

[5] Baumann H, Tillman A-M. The hitch hiker's guide to LCA. Lund; Sweden: Studentlitteratur; 2004.

[6] EU. Directive 2008/98/EC of the European parliament and of the council of 19 November 2008 on waste and repealing certain directives. Brussels, Belgium: European Commission; 2008.

[7] European Commission. Closing the loop – an EU plan for the circular economy. 2015 [Brussels, Belgium].
[8] Tchobanoglous G, Theisen H, Vigil SA. Integrated solid waste management: engineering principles and management issues. New York, USA: McGraw-Hill; 1993.
[9] Xevgenos D, Papadaskalopoulo C, Panaretou V, Moustakas K, Malamis D. Success stories for recycling of MSW at municipal level: a review. Waste and Biomass Valorization 2015;6(5):657–84.
[10] Rousta R, Ordoñez I, Bolton K, Dahlén L. Support for designing waste sorting system: a mini-review. Waste Management & Research 2017;35(11):1099–111.
[11] Engblom GM. Waste management, the Swedish experience. Stockholm; Sweden: The Swedish Environmental Protection Agency, Graphium Norstedts; 1999.
[12] SFS. Miljöskyddslagen, SFS. 1969:387 [environmental protection law, SFS. 1969:387]. Stockholm, Sweden (in Swedish): The Department of the Environment; 1969.
[13] SFS. Miljöbalken, SFS. 1998:808 [environmental code, SFS. 1998:808]. Stockholm, Sweden (in Swedish): The Department of the Environment; 1998.
[14] SFS. Förordningen om producentansvar för förpackningar, SFS. 1994:1235 [The ordinance of producer responsibility for packaging, SFS. 1994:1235]. Stockholm, Sweden (in Swedish): Swedish legislation; 1994a.
[15] SFS. Förordningen om producentansvar för returpapper, SFS. 1994:1205 [The ordinance of producer responsibility for waste paper, SFS. 1994:1205]. Stockholm, Sweden (in Swedish): Swedish legislation; 1994b.
[16] Milios L. Municipal waste management in Sweden. European Environment Agency; 2013.
[17] Pantamera. Pantamera 2018. 2018. Cited March 12, 2018. Available from: http://pantamera.nu/pantsystem/statistik/sa-pantar-ni-i-din-kommun/.
[18] Rousta K, Bolton K, Dahlén L. A procedure to transform recycling behavior for source separation of household waste. Recycling 2016;1(1):147–65.
[19] Swedish Waste Management Association. Hushållsavfall i siffror – kommun- och länsstatistik 2017 [Household waste in numbers – municipal and region statistics 2017]. [Malmö, Sweden (in Swedish)]. 2017.
[20] Rousta K. Municipality solid waste management: an Evaluation on the Borås system. Sweden: University of Borås; 2008.
[21] Borås Energi och Miljö. Borås waste management plan 2012–2020. Borås, Sweden: Borås Energi och Miljö; 2011.
[22] Swedish goals. 2015. cited March 9, 2018. Available from: http://www.regeringen.se/regeringens-politik/globala-malen-och-agenda-2030/.
[23] Abila N. Managing municipal wastes for energy generation in Nigeria. Renewable and Sustainable Energy Reviews 2014;37:182–90.
[24] Akinbomi J, Brandberg T, Sanni SA, Taherzadeh MJ. Development and dissemination strategies for accelerating biogas production in Nigeria. BioResources 2014; 9:5707–37.
[25] Akinrinola FS, Darvell LI, Jones JM, Williams A, Fuwape JA. Characterization of selected Nigerian biomass for combustion and pyrolysis applications. Energy & Fuels 2014;28:3821–32.
[26] Rousta K, Dahlén L, Bolton K, Lundin M. Quantitative assessment of distance to collection point and information effects in source separation of household waste. Waste Management 2015;40:22–30.
[27] Bolton K, De Mena B, Schories G. Sustainable management of solid waste. In: Taherzadeh MJ, Richards T, editors. Resource recovery to approach zero municipal waste. Boca Raton, USA: CRC Press; 2016. p. 23–40.
[28] Brancoli P, Rousta K, Bolton K. Life cycle assessment of supermarket food waste. Resources, Conservation and Recycling 2017;118:39–46.
[29] Garnett T. Three perspectives on sustainable food security: efficiency, demand restraint, food system transformation. What role for life cycle assessment? Journal of Cleaner Production 2013;73:10–8.
[30] FAO. Food wastage footprint. Impacts on natural resources – summary Report. Rome: Food and Agriculture Organization of the United Nations; 2013.
[31] FAO. Global food losses and food waste – Extent, causes and prevention. Rome: Food and Agriculture Organization of the United Nations; 2011.
[32] FAO, IFAD, WFP. The State of Food Insecurity in the World. Meeting the 2015 international hunger targets: taking stock of uneven progress. Rome: Food and Agriculture Organization of the United Nations; 2015.
[33] FAO. How to feed the world in 2050. Rome: Food and Agriculture Organization of the United Nations; 2009.

CHAPTER 5

Influential Aspects in Waste Management Practices

KARTHIK RAJENDRAN, PHD • RICHEN LIN, PHD • DAVID M. WALL, PHD • JERRY D. MURPHY, PHD

INTRODUCTION

Between 1880 and 2017, global surface temperatures have increased by 1°C. Similarly, the concentration of carbon dioxide has witnessed a surge to greater than 400 ppm, the highest level in the past 650,000 years. This increase in temperature will elevate sea levels, which will ultimately affect the livelihood of millions of people across the world. The average global sea levels are rising by 3.2 mm/a [1]. Most climate change models predict that limiting global warming to less than 1°C may not be possible by the end of this century. Global warming is caused by activities such as industrialization, use of fossil fuels, which release greenhouse gases (GHGs) and contribute to global warming [2].

Human-led industrialization has caused a surge in the economy that has radically transformed our consumption patterns. This changing pattern has left a human trail across the globe in the form of waste. Today, the global solid waste generation has reached about 2 billion tonnes/a [3]. A strong correlation between economic growth and waste generated can be observed, highlighting the changes in consumption patterns. However, only a few countries in the world manage their waste disposal and recovery efficiently.

Several components within the wastes generated are valuable resources that are typically unused and dumped, polluting our environment, rivers, and oceans. For example, plastic bags are produced from fossil sources that leave a significant trail in the oceans affecting the flora and fauna of the aquatic environment. If plastic bags are recycled efficiently, the need for fossil sources may decrease, reducing our emissions and strengthening our fight against climate change [4]. On the other hand, waste generation is inevitable. Thus a fine balance between waste production and treatment/recovery is necessary.

There are several aspects in providing effective waste management practices. These include the use of the appropriate technology, which is economical, backed by the government in the form of policy and has gained support from the public. These interventions affect waste management practices across the globe. However, the complexity in this practice is the interactional effects of these factors that can adversely affect waste management [5].

This chapter attempts to address the influential aspects of an effective waste management practice, including an overview of global waste facts, technologies available, technical issues in waste management, and economic, sociocultural, and political factors. Finally, a case study that compares the waste management practices in different countries is presented. This chapter highlights issues to be considered that go beyond the conventional and technical side of waste management practices.

GLOBAL WASTE—FACTS AND FIGURES

Global solid waste generation is on the rise, with more than 2 billion tonnes/a of waste produced. This increase in waste generation is influenced by several factors including expanding economies, new products, change in mind-set among the public, consumerism, increases in income, increase in population, etc. [6]. A global heat map of per capita solid waste generation per day is highlighted in Fig. 5.1 [3,7]. It is evident that the developed countries exceed the developing countries in terms of waste generation. This is primarily due to the socioeconomic status and the capacity of purchasing

Sustainable Resource Recovery and Zero Waste Approaches. https://doi.org/10.1016/B978-0-444-64200-4.00005-0

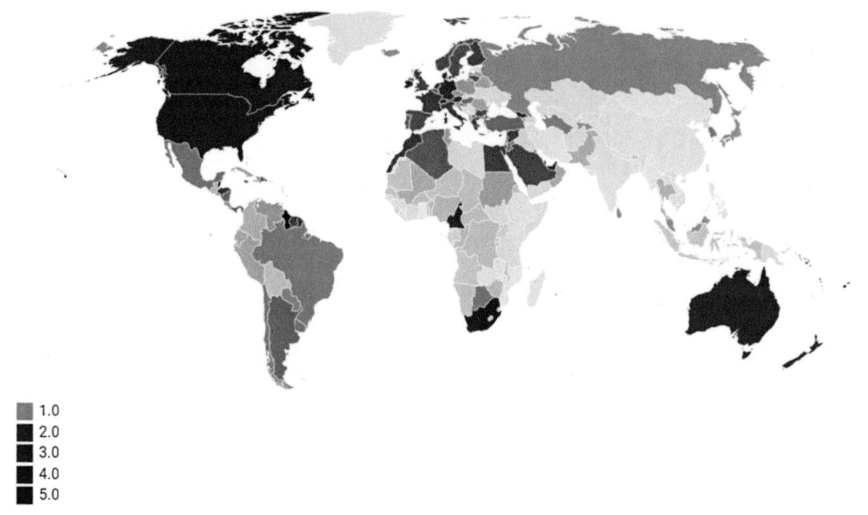

FIG. 5.1 Global heat map of per capita solid waste generation per day (kg/capita/d).

power among the people in these countries. The developed countries, including the United States, Canada, the European Union, and Australia, produce significantly more waste than the third-world countries [6].

In the past, products have been developed with the intention for use, but not designed to include for disposal/treatment, i.e., the end of a product's life cycle. This design approach has caused difficulties environmentally in terms of global warming and climate change. It is necessary to change this consumption pattern to reflect sustainability into the consumption cycle and for the product waste to become a part of it. Certain countries have established a polluter pays principle and also indicated the producer responsibility to reflect this change in a product cycle [8]. However, these laws are enacted mainly in developed countries. There are several factors that affect global solid waste generation and its treatment patterns. This includes access to affordable technology, economic conditions, sociocultural influences, and political directives. There is a strong correlation between the gross domestic product and the waste generated in countries. This is an almost linear correlation (Fig. 5.2). As the economy of many countries continues to expand, this correlation poses a serious threat how the wastes generated will be treated and processed.

As products are produced, and utilized, at a faster rate, it is necessary to develop suitable technologies for waste treatment. With more products being produced, fewer resources will be available in the future to sustain these requirements. It is becoming extremely important that the technologies developed for waste management should consider sustainable resource recovery mechanisms.

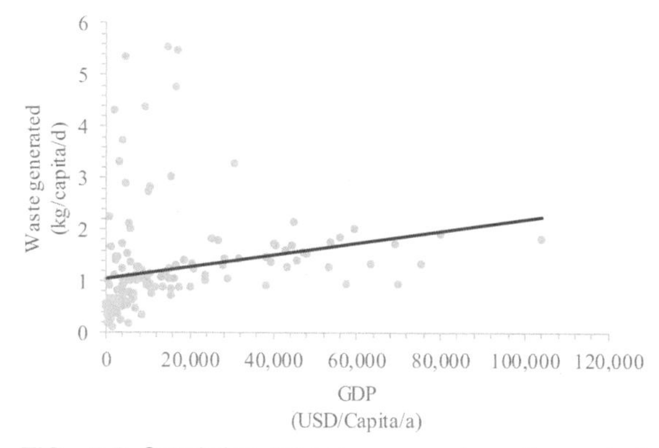

FIG. 5.2 Correlation between gross domestic product (GDP) in USD versus waste generated (kg/capita/d).

WASTE MANAGEMENT TECHNOLOGIES

There are a variety of waste treatment technologies; the long-established technology is simply landfilling. Other available technologies include composting and recycling. These technologies have been widely employed to treat various types of wastes. Alternatively, advanced waste treatment technologies have attracted increasing attention and have been explored for the purpose of waste-to-energy conversion. These advanced technologies include biological (e.g., anaerobic digestion and fermentation) and thermal/thermochemical technologies (e.g., incineration, pyrolysis, and gasification), which are outlined in Table 5.1 in terms of the conversion process, conditions, main products, and by-products. These technologies are not necessarily more complex than the established technologies, but

TABLE 5.1
Advanced Waste Treatment Technologies.

Conversion Process	Conditions	Main Products	By-Products
Anaerobic digestion	35–55°C, anaerobic environment, pH 6.5–7.5	Biogas (around 60% CH_4 and 40% CO_2)	Digestate (used as fertilizer/for soil amendment)
Hydrogen fermentation	35–55°C, anaerobic environment, pH 5.5–6.5	Biohydrogen (around 60% H_2 and 40% CO_2)	Volatile fatty acids (used for downstream chemicals)
Ethanol fermentation	30–35°C, anaerobic environment, pH 4.5–6.0	Ethanol and CO_2	Remaining feedstock (used as animal feed)
Incineration	800–1000°C, air, oxygen	Heat, electricity	Ash
Gasification	800–900°C; air, oxygen, or steam; 1–30 bar	Syngas (CO, CH_4, N_2, H_2, CO_2)	Ash
Pyrolysis	400–1200°C, the absence of oxygen	Syngas, bio-oil	Biochar (used for soil amendment, activated carbon)

they exhibit many advantages with respect to the reduction of volume and destruction of toxic organic compounds, and energy recovery. However, the economics need to be carefully evaluated when using the advanced waste treatment technologies. This section will discuss the advantages and the major challenges associated with these advanced waste treatment technologies.

Anaerobic Digestion

Anaerobic digestion is a series of biological processes in which diverse microorganisms break down biodegradable materials in the absence of oxygen. Wet organic wastes such as food wastes, animal slurries, and agricultural wastes are preferable feedstocks for anaerobic digestion. One of the end products is biogas (mainly containing 60% methane and 40% carbon dioxide), which is combusted to generate electricity and heat, or can be upgraded into renewable natural gas and transportation fuels. The other product is nutrient-enriched digestate, which can be used as soil conditioners or fertilizers. Anaerobic digestion has many environmental benefits including the production of a renewable energy platform, the possibility of nutrient recycling, and the reduction of waste volumes [9]. As a result, anaerobic digestion has, in the recent years, received increasing attention in a number of countries. For example, the total number of biogas plants in Europe was 14,572 as of 2013 and Germany has the most developed biogas industry, with around 9000 plants in operation [10]. In Asian countries, the biogas industry is booming; for example, the biogas production in 2015 in China reached 19 billion m^3, with an average annual growth rate of 6.3% since 2010.

When considering the application of anaerobic digestion systems, the feedstock is a key factor affecting the performance of digestion. Wet organic wastes such as food waste and animal slurries typically contain abundant biodegradable components that can be used for biogas production. However, the individual feedstock may have a suboptimal carbon-to-nitrogen (C/N) ratio, which can lead to an unstable digestion process. Carbohydrates are more favorable substrates than proteins. The excess protein content in feedstocks can lead to a low C/N ratio (typically below 10); this would be a critical issue for long-term monodigestion. It is known that anaerobic degradation of protein compounds produces ammonia. High ammonia concentration in the digester can inhibit methanogens, causing volatile fatty acid accumulation and digestion failure. To address this issue, codigestion of carbohydrate-rich and protein-rich feedstock has been carried out to achieve a balanced C/N ratio, thus increasing the activity of microorganisms in digestion associated with an increased biogas yield. Research has shown that codigestion of municipal waste and food waste can help improve biogas production by up to 40%–50% compared with monodigestion of food waste alone [11]. It has been demonstrated that biogas yield from the mixture of wastewater sludge and food waste increased linearly with an increased fraction of food waste, and addition of 35% of food waste in the mixture exhibited not only a higher methane yield but also an accelerated methane production [12].

Another issue that may affect the application of anaerobic digestion is the relatively long digestion time (typically 20–40 days) due to the long duration of the microbial reactions. The major biological processes involved in anaerobic digestion include four steps: hydrolysis, acidogenesis, acetogenesis, and methanogenesis (Fig. 5.3), and each step requires certain types of microorganisms. Most researchers have reported that the rate-limiting step for complex organic substrates is the hydrolysis step, which is ascribed to the formation of toxic by-products or non-desirable volatile fatty acids formed during the hydrolysis step. While methanogenesis is the rate-limiting step for easily biodegradable substrates, due to the relatively low growth rate of methanogens. To accelerate the hydrolysis and enhance subsequent methane productivity, a variety of pretreatment options, such as mechanical, thermal, chemical, or biological processes, or a combination of these, have been developed at laboratory or pilot scale with various levels of success [13–15]. However, a systematic assessment of different pretreatment options is quite necessary for deciding which one would be the most suitable from an industrial point of view.

Fermentation

Fermentation technologies can be employed to produce either biohydrogen or bioethanol from wet organic wastes, such as food waste, agricultural waste, sewage sludge, using different microorganisms. Before fermentation, some wastes require saccharification or hydrolysis for converting carbohydrates (such as cellulose and starch) into sugars.

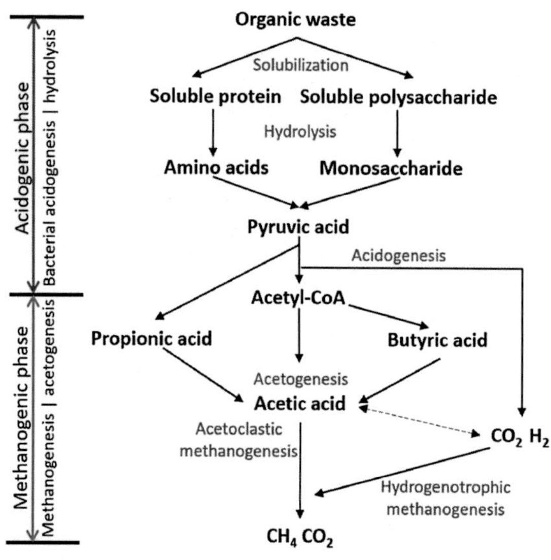

FIG. 5.3 Metabolic pathway for methane production via anaerobic digestion.

Biohydrogen is usually produced through dark hydrogen fermentation, during which hydrogen-producing bacteria, such as *Clostridium* and *Enterobacter*, can convert fermentable sugars to hydrogen and volatile fatty acids. Hydrogen is a potentially versatile energy carrier that could alter the use of liquid fossil fuels because hydrogen has a high energy density per unit mass of 122 kJ/g, which is 2.8-fold higher than that of hydrocarbon fuels [16]. In addition, the combustion of hydrogen produces only water as a by-product, contributing to a favorable outcome for the reduction in GHG emissions. Biohydrogen production through dark hydrogen fermentation is still in its infancy and most studies are based on pilot scales. The critical challenges of hydrogen fermentation lie in the low hydrogen conversion efficiency and unstable hydrogen production, partly because of the formation of various by-products. The theoretic yield of hydrogen through dark fermentation is 4 mol/mol of glucose ($C_6H_{12}O_6 + 2H_2O = 2CH_3COOH + 2CO_2 + 4H_2$) [17]. However, the reported data are typically below 2.5 mol/mol glucose in the state-of-the-art literature [17–19]. For example, the hydrogen yields of wild *Enterobacter aerogenes* (a typical species of hydrogen-producing bacteria) are reported as approximately 1.0–1.8 mol/mol of glucose [20]. The low yields are due to the fact that the bacterial strain is sensitive to and inhibited by operational parameters such as particular pH ranges, accumulated hydrogen, and volatile fatty acids.

With regard to bioethanol production, this fermentation process mainly converts sugars to bioethanol. As compared with biohydrogen production, bioethanol production is a more mature technology. The basic steps for large-scale production of bioethanol are fermentation of sugars, distillation, dehydration, and denaturation (optional). For bioethanol production from the cellulosic materials of wastes, effective pretreatment and enzymatic hydrolysis are required to produce a high concentration of glucose. Fermentation can then convert glucose to ethanol by microbes, such as *Saccharomyces cerevisiae, Escherichia coli, Zymomonas mobilis, Pachysolen tannophilus*, and *Candida shehatae*. Ethanol is broadly used as a liquid biofuel for transportation and has a great potential as a substitute of gasoline in the transport fuel market. But the cost of bioethanol production is higher than that of fossil fuels. To address this issue, progressive research is needed to reduce the cost of enzymes and to select robust microorganisms with high tolerance to inhibitory compounds [11].

Incineration

Incineration is a relatively mature waste treatment technology that involves the combustion and conversion of wastes into heat, flue gas, and ash [21]. It is the thermal degradation and decomposition of wastes in the presence of oxygen at temperatures of 800−1000°C. The heat from the combustion process can be used to operate steam turbines for energy production, or for heat exchangers in industry. The ash is mostly formed by the inorganic composition of the wastes and may be carried by the flue gas in the form of solid particulates. The flue gases must be cleaned to remove the gaseous and particulate pollutants before they are dispersed into the atmosphere. Incinerators are capable of reducing the volume of original solid wastes by up to 80−85%, and thus they significantly reduce the necessary volume for disposal [11]. In addition, incineration has particular benefits for the treatment of certain waste types such as clinical wastes and certain hazardous wastes where pathogens and toxins can be completely destroyed by high temperatures [22].

The major issue with incinerators is the potential pollution associated with incineration of waste. Emissions include the following elements and compounds: sulfur, chlorine, fluorine, N_2, CO, CO_2, NO_x, SO, polychlorinated dibenzodioxine, furan, methane, ammonia, hydrochloric acid, and hydrogen fluoride [23]. Extensive efforts have been made to control the pollutant emissions, such as by modifying the fuel composition, the moisture content of the fuel, the particle size of the fuel, and the incinerator configuration. For example, an electrostatic precipitator was used to remove dioxin in an incinerator, resulting in removal efficiencies greater than 90% for all congeners and homologs of dioxins [24].

Gasification

As another thermal approach for waste treatment, gasification is similar in principle to incineration. The energy produced from incineration is high-temperature heat, whereas combustible gas is often the main energy product from gasification. Gasification converts wastes into a combustible gas mixture by partially oxidizing wastes at high temperatures, typically in the range of 800−900°C. The low-calorific-value gas produced can be burned directly or used as a fuel for gas engines and gas turbines. The product gas (a mixture of CO, H_2, CO_2, and H_2O) can be used as a feedstock in the production of chemicals (such as methanol). The gasification of solid waste includes a sequence of successive endothermic and exothermic steps, with respect to the reactants and products (Fig. 5.4).

Gasification of municipal solid waste (MSW) and biomass as an energy recovery method has been widely studied all over the world. Several types of research on biomass gasification technologies have been conducted in the European Union and the United States and most of them are in the laboratory or demonstration phase [25]. Gasification has several advantages over the traditional incineration technology, primarily in terms of

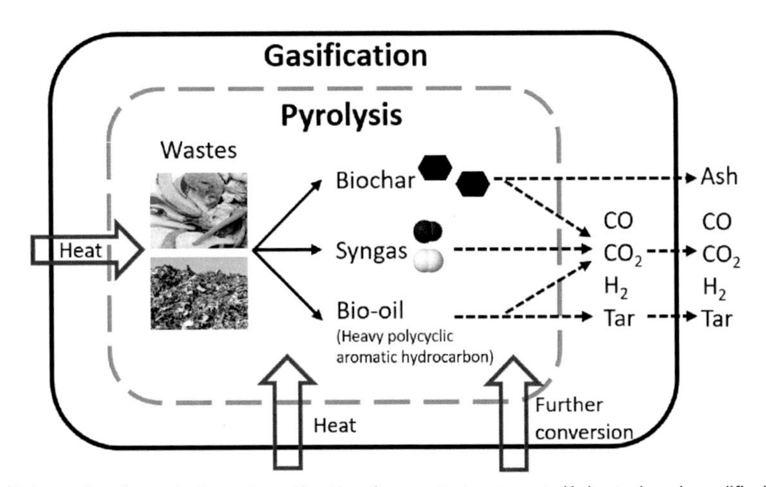

FIG. 5.4 Schematic of pyrolysis and gasification for waste treatment. (Adapted and modified from [25].)

more acceptable cost and the flexibility of coupling the operating conditions (such as temperature and equivalence ratio) and the reactor configurations (such as fixed bed, fluidized bed, entrained bed, vertical shaft, moving grate furnace, rotary kiln) to obtain syngas, which is suitable for use in different applications [11]. However, the variable characteristics of various wastes tend to make gasification much more challenging; the variation of feedstock properties has a major impact on the design, performance, maintenance, and cost of gasification, and ultimately on its feasibility [26]. Even if a number of significant applications do exist, the extreme challenges of waste gasification prevent it from consideration as an established commercial option: operating experience is limited and data on actual performance, reliability, and costs are incomplete as well, thus making comparisons with conventional technologies very difficult.

Pyrolysis

Pyrolysis is a technology that breaks down organic materials in the absence of oxygen to produce liquid (bio-oil), gaseous (syngas), and solid (biochar) products, as illustrated in Fig. 5.4. Syngas comprises mainly of CO and H_2 (together 85%) with a small proportion of CO_2 and CH_4. The bio-oil produced through pyrolysis typically has a heating value of around 17 MJ/kg. The pyrolysis process can occur in the temperature range of 400–1200°C. Although the product yield depends on various operating parameters, generally low temperature and high residence time favor biochar production [27]. Pyrolysis has been investigated as an attractive alternative to incineration for waste disposal. The pyrolysis process conditions can be optimized to produce a solid char, gas, or liquid/oil product, indicating that a pyrolysis reactor can act as an effective waste-to-energy converter. When compared with the conventional incineration plant that runs in the capacity of kilotonnes (kt) per day, the scale of the pyrolysis plant is more flexible and the output of pyrolysis can be integrated with other downstream technologies for product upgrading. The existing pyrolysis technologies seldom run alone with gas, bio-oil, and biochar output as end products, most of them are combined with gasification, combustion, and smelting. The combination with gasification produces fuel gas of moderate calorific value, and this will be a competitive choice in the future. However, at the same pyrolysis-based technologies are expensive and may not be affordable compared to commercial waste treatment methods.

FACTORS AFFECTING WASTE MANAGEMENT

There are several factors that affect sustainable waste management practices. The problems are typically different depending on the country and thus it is a multidimensional localized problem that cannot be solved by a single solution. A combination of different aspects needs to be assessed to reach sustainable solutions. There are four distinct aspects of interest in waste management: technology, economics, sociocultural aspects, and politics.

Technology

The developed world has access to high-end technologies because of the increased efforts in innovation, research, and development. However, one technology that works in a particular country may not be as effective in another because of the local conditions and requirements. For example, Japan has limited land availability that drives incineration as the preferred waste treatment technology over other forms of treatment methods [3]. Waste management technologies should be developed to reflect the 5R principles (reduce, reuse, recycle, recover, and refuse) [28]. Yet the bottom-up approach has been practiced more than the top-down approach (Fig. 5.5). Many countries landfill/dump waste, and some developed countries use energy recovery and recycling. For a sustainable future, more emphasis needs to be placed on the reduce and reuse concepts.

Conventional waste management technologies have less scope in the market in terms of recycled materials. Most present-day technologies can sort similar materials but not a composite such as food packaging [29]. Other technologic aspects such as waste collection systems need to be simplified particularly for third-world countries. In most western countries, wastes are sorted into multiple fractions including organics, recyclables, glass, metals, and combustibles [30]. In the developing countries, waste collection systems are not efficient and the collection rate is less. To tackle this issue, simplified collection methods are necessary or a boost in the form of incentives is necessary, along with adequate training/education programs on the importance of waste segregation.

Economics

The key hindrance to sustainable waste management is the economic feasibility of the waste treatment methods. Most waste management systems are not economically viable unless the government provides a

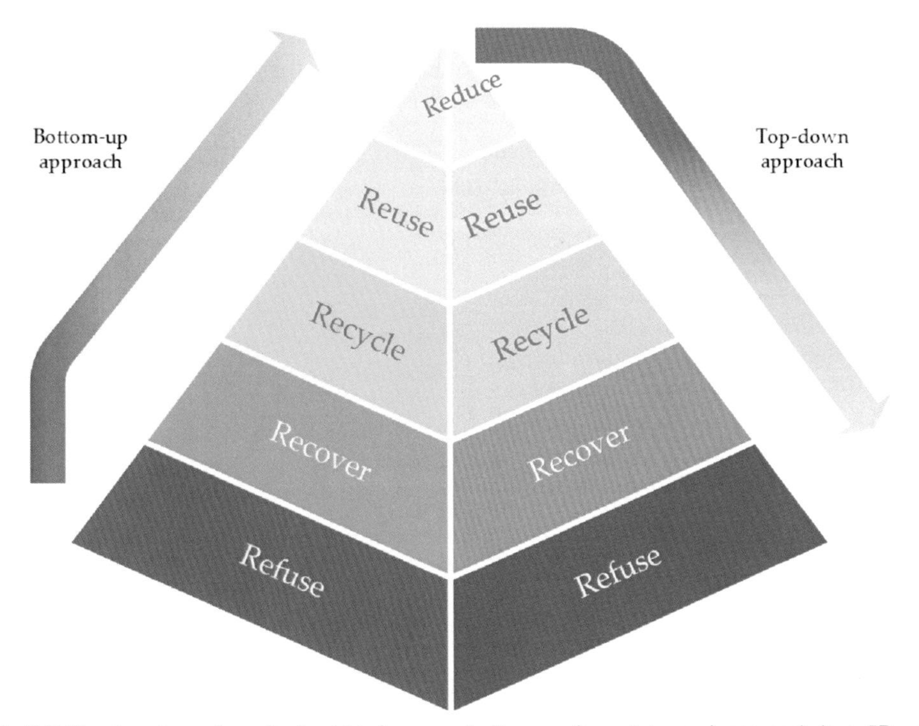

FIG. 5.5 The top-down (how it should be) versus bottom-up (how it is now) approach in a 5R waste management system.

subsidy. This is why landfilling has been preferred in developing countries rather than waste treatment and energy recovery options. Most developing countries require not only financial support to establish these technologies [6] but also a change in the business model from how conventional businesses operate.

Furthermore, the technologies need to be cost-effective for companies to operate and expand. As mentioned earlier, current waste management technologies require incentives/support systems from the government, at least in the initial phase [31]. Incentives are required in high-income countries, and for developing countries, this situation is even more challenging. Often, a change in taxation is required for the uptake of certain waste management technologies. For example, Germany provided incentives that accelerated the use of renewable energy technologies, accelerated waste management, and reduced GHG emissions [32]. If a government starts to provide tax reductions and subsidies, then private/semistate companies may be willing to enter the waste management sector. Assurances are needed in some form for companies to invest and thereby make profit in the waste management sector.

Current revenue streams in waste management systems include the polluter pays principle, producer responsibility, and gate fees for organics. This needs to be enhanced to generate further income through the selling of products. There are certain macro- and microeconomic aspects attached to this. For example, if either the crude oil price or carbon tax for fossil fuels increases, most energy recovery waste management systems may become more profitable.

Sociocultural Aspects

Waste management is also a societal issue. This is where the sociocultural aspect gains sufficient importance, equivalent to technology or economic feasibility. No waste treatment technology can be feasible unless people are willing to accept and abide by it [33]. In western countries, from an early age, children are taught how to segregate waste. When they become adults, they pass this trait on to future generations and so the tradition is carried forward. In the developing world, the awareness and education among adults is minimal, causing a significant challenge in overcoming this barrier.

Often, people have a resistance to change from their current behavior. This is one of the main facets of a sociocultural hindrance. It requires personal motivation to adhere to the change and a strong policy might provide the kick start that is required. A strong road map, alongside adequate training and awareness, is essential to overcome these barriers (Fig. 5.6).

Policy and Political Aspects

Policy is one of the key drivers that can bring all other factors to work in harmony. Good policy can drive a system. For example, Germany became the leader in the biogas industry primarily because of the policy support from the government in the form of incentives and assured tariffs. The government helped industries to innovate and reduce costs and emissions by providing additional incentives. Besides specific policies, there are other sides of political aspects that need to be considered for a sustainable waste management system. These include transparency in governance, measures to reduce corruption, etc. In developing countries, the corruption rates can be high and this is one of the reasons why there may be higher investment costs. These higher investment can costs make the technology unfeasible. More open markets and transparency in governance, along with strong policies, will help toward a transition to sustainable waste management.

FIG. 5.6 Multidimensional factors for an efficient waste management system.

CASE STUDIES

Sweden

Sweden, with 9.9 million inhabitants, has exemplified great success from a waste management perspective. Quantities of waste generated in recent years have been on the decline and recycling of materials is now common practice. Sweden is located in the cold Nordic region and thus heating is required for several months during the winter, whereas air-conditioning is necessary for shopping centers and hospitals during the summer. Consequently, energy recovery from waste is a very practical solution in Sweden. Methods include incineration (combustion) of residual waste and anaerobic digestion that can produce biogas, which can be used for electricity or, if cleaned, as a transport fuel, and digestate, a biofertilizer.

The Swedish national strategy for waste management is based on a hierarchy in which waste prevention is the first stage [34]. From 2015 to 2016, the quantity of household waste generated decreased by 11 kg per person to 467 kg household waste per person and the total quantity of household waste produced in Sweden was 4.67 million tonnes (Mt) [34]. The breakdown of treatment of household waste was as follows: 34.6% material recycling (equivalent to 1.62 Mt), 16.2% biological treatment (equivalent to 0.76 Mt), 48.5% energy recovery (equivalent to 2.26 Mt), and the remaining 0.7% sent to landfill [34]. Hence, little burden to the environment is seen through the generation of waste in Sweden. The effective utilization of waste has been possible by changing the waste production patterns, implementing corrective legislation since the 1960s and designing less material for packaging [35].

Sweden is divided into a number of municipalities for waste management. Each municipality has its own waste management strategy and regulations, and often municipalities collaborate in the development of waste management plans. It is the responsibility of the households to separate the wastes at source and follow the plans set by the municipality. Both collection of waste and waste treatment can be carried out by the municipalities or private contractors or a combination of both. Municipalities in Sweden primarily use a volume-based tariff for the collection, transport, recovery, and disposal of waste; however, a smaller proportion of municipalities operate off a weight-based tariff [36]. The money accrued by the municipality through charging customers for the collection of waste only covers the total cost of municipal waste management. In essence, the revenue will not exceed these costs. The cost on the consumer varies by location, by dwelling, and whether the household source separates food waste (which can result in lower costs). As of 2016, the average annual waste collection charge for a Swedish household was SEK 2094 and the total average annual cost in municipalities was SEK 787 per person excluding VAT [34].

The recycling of material plays a pivotal role in Sweden's plans for a sustainable society. Citizens are encouraged to separate the waste in two ways: one is by educating them about the waste and its importance and the other is by providing economic incentives. For instance, a deposit system named PANT was created to increase recycling of PET and glass bottles and aluminum cans. The customers pay a deposit of SEK 1–2 for each can or bottle and get it back when returning the bottles/cans to any shop or supermarket in Sweden. As a result of this method, more than 88% of these materials are recycled in a relatively pure form [37]. In Sweden the governmental policies were formulated to enhance the extended producer responsibility (EPR) for packaging, waste paper, refrigerators, printers, etc. The EPR shifts the responsibility to the producer to collect waste after use of a product and dispose of the waste properly [38]. The collection of packaging and paper is undertaken at recycling centers and processed to become new products. Further products for reuse are also recycled, such as plastic, glass, textiles, and construction materials. Recycling centers are available in which households can also dispose of bulky and electric waste; 580 of these centers currently exist [34]. The more the waste is separated, the more value it has and it can be easily recycled or converted into other materials or energy. Recycling centers are made convenient for the public and typically located no more than 300m from residential areas [34].

Food and residual wastes are typically separated into organic and combustible bins in Sweden, while the collection of paper and packaging is becoming more common for households. From the Swedish perspective, biological treatment of organic waste is achieved through anaerobic digestion or composting, the latter becoming less popular. For 2018, a goal has been set that for 40% of all food waste collected, both energy and nutrients, will be recovered [34]. Materials that do not fall under the category of recycling or biological treatment in Sweden are sent for energy recovery through incineration; there are 34 incineration plants in the country [34]. For 2016, 18.1 TWh of electricity was produced through energy recovery, accounting for more than any other country in Europe [34]. Additionally, Sweden imports residual waste from other countries, enhancing the country's overall fuel supply. Studies from Sweden have indicated that the separate collection of food waste can reduce incineration by reducing the collection of residual waste while simultaneously increasing recycling and biological recovery of waste, thereby offering a successful policy instrument [36]. Furthermore, valorization of food waste is common in Sweden through anaerobic digestion or food waste could potentially be used for the production of bio-based chemicals in the future [39]. Landfills are still required for wastes that do not fall under the categories of material recycling, biological treatment, and energy recovery; however, waste volumes decreased by 19% in just one year from 2015 to 2016 [34].

United States

The United States is one of the most industrialized countries in the world. The country generates a very high quantity of MSW per capita; according to the Organisation for Economic Co-operation and Development, the latest figure stands at 738 kg per person per year [40]. With a population of 319 million, the total generation of MSW in the United States was calculated in excess of 262 Mt as of 2015 [40,41]. Of this MSW, approximately 68 Mt are recycled, 2 Mt are composted, 33 Mt are combusted for energy recovery, and 137 Mt are landfilled [41]. Thus over half of all MSW generated in the United States is still sent to landfill disposal; however, this figure has reduced significantly from 94% of all MSW in the 1960s [42]. The typical makeup of MSW in the United States constitutes 29.7% containers and packaging, 20% durable goods, 20% nondurable goods, 15.1% food waste, 13.2% yard trimmings, and 1.5% other wastes. Since 1960, the generation of MSW has increased almost threefold, with a daily generation rate of 2.03 kg per person in 2015 [42].

The United States initiated its waste management measures back in 1895 in New York City. It was started with unit operations approach, control of waste management, collection and transportation, processing, incineration, and landfilling [43]. The evolution of waste management in the United States has instigated a rise in recycling, composting, and energy recovery in the recent years. However, the country can be considered to be starting from a low base, as landfilling was the dominant waste management method for decades. The tipping fee for landfills has seen an evident increase in the past 35 years in the United States. The fee has increased 2.5-fold from approximately $19 (in 1980) to $48 (in 2015), with the use of landfills declining in this period [41]. The number of active MSW landfills in the United States has decreased from approximately 7900 in 1988 to 1900 in 2009 [44]. A factor in the significant use of landfills can be attributed to the strong private sector responsible for the waste management systems in the United States, which have been somewhat resistant to change. The US Environmental Protection Agency (EPA) guidelines suggest a preferential waste treatment hierarchy that is headed by source reduction and reuse, followed by recycling/composting and energy recovery, with the least preferred option being treatment and landfill disposal [42].

Collection of household waste in the United States varies between local governments and private collectors, depending on the jurisdiction. It has been previously estimated that collection and transportation costs for waste in the United States exceeds the revenues generated from waste treatment and disposal [43]. Traditionally, households pay for waste disposal through property taxes or on a fixed fee basis. However, in some communities, pay-as-you-throw (PAYT) programs have been established. This is a variable-rate pricing system where the customer is charged based on the weight of waste they are disposing of or on a per bag basis [45]. The advantage of such an initiative is the greater incentive for household recycling and production of less waste. The benefits of recycling have been illustrated in the 2016 Recycling Economic Information study that suggested that as of 2007, recycling and reuse activities in the United States accounted for 757,000 jobs, $36.6 billion in wages, and $6.7 billion in tax revenues [41]. An example scheme is the container deposit laws set in 10 US states (California, Connecticut, Hawaii, Iowa, Maine, Massachusetts, Michigan, New York, Oregon, and Vermont), which carry deposit refund systems for beverage containers. Depending on the state, the deposit will be in the range of $0.05–$0.15. In essence, the consumer pays a deposit on the purchase of the beverage. The purpose of such a system is to shift the responsibility of packaging (and waste) to the manufacturer from the consumer. Through this, many benefits are obtained, including increased recycling rates through financial incentive and job creation.

Within the US waste management strategy is the food recovery hierarchy that includes for source reduction, food donation, animal feed, industrial use, and composting. Food waste reduction is directed at both businesses and individuals to calibrate the quantity of food that might be required, with the aim of reducing waste. The second tier of the hierarchy proposes that any nonperishable and unspoiled perishable foods can be donated to local food banks, soup kitchens, pantries, and shelter food donation [46]. Beyond this, remaining scraps can be fed to livestock as a cheaper alternative to landfill disposal. Only after these initial steps is food waste considered as a means of generating energy (fourth tier on the hierarchy). Anaerobic digestion of fats, oil, and grease is particularly befitting, as disposal of these wastes can be difficult [46]. As of 2016, there were 77 waste-to-energy facilities operating in the United States (in 22 states), with a total daily throughput of approximately 95,000 t per day [47].

Ireland (Republic of Ireland)

Ireland's waste management strategy is underpinned by the Waste Management Act 1996 and the EU Waste Framework Directive. The Directive specifies the waste management hierarchy for Ireland under which the pinnacle is the prevention of waste, followed by reuse, recycling, recovery, and disposal. Under the Waste Management Act 1996, all local authorities in Ireland must facilitate the collection of household waste in their allocated area and also provide for the provision of disposal and recovery facilities. Households typically have their waste collected once a week via private operators and it is common for operators to collect different types of wastes every other week. The collection of waste is typically a central issue in Irish policy. Under the most recent framework introduced, waste collectors can offer a range of pricing options that include for standing charges, charges per lift, charge per kilogram, charge by weight, or a combination of these options. The old structure of a flat rate charge for wastes is now being phased out for customers. Per capita, Ireland is among the highest municipal waste producers in Europe. As of 2014, Ireland had the sixth highest level of municipal waste per capita in the European Union at 586 kg, which was higher than the EU average of 474 kg per capita per year [48]. According to the latest survey available, the total quantity of waste generated in Ireland was 19.8 Mt [49].

For the purposes of future waste management planning Ireland has been split into three regions, namely, Southern Region, Eastern Midlands Region, and the Connacht Ulster Region. The national government, local government, and EPA establish how the waste management hierarchy is being achieved within these regions [50]. The Irish government typically applies the polluter pays principle in which the producer of the waste must assume the responsibility in ensuring the waste is correctly disposed of. In the past two decades, Ireland has seen an evident change in terms of its waste management strategy. This is exemplified by the avoidance of landfill, which has been directly targeted, driven by both national and EU legislations. In Ireland, only five landfills remain in operation; this is a reduction from 22 sites in 2010. From 2010 to 2019 the landfill levy increased from €30 per tonne to €75 per tonne [51].

In Ireland the MSW to be treated is described as a combination of household and commercial wastes that include dry recyclables, residual waste, food waste, and garden waste [51]. Three options are prescribed for the treatment of municipal waste: recycling, recovery through fuel production, and disposal to landfill. Latest figures suggest that as much as 79% of municipal waste is being recovered (approximately 1.94 Mt through recycling and fuel production), whereas the remainder is sent to landfill (approximately 0.54 Mt) [51]. As of 2014, Ireland had the ninth highest rate in the European Union for sending municipal waste to landfill (223 kg per capita) [48]. However, the recovery of fuel waste is expected to reduce quantities further, largely in part to a second incineration facility that became operational in 2017.

Ireland has succeeded in certain areas of waste management and a shift toward a circular economy is now the focus. For example, recovery of specific packaging wastes (cardboard, paper, glass, plastic, steel, aluminum, and wood) reached 88% and has been consistently above the targets set in this area [48]. Food waste has specifically been targeted in Ireland with regard to prevention and treatment. By 2016, the quantity of biodegradable MSW that may be sent to landfill was capped at 35% of the baseline year 1995 [52], and the 2020 target for biodegradable MSW is below 427,000 t; as of 2017, Ireland disposed of approximately 307 kt of biodegradable waste [51]. The reduction measures put in place for biodegradable waste were implemented to reduce fugitive methane emissions, a harmful GHG. Composting and anaerobic digestion are the biological treatment options available in Ireland, the latter expected to become a more sizable industry in the coming years. Currently composting is dominant and accounts for 79% of all tonnage treated [51] (Table 5.2). As of 2016, a total of 231 kt of biodegradable municipal waste is treated and this figure was reached through the introduction of productive legislation such as the Commercial Food Waste Regulations (2010) and the Household Food Waste Regulations (2013). In particular, the Commercial Food Waste Regulations (2010) required households to have a separate collection of food and biowastes and to segregate their food waste before its collection. Approximately 640,000 households in Ireland had an organic bin kerbside collection service in 2016. To achieve this, all waste collectors offered the collection service in population agglomerations of greater than 500 persons [51,53]. A trend has developed in Ireland where much

TABLE 5.2
Different Countries and Their Waste Management Activities.

Countries	GDP (USD/ capita/a)	Population (millions)	MSW Generation (Mt/a)	Recycling (%)	Incineration (%)	Compost/ Biological Treatment (%)	Landfilling (%)
Sweden	53,442	9.9	4.6	35	49	16	0.7
Ireland[a]	69,330	4.8	2.6	34	35	7	21
USA	59,531	325	262	26	13	9	53

GDP, gross domestic product; *MSW*, municipal solid waste.
[a] Waste export accounts for fraction of waste treatment.

of the organic bin waste collected (up to 32%) is transported to Northern Ireland for treatment in anaerobic digestion facilities [51].

Infrastructure planning is seen as a key step in Ireland's future waste management plans. The deficits in the waste management strategies have been recognized, such as the unavailability of a hazardous waste landfill and the high exportation rate of recyclable waste to other countries. However, three types of permanent facilities for the recycling of materials are available nationally. Bring banks are established for the recycling of materials such as glass bottles, aluminum cans, and unwanted clothes. Civic amenity sites are staff-run centers that open at specific hours and tend to accept much of the same wastes but with more variety, including garden wastes at specific sites. Finally, recycling centers are for the civic amenity sites (with staff and specific opening hours), typically located at local authority depots but tend to not accept very bulky items.

CONCLUSIONS AND PERSPECTIVES

Sustainable waste management practices are not limited to developing affordable technologies but beyond. A coherent intervention including various stakeholders such as government, public, industries, scientists, and nongovernmental organizations is needed to solve this problem. Waste management is a multidimensional localized problem that needs localized solutions. The important aspect of a sustainable waste management practice is the will to do it rather than passing the responsibility to other stakeholders. Several countries have successfully implemented waste management technologies and are on the verge of developing zero-waste cities. It is necessary for those countries to transfer that gained knowledge and experience to other countries for a sustainable world in the future.

REFERENCES

[1] NASA. USA, https://climate.nasa.gov/; 2018.
[2] Weart SR. The discovery of global warming. Harvard University Press; 2008.
[3] Taherzadeh MJ, Rajendran K. Factors affecting development of waste management. Waste management and sustainable consumption: reflections on consumer waste. Routledge; 2014. p. 67–87.
[4] Jambeck JR, Geyer R, Wilcox C, Siegler TR, Perryman M, Andrady A, et al. Plastic waste inputs from land into the ocean. Science 2015;347:768–71.
[5] Mihai F-C, Taherzadeh MJ. Introductory chapter: rural waste management issues at global level. Solid waste management in rural areas. InTech; 2017.
[6] Hoornweg D, Bhada-Tata P. What a waste: a global review of solid waste management. 2012.
[7] Kawai K, Tasaki T. Revisiting estimates of municipal solid waste generation per capita and their reliability. Journal of Material Cycles and Waste Management 2016;18: 1–13.
[8] McAllister J. Factors influencing solid-waste management in the developing world. 2015.
[9] Xu F, Li Y, Ge X, Yang L, Li Y. Anaerobic digestion of food waste – challenges and opportunities. Bioresource Technology 2018;247:1047–58.
[10] Torrijos M. State of development of biogas production in Europe. Procedia Environmental Sciences 2016;35: 881–9.
[11] Pham TPT, Kaushik R, Parshetti GK, Mahmood R, Balasubramanian R. Food waste-to-energy conversion technologies: current status and future directions. Waste Management 2015;38:399–408.
[12] Koch K, Helmreich B, Drewes JE. Co-digestion of food waste in municipal wastewater treatment plants: effect of different mixtures on methane yield and hydrolysis rate constant. Applied Energy 2015;137:250–5.
[13] Zhen G, Lu X, Kato H, Zhao Y, Li Y-Y. Overview of pretreatment strategies for enhancing sewage sludge disintegration and subsequent anaerobic digestion: current advances, full-scale application and future perspectives. Renewable and Sustainable Energy Reviews 2017;69: 559–77.

[14] Ariunbaatar J, Panico A, Esposito G, Pirozzi F, Lens PNL. Pretreatment methods to enhance anaerobic digestion of organic solid waste. Applied Energy 2014;123:143–56.
[15] Harris PW, McCabe BK. Review of pre-treatments used in anaerobic digestion and their potential application in high-fat cattle slaughterhouse wastewater. Applied Energy 2015;155:560–75.
[16] Chandrasekhar K, Lee Y-J, Lee D-W. Biohydrogen production: strategies to improve process efficiency through microbial routes. International Journal of Molecular Sciences 2015;16:8266–93.
[17] Lin R, Cheng J, Ding L, Song W, Liu M, Zhou J, et al. Enhanced dark hydrogen fermentation by addition of ferric oxide nanoparticles using Enterobacter aerogenes. Bioresource Technology 2016;207:213–9.
[18] Liu Z, Zhang C, Lu Y, Wu X, Wang L, Wang L, et al. States and challenges for high-value biohythane production from waste biomass by dark fermentation technology. Bioresource Technology 2013;135:292–303.
[19] Ren N-Q, Zhao L, Chen C, Guo W-Q, Cao G-L. A review on bioconversion of lignocellulosic biomass to H 2: key challenges and new insights. Bioresource Technology 2016;215:92–9.
[20] Zhang C, Lv F-X, Xing X-H. Bioengineering of the Enterobacter aerogenes strain for biohydrogen production. Bioresource Technology 2011;102:8344–9.
[21] Pan S-Y, Du MA, Huang IT, Liu IH, Chang EE, Chiang P-C. Strategies on implementation of waste-to-energy (WTE) supply chain for circular economy system: a review. Journal of Cleaner Production 2015;108:409–21.
[22] Rodriguez-Garcia G, Frison N, Vázquez-Padín JR, Hospido A, Garrido JM, Fatone F, et al. Life cycle assessment of nutrient removal technologies for the treatment of anaerobic digestion supernatant and its integration in a wastewater treatment plant. The Science of the Total Environment 2014;490:871–9.
[23] Tabasová A, Kropáč J, Kermes V, Nemet A, Stehlík P. Waste-to-energy technologies: impact on environment. Energy 2012;44:146–55.
[24] Karademir A, Bakoğlu M, Ayberk S. PCDD/F removal efficiencies of electrostatic precipitator and wet scrubbers in IZAYDAS hazardous waste incinerator. Fresenius Environmental Bulletin 2003;12:1228–32.
[25] Arena U. Process and technological aspects of municipal solid waste gasification. A review. Waste Management 2012;32:625–39.
[26] Consonni S, Viganò F. Waste gasification vs. conventional Waste-To-Energy: a comparative evaluation of two commercial technologies. Waste Management 2012;32:653–66.
[27] Tripathi M, Sahu JN, Ganesan P. Effect of process parameters on production of biochar from biomass waste through pyrolysis: a review. Renewable and Sustainable Energy Reviews 2016;55:467–81.
[28] Tam VW, Tam CM. A review on the viable technology for construction waste recycling. Resources, Conservation and Recycling 2006;47:209–21.
[29] Ruj B, Pandey V, Jash P, Srivastava V. Sorting of plastic waste for effective recycling. International Journal of Applied Science and Engineering Research 2015;4: 564–71.
[30] Chu Z, Wang W, Wang B, Zhuang J. Research on factors influencing municipal household solid waste separate collection: bayesian belief networks. Sustainability 2016;8:152.
[31] Wagner J. Incentivizing sustainable waste management. Ecological Economy 2011;70:585–94.
[32] IEA. Renewable energy sources act 2014. 2014. https://www.iea.org/policiesandmeasures/pams/germany/name-145053-en.php.
[33] Petit-Boix A, Leipold S. Circular economy in cities: reviewing how environmental research aligns with local practices. Journal of Cleaner Production 2018;195: 1270–81.
[34] Sverige A. Swedish waste management. 2017. https://wwwavfallsverigese/fileadmin/user_upload/Publikationer/Avfallshantering_2017_eng_lowpdf.
[35] Hultman J, Corvellec H. The European Waste Hierarchy: from the sociomateriality of waste to a politics of consumption. Environment and Planning-Part A 2012; 44:2413.
[36] Andersson C, Stage J. Direct and indirect effects of waste management policies on household waste behaviour: the case of Sweden. Waste Management 2018;76: 19–27.
[37] Rajendran K, Björk H, Taherzadeh MJ. Borås, a zero waste city in Sweden. Journal of Development Management 2013;1:3–8.
[38] Pires A, Martinho G, Chang N-B. Solid waste management in European countries: a review of systems analysis techniques. Journal of Environmental Management 2011;92:1033–50.
[39] Brunklaus B, Rex E, Carlsson E, Berlin J. The future of Swedish food waste: an environmental assessment of existing and prospective valorization techniques. Journal of Cleaner Production 2018;201:1–10.
[40] OECD. Municipal waste data. 2018. https://dataoecdorg/waste/municipal-wastehtm.
[41] US EPA. Advancing sustainable materials management: 2015 fact sheet. 2018. https://wwwepagov/sites/production/files/2018-07/documents/2015_smm_msw_factsheet_07242018_fnl_508_002pdf.
[42] US EPA. National overview: facts and figures on materials, wastes and recycling. 2018. https://wwwepagov/facts-and-figures-about-materials-waste-and-recycling/national-overview-facts-and-figures-materials.
[43] Kollikkathara N, Feng H, Stern E. A purview of waste management evolution: special emphasis on USA. Waste Management 2009;29:974–85.
[44] US EPA. Municipal solid waste landfills: economic impact analysis for the proposed new subpart to the new source performance standards. 2014. https://www3epagov/ttnecas1/regdata/EIAs/LandfillsNSPSProposalEIApdf.

[45] US EPA. Pay-as-you-throw. 2016. https://archiveepagov/wastes/conserve/tools/payt/web/html/indexhtml.
[46] US EPA. Sustainable management of food. 2018. https://wwwepagov/sustainable-management-food/food-recovery-hierarchy.
[47] Energy Recovery Council. 2016 Directory of waste-to-energy facilities. 2016. http://energyrecoverycouncilorg/wp-content/uploads/2016/05/ERC-2016-directorypdf.
[48] CSO. Environmental indicators Ireland. 2016. https://wwwcsoie/en/releasesandpublications/ep/p-eii/eii2016/waste/.
[49] EPA. Waste in Ireland. 2015. http://wwwepaie/pubs/reports/indicators/epa_factsheet_waste_v2_fapdf.
[50] Department of the Environment Community and Local Government. Waste management policy in Ireland: a resource opportunity. 2012. http://wwwepaie/pubs/reports/waste/plans/Resource_Opportunity2012pdf.
[51] EPA. EPA waste data release: MUNICIPAL waste statistics for Ireland. 2017. http://wwwepaie/nationalwastestatistics/municipal/.
[52] Council of the European Union. Council directive 1999/31/EC, EU landfill directive. 1999.
[53] EU. EU (Household food waste and bio-waste) regulations. 2015. https://wwwdccaegovie/en-ie/environment/legislation/Documents/21/SI%20No%20430_2015%20Household%20food%20waste%20and%20Bio-Wastepdf.

CHAPTER 6

Sustainable Management of Solid Waste

MUKESH KUMAR AWASTHI, PHD • JUNCHAO ZHAO, MASTER • PARIMALA GNANA SOUNDARI, PHD • SUMIT KUMAR, PHD • HONGYU CHEN, MASTER • SANJEEV KUMAR AWASTHI, PHD • YUMIN DUAN, MSC • TAO LIU, PHD • ASHOK PANDEY, PHD • ZENGQIANG ZHANG, PHD

WASTE MANAGEMENT AND SUSTAINABILITY: AN INTRODUCTION

Rapid population and economic growth, urbanization, industrialization, and changing socioeconomic conditions have resulted in increase in solid waste (SW) generation. SW management is a serious issue, as there are environmental, public health, and aesthetic concerns associated with its safe disposal. SWs generate a large amount of greenhouse gases that have a significant influence on global warming. Contamination of water and soil, generation of unpleasant odor, and spread of diseases are other noteworthy concerns associated with their management. Despite the many efforts made toward their handling, much of their chunks end up in landfills. Landfills require much land area as well as money and manpower for their successful operation. In short, SW management is a problem at a global scale and it takes an economic as well as environmental toll on our society. Hence, we should look toward sustainable waste management and mitigation of waste generation.

Viable SW management requires public awareness and participation, stringent laws and policies at the local body and government levels, and scientific solutions. Different studies have been undertaken in diverse parts of the world for the successful handling of SWs. A study in Brazil has highlighted that good sanitation (including drinking water, sewage, and SW management) can be achieved by public participation combined with private enterprises and local communities [1]. Another study in São Paulo, Brazil, has revealed that socioeconomic aspects influence municipal solid waste (MSW) generation, and thus social aspects and inequality must be considered while planning for MSW management [2].

A research in Saguenay, Quebec (Canada), has emphasized that information on the quantity and type of residual household waste generated by inhabitants can help in successful planning of its management [3]. Analysis of household solid waste (HSW) in Okayama and Otsu cities of Japan has found that HSW generation and composition is influenced by the individual consumption expenditure. Greener lifestyle, intensive recycling, producer responsibility, separation at source, and composting should be practiced for better handling of HSW [4]. Landfilling, composting, and incineration are the widely practiced methods to handle MSW in China and they result in air pollution. A review on air pollution and greenhouse gas emission associated with MSW management in China has emphasized that reduction, sorting, recycling of waste, advanced disposal technologies, effective laws/regulations, public awareness, and education are approaches to control air pollution associated with MSW clearance [5].

Contemporary researches highlight that for sustainable management of SWs, we should practice the four *R*'s Reduce, Reuse, Recycle, and Recover. A lot of studies conducted on this topic emphasize establishment of a circular economy during the handling of MSWs. Wastes should be seen as a resource, and a lot of economic benefits can be harvested from them if they are disposed properly. Investigation of three case studies, namely, San Francisco's zero waste programs, Flanders's sustainable materials management initiative, and Japan's sound material-cycle society plan, showed that the present transition to view waste as a resource for the establishment of a circular economy is at the initial stage and additional advancement in government policies, planning, and behavior is needed [6].

Sustainable Resource Recovery and Zero Waste Approaches. https://doi.org/10.1016/B978-0-444-64200-4.00006-2

Vermicomposting is an age-old process that converts organic wastes to useful manure by the synergistic action of earthworms and microbes. This process has also been used to detoxify industrial waste to produce manure for agriculture [7]. Microbial anaerobic digestion is an additional approach to convert animal waste to renewable energy (biogas) and nutrient-rich biofertilizer [8]. Production of lactic acid from food waste [9], generation of activated carbon from SW [10], and recovery of cement, salt, metals, and rare earth metals from MSW ash after incineration [11] are some of the examples of production of value-added products from wastes. To sum up, the change in the societal mind-set to view waste as a resource for the production of useful products will lead to a circular economy and simultaneously result in sustainable management of SW.

WASTE CHARACTERISTICS AND GENERATION

Understanding of the quantity and characteristics of SWs is prerequisite for their sustainable management. Wastes are characterized based on their source of generation, properties, and geographic locations. There are some broad types of waste classification. *Biodegradable* wastes such as organic food wastes can be converted by the action of microbes into useful products, whereas *nonbiodegradable* wastes such as most types of plastics are tough materials that can be recycled or end up in landfills. *Hazardous* and *nonhazardous* are the other waste categories. Hazardous waste can pose physical, chemical, radioactive, and infectious hazards after exposure and needs careful handling and disposal by experts under strict guidelines. Wastes are also largely distinguished into *recyclable* and *nonrecyclable* based on their feasibility of recovery. SWs are characterized by the biowaste, combustible, and noncombustible fractions. Determination of moisture content, organic fractions, calorific value, and chemical composition of different fractions of wastes helps in their sustainable management. Some of the widely generated wastes are discussed in the following.

MSW is a major waste generated by household and business enterprises. The municipality, local body, and private enterprises have the responsibility to collect this category of waste. This is a most heterogeneous waste with a major portion of it being food waste. MSW can also contain hazardous wastes such as discarded batteries containing mercury. Plastic, paper, wood, glass, textile, rubber, leather, and metals are the common constituents of MSW worldwide [12]. *Construction and demolition (C&D)* wastes are created during construction, renovation, and demolition of buildings, roads, and bridges. This category of waste consists mainly of inert materials such as concrete, brick, metals, and woods and there is emphasis to recycle most of its fractions. Asbestos and particulate matters are hazardous components of C&D wastes. Like any other SW, quantification is necessary for successful handling of C&D wastes, and in this direction a review has explained and evaluated 57 papers on its quantification techniques [13]. Advances in technology have resulted in varied electronic products such as desktop, laptop, cell phone, smartphone, tablet, digital camera, and flat television. The wastes associated with these products are called *electronic waste (e-waste)*. Growing demand, greater affordability, societal needs, and shorter use time of these products have resulted in a sharp increase in generation of e-wastes. The presence of heavy metals in them poses grave environment pollution concerns. Recycling of e-waste is priority in the present scenario, and its various portions such as plastic, glass, metals, rare earth elements, and minor metals can be recycled [14].

Healthcare wastes are generated by medical facilities such as hospitals, health centers, and clinics. Aging population, prevalence of lifestyle diseases, and affordability of medical treatments and use of disposable items have led to the increase in healthcare wastes. Syringes, bandages, cotton, gloves, medicines, pathologic waste, chemicals, placenta, plastic, and paper are some common constituents of this kind of wastes. Some specific wastes among them must be compulsorily incinerated due to strict regulations, whereas some components posing biological risks must be treated before disposal. Healthcare wastes are broadly classified into infectious and noninfectious wastes. Yellow bags are used to collect infectious, pathologic, and sharp objects requiring greater attention, whereas paper and other common wastes are collected in black bags. The healthcare waste generation varies among different wards of hospitals. Healthcare waste shows variability in its constituents' characteristics, such as bulk density, calorific value, and moisture content [15].

WASTE STORAGE, SEGREGATION, AND COLLECTION

Waste collection at source, its segregation into different components, and storage are important aspects while planning sustainable waste management. Collection of waste is generally done by door-to-door collection, roadside bins, and common collection points. Waste collection responsibility mainly lies on local government such as municipality/local body. In some cities, roadside bins are segregated to collect biodegradable and nonbiodegradable wastes. Proper waste segregation is most important from a recycling point of view and for decreasing the waste load for further disposal. Responsibility of waste segregation lies on individual citizens as well as on dedicated staff for waste collection and segregation. Segregation of plastic, metals, woods, papers, and biowastes not only provides economic advantage due to recycling but also saves money and efforts for their further treatment or disposal. Public awareness is most important for proper waste segregation. Socioeconomic background is also crucial for public behavior in proper waste sorting. The local government should take care to raise public awareness for waste sorting through different kinds of engagements emphasizing its economic and environmental benefit. Although urban areas and cities are well covered for waste collection, sorting, and storage, rural areas are mostly neglected, resulting in open disposal of wastes. Collected wastes are stored at specific points from where they are transported for treatment or disposal at the landfill. Some wastes such as e-wastes need separate storage facility and their storage facilities are not well developed till now.

Although mainly carried out on similar lines, the planning, modalities, and behavior of different countries/regions differ in the collection, segregation, and storage of wastes. A study on Guangzhou, China, has pointed out that for integrated waste management, sorting of waste by individuals and sanitation workers is the most practical approach. This approach helped recycle a significant amount of plastic, paper, and metal, resulting in economic and environmental benefits [16]. Socioeconomic conditions were found to affect waste recycling in Hong Kong. Privately managed properties inhabited by well-off families recycled more wastes [17]. A review on waste sorting has underlined that suitable infrastructure, less distance to collection point, and engagement with public are key for a successful sorting system [18]. In India, MSW is mainly disposed of at nearby common collection centers by individuals and further sorted by sanitation workers. A few cities, such as Delhi and Bangalore, have house-to-house waste collection facility. The average collection efficiency of MSW is 70% in most of the cities and states of India [19]. In Dhaka, Bangladesh, the Dhaka City Corporation (DCC) is responsible for the collection of MSW. The DCC can collect only 50% of the waste and the rest is discarded in open spaces, resulting in drain clogging, bad odor, water pollution, and mosquitoes [20].

WASTE PREVENTION

Prevention of waste is a vital part of integrated waste management and is at the top of hierarchy. Prevention of waste is immensely beneficial from an environment protection and economy point of view. Prevention of waste is advantageous from any waste management strategy such as energy recovery, recycle, and landfill because the production of material that becomes waste as well as its treatment is circumvented. The definition of waste prevention includes avoidance of waste-creating products, waste reduction at source, increasing the life cycle of a product, and reuse. Although recycling is excluded in this definition, it is sometimes referred as the waste prevention approach by some researchers. Much of the onus for waste prevention lies on individuals and policy makers. Engagement with the public can be through public campaigns, explaining the environmental and monetary benefits of waste prevention. Individuals should be encouraged for a greener and minimalistic lifestyle owing to our moral responsibility to conserve the environment for future generations. Food waste is one waste component that can be avoided by households. Individuals should be made aware that food security is a global problem and thus we should strive toward zero food wastage. We can also share surplus food with people in need of it. In this direction, some nongovernmental organizations (NGOs) collect surplus food from restaurants and parties and distribute it to the people in need. Waste prevention should be inculcated as a social responsibility and needs a behavioral change in individuals.

Policy making also has a great role to play in the prevention of waste generation. Encouraging the use of secondhand/refabricated items with providing incentives for them is one approach for reuse. Some government policies, such as ban on plastics, have an immense role to play in waste prevention. The government should also take initiatives for recycling of bulky items. Payment by residents and business ventures for waste management is also an approach toward waste prevention and sustainable waste management. The policy that the more you generate waste, the more you have to pay will encourage people in prevention of waste generation. Waste prevention is an interdisciplinary approach and needs scientific innovations. Several recent innovations, such as use of plastics in road making, are contributing to waste prevention and reduction. Several research papers have highlighted the importance of waste prevention in integrated waste management. Zacho and Mosgaard [21] have suggested the need for incorporation of waste prevention in local waste management. They also suggest setting quantitative targets for waste prevention rather than just the announcement of intention. The quantitative target has more binding and we work to achieve that. A review on household waste prevention has suggested targeting for waste prevention, producer accountability, householder charging, public sector financial support for pilot projects, and the partnership among public, private, and other organizations backed by strong public engagement and campaigns as policy measures [22]. Johansson and Corvellec [23] have underlined that the European and Swedish waste prevention plans are flawed, as they pay no attention to consumption, the primary reason of waste generation. Their waste prevention plans are more about handling the created wastes and are also not clear on incentives and sanctions. Instead of focusing on household waste prevention, the spotlight should be upstream on the consumption and manufacture of goods with elevated environmental toxicity.

MATERIAL RECYCLING AND RESOURCE RECOVERY

SW separation system, material recycling, and resource recovery systems of various kinds of SWs and the legal framework for SW sustainable management are the main parts for sustainable management of SW. Furthermore, the awareness of SW management among every individual should also be increased; thus the goal of sustainable development of SW management would be achieved finally. Resource recycling can achieve good economic, social, and environmental benefits. It can not only save natural resources, reduce energy consumption, transfer labor, and expand employment but also help curb waste and reduce pollutant emissions. Recycling of SWs makes it possible for the conversion of useless waste into useful materials, turning hazardous waste into harmless resources, and promoting waste utilization. The recycling of SW is of great significance in terms of resource recovery and saving energy, as well as reducing risk to the environment. The SW separation and recovery system is gaining importance because it not only resulted in reducing waste accumulation but also helped to make profits out of it. The household wastes can be collected by separating paper, plastics, scrap metals, etc., which will substantially reduce the amount of waste accumulation. Similarly, the recovery of biogas and heat energy from landfills and incineration plants, respectively, will reduce waste generation and help in the appropriate reutilization of resources. The proper recovery of precious metals from e-wastes can slow down the depletion of natural mines, developing and encouraging the utilization of urban mines. Resource utilization of livestock and manure, agricultural waste, domestic sewage sludge, and other organic SWs during aerobic composting and anaerobic digestion and then recycling the organic substances and nutrients, etc. are some of the efficient ways to realize SW resources and materials recovery systems.

PUBLIC ENGAGEMENT FOR THE IMPLEMENTATION OF WASTE REDUCTION AND RECYCLING POLICIES

Public Engagement System

Public engagement is a social activity in which the social organizations, units, or individuals serve as the subject and participate in the project, planning, policy formulation, and evaluation procedures in a targeted manner within its rights and obligations. The idea of public engagement was formed in the 1970s. In 1970 the requirements for public engagement in the National Environmental Policy Law were clearly stated in the United States. In 1978 the United Nations Environment Programme (UNEP) explicitly proposed the concept of public engagement in its basic procedures for environmental impact assessment. Furthermore, the World Bank implemented public engagement as a World Bank policy in October 1981, which is clearly defined in its work order: when borrowers were going to design and implement projects, especially in the process of environment evaluation, the World Bank expects that the opinions of the affected groups and NGOs should be fully considered by borrowers. The public engagement mechanism in environmental protection is in the public interest, which represents that citizens have the right to take part in environment-related decision-making links through certain procedures or methods.

The Characteristics of the Public Engagement System

The smooth implementation of social behaviors must rely on the maximized identity, support, and participation of the public and the community, which also needs to have a broad mass base. The public should not only participate in the formation of relevant decisions but also need to supervise their implementation process. Therefore it is necessary to establish a sound supervision mechanism for the public and social groups. Whether it is a nonprofessional general public or a technical expert, regardless of the field of expertise, the public can provide a completely different perspective, which is conducive to think about the problem in all aspects and to make the policy plan more comprehensive. The awareness of the role of public participation would be increased by a sound public participation mechanism and would consequently protect the public's own interests, which in turn will promote the consciousness of public participation. Conducive to policy-making by government are more in line with popular opinion and actual situation. The public engagement system can increase the transparency of a government, thus making the administration more open and rational by reflecting the public opinion in government decision-making.

Public Engagement for the Implementation of Waste Reduction and Recycling Policies in Different Countries

United States

The United States has reached the forefront in SW pollution legislation and prevention (Ellen MacArthur Foundation, 2016) in the world. The US environmental legislation on SW started earlier and established a relatively complete, rigorous, and scientific legal system. As early as 1976, the United States enacted the "Solid Waste Disposal Act," which also can be said of the "Resource Conservation and Recovery Act" (RARC). The Solid Waste Disposal Act was enacted to comprehensively control SW from causing land pollution in the United States. In 1984 the US Congress amended the law and the revised Solid Waste Disposal Act has been in use ever since, which has the legislative purpose to improve the public health and the environment as well as save essential materials and energy. There are two basic principles: first reduce or eliminate the generation of hazardous waste as soon as practicable and the second is to reduce the environmental risk to public health during the disposal and storage of SW. The public participation mechanism mainly reflects the following points:

1. The legislation stipulates the obligations of SW emitters and establishes an extension system for producer responsibility. The US Environmental Protection Agency believes that different products require different producer responsibility extension systems. The government tends to use the power of the market to implement a producer responsibility extension system, which also supports the state governments to explore various management approaches for e-waste. The current US policies at the federal level to encourage the extension of producer responsibility include partnership agreements, voluntary product environmental information, mandatory disclosure of environmental information, and mandatory identification of product content.
2. Application of market mechanism: As for the management of SW, the US government has adopted the strategy of introducing market mechanisms. The overall goal and stage objectives should be established by the federal government for SW. Companies, which achieve that objective in the most cost-effective way, can get the chance to achieve the government's stated goals through the licensing system. These companies can also reduce waste emissions by selling additional credit to those who have difficulty achieving this goal.
3. The use of economic measures, such as the charges for SW. First measure is the user charge system that requires the person responsible for producing SW to pay a certain fee to assume its responsibility to the society. Second is the landfill tax, which includes taxes collected for the waste for the process of landfilling, to improve the comprehensive utilization efficiency of SW.

Japan

Japan emphasizes the classification of SW. The "Waste Disposal Act" in Japan divides waste into household and industrial wastes. Household waste is disposed of by the government and industrial waste is disposed of by the enterprise. In the 1990s the Japanese government formulated some laws related to waste reduction and recycling polices, namely, "Recycling," "Basic Law on the Environment," "Containers Packaging Recycling," and "Construction Waste Material Recycling." In 2000 the "Basic Law for the Construction of Recycling Society" was promulgated to promote the recycling of resources. Consequently, Japan had established a legal system for industrial waste recycling. The waste reduction and recycling policies in Japan require companies to not only reduce the generation of industrial waste but also recycle industrial waste. In the recent years, Japanese companies have transformed resources and energy through effective and organized cooperation among various industrial sectors, thereby minimizing the amount of waste generated and its environmental risk.

United Kingdom

The "Control of Pollution Act" of 1974 and the "Environmental Protection Act" of 1990 promulgated by the British government regulated the prevention and control of SW pollution. The "Waste Management

License Regulations" of 1994 and the "Waste Management Document" of 2003 by the British government strengthened the disposal of SW and formed a relatively complete SW pollution control system, which was consisted by three main regulations.

1. Waste management permit system. The waste disposal, recycling, and other actions were restricted through administrative licensing according to this system, which ensured that these actions do not cause environmental pollution and human damage. The waste management permit can be transferred with the approval of the UNEP, but the obligation to landfill greening and monitoring cannot be exempted from the licensee when the license is revoked and temporarily detained for any reason. If the license is transferred in violation of the regulations or the statutory obligation is violated, the licensee may be subject to a higher penalty or imprisonment.
2. Producer extended liability system. The system of this project is similar to that in the United States. The producer extension responsibility mainly refers to the responsibility of the producer to be responsible for the recovery, recycling, and disposal of the product after use. This strategy is to attribute the responsibility of the product abandonment stage to the producer.
3. The United Kingdom also imposed a landfill tax on industrial SW and established a waste recycling fund management system to prevent the industrial SW pollution. Similar cases were also founded in the United States that employ economic measures to prevent pollution.

THERMAL TREATMENT TECHNIQUES: INCINERATION, GASIFICATION, AND PYROLYSIS

Incineration

Incineration refers to the oxidative combustion reaction between the combustible components in the waste and oxygen under high temperatures and aerobic conditions and then heat is released, with conversion to high-temperature combustion flue gas and a stable solid residue [24]. Besides, the volume is greatly reduced. The combustion flue gas is recycled in the form of energy (such as power generation and heating), and the stable residue can be disposed of in landfill or used as building material [25]. During incineration, many toxic and harmful chemicals and pathogens are oxidatively decomposed and destroyed at high temperatures [26].

Main phase during incineration process

The combustion of SW can be divided into three steps: drying phase, combustion phase, and burnout phase.

1. Drying phase: Water in SW needs to be vaporized during drying phase, which is of great significance for incineration. In addition, a great amount of energy and heat are consumed. The drying time is positively consistent with the moisture content of the organic waste. If the moisture content is too high, the drying time could be prolonged and then it is difficult for combustion.
2. Combustion phase: After the drying phase, the organic waste could be burned at high temperatures. Macromolecular organic compounds are decomposed into combustible gases including CO, CH_4, H_2, and so on during the combustion phase, which could be easier to burn than the solid.
3. Burnout phase: After the incineration phase, the concentration of the materials participating in this reaction is reduced. The inert substance formed by the reaction increases the generation of CO_2, H_2O, and solid ash, which makes it difficult for the remaining oxidant to react with the internal combustible component under the ash layer. Therefore to burn off the remaining combustible components of the material, it is necessary to have sufficient burnout time by extending the entire incineration process.

Key factors during incineration

In general, four factors are needed to control incineration, so as to maintain a comprehensive combustion [27,28]. The factors are as follows.

1. Temperature: The incineration temperature significantly affects the oxidization and decomposition of harmful components in waste. The incineration temperature is referred to as the highest temperature that can be achieved by combustible solids. Increasing the incineration temperature is beneficial to the decomposition and destruction of organic poisons in the waste and to suppress the generation of black smoke. However, excessively high incineration temperatures not only increase fuel consumption but also increase the amount of metal volatilized in the waste and the amount of nitrogen oxides, which is responsible for secondary pollution. Therefore it is necessary to control the appropriate temperature. Different types of organic matter have different optimum temperatures. For example, the optimum temperature for chlorine-containing substances is 800–850°C, whereas the optimum temperature for cyanide-containing substances is 800–950°C.

2. Residence time: The residence time has two meanings. One is the residence time of the combustible SW in the incinerator, which refers to the time required for the combustible SW to be discharged from the furnace to the incinerator. The other meaning is the residence time of waste incineration flue gas, which refers to the time required for the flue gas generated by the incineration of combustible SW to be extracted from the combustible SW to the discharge incinerator, mostly in 1–3 s.
3. Turbulence: It is an indicator of the mixing ratios of combustible SW and air. The higher the turbulence, the more the possibility for the organic combustibles to obtain the required amount of oxygen on time, thus ensuring the efficiency of incineration.
4. Excess air rate: According to the composition of the combustibles and the stoichiometric equation, the amount of oxygen required to burn a unit mass of solid fuel corresponds to the amount of air called the theoretic amount of air. In order to ensure complete waste incineration, it is usually necessary to supply more air than the theoretic air volume, that is, the actual air volume. The actual air volume to theoretic air volume ratio is called the excess air coefficient. Excessive air ratio has a great impact on waste incineration. Supplying proper excess air is a necessary condition for complete combustion of organic matter. Increasing the air ratio not only provides excess oxygen but also increases the turbulence in the furnace, which is beneficial for incineration. But the excessive air ratio of nitrogen may lower the temperature inside the furnace, affecting incineration.

Pyrolysis

The pyrolysis of solid organic waste means that the solid organic waste is chemically decomposed by heating under anaerobic conditions, finally forming liquid bio-oil, chemical gas, and biochar [29]. The ratio of these compounds in the final product depends on the reaction conditions. Generally, in low-temperature slow pyrolysis (<500°C) the product obtained is mainly biochar, in high-temperature flash pyrolysis (700–1100°C) the product obtained is mainly flammable gas, and in medium-temperature rapid pyrolysis (500–700°C) the product obtained is mainly liquid bio-oil [30,31].

Key factors during pyrolysis

1. Temperature: The product distribution of biomass pyrolysis and the composition of combustible gases are significantly affected by temperature. In general, low-temperature, long-term retention slow pyrolysis is mainly used to maximize carbon production, and its mass yield and energy yield can reach 30% and 50%, respectively. When the temperature is less than 600°C, a medium reaction rate is employed, and the yields of bio-oil, noncondensable gas, and carbon are substantially equal. The pyrolysis temperature in the range of 500–650°C is mainly used to increase the production of bio-oil. The bio-oil yield can reach 80% under these temperature conditions. A very high reaction rate and short gas phase retention period can be established if employing a higher temperature range of above 700°C; at this temperature range, the yield can reach up to 80%, mainly for the production of gaseous products.
2. Time: At a given particle size and reactor temperature, a small solid holdup period is required to completely convert the biomass. In order to achieve maximum bio-oil production, the volatile products produced during pyrolysis should leave the reactor quickly to reduce the time required for secondary tar cracking. Therefore in order to obtain the maximum bio-oil yield, the gas phase retention period is a key parameter [32].
3. The characteristic of biomass materials: The characteristics of biomass, particle size, shape, and particle size distribution have an important influence on biomass pyrolysis behavior and product distribution. For example, when the particle size is less than 1 mm, the pyrolysis process is controlled by the reaction kinetic rate, whereas when the particle size is larger than 1 mm, the particles will become a limiting factor for heat transfer [33]. The magnitude of the pressure will affect the gas phase retention period, which will affect the secondary cracking and ultimately affect the pyrolysis product yield distribution. Besides, the lower heating rate is beneficial to the formation of char, which is not conducive to tar production. Therefore flash cracking for the purpose of producing bio-oils uses a higher heating rate.

Gasification

Oxygen or oxygenated substances (such as water vapor) are used as the gasifying agent during biomass gasification. Water vapor is generated through the reaction of carbon and air, which also generates CO_2 and CO. Water vapor reacts through a red carbon layer in the reduction zone and is converted into a combustible gas containing components such as CO, H_2, and a small amount of CH_4 under the high temperature. The gasification process includes drying of the solid fuel, dry distillation (pyrolysis), and oxidation and reduction processes.

The principle of gasification

The gasification agent (water vapor, oxygen) rising in the gasifier performs complex multiphase physical and chemical reactions with the reverse biomass. After the wet material is added from the top, it is dried by the rising hot gas stream to discharge the water vapor. When the dried raw material is dropping, it is heated and thermally decomposed by the hot gas stream, and the volatile matter is released and the remaining carbon continues to fall with the rising CO_2 and water. Water vapor, CO_2, and H_2O are reduced to CO, H_2, etc. and the remaining carbon is oxidized by the air entering from the bottom, and the released heat of combustion supplies heat to the entire gasification process [34,35].

ANAEROBIC DIGESTION OR CODIGESTION FOR SUSTAINABLE SOLID WASTE TREATMENT/MANAGEMENT

Anaerobic digestion is an environment-friendly technology that utilizes anaerobic microorganisms that efficiently degrade and utilize organic waste, producing clean energy gas. It not only meets the needs of energy in today's society but also effectively solves the problem of organic waste disposal [36]. The early domestic anaerobic fermentation technology is mainly used for the treatment of waste sludge and high-concentration organic matter wastewater. It has less application in the treatment of organic matter [37].

The Principle of Anaerobic Digestion

The stages of anaerobic microbial degradation of macromolecular organic matter, namely, hydrolysis, acidification, and methanation, are listed in Fig. 6.1 [38]. In anaerobic fermentation, organic matter such as carbohydrates, proteins, and fats in organic waste is first hydrolyzed by microorganisms into low-molecular-weight substances such as polysaccharides, peptides, and long-chain fatty acids. The hydrolysis phase of particulate organic matter has been confirmed by most researchers as the rate-limiting step. Because of the hydrolysis process, hydrolase or coenzyme first needs to be attached to the organic particles, and the long-chain polymer is decomposed by the action of the enzyme [39]. Therefore the rate of degradation of the entire anaerobic fermentation depends mainly on the speed of the hydrolysis process. The resulting short-chain organic matter is further degraded by acid-producing bacteria during the acidification stage, forming small molecular substances such as glucose, amino acids, and short-chain organic acids, and this process also produces secondary metabolites such as NH_3 and H_2S. The third stage is the acetogenic process. Glucose and amino acids are used by acetogens to generate volatile fatty acids, such as acetic acid and butyric acid, as well as H_2 and CO_2. Finally, methanation by methanogens will volatilize and then fatty acids and H_2/CO_2 are converted into methane [40]. During the whole fermentation process, most of the energy is stored in methane and a small part is needed for the growth of microorganisms.

Influencing Factors

Anaerobic fermentation is a complex biochemical process that is influenced by many factors and these factors are usually interacting.

1. Temperature: Temperature is one of the important factors affecting the life activities of microorganisms. Sousa et al. [41] found that anaerobic fermentation can be carried out normally and the gas yield of rice straw increases with the increase in temperature within a certain temperature range.

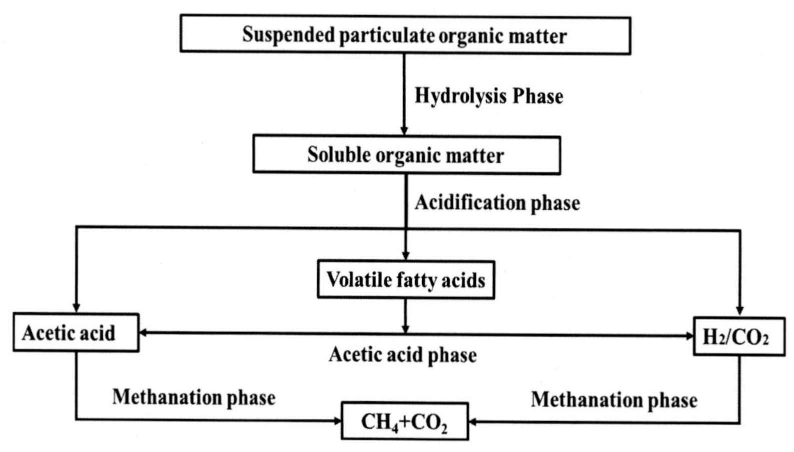

FIG. 6.1 Subsequent steps in the anaerobic digestion process.

2. pH: The fluctuation of the pH value has a great influence on the anaerobic fermentation system, which directly affects the survival and metabolic activities of the microorganisms in the system. For example, the appropriate range (6.5 to 7.5) for formazan-producing bacteria will be altered if the pH gets altered and similarly, for some of the fungi the production will be interfered if pH becomes higher or lower than the appropriate range. In general, pH value affects gas production [42].
3. The carbon-to-nitrogen ratios of fermentation materials: The metabolic activities of microorganisms during anaerobic fermentation require proper and balanced nutrients. Among the agricultural wastes, some organic waste such as rice straw is a high-carbon raw material, and the nitrogen content of the livestock manure is relatively high. Therefore proper carbon-to-nitrogen ratios could be adjusted by different mixing ratios of these wastes, and the fermentation gas production effect can be improved.

BIOHYDROMETALLURGICAL PROCESSING OF METALLIC COMPONENTS OF ELECTRONIC WASTES

The electronic information technology and application fields have developed so much, which consequently accelerated the speed of upgrading of electric and electronic products, thus the e-waste has become one of the fastest growing SW streams in the world [43] (Fig. 6.2). Biohydrometallurgy, including bioleaching and biosorption, is one of the most promising technologies in the metallurgical field. In the recent years, it has been increasingly used for the recovery of precious metals in e-waste.

FIG. 6.2 Large amount of electronic waste discharged into the environment.

Bioleaching refers to the process in which precious metals are transformed from e-waste to liquid solution. The pretreated e-waste and the metabolite hydrogen cyanide, which is generated by some anaerobic bacteria (such as *Pseudomonas aeruginosa*, *Pseudomonas fluorescens*, and *Bacillus megaterium*), undergo cyanidation to form water-soluble cyanide complex such as $[Au(Cn)_2]^-$, $[Ag(CN)_2]^-$, and $[Pt(CN)_4]^-$. Metal cyanide is separated by chromatography and then recovered by the adsorption of activated carbon [44]. Biosorption (bioreduction, bioaccumulation, or bioconcentration) refers to the approach to enrich metals through the use of algae, bacteria (such as *Desulfovibrio*), yeast, and fungi [45]. Bacteria used for biosorption have the characteristics of high specific surface area and metal affinity. The biosorption mechanism of adsorbed metals in solution is divided into chemical and physical adsorption mechanisms. The former mechanism includes complexation, chelation, microbial precipitation, and microbial reduction, and the latter mechanism typically includes electrostatic forces and ion exchange.

There are many kinds of microorganisms involved in the biohydrometallurgy of e-waste, which mainly include acidophilic inorganic autotrophic bacteria, such as *Thiobacillus ferrooxidans*, *Thiobacillus thiooxidans*, and *Leptospirillum ferrooxidans*. These microorganisms can survive in a low-pH inorganic salt medium and can withstand high concentrations of heavy metals. The two main functions of these microorganisms are to oxidize Fe^{2+} to Fe^{3+} and S^0 to H_2SO_4 during leaching. According to the tolerance temperature, these acidophilic bacteria are divided into three categories: medium-temperature bacteria, medium-thermophilic bacteria, and extreme thermophiles. *T. ferrooxidans*, *T. thiooxidans*, and *L. ferrooxidans* are the key medium-temperature bacteria with the appropriate temperature ranges from 28 to 38°C. In addition to the leaching function of acidophilic autotrophic bacteria, some cyanogenic bacteria can also dissolve solid precious metals from gold-containing printed circuit boards and jewelry with Ag and Pt [46]. Some of the microorganisms having the ability to extract SW are listed in Table 6.1. In short, as the research progresses, more microorganisms with the ability to leach metal will be discovered and more microorganisms will be applied to the recycling of e-waste metals.

In general, microorganisms are dominated by the oxidation of metal elements during leaching. Some researchers believe that the "contact leaching" of microorganisms during dissolution plays an important role in the dissolution of metals. The microbial dissolution of metal occurs at the metal-bacteria-solution interface,

TABLE 6.1
Bioleaching of Metals From Solid Waste.

Waste	Microorganisms	Leaching Efficiency	Leaching Condition
Ash from a copper smelter	*Thiobacillus ferrooxidans*, *Thiobacillus thiooxidans*	Cu, 87%	Shake flask, 22 days; pH = 1.8; ash content, 50 g/L
Copper converter slag	*T. ferrooxidans*, *T. thiooxidans*	Cu, 86% Ni, 50% Co, 64%	Shake flask, 25 days; pH= 1.7; ash content, 50g/L
Zinc smelting slag	Moderately thermophilic bacteria	Al, 82%–84% As, 85%–91% Cu, 86%–88% Zn, 95%–97%	Shake flask, pH = 1.5, smelting slag, 50 g/L
Industrial waste	*Penicillium simplicissimum*	Al, 75%	Shake flask, content, 50 g/L
Copper-containing sludge	*Sulfolobus acidocaldarius*	Cu, 99%	The bioreactor was leached for 12 h
Spent lithium-ion batteries	*Aspergillus niger*	Co, 65% Li, 10%	Shake flask, 18 days; content, 100 g/L

and the microbe adsorbs on the biofilm formed on the solid surface. The formation of the biofilm is closely related to the dissolution effect.

Microbial dissolution of metals in e-waste is a complex multi-interface process. It involves both the oxidation of metals by microorganisms and the formation of microbial biofilms to promote dissolution, which can be considered as a complex process of various biochemical interactions, for example, the participation of enzymes in the electron transport process from the metal surface to the inside of the bacterial cells and the adhesion of the extracellular polysaccharide layer secreted by the bacteria to the metal surface. Microbial leaching of metals is a complex biochemical process, which would be clearly presented in the near future.

HEALTHCARE WASTE MANAGEMENT

Healthcare Waste

Healthcare wastes come from the hospitals, research institutes, and veterinary institutes because of the diagnosis, treatment, and/or immunization of humans as well as animals; they are no different in generating wastes as like the other industries. Healthcare wastes threaten the ecosystem because they are infectious, toxic substances that may be genotoxic and radioactive. If the proper disposal or handling of the medical waste is not ensured, it may pose a serious risk of infection to the one handling it and also sometimes may be the cause of an epidemic or endemic outbreak that is of public concern [47]. All medical wastes should be collected, processed, and disposed in a proper way ensuring safety to the public as well as environmental health, which can be achieved by practicing the rules and regulations right from their collection to disposal [48].

In many of the developing countries, medical wastes are not treated properly but are dumped along with the domestic waste, thus threatening the health of the environment and public. Dumping is one of the cost-effective treatments, but it creates uncontrollable negative impacts on the public and environment and needs to be addressed for its proper implementation [49]. The healthcare wastes could pose a direct microbial threat or indirect risks to the public. The indirect risks may be from sharp objects such as syringes, needles, probes, scalpels, and products of blood and internal human wastes such as embryo tissue, placenta, and other excavated tissues from the patients. The other wastes that are of great concern include expired or used medicines and their containers, urine or fecal samples from patients, and the remnants of cytotoxic compounds used for treatment and analysis, such as radiographic films as well as some radioactive substances used in nuclear medicines [50].

Waste Management

Several healthcare waste treatment methods, including sterilization, sanitation, and disinfection using chemicals, microwaves, and dry heat, as well as by superheat steams and incineration, may be used [51]. Management should be started from the beginning of collection toward the end of disposal in each and every stage to ensure prominent management; small differences in any of the operations at any stage may complicate the success rates. The very beginning stage is the collection and segregation of the wastes. Unlike the other wastes, medical waste is infectious, so the workers have to be advised and have to act in a proper manner without getting infected. The wastes should be segregated according to their nature and types. The color codes of the respective bags for waste segregation, such as green for anatomic wastes, should be known by workers before collecting the wastes. So the proper collection and segregation itself narrows down the shortcomings by 50%.

Handling and transport is the second stage. In many countries, medical waste has been disposed along with the municipal waste stream without realizing the disaster caused by it. Mishandling threatens the environment and public, and the most common way for getting infection is by direct contact, vector transmission, and sources of water and air. The persons who are picking up wastes for recycling, drug addicts who are using previously used syringes and needles from dumpsites, low-level hospital staff, and many others who are seeking this job for additional income are at high risks of being infected [52]. Hence, the workers from municipal and healthcare sectors should be provided with the equipment that could safeguard them and also with the basic awareness programs, which make them avoid infections.

Treatment and disposal is the final stage and various treatment methods are available. One needs to understand the category of waste and its volume before selecting the treatment method. The infectious wastes can be treated according to their classification of being and not being burnable, and every treatment has its own pros and cons. The most adopted treatment method is incineration in which, under controlled conditions, the waste undergoes combustion. During combustion, there occurs a significant reduction in the volume of waste and killing or destruction of pathogens and organics, even the radioactive wastes [53], but at the same time, the release of toxic ashes and the gases during this process has serious implications on the environment and the public. Thus new technologies such as immobilization or stabilization of these toxic ashes to recyclable products such as concretes, ceramics, glass, and other durable matrices of economic value should be followed to overcome the abovementioned cons.

Role of Stakeholders

The various facets of healthcare waste management can be achieved by proper cooperation between internal and external stakeholders at all the levels, starting from the patient, caretaker, concerned staff to the government, law, various policy-making authorities, academia, and service providers. The internal stakeholders from the top level of hospital management should prepare and implement management programs to maintain targets via monitoring and following. The employees, patients, and caretakers should cooperate and ensure active participation to achieve waste minimizations. Policy-makers, along with the national and international funding bodies, should provide political as well as financial support that is required for steady improvement to manage the necessary funds to implement the healthcare waste management program.

CONSTRUCTION AND DEMOLITION WASTE MANAGEMENT

The construction, renovation, and demolition of infrastructure and building activities generate many substances that are left abandoned and are collectively known as C&D wastes [54]. According to their chemical nature, they are of two categories: (1) inert materials that hardly participate in chemical reactions, such as bricks, soil, and concrete, and (2) noninert materials such as wood and rebar, which react readily (Table 6.2). These wastes are unavoidable as a result of construction and pose negative environmental impacts. However, their effective management can minimize their

TABLE 6.2
Composition of Construction and Demolition Wastes.

Types of Waste Materials	Maximum Range (%)
Masonry	54
Concrete	40
Wood	4
Gypsum	0.4
Plastics	2
Minerals	9
Miscellaneous	36

Modified from Gálvez-Martos, 2018 Gálvez-Martos JL, Styles D, Schoenberger H, Zeschmar-Lahl B. Construction and demolition waste best management practice in Europe. Resources, Conservation and Recycling 2018;136:166–78.

generation; even they have a great possibility of being recycled to a useful resource if they underwent proper treatments. In case of waste management after natural disasters', such as earthquake, a few of the following key factors should be considered before implementing the management strategies: identifying proper disposal sites and recyclable materials, composition and nature of waste, handling and processing area and its holding capacity, reuse and reconstruction basis, and finally, authority structures of a government [55].

As reported by Lu and Yuan [57], we can divide the management practices under two heads: technical measures that involve construction and recycling and managerial measures that involve on-site sorting and disposal. The waste management also needs industrial guidelines for the crucial assessment processes; for example, as reported by Wu et al. [58], the C&D waste management was found to be at 10%, 8.16%, and 11.84% in the United States, the United Kingdom, and China, respectively.

Controlling the waste stream could be achieved by better understanding the factors that generate wastes, which may be a basic step for both policy-makers and decision-makers to plan, change, and implement proper waste management. Ortiz et al. [59] and Mohamed [60] reported that the greener way of treating C&D wastes is recycling followed by incineration and landfilling after careful assessment of these three methods. The three *R*'s "Reduce, Reuse, and Recycle" provide the best opportunities for reducing these wastes (Fig. 6.3). Reduce gains the top priority; using less material costs less, reduces pollution from its manufacture and transportation, saves energy and water, and keeps material out of landfills. Reusing, the second approach, extends the life of existing materials and decreases the amount of new resources needed. Recycling, the third approach, again conserves resources and diverts materials from landfills [61].

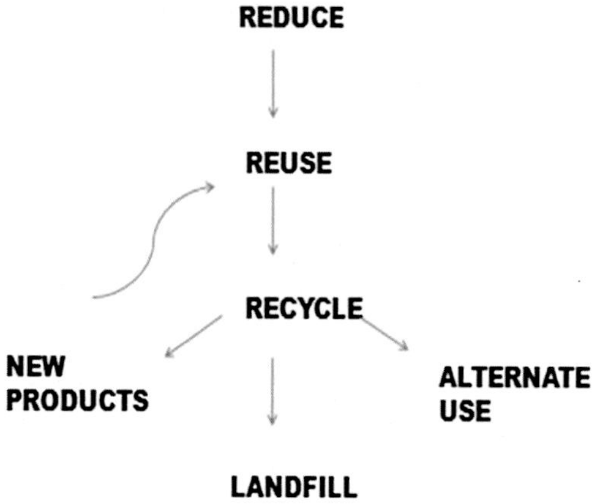

REDUCE

REUSE

RECYCLE

NEW PRODUCTS

ALTERNATE USE

LANDFILL

FIG. 6.3 The three *R*'s principle.

A concept called circular economy can be opted for better efficiency of sustainable C&D waste management. By definition, it is "an industrial system that is restorative or regenerative by intention and design. It replaces the end-of-life concept with restoration, shifts toward the use of renewable energy, eliminates the use of toxic chemicals impairing reuse, and aims at eliminating waste through the superior design of materials, products, systems, and business models" [62]. It mainly depends on the stakeholders and professionals by their unfailing plans to make it a great success. The recycled concrete, blocks, and aggregates can be obtained from wastes generated during building processes, which may be a better option to reuse and recycle them within their industries. The recycled building wastes can also be used in other industries, as apart from bricks, mortar, concrete, and SW, they also contain scraps of glasses, woods, clay, metal, and even shale slabs [63].

Once after smelting, scraps of metals can be recycled and reused by metallurgical industries; they can be very well used for manufacturing electric, automobile, and plumbing appliances, as well as different types of containers. After crushing and melting, the waste glass can be utilized by the glass industry; it can also replace abrasives such as alumina and silica in preparing the surfaces of equipment parts and bridges and even in ship constructions because it helps avoid adverse health effects from constant exposure to other abrasives [64]. The paper industries can effectively utilize the woods for paper transitions. The ceramic bricks can be manufactured using clay and shale bricks as raw inputs. The crushed waste bricks can be used as filler in paints; even in rubber plastic composites once been organically transformed [65].

TREATMENT AND USE OF ASHES FROM SOLID WASTE PROCESSING

Rapid industrialization and continuous operations are associated with the increase in the generation of wastes, such as generation of tons of ashes including fly ashes and bottom ashes from coal as well as municipal waste; ashes of biomass, sludge, and co-combustion; and even slags of blast furnaces. The common practice of disposing is usually landfilling because it is economical but at the same time it threatens the environment striking its well-being. This necessitated the reuse and recycling of ashes as the source of useful materials that increase the economic development; some wastes require the knowledge of their properties and nature for their different applications. The bottom ash contributes about 15%–30% of the mass of the input wastes [66]. Metals and metal aggregates are the two types of resources found in ash.

A report by Crillesen and Skaarup [67] stated that after incineration, many of the larger pieces of metal scraps such as magnetic metals are being removed from basal ashes, whereas the devoid in the utilization of aggregated, rare, valuable metals from them leads to resource losses, thus lowering their performance capacities. Based on magnetism, scrap metals are of two types: ferrous and nonferrous. Ferrous scraps are separated by employing magnets, whereas eddy current and inductive sorting are widely employed for separating nonferrous aluminum and other conductive metals such as stainless steel. A report by Vassilev et al. [68] studied some of the biomass ashes in comparison with better known carbon ashes to predict their chemical combinations, as the huge amount of alkalis and chlorides in them limited their viability in reutilization as like other carbon wastes. The potential recovery of minerals from ashes of petroleum industries has been studied by Park et al. [69] by effectively incorporating the mineral carbonation process.

Like natural processes, the sequestration of carbon dioxide can be achieved by combining carbon dioxide with minerals, which was effectively studied and reported by Seifritz [70]. Some of the mineral ions such as calcium, magnesium, and barium can be stabilized by the formation of $CaCO_3$, $MgCO_3$, and $BaCO_3$, respectively, when they constantly react with atmospheric carbon dioxide. Mineral carbonation includes direct processes employing raw minerals and indirect processes employing acids or alkalis; both are reported to be energy-consuming processes. As reported by Bhattacharyya et al. [71], chemical stabilization is economical with low capital investment; it stabilizes heavy metal contaminants of ashes either by altering to insoluble or less soluble forms [72] by or making them stable by reducing their leachability [73]. Wet milling, similar to chemical stabilization, stabilizes metals by enhancing their reactivity and convertibility [74]. The utilization of waste ashes for cement production gained attention worldwide, thus reducing land filling as well as generating profits by producing important secondary materials. Because of being rich in silicon, iron, sulfur, calcium, and aluminum, waste ashes are said to be pozzolanic, which forms substances with properties of complex cementitious materials after their chemical reaction [75]. Roethel and Breslin [76] in a study reported the positive replacement of these cementitious blocks prepared from waste ashes over the normally used concrete blocks in boathouse constructions; even these cementitious blocks were found to be satisfactory meeting the criteria.

LANDFILL DESIGN AND OPERATION

Methods such as landfilling, composting, grinding, milling, compaction, open dumping, and anaerobic digestion were commonly employed for SW disposal [77]. Above all, landfills gained attention because they are economical when compared with their counterparts and they can hold a maximum capacity of handling before wastes are being stabilized. Factors such as economic development, culture, climate, and energy resources of a particular landscape, along with its waste generation capacity, decided the composition of wastes. Landfills are generally engineering works under continuous development processes; their optimal design is almost complex and unresolved till date. The various design processes and their operations used to depend on several variables such as height, thickness, layers and kind of waste material, and generation of leachates.

A generalized landfill should possess a base lining system may be with some claylike particles or high–density polyethylene sheets to prevent the direct contact of wastes from the surrounding environment; this can be achieved by a single base liner or composite lining system. To overcome the leachate accumulation, sufficient drainage pipes can be used. The operations such as measurement and record keeping, loading and unloading of wastes, removal of leachates, waste compaction, covering applications to prevent bad odor, landfill gas management, emission monitoring, final cover applications, and postclose care help in stabilizing the landfills.

For the past few decades, the designing and technologies for landfill construction are rapidly growing combating their constraints and their demands, as there is increase in the rate of waste generation; the landfill construction technologies even offer flexibility in the expansion of already constructed landfills depending on the investment over time. Nowadays, landfill construction is constrained by phase programs, which provide room for initiating expansion or construction of new units parallel to the existing ones, undertaking first phase for the construction of second–phase buildings by using maturity benefits as the investment for the latter. The multibarrier system is of great concern, nowadays, to reduce impacts to the environment in active as well as closed landfills [78]. Leachate is one of the main barriers, as it needs separate collection systems, which can reduce its hazardous potency before directly reaching the ground; this can be easily achieved by employing capping of impermeable membranes at the top and bottom of landfills. The requirement for water and soil protection used to be the foremost criterion to be met while designing landfills [79].

Many factors are to be considered before formulating a successful experimental design. The traditional methods of analyzing all the factors, experimentations, and prototyping them need extra efforts and also used to be time consuming; scientific computing, in its way, allowed us to evaluate a large number of factors with minimal time and ensured effectiveness [80]. The main processes of the selected landfills such as hydrology, degradation, and settlement have to be considered and they are closely linked to each other in operations. If one variable is affected, it will influence the rest, so one needs to analyze each and every variable in optimizing the designs. For example, if one focuses on the landfill volume, the other variables such as height of the waste layer, pollutants to be emitted, and power to be generated or utilized for the processes have to be considered, along with the wind force of that geographic location, so that the plan and design could be effective [81]. Traditionally, manual analysis and design are used but the drawbacks in simultaneously correlating all the variables necessitated the computing techniques. Many proposed mathematic models for achieving appropriate processes of landfill designing; hydrologic models such as HELP (Hydrologic Evaluation of Landfill Performance) [82], degradation models [83], biogas generation models [84], and settlement models [85]. Opting such models (general-use software packages) ensures the qualitative and quantitative analysis of each variable in each design as well as one can effectively design decision support systems to get optimal results.

LANDFILL LEACHATE COLLECTION AND TREATMENT

As there is a serious ecological issue with landfilling, different counterparts were adopted because there were always problems with the generation of leachates. The age of the landfill determines the composition of landfill leachates; many of which contains mainly biodegradable, a few refractory too, chlorinated salts, metals, ammonia, nitrogen, and humic substances [86,87]. The important reason behind leachate collection and treatment is its high toxicity and high chemical oxygen demand (COD). The variation in leachate volume and composition challenges the most efficient treatment methods rather than conventional treatment methods because of being difficult in supplying the discharge standards [88]. The adoption of treatment methods such as physiochemical, biological, and hybrid and/or combined treatment methods mainly depends on the leachate's chemical constituents, biological oxygen demand (BOD)/COD ratio, and age.

The physiochemical treatment is mainly used to remove the suspended as well as dissolved solids from leachates. Air stripping, coagulation-flocculation, chemical precipitation, chemical oxidation, and adsorption are the most preferred and usually adopted treatment techniques. The air stripping process ensures the effective stripping of the ammonia gas from the leachate, with enough oxygen availability to the downstream processing. Alum and lime are used for successful coagulation-flocculation processes before subjecting the leachates to biological treatments. Electrocoagulation can also be employed using electric potential along with chemicals. It can efficiently remove organics and solids, only requires limited usage of chemicals, and generates less sludge quantity; however, it requires more energy and high cost for the treatment. Based on the ratio of BOD/COD in the leachates the aerobic or nonaerobic methods of biological treatments are determined. Microorganisms play a vital role in both the processes via their growth by utilizing nutrients in the leachate. The biological methods of treatment process are gaining special interest due to their cost-effectiveness, the simplicity in their operational methods, and their maintenance process. Like every other method, this method is also limited because of the complexity of metabolic pathways and growth characteristics of the microorganisms, in addition to their culture conditions, interactions, and survival rates.

Different combinations or hybrid processes are usually followed to get better results because the leachate is complex and the adoption of only a single technique failed to produce the expected outcomes to meet the desired standards. Various combinations of physiochemical and biological treatments were reported to be satisfactory by many researchers. For example, as reported by Millar et al. [89], the combination of electrochemical coagulation treatments was found to be best over its chemical-alone counterpart because of the following reasons: prompt abilities in treating the colloidal particles, even smaller volumes of sludge can be treated with less chemical addition, less power consumption, and being cheaper with high output results. After many attempts of these complex treatments in treating landfills for more than a few decades, minimizing their environmental impacts is still important to the researchers, thus gaining further research attention. To ensure hazard removal and potential water recovery with purity, reverse osmosis (RO) can be adopted as a single treatment or paired with other treatments and was simply found to be successful [90]. Instead of using a single treatment process, the multistage RO proved to be efficient in water recovery and also reduced specific energy consumption, as reported by Judd [91]. The functional ability and sustained treatments of the RO systems are acknowledged usually with the process of brine disposal; more than 30% volume of raw leachate is found to be returned back to the landfill kindling an osmotic pressure hype [92]. As reported by Joo and Tansel [93], optimal performances of landfills can be achieved by coupling more than two stages of reaction process, which may leave less concentrated water as well as good water recovery rate.

LANDFILL AFTERCARE AND MAINTENANCE

The major geotechnical aspect is the settlement and restoration of the closed landfill sites. MSW settlement causes many problems such as reverse flow of drain, sagging of channels (even ground channels), damage to buildings and distortions, water ponding, cracking, and rupture of drain pipes, which need further attention for proper settlement processes. The settlement is nowadays irregular as well as excessive, but it could be handled relatively easily, as it needs differential settlement processes [94]. The total and differential settlements necessitated the study and proper estimation of settlement to the landfill sites according to MSW treatment.

The MSW landfill settlement is a complex process including short-term changes and long-term degradations, ensuring proper results [95]. It can be achieved by five-stage settlement processes: the initial stage dealing with mechanical compression of waste components, second stage involves primary settlement due to continuous compression and reorientation of wastes in the landfill, the third stage is the secondary settlement where the initial decomposition of organic substances after compression is started, through the fourth stage the organic decomposition is achieved, and in the final stage, residuals are treated and deformed. The first two stages of the settlement process are estimated to be achieved within 3 months from the closure of landfills, whereas the rest of the stages are more dependent on time, as there are various stages of biodegradation processes, mainly the fourth decomposition stage is of major concern because it emits several types of gases and even liquids that need further attention; hence, thoroughly the settlement process is a time-dependent variable. There used to be an imposed equilibrium during aftercare due to active management and controls through engineering works; thus arises a concept called hydraulic equilibrium, which needs further attention. There will be an establishment of slow rising of leachates in the landfills due to the drop of the active as well as engineering control systems retrieved. For completion to occur, the regulator must be satisfied that future fluxes to the environment will be acceptable under a range of hydraulic equilibrium situations.

LEGAL AND INSTITUTIONAL FRAMEWORK FOR SOLID WASTE MANAGEMENT

The growing per capita waste generation and also failure in managing it in a manner safe to the environment leads to the cause severe health issues in the environment and public. Fast-growing megacities challenge waste management to come up with innovative and environmentally sustainable practices that can reduce, recycle, and dispose the increasing wastes, ensuring safety. The developing and transitional nations are therefore in need of preparing effective policy resources and legal and institutional framework for coping up with the lurking environmental problems with their effective SW management [96]. Legal and institutional framework is a broad concept and it concerns the functions of governments, private enterprises, political jurisdictions, judicial systems, legislative bodies, and regulatory agencies. More specifically, legal framework incorporates laws, amendments, treaties, acts, ordinances, mandates, regulations, and their enforcement mechanisms. Institutional system refers to the structure of government and its agencies, independent think tanks, and private sector services. A well-defined legal framework articulates the purpose for SW prevention, reduction, segregation, recycling, combustion, and landfilling. The role of the government in crafting a legal system varies with different countries. The US government is concerned much about landfills: it employed several regulations for landfilling, waste discharge management, and gas and ash emissions and their conversion into energy products, whereas there are no such regulations in practice by the Canadian government [97].

SW laws and regulations are aimed at building institutions to address direct services for waste pretreatment from collection to safe disposal and indirect or support services for technical services and financing [98]. NGOs and the private sector are playing an increasingly important role in waste management; the involvement of both government and NGOs in effective building up of institutions is said to be the primary goal in a regulatory statutory framework [99]. The legal and judicial infrastructure of the industrializing nations must address company law, joint ventures, and public-private partnerships, as well as formulate fair and predictable tax laws for the infusion of foreign capital in their economies [100]. The existence of rule framework and protection of property rights correspond the economic development of a developing country by reducing other influences against it [101]. The policies and legal institutional framework used to be different in different countries. SW disposal act, resource recovery act, and resource conservation and recovery act are passed with respect to SW management by the United States. In 1945 the U.S. Department of Public Health (USDPH) was charged at the wake of the debate over what to do with the massive generation of refuse. The first two acts are for efficient disposal of hazardous materials and recovery of useful resources, respectively. The legacy of EU waste policy and legislative framework goes back to the 1970s, but the key features of the current strategy can be found in the Directive 2008/98/EC, which has seven chapters divided into 42 articles, with "extended producer responsibility" as one of the major highlights.

South Asian countries may give an illusion of similarity in their physical and economic characteristics but a closer observation will reveal variations. The establishment of the Ministry of Environment in 1980, later renamed Ministry of Environment and Forests (MoEF), marked the beginning of environmental regulation in India. SW management in India is primarily a local affair. Pakistan and Bangladesh have passed environmental laws and set up environmental ministry at the national level. Under the ministry of housing works, environment and urban affairs division was established by Pakistan in 1974. The Water Pollution Control Ordinance, adopted in 1970, is the first major environmental law in Bangladesh (during erstwhile East Pakistan), which was repealed by the passage of the Environmental Pollution Control Ordinance in 1977. Subsequently, the Environmental Conservation Act (ECA) was passed in 1995, replacing the 1977 ordinance and was followed by the Environmental Conservation Rules in 1977. The Biological Diversity Act was enacted in 2012 and the ECA was amended in 2010 [102]. The 5-year plan (FYP) by the government of China [103] were established to address the environmental issues. The local and central governments have issued several regulations and policies and many laws to regulate the issues of SW (Table 6.3).

TABLE 6.3
National Environmental Protection Plans of China.

Year	Policies and Plans for Solid Waste	Major Concerns
2003	11th FYP	To reduce major air pollutants
2009	12th FYP	To reduce pollution and emissions
	National total emission control	To improve environmental conditions in metropolitans
2011	12th FYP for national environmental protection	Governing industrial solid waste disposal
2015	Law of solid waste pollution prevention and control	To encourage circular economy in solid waste disposal

FYP, 5-year plan.
Modified from Zhou B, Sun C, Yi H. Solid waste disposal in Chinese cities: an evaluation of local performance. Sustainability 2017;9: 2234.

Like the above-discussed countries, the rest of the world is also running with many rules that enact the efficient management of SW.

LIFE CYCLE ASSESSMENT FOR DECISION-MAKING IN SOLID WASTE MANAGEMENT

Economy of a country and its climatic conditions decide the composition, nature, and rate of MSW generation [104]. The high gross domestic product (GDP) and high per capita income of a country are directly proportional to the generation of nonbiodegradable wastes, such as plastics from packaging wastes, when compared with the low-GDP countries where wastes are generally biodegradable. The depletion of natural resources, climatic changes, and hazardous environmental as well as human health are due to the underdevelopment of or failure in MSW management [105]. Different management technologies are compared with the tool called life cycle assessment (LCA) in determining the environmental impact and acceptable management options [106]. The local specificities while modeling the waste management systems yielded potential LCA [107]. The methodologies of each and every process including from the start till the end used to ensure good LCA in any study [108]. The developing Asian countries usually followed open dumping and landfilling as major waste management practices, whereas the developed Asian countries followed incineration as the prime technique.

The release of hazardous substances into the ecosystem, extending to soil, air, and water pollution, is due to the unsegregated wastes and their improper open burning, dumping, and landfilling. The developed countries such as Japan and Singapore mainly allocate their budgets for disposal of wastes thus achieving efficient management systems, whereas many of the developing countries utilize their budget only for waste collection rather than for its disposal; thus, one needs to evaluate and identify the impacts of management practices, for which LCA is used and is promising for solving many issues [109]. The LCA method can be applied in many of the management strategies such as in natural resource exploitation and energy consumption, comparing acidification potential, toxicity evaluation, greenhouse gas emission, and various other strategies, as reported by Laurent et al. [105].

Material flow analysis (MFA), along with LCA, allows one to assess several environmental issues, providing consistency between the variables, transfer coefficients, and waste resource availability, along with capacity restrictions [110]. It also enables to figure out process performance for MSW management [111] and for the secondary utilization of remnants such as ashes in other industries, including cement industries [112]. The main influencing factors of LCA reported by Laurent et al. [105] are composition, nature of wastes, and the surrounding energy systems. The important research gaps may be outlined by the use of geographic status in the life cycle inventories. The models of geographic interests allowed one to model the material substitution accurately by outlining the reusable and recyclable product output. The coupling of MFA and LCA requires inventory, thus identifying all the processes to be carried out in a specific study region. The development of site-specific models improves the efficiency of the waste management process [113]. The development of LCAs and several efficient tools for waste management is by the specific regional assessment [114]. The attribution LCA (ALCA) and consequential LCA (CLCA) approaches offered different results and messages to the users [115]; the inputs and outputs in ALCA are attributed to functional units by a system-modeling approach. CLCA is change-oriented that quantifies effects that associated changes in LCA of a system by decision [116]. Thus the consequential approaches took environmental assessment a step further by analyzing its burdens in response to market implications and even beyond foreground systems.

CONCLUSIONS AND PERSPECTIVES

Sustainable waste management is an integrative and interdisciplinary approach requiring proper planning and execution of each aspect involved in it, i.e., waste prevention, collection, segregation, storage, recycle/recovery, treatment, and disposal. Individual citizens and local government/policy-makers are the major stakeholders, who need to work in harmony to achieve the goals of waste handling. There should be a behavioral mind-set change at the individual level to take moral responsibility for lesser waste generation to protect the environment and make this world a better livable place. A greener lifestyle, minimalistic attitude, and following the guidelines for waste management will be further helpful. The government's and policy-maker's roles should be toward framing policies for greater public engagement, providing incentives for reuse/recycle, charging for waste, and developing scientific innovations for waste management. Policies and approach for waste handling should differ in different provinces taking into account the cultural and socioeconomic background and the waste generation patterns of different regions. We should endeavor toward considering waste as a resource and establish the circular economy by harnessing value-added products from it. At last, for unavoidable wastes, we should employ scientific innovations for their safe disposal in a way resulting in minimal environmental impact.

ACKNOWLEDGMENTS

The authors are grateful for the financial support from the Research Fund for International Young Scientists from the National Natural Science Foundation of China (Grant No. 31750110469), China, and The Introduction of Talent Research Start-up Fund (No. Z101021803), College of Natural Resources and Environment, Northwest A&F University, Yangling, Shaanxi Province 712100, China. We are also thankful to our all laboratory colleagues and research staff members for their constructive advice and help.

REFERENCES

[1] Dias CMM, Rosa LP, Gomez JMA, D'Avignon A. Achieving the sustainable development goal 06 in Brazil: the universal access to sanitation as a possible mission. Anais da Academia Brasileira de Ciências 2018;90(2):1337–67.

[2] Vieira VHAM, Matheus DR. The impact of socioeconomic factors on municipal solid waste generation in São Paulo, Brazil. Waste Managment Research 2018; 36(1):79–85.

[3] Guérin JÉ, Paré MC, Lavoie S, Bourgeois N. The importance of characterizing residual household waste at the local level: a case study of Saguenay, Quebec (Canada). Wastes Management 2018;77:341–9.
[4] Gu B, Fujiwara T, Jia R, Duan R, Gu A. Methodological aspects of modelling household solid waste generation in Japan: evidence from Okayama and Otsu cities. Waste Managment Research 2017;35(12):1237–46.
[5] Tian H, Gao J, Hao J, Lu L, Zhu C, Qiu P. Atmospheric pollution problems and control proposals associated with solid waste management in China: a review. Journal of Hazardous Materials 2013;252–253:142–54.
[6] Silva A, Rosano M, Stocker L, Gorissen L. From waste to sustainable materials management: three case studies of the transition journey. Wastes Management 2017;61:547–57.
[7] Bhat SA, Singh S, Singh J, Kumar S, Bhawana, Vig AP. Bioremediation and detoxification of industrial wastes by earthworms: vermicompost as powerful crop nutrient in sustainable agriculture. Bioresource Technology 2018;252:172–9.
[8] Manyi-Loh CE, Mamphweli SN, Meyer EL, Okoh AI, Makaka G, Simon M. Microbial anaerobic digestion (Bio-Digesters) as an approach to the decontamination of animal wastes in pollution control and the generation of renewable energy. International Journal of Environmental Research and Public Health 2013;10(9):4390–417.
[9] Li J, Zhang W, Li X, Ye T, Gan Y, Zhang A, Chen H, Xue G, Liu Y. Production of lactic acid from thermal pre-treated food waste through the fermentation of waste activated sludge: effects of substrate and thermal pre-treatment temperature. Bioresource Technology 2018;247:890–6.
[10] Adelopo AO, Haris PI, Alo B, Huddersman K, Jenkins RO. Conversion of solid waste to activated carbon to improve landfill sustainability. Waste Managment Research 2018;36(8):708–18.
[11] Quina MJ, Bontempi E, Bogush A, Schlumberger S, Weibel G, Braga R, Funari V, Hyks J, Rasmussen E, Lederer J. Technologies for the management of MSW incineration ashes from gas cleaning: new perspectives on recovery of secondary raw materials and circular economy. The Science of the Total Environment 2018;635:526–42.
[12] Castaldi MJ. Perspectives on sustainable waste management. Annual Review of Chemical and Biomolecular Engineering 2014;5:547–62.
[13] Wu Z, Yu AT, Shen L, Liu G. Quantifying construction and demolition waste: an analytical review. Wastes Management 2014;34(9):1683–92.
[14] Tansel B. From electronic consumer products to e-wastes: global outlook, waste quantities, recycling challenges. Environment International 2017;98:35–45.
[15] Diaz LF, Eggerth LL, Enkhtsetseg S, Savage GM. Characteristics of healthcare wastes. Wastes Management 2008;28(7):1219–26.
[16] Tang J, Wei L, Su M, Zhang H, Chang X, Liu Y, Wang N, Xiao E, Ekberg C, Steenari BM, Xiao T. Source analysis of municipal solid waste in a mega-city (Guangzhou): challenges or opportunities? Waste Management Research 2018. https://doi.org/10.1177/0734242X18790350.
[17] Lo AY, Liu S. Towards sustainable consumption: a socio-economic analysis of household waste recycling outcomes in Hong Kong. Journal of Environmental Management 2018;214:416–25.
[18] Rousta K, Ordoñez I, Bolton K, Dahlén L. Support for designing waste sorting systems: a mini review. Waste Managment Research 2017;35(11):1099–111.
[19] Sharholy M, Ahmad K, Mahmood G, Trivedi RC. Municipal solid waste management in Indian cities – a review. Wastes Management 2008;28(2):459–67.
[20] Yasmin S, Rahman MI. A review of solid waste management practice in Dhaka City, Bangladesh. International Journal of Environmental Protection and Policy 2017;5(2):19–25.
[21] Zacho KO, Mosgaard MA. Understanding the role of waste prevention in local waste management: a literature review. Waste Managment Research 2016;34(10):980–94.
[22] Cox J, Giorgi S, Sharp V, Strange K, Wilson DC, Blakey N. Household waste prevention - a review of evidence. Waste Managment Research 2010;28(3):193–219.
[23] Johansson N, Corvellec H. Waste policies gone soft: an analysis of European and Swedish waste prevention plans. Wastes Management 2018;77:322–32.
[24] Zha J, Huang Y, Xia W, Xia Z, Liu C, Dong L, Liu L. Effect of mineral reaction between calcium and aluminosilicate on heavy metal behavior during sludge incineration. Fuel 2018;229:241–7.
[25] Hong JL, Chen YL, Wang M, Ye LP, Qi CC, Yuan HR, Zheng T, Li XZ. Intensification of municipal solid waste disposal in China. Renewable and Sustainable Energy Reviews 2017;69:168–76.
[26] Shen F, Liu J, Dong Y, Gu C. Insights into the effect of chlorine on arsenic release during MSW incineration: an on-line analysis and kinetic study. Wastes Management 2018;75:327–32.
[27] Wang Y, Zhang X, Liao W, Wu J, Yang X, Shui W, Deng S, Zhang Y, Lin L, Xiao Y, Yu X, Peng H. Investigating impact of waste reuse on the sustainability of municipal solid waste (MSW) incineration industry using energy approach: a case study from Sichuan province, China. Wastes Management 2018;77:252–67.
[28] Makarichi L, Jutidamrongphan W, Techato K. The evolution of waste-to-energy incineration: a review. Renewable and Sustainable Energy Reviews 2018;91:812–21.
[29] Czajczynska D, Nannou T, Anguilano L, Krzyzynaska R, Ghazal H, Spencer N, Jouhara H. Potentials of pyrolysis processes in the waste management sector. Procedia Engineering 2017;123:387–94.

[30] Arni SA. Comparison of slow and fast pyrolysis for converting biomass into fuel. Renew Energy 2017;124:197–201.

[31] Choi MK, Park HC, Hang SC. Comprehensive evaluation of various pyrolysis reaction mechanisms for pyrolysis process simulation. Chemical Engineering and Processing: Process Intensification 2018;130:19–21.

[32] Gong Z, Wang Z, Wang Z, Fang P, Meng F. Study on pyrolysis characteristics of tank oil sludge and pyrolysis char combustion. Chemical Engineering Research & Design Impact 2018;135:30–6.

[33] Senneca O, Cerciello F, Heuer S, Ammendola P. Slow pyrolysis of walnut shells in nitrogen and carbon dioxide. Fuel 2018;225:419–25.

[34] Lan W, Chen G, Zhu X, Wang X, Liu C, Xu B. Biomass gasification-gas turbine combustion for power generation system model based on aspen plus. The Science of the Total Environment 2018;628–629:1278–86.

[35] Zhang WW, Chen XL, Wang FC. Process simulation of biomass entrained flow gasification based on aspen plus. Acta Energiae Solaris Sinica 2007;28:1360–4.

[36] Ebner JH, Labatut RA, Rankin MJ, Pronto JL, Gooch CA, Williamson AA, Trabold TA. Lifecycle greenhouse gas analysis of an anaerobic co-digestion facility processing dairy manure and industrial food waste. Environmental Science and Technology 2015;49:11199–208.

[37] Siddique MNI, Wahid ZA. Achievements and perspectives of anaerobic co-digestion: a review. Journal of Cleaner Production 2018;194:359–71.

[38] Fagbohungbe MO, Herbert BM, Hurst L, Ibeto CN, Li H, Usmani SQ, Semple KT. The challenges of anaerobic digestion and the role of biochar in optimizing anaerobic digestion. Wastes Management 2016;61:236–49.

[39] Verstraete W, Van de Caveye P, Diamantis V. Maximum use of resources present in domestic "used water". Bioresource Technology 2009;100(23):5537–45.

[40] Batstone DJ, Hülsen T, Mehta CM, Keller J. Platforms for energy and nutrient recovery from domestic wastewater: a review. Chemosphere 2015;140:2–11.

[41] Sousa DZ, Salvador AF, Ramos J, Guedes AP, Barbosa S, Stams AJM, Alves MM, Pereira MA. Activity and viability of methanogens in anaerobic digestion of unsaturated and saturated long-chain Fatty acids. Applied and Environmental Microbiology 2013;79(14):4239–45.

[42] Calicioglu O, Brennan RA. Sequential ethanol fermentation and anaerobic digestion increases bioenergy yields from duckweed. Bioresource Technology 2018;257:344–8.

[43] Li J, Zeng X, Chen M, Ogunseitan OA, Stevels A. "Control-Alt-Delete": rebooting solutions for the E-waste problem. Environmental Science and Technology 2015;49(12):7095–108.

[44] Faramarzi MA, Stagars M, Pensini E, Krebs W, Berndl H. Metal solubilization from metal-containing solid materials by cyanogenic *Chromobacterium violaceum*. Journal of Biotechnology 2004;113:321–6.

[45] Creamer NJ, Baxter-Plant VS, Henderson J, Potter M, Macaskie LE. Palladium and gold removal and recovery from precious metal solutions and electronic scrap leachates by *Desulfovibrio desulfuricans*. Biotechnology Letters 2006;28(18):1475–84.

[46] Lee JC, Pandey BD. Bio-processing of solid wastes and secondary resources for metal extraction - a review. Wastes Management 2012;32(1):3–18.

[47] Mbongwe B, Mmereki BT, Magashula A. Healthcare waste management: current practices in selected healthcare facilities, Bostswana. Waste Management 2008;28:226–33.

[48] Bdour A, Altrabsheh B, Hadadin N, Al-Shareif M. Assessment of medical wastes management practice. A case study of the northern part of Jordan. Wastes Management 2007;27:746–59.

[49] Diaz LF, Savage GM, Eggerth LL. Alternatives for the treatment and disposal of healthcare wastes in developing countries. Wastes Management 2005;25:626–37.

[50] Bendjoudi Z, Taleb F, Abdelmalek F, Addou A. Healthcare waste management in Algeria and Mostaganem department. Wastes Management 2009;29:1383–7.

[51] Yong-Chul J, Cargro L, Oh-Sub Y, Hwidong K. Medical waste management in Korea. Journal of Environmental Management 2006;80:107–15.

[52] Appleton J, Ali M. Healthcare or health risks? Risks from healthcare waste to the poor. Leicestershire, UK: Water Engineering and Development Centre, Loughborough University; 2000.

[53] Stoch P, Ciecińska M, Stoch A, Kuterasiński L, Krakowiak I. Immobilization of hospital waste incineration ashes in glass-ceramic composites. Ceramics International 2017;44:728–34.

[54] Wu Z, Yu ATW, Shen L. Investigating the determinants of contractor's construction and demolition waste management behavior in Mainland China. Wastes Management 2017;60:290–300.

[55] Karunasena G, Amaratunga D, Haigh R, Lill I. Post disaster waste management strategies in developing countries: case of Sri Lanka. International Journal of Strategic Property Management 2009;13(2):171–90.

[56] Gálvez-Martos JL, Styles D, Schoenberger H, Zeschmar-Lahl B. Construction and demolition waste best management practice in Europe. Resources, Conservation and Recycling 2018;136:166–78.

[57] Lu WS, Yuan HP. A framework for understanding waste management studies in construction. Wastes Management 2011;31(6):1252–60.

[58] Wu Z, Yu ATW, Wei Y. Predicting contractor's behavior toward construction and demolition waste management. In: Proceedings of the 19th international symposium on advancement of construction management and real estate. Springer; 2015. p. 869–75.

[59] Ortiz O, Pasqualino JC, Castells F. Environmental performance of construction waste Comparing three scenarios from a case study in Catalonia, Spain. Wastes Management 2010;30:646–54.

[60] Mohamed M. Environmental and economic impact assessment of construction and demolition waste disposal using system dynamics. Resources, Conservation and Recycling 2014;82:41–9.

[61] Samton G. Construction & demolition waste manual. LLP with City Green Inc., Prepared for NYC Department of Design & Construction; 2003.

[62] Ellen MacArthur Foundation. Circular economy – UK, Europe, Asia. South America & USA. Ellen MacArthur Foundation; 2016. http://www.ellenmacarthur foundation.org/circular economy.

[63] Ying Y, Xu HZ. Study on the countermeasures of building waste resource based on 3R principle. Business 2013; 15:270.

[64] Li Y, Zheng Y, Chen JL. Policy study on construction waste reclamation in Beijing. Build Science 2008; 24(10):4–10.

[65] Huang B, Wang X, Kua H, Geng Y, Bleischwitz R, Ren J. Construction and demolition waste management in China through the 3R principle. Resources, Conservation and Recycling 2018;129:36–44.

[66] Hjelmar O, Johnson A, Comans R. Incineration: solid residues. In: Christensen TH, editor. Solid waste technology & management, vols. 1 and 2. Chichester, UK: John Wiley & Sons Ltd; 2010.

[67] Crillesen K, Skaarup J. Management of Bottom Ash from WTE Plants: an overview of management options and treatment methods. Report by ISWA-WG Thermal Treatment. Subgroup Bottom Ash from WTE-Plant; 2006.

[68] Vassilev S, Baxter D, Andersen L, Vassileva C. An overview of the composition and application of biomass ash. Part 1. Phase–mineral and chemical composition and classification. Fuel 2013;105:40–76.

[69] Park S, Song K, Jeon CW. A study of mineral recovery from waste ashes at an incineration facility using the mineral carbonation method. International Journal of Mineral Processing 2016;155:1–5.

[70] Seifritz GW. CO_2 disposal by means of silicates. Nature 1990;345:486.

[71] Bhattacharyya S, Donahoe RJ, Patel D. Experimental study of chemical treatment of coal fly ash to reduce the mobility of priority trace elements. Fuel 2009;88: 1173–84.

[72] González Corrochano B, Alonso Azcárate J, Rodas M. Effect of thermal treatment on the retention of chemical elements in the structure of lightweight aggregates manufactured from contaminated mine soil and fly ash. Construction and Building Materials 2012;35: 497–507.

[73] Huang SJ, Chang CY, Chang FC, Lee MY, Wang CF. Sequential extraction for evaluating the leaching behavior of selected elements in municipal solid waste incineration fly ash. Journal of hazardous materials 2007;149:180–8.

[74] Ke Y, Chai LY, Liang YJ, Min XB, Yang ZH, Chen J. Sulfidation of heavy metal containing metallurgical residue in wet-milling processing. Minerals Engineering 2013; 53:136–43.

[75] Schneider M, Romer M, Tschudin M, Bolio H. Sustainable cement production—present and future. Cement and Concrete Research 2011;41:642–50.

[76] Roethel FJ, Breslin VT. Municipal solid waste (MSW) combustor ash demonstration program "the boathouse". EPA/600/SR-95/129. Ohio, USA: US Environmental Protection Agency; 1995. 8 pp.

[77] Aziz SQ, Aziz HA, Yusoff MS, Bashir MJ, Umar M. Leachate characterization in semi-aerobic and anaerobic sanitary landfills: a comparative study. Journal of Environmental Management 2010;91:2608–14.

[78] Cossu R. The multi-barrier landfill and related engineering problems. In: Proceedings sardinia 95, fifth international landfill symposium, vol. 2. Cagliari: CISA; 1995. p. 3–26.

[79] Dajić A, Mihajlović M, Jovanović M, Karanac M, Stevanović D, Jovanović J. Landfill design: need for improvement of water and soil protection requirements in EU Landfill Directive. Clean Technology Environment Policy 2016;18:753–64.

[80] Denning PJ. Computer science: the discipline. Encyclopedia of Computer Science; 2000.

[81] Cuartas M, López A, Pérez F, Lobo A. Analysis of landfill design variables based on scientific computing. Wastes Management 2017;71:287–300.

[82] Berger KU. On the current state of the hydrologic evaluation of landfill performance (HELP) model. Wastes Management 2015;38:201–9.

[83] Gawande NA, Reinhart DR, Yeh GT. Modeling microbiological and chemical processes in municipal solid waste bioreactor, part II: application of numerical model BIOKEMOD-3P. Wastes Management 2010;30:211–8.

[84] Kamalan H, Sabour M, Shariatmadari N. A review on available landfill gas models. Journal of Environmental Science and Technology 2011;4(2):79–92. https://doi.org/10.3923/jest.2011.79.92.

[85] Babu GS. Prediction of long-term municipal solid waste landfill settlement using constitutive model. Practice Periodical of Hazardous, Toxic, and Radioactive Waste Management 2010;14:139–50.

[86] Cui Y, Wu Q, Xiao S, An X, Sun J, Cui F. Optimum ozone dosage of preozonation and characteristic change of refractory organics in landfill leachate. Ozone: Science and Engineering 2014;36:427–34.

[87] Zhang QQ, Tian BH, Zhang X, Ghulam A, Fang CR, He R. Investigation on characteristics of leachate and concentrated leachate in three landfill leachate treatment plants. Wastes Management 2013;33:2277–86.

[88] Top S, Sekman E, Hoşver S, Bilgili MS. Characterization and electrocoagulative treatment of nanofiltration concentrate of a full-scale landfill leachate treatment plant. Desalination 2011;268:158–62.

[89] Millar GJ, Lin J, Arshad A, Couperthwaite SJ. Evaluation of electrocoagulation for the pre-treatment of coal seam water. Journal of Water Process Engineering 2014;4: 166–78.

[90] Renou S, Givaudan JG, Poulain S, Dirassouyan F, Moulin P. Landfill leachate treatment: review and opportunity. Journal of Hazardous Materials 2008; 150(3):468–93. https://doi.org/10.1016/j.jhazmat.2007.09.077.

[91] Judd SJ. Membrane technology costs and me. Water Research 2017;122:1–9. https://doi.org/10.1016/j.watres.2017.05.027.
[92] Li F, Wichmann K, Heine W. Treatment of the methanogenic landfill leachate with thin open channel reverse osmosis membrane modules. Wastes Management 2009;29:960–4.
[93] Joo SH, Tansel B. Novel technologies for reverse osmosis concentrate treatment: a review. Journal of Environmental Management 2015;150:322–35.
[94] Wong CT, Leung MK, Wong MK, Tang WC. Afteruse development of former landfill sites in Hong Kong. Journal of Rock Mechanics and Geotechnical Engineering 2013;5:443–51.
[95] Liu CN, Chen RH, Chen KN. Unsaturated consolidation theory for the prediction of long-term municipal solid waste landfill settlement. Journal of Waste Management Research 2006;24:80–91.
[96] Shekdar AV. Sustainable solid waste management: an integrated approach for Asian countries. Wastes Management 2012;29(4):1438–48.
[97] Hickman HL. Principles of integrated solid waste management. Annapolis, MD: American Academy of Environmental Engineers; 1999.
[98] Mahar RB. Legal framework and financing mechanisms for waste agricultural biomass (WAB) solid waste in district Sanghar, Pakistan in converting waste agricultural biomass into energy source. Osaka, Shiga: United Nations Environmental Programme, International Environmental Technology Centre; 2010.
[99] Schübeler P, Wehrle K, Christen J. Conceptual framework for municipal solid waste management in low-income countries. UNDP/UNCHS/World Bank; 1996.
[100] Perry A. Legal systems as a detriment of FDI: lessons from Sri Lanka. Hague: Kluwer Law International, Alphen aan den Rijn; 2001.
[101] Beck T, Clarke G, Groff A, Keefer P, Walsh P. New tools in comparative political economy: the database of political institutions. The World Bank Economic Review 2001;15(1):165–76.
[102] Naureen M. Development of environmental institutions and laws in Pakistan. Pak. J. His. Cul. 2009;30(1):93–112.
[103] Zhou B, Sun C, Yi H. Solid waste disposal in Chinese cities: an evaluation of local performance. Sustainability 2017;9:2234.
[104] Yadav P, Samadder SR. Assessment of applicability index for better management of municipal solid waste: a case study of Dhanbad, India. Environmental Technology 2017;37:1–36.
[105] Laurent A, Bakas I, Clavreul BA, Niero M, Gentile E, Hauschilda MZ, Christensen TH. Review of LCA studies of solid waste management systems-Part I: lessons learned and perspectives. Wastes Management 2014;34:573–88.
[106] Damgaard A, Riber C, Fruergaard T, Hulgaard T, Christensen TH. Lifecycle-assessment of the historical development of air pollution control and energy recovery in waste incineration. Wastes Management 2010;30(7):1244–50.
[107] Banar M, Cokaygil Z, Ozkan A. Life cycle assessment of solid waste management options for Eskisehir Turkey. Wastes Management 2009;29:54–62.
[108] Winkler J, Bilitewski B. Comparative evaluation of life cycle assessment models for solid waste management. Wastes Management 2007;27:1021–31.
[109] Rebitzer G, Ekvall T, Frischknecht R, Hunkeler D, Norris G, Rydberg T, Schmidt S, Suh WP, Weidema BP, Pennington DW. Life cycle assessment: Part 1: framework, goal and scope definition, inventory analysis, and applications. Environment International 2004;30(5):701–20.
[110] Haupt M, Kägi T, Hellweg S. Modular life cycle assessment of municipal solid waste management. Wastes Management 2018. https://doi.org/10.1016/j.wasman.2018.03.035.
[111] Boesch ME, Vadenbo C, Saner D, Huter C, Hellweg S. An LCA model for waste incineration enhanced with new technologies for metal recovery and application to the case of Switzerland. Wastes Management 2014;34:378–89.
[112] Vadenbo CO, Boesch ME, Hellweg S. Life cycle assessment model for the use of alternative resources in ironmaking. Journal of Industrial Ecology 2013;17:363–74.
[113] Gheewala SH. LCA of waste management systems—research opportunities. The International Journal of Life Cycle Assessment 2009;14:589–90.
[114] Clavreul J, Baumeister H, Christensen TH, Damgaard A. EASETECH – an environmental assessment system for environmental technologies. Environmental Modelling & Software 2014;60:18–30.
[115] Thomassen MA, Dalgaard R, Heijungs R, DeBoer I. Attributional and consequential LCA of milk production. The International Journal of Life Cycle Assessment 2008;13:339–49.
[116] Curran MA, Mann M, Norris G. The international workshop on electricity data for life cycle inventories. Journal of Cleaner Production 2005;13(8):853–62.

CHAPTER 7

Law and Public Management of Solid Waste in Brazil: A Historical-Critical Analysis

MARIA CECÍLIA LOSCHIAVO DOS SANTOS, PHD • TERESA VILLAC, PHD

INTRODUCTION

Brazil established a national public policy for the management of solid waste only in 2010, after 20 years of legislation proceeding. Previously, in addition to the cities' public services responsible for urban cleaning, collectors were the real agents responsible, for years, for the recycling of urban solid waste.

Currently, almost 10 years after the definition of a national policy (Act 12,305 of 2010), solid waste management in Brazil is still subject to harsh criticism, and the situation of recyclable waste collectors remains precarious.

The objective of this chapter is to analyze the National Policy for Solid Waste and the ruptures that it brought, in the letter of the law, to the public management of waste in Brazil. We also address the difficulties for its implementation and identify the reasons for the little progress in terms of effectiveness.

THE NATIONAL POLICY FOR SOLID WASTE AND THE MANAGEMENT OF SOLID WASTE IN BRAZIL

Before the establishment of a national policy for the management of solid waste in Brazil, there were only specific national regulations per the type of waste, and in some Brazilian states, there were general rules for their local implementation.

Only in 1989, the debate on the formulation of a national law began, based on a bill that initially addressed the management of health services waste. Approved in one of the Legislative Houses (the Federal Senate), it was referred for vote in the House of Representatives in 1991, and an environmentalist, Fabio Feldmann, proposed that the bill should address solid waste management in general.

Progress took years in the Legislative Houses, with more than a 100 new bills added, the constitution of commissions, and a public hearing, until the approval in 2010. The efforts of the National Movement of Recyclable Materials Collectors in favor of a legislation that considered them as relevant social actors in the management of solid waste are worth mentioning.

The principles of the National Policy for Solid Waste were innovative in the Brazilian scenario, highlighting the need for a systemic view on waste management that takes into account the environmental, social, cultural, economic, technologic, and public health variables. In addition, there was a legal recognition that solid waste has an economic value and that waste collectors must take part in its management and there was a guideline on the need for cooperative management among the public power, the business sector, and other segments of the society.

In effect, the search for eco-efficiency in solid waste management considers not only environmental and economic aspects, but also cultural, through the valorization of environmental education and consumption decrease; technological, aiming to improve clean technologies, and public health factors, with the eradication of dumps; and human aspects, that bring quality of life.

The sustainable development model pursued by the Brazilian National Policy for Solid Waste is not limited to the tripod of sustainability and expands with the human right to a healthy environment, a greater cooperation between social agents, and the empowerment of waste collectors, in line with Amartya Sen's [1] thinking that emphasizes sustainable development as the guarantor of human freedoms.

Thus we highlight the legally foreseen mechanism of integrated waste management, which consists of a set of actions directed toward the search of solutions for solid

Sustainable Resource Recovery and Zero Waste Approaches. https://doi.org/10.1016/B978-0-444-64200-4.00007-4

waste management, by considering political, economic, environmental, cultural, and social dimensions, with social control and under the assumption of sustainable development.

In addition, Act 12,305 provides for the social inclusion and economic emancipation of collectors of reusable and recyclable materials by integrating them into actions that involve shared responsibility for the products' life cycle, as well as selective collection. Reverse logistics must necessarily add social parameters and be an instrument of economic and social development, through a set of actions designed to enable the collection and restitution of solid waste to the business sector, for reuse in its cycle or in other production cycles, or other environmentally appropriate final destinations.

COLLECTORS' VISIBILITY

Government's incentive for the creation and development of cooperatives and associations of collectors of reusable and recyclable materials was provided for in the National Policy and in the Decree that regulated it (Decree 7.404 of 2010). It aimed to give social visibility and strength to this segment, which comprises socially excluded persons in situation of human vulnerability, with the recovery of their dignity through a decent work.

Accordingly, there was a plan for the elimination of dumps, which are open-air sites for improper environmental disposal of urban solid waste, with harmful environmental consequences, and where people in a situation of absolute social exclusion, including children, work daily by collecting recyclable waste.

In addition to the goal of eradication of dumps, plans for municipal solid waste management that implement selective collection with the participation of waste collectors' cooperatives, formed by individuals of low income, have priority access to financial resources from the Federal Government.

Participation of collectors' cooperatives was also foreseen as a component of the reverse logistics systems. These should be implemented in the country through the return of products by the consumer after their use, independent of the public service of urban cleaning and solid waste management.

Considering that urban cleaning and waste management services in cities are under the responsibility of Brazilian municipalities, it was anticipated that selective collection systems of recyclable solid waste would prioritize the participation of collectors' cooperatives by the elaboration of municipal plans for the integrated management of solid waste.

Public policies oriented toward collectors of reusable and recyclable materials must include (1) the possibility of waiver of public bidding for hiring these collectors' cooperatives or associations; (2) the promotion of training and incubation and institutional strengthening of cooperatives; (3) the promotion of research aimed at integrating the policies into actions that involve shared responsibility for a products' life cycle; and (4) the improvement of the collectors' working conditions.

The Pro-Collector Program (Decree 7.405 of 2010) was established to improve the working conditions and opportunities for social and economic inclusion of collectors of reusable and recyclable materials. It involves training, research, incubation of cooperatives and solidary social ventures, credit lines, acquisition of equipment and improvements in the physical structure of cooperatives, organization of marketing networks, and strengthening the collector's participation in recycling chains.

CONCLUSIONS AND PERSPECTIVES

Almost 10 years after the launching of the national policy, which took 20 years to be approved, the situation of solid waste management in Brazil is critical. According to data from the Brazilian Institute of Geography and Statistics on the Profile of Brazilian Municipalities [2], in 2017, only 54.8% of the 5570 Brazilian municipalities had an integrated plan for solid waste management and 41.9% had selective collection of household solid waste. The deadline for the eradication of dumps, provided for in the National Policy for Solid Waste as until 2014, was not met yet, and there is a bill underway for further extension.

The low effectiveness in policy implementation stems from a combination of factors, of which we highlight the Brazilian development model, which historically focused on economic growth. The emergence of a democratic constitution in 1988, with values and principles for the reduction of social inequalities and preservation of the environment, was not sufficient to break and overcome this economic cycle with a development model based on human freedoms.

Regarding the National Policy for Solid Waste, if we consider the economic value attributed to waste and the significant changes in management that it brought, which affect the whole society on the shared responsibility for products' life cycle, the conflicts involve a wide range of social actors, such as the production sector, suppliers, recycling industry, public power, waste collectors, importers, individuals, and companies.

In addition, despite the reference to environmental education and the need to reduce consumption in Brazil, there was no mention in the policy for the empowerment of the citizen, who was reduced to the consumer role.

Environmental education, little developed in the country and restricted to specific sectors, is one of the paths through which a citizen, individually and socially, can realize the ethical responsibility of his/her consumption pattern and the unconscious way of waste disposal by organizing and acting socially and politically for public policy implementation.

The moment is critical, in a scenario of environmental legislative setbacks in other areas in Brazil, and collectors' social movements did not achieve visibility and active participation in the formulation of strategies for the policy's effectiveness.

Therefore the relationships among consumption, inappropriate disposal, and collecting activity persist in Brazilian cities, revealing an ethical crisis that refers to human responsibility for the other and for the future.

REFERENCES

[1] Sen A. Desenvolvimento como liberdade [Development as freedom]. São Paulo: Companhia das Letras; 2009.

[2] Instituto Brasileiro de Geografia e Estatística. Perfil dos municípios brasileiros – Munic 2017 [Profile of Brazilian municipalities 2017]. Rio de Janeiro: IBGE; 2017.

ADDITIONAL REFERENCES

Santos MCL. Design wisdom: beyond the frontiers of reductionist discourses. In: Paper presented at the 5th European Academy of Design Conference, Barcelona, Spain; 2003. Retrieved from: http://www.closchiavo.pro.br/site/pdfs/design_wisdom_beyond.pdf.

Santos MCL. Re-shaping design. A teaching experience at COOPAMARE. Listen to the recyclable collector's voice. Cumulus Working papers. Helsinki: University of Art and Design; 2005. Retrieved from: http://www.cumulusassociation.org/wp-content/uploads/2015/09/WP_Utrecht-13_04.pdf.

Santos MCL. Consumo, descarte, catação e reciclagem: notas sobre design e multiculturalismo [Consumption, disposal, collecting and recycling: notes on design and multiculturalism]. In: Moraes D, org. Cadernos de estudos avançados em design: multiculturalismo (Caderno I. Belo Horizonte: Centro de Estudos, Teoria, Cultura e Pesquisa em Design, vol. 1). Universidade Estadual de Minas Gerais; 2008.

Santos MCL, coord. Design, resíduo & dignidade [Design, waste & dignity]. São Paulo: Olhares; 2014.

Santos MCL. Lições das cidades de plástico e papelão: resíduos, design e o panorama visto da margem [Lessons from the plastic and carton cities: waste, design and the panorama seen from the edge]. In: Santos MCL, coord. Design, resíduo & dignidade. São Paulo: Olhares; 2014. p. 41–55.

Villac T. A construção da Política Nacional de Resíduos Sólidos [The construction of the National Policy for Solid Waste]. In: Santos MCL, coord. Design, resíduo & dignidade. São Paulo: Olhares; 2014. p. 149–59.

Villac T. Sustentabilidade e contratações públicas no Brasil: Direito, Ética Ambiental e Desenvolvimentto. PhD thesis Thesis Advisor: Santos, MCL. São Paulo: Instituto de Energia e Ambiente, Universidade de São Paulo; 2017. Retried from: http://www.teses.usp.br/teses/disponiveis/106/106132/tde-08112017-141101/.

CHAPTER 8

Sorting Household Waste at the Source

KAMRAN ROUSTA, PHD • KIM BOLTON, PHD

INTRODUCTION

It is expected that the quantity of municipal solid waste (MSW) will increase due to the rise in population growth, increased rate of urbanization, and increased economic power of the developing countries [1]. The world generates 2.01 billion tons of MSW per year and it is expected to increase to 3.40 billion tons by 2050 [1]. It is estimated that solid waste generated 1.6 billion tonnes of carbon dioxide equivalents in 2016, which corresponds to 5% of the global emission [1]. This is estimated to increase to 2.6 billion tonnes carbon dioxide equivalents in 2050 [1]. Development of efficient and robust waste management systems is required to limit the increase in these impacts, as well as to prevent other negative aspects associated with uncontrolled landfill or dumping, such as health and social issues [1,2]. These issues will also lead to increased economic costs, such as in healthcare.

One of the unique challenges that arises when managing MSW is that it contains many different fractions including food, plastic, metal, and glass wastes. These different fractions need to be separated from each other so that each fraction can be treated in a way that is best for that specific fraction. For example, food waste can be treated biologically to produce biofuel or compost, whereas plastic, metal, and glass wastes should be recycled into new plastic, metal, and glass products. Contamination of the pure, separated fractions with waste from other fractions decreases the efficiency of waste treatment and yields recycled products and materials that have a lower quality. For example, contamination of the food waste fraction with other fractions will decrease the efficiency of the biological treatment process and the quality of the compost. Similarly, contamination of the plastic waste fraction with food waste (or any waste fraction other than plastic) will decrease the efficiency of the recycling process and the quality of the recycled plastic. Hence, sorting the waste into various fractions is the key for achieving efficient recycling processes and for obtaining high-value products.

Sorting and temporary storage of the waste fractions may be done at the interface between the actors that generate the waste, such as households and restaurants, and the technical system that treats the waste (recycling, incineration), as illustrated in Fig. 4.3 in Chapter 4. In spite of the importance of this interface, many studies focus only on the households or only on the technical system, rather than the interplay between the households and the technical system.

For example, a review by Rousta et al. [3] revealed that most studies that offer design guidelines for waste management systems focused only on the technical system. This can be questioned because the needs of the households are key when designing the system. The same technical system may not be relevant for different contexts and cultures. Similarly, most of the studies that focused on households did not consider ways to develop the technical system to fit the needs of the households. Rousta et al. [3] suggest that multidisciplinary studies that combine user behavior with technical aspects are needed in order to focus on the interplay between these areas. These types of studies would also combine the different research methods used by social scientists and engineers. Only 2 of the 51 studies reviewed by Rousta et al. [3] were multidisciplinary, the remainder were from either the social sciences (26) or the engineering (23) fields.

Note that sorting of the waste into the different fractions does not have to be done at, or by, the households. It can also be done at a material recovery facility (MRF). Commingled waste or a mixture of commingled and sorted fractions are transported to the MRF, where manual and/or mechanical sorting can separate the different fractions. It is important to note that there is currently no available equipment that sorts all the fractions in a commingled waste stream. Therefore besides large investments for mechanical sorting at MRFs, manual sorting is also needed, even when there is no source separation system in place. Moreover, the quality of the sorted material in mechanical sorting

Sustainable Resource Recovery and Zero Waste Approaches. https://doi.org/10.1016/B978-0-444-64200-4.00008-6

of mixed wastes is not as good as when waste is sorted into different fractions at the source [4]. When the waste fractions are not separated at the source and are mixed together, they become a comingled pulp, which is difficult to treat in the downstream process [5]. However, the focus of this chapter is waste sorting at the source, which means that households are responsible for sorting the waste that they generate in their homes.

This chapter is structured as follows. The role of the households in waste sorting at the source, as well as the factors that influence this sorting, is presented before discussing and giving examples of technical systems that are used when sorting at the source. Following this discussion, we return to the system perspective when presenting the recycling behavior transition (RBT) procedure [6] that, based on the needs of the households, can be used to identify, implement, and analyze interventions to improve the interface between the households and the technical part of any waste management system. Finally, we explain the impacts of source separation in a sustainable society and give the conclusions and perspectives.

THE ROLE OF HOUSEHOLDS IN WASTE SEPARATION AT THE SOURCE

Waste management should be seen as a component of any city's infrastructure. Similar to other types of infrastructure, such as access to clean water, electricity, and healthcare, there should be a political agenda for long-term development and improvement of the waste management system. This should be the responsibility of the politicians, e.g., at the municipal level, and can be implemented by the municipalities or in cooperation with private companies. When implemented by municipalities, the costs associated with these types of infrastructures are often covered by taxes and tariffs.

In Sweden, households pay tariffs to the municipalities for collection and treatment of their waste. The tariffs are not the same for all municipalities, but they are usually aimed at minimizing household waste and motivating households to sort their waste correctly. Hence, as for any other infrastructure, households pay for this service and therefore expect a high-quality waste management system. However, in contrast to many other infrastructures, households in Sweden are expected to actively participate in the system. They are expected to sort their waste, usually without payment or with a very small reduction in tariffs. This may cause some confusion because households are paying for a service and at the same time are expected to work for the service provider. The importance of households in both these roles should be rational and clear to the households. In addition, the barriers to their (unpaid) work in the system must be as small as possible. The factors that motivate and enable households to engage in the system must be understood and enforced. These factors depend on the local context of the households.

Table 8.1 shows the typical compositions of MSW in countries with different income levels [1]. The data is from 2016. The lists of countries that are considered to be at the levels of low income, lower-middle income, upper-middle income, and high income are given by Kaza et al. [1]. The waste fractions shown in the table are organic (food, garden refuse, etc.), paper (paper and cardboard), plastic, glass, metal, and others (textiles, rubber, wood, electronic waste, etc.). It is evident that the composition of the waste differs in different countries. For example, countries with lower income typically have a larger fraction of organic waste and countries with higher income typically have a larger fraction of paper, metal, and glass wastes. Irrespective of the relative amounts of each waste fraction, these fractions need to be separated so that they can be treated in the most efficient and relevant manner.

Factors That Influence Household Participation in Sorting Waste at the Source

Ajzen and Fishbein [7] developed the theory of reasoned action to understand the relation between a person's attitude and his/her behavior. According to

TABLE 8.1
Composition of Municipal Solid Waste in Countries With Different Income Levels.

Income	Organic (%)	Paper (%)	Plastic (%)	Glass (%)	Metal (%)	Others (%)
Low income	56.4	7.4	6.6	1.3	1.6	26.7
Lower-middle income	52.8	12.6	11.2	3.4	2.4	17.6
Upper-middle income	53.6	10.7	10.4	4.0	2.2	18.7
High income	31.5	25.2	12.5	5.5	6.4	18.9

Source: Data from 2016.

this theory, the most important aspects that determine behavior are intentions. Intentions are influenced by the person's attitudes to behave in a certain way and can be based on the individual beliefs. It is also influenced by subjective norms that are based on how the person perceives other people behaving, which can also affect his/her behavior. After several studies, Ajzen [8] added a third aspect to the theory. This was the person's perception regarding the difficulty and chance of success when behaving in a given way.

Ölander and Thøgersen [9] extended this theory to the motivation-ability-opportunity-behavior model. According to this model, intention is the aspect that ultimately results in a certain behavior. Intentions are influenced by attitudes and social norms, which together build the motivation for the behavior. However, the behavior will not be carried out unless the person perceives that he/she has the ability and opportunity to perform the behavior. Ability is partly based on a person's knowledge of his/her ability to successfully perform a task and his/her previous experiences from similar tasks. There must also be an opportunity, such as nearby and convenient sorting facilities, to enable the person to perform the behavior.

Rousta and Dahlén [5] divided the factors that influence household participation in waste separation schemes into internal, external, and sociodemographic factors. Internal factors are those that are intrinsic to each individual, and they affect an individual's participation in the waste separation scheme. These include the individual's motivation, attitude (including health concerns associated with poorly managed waste), and intention toward recycling. For example, certain people may exercise a responsibility to participate, either because there are laws that govern participation or because they are concerned about the environmental impact of waste.

External factors are extrinsic to the individual. They facilitate or hinder the individual's participation in the recycling scheme [5]. One example is the availability of a waste separation scheme, and this scheme is easy and convenient to use. Having too many waste fractions may cause inconvenience, as this may require more work to sort the fractions and more space in households to temporarily store the sorted or partially sorted fractions. The recycling stations should not be too far from the households, so that it is convenient for the households to discard their waste on a regular basis. The second example of an external factor is information that provides knowledge about sorting and recycling and the benefits of these activities. Rewards and positive feedback are the other examples of external factors. The importance of two of these external factors, availability of a waste separation scheme and proper information, is discussed in the Section The Recycling Behavior Transition Procedure.

Sociodemographic factors are variables such as culture, age, gender, level of education, and income. There have been numerous studies, reviewed by Rousta and Dahlén [5], on the effect of these factors on an individual's participation in the waste separation scheme. For example, Ando and Gosselin [10] and Beigl et al. [11] stated that sociodemographic factors play a major role in recycling behavior, but how they affect this behavior is still unclear. In a meta-analysis of various studies, Miafodzyeva and Brandt [12] also identified that there are some doubts on how sociodemographic factors influence recycling behavior. It is important to note that most of the research in this field was based on case studies in a limited geographic and cultural area. Therefore it is not simple to generalize the influence of the different factors on recycling behavior. However, these studies give a broad understanding of sociodemographic factors on recycling behavior and also report that they are important in determining the recycling behavior in households.

The interaction between the internal, external, and sociodemographic factors that influences household participation in the waste sorting system, called recycling behavior, is shown in Fig. 8.1. Even though it is uncertain as to how different sociodemographic factors influence the sorting of household waste, there is agreement that they are an important factor [10,11] and are therefore included in the figure. It is important to note that it is not simple to separate the factors into isolated categories. The factors in each category can directly influence recycling behavior or influence the factors in another category. This relation is shown with dashed lines in Fig. 8.1.

As mentioned earlier, Rousta et al. [3] reviewed previous studies that included interaction of households with the technical part of the waste management system. They drew the following conclusions:

- The perceived convenience of the technical infrastructure is vital for increasing sorting into different waste fractions. For example, the volumes of sorted materials have consistently increased when reducing the distance to the sorting infrastructure. The waste management service for sorted fractions should be at least as convenient as that for the unsorted (residual) fraction.
- Information that encourages household participation should be provided at regular intervals and in an engaging way. This can be achieved through several communication channels.

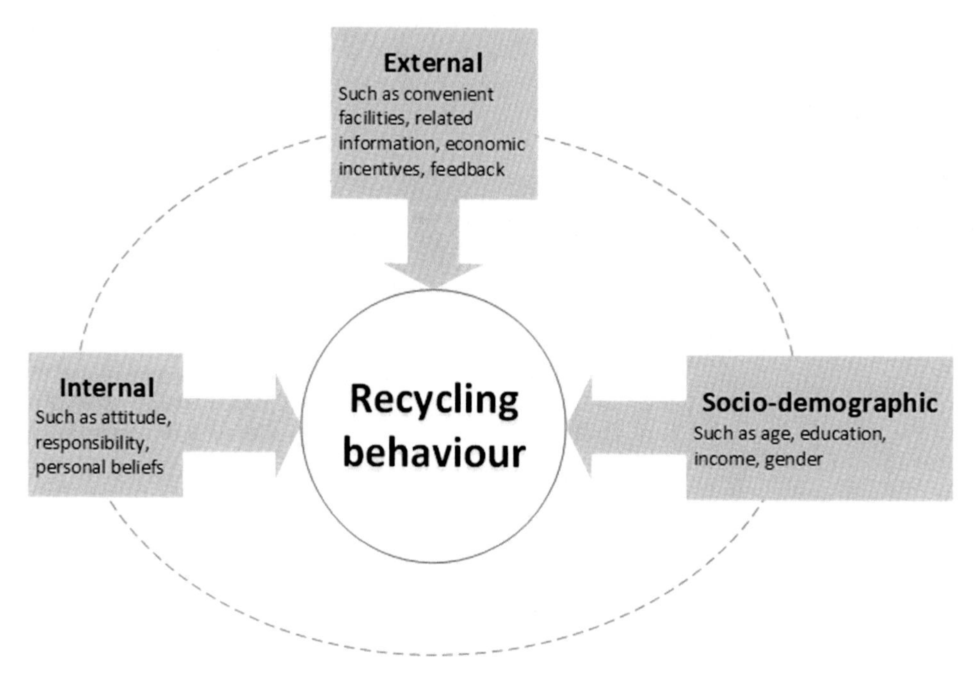

FIG. 8.1 Factors that influence recycling behavior.

- Economic incentives do not appear to be a productive way to encourage long-term engagement of households in sorting waste, as long as they receive other benefits. In contrast, economic incentives may be relevant when introducing a new, or changing an existing, waste separation system.
- Although sociodemographic factors appear to influence waste generation and sorting, there is no consensus on how they affect these behaviors.
- Sorting of waste is influenced by social norms. Infrastructures that increase social contact may therefore improve sorting rates.
- The local context, for example, of sorting in rural or urban areas must be considered when developing a waste management system.

INFRASTRUCTURE FOR COLLECTING SEPARATED HOUSEHOLD WASTE

There is no single infrastructure for collecting household waste that is best for all households and contexts. It is important that the infrastructure that is chosen for a specific household encourages participation. This also implies that infrastructures may change over time, reflecting the changing consumption patterns, needs, and desires of households.

The infrastructure that is selected for household waste is a crucial part of the waste management system, as it must be accessible to the households as well as be suitable for the downstream waste management processes. For example, pure waste fractions are needed for efficient recycling. This would indicate that the infrastructure should have at least as many containers as there are recyclable fractions. Assuming that households sort correctly, waste collected in these containers could be sent directly to recycling companies, without the need for further sorting at MRFs. However, multiple containers also require a more complex logistic system, which may not be economically viable in areas that have low population densities. Similarly, depending on the internal, external, and sociodemographic factors discussed earlier, households may not be motivated to sort into many different fractions, especially if it is difficult to identify the type of waste (e.g., whether to sort combined paper and plastic packaging into the paper or the plastic fraction). In this case, the waste may incorrectly be sorted in another fraction, such as the residual waste fraction. It may be more convenient for households to partially separate the waste into fewer waste fractions (e.g., comingled paper and plastic packaging), which would then require subsequent sorting at an MRF.

In the following section, examples of different types of infrastructures, commonly used in Sweden, are presented.

Recycling Centers

Fig. 8.2 shows an example of a recycling center. In Sweden, these recycling centers are usually used in conjunction with the other types of infrastructures discussed later, and hence, waste that is sorted in the other infrastructures (smaller packaging waste, food waste, and residual waste) is not typically accepted at recycling centers. Instead, recycling centers accept fractions such as bulky waste, garden waste, furniture waste, textile waste, building waste, large packaging waste, hazardous waste, light bulbs, and electronic waste. At some recycling centers, there are also containers for reusable waste, e.g., waste that can be given or sold to other households to be reused.

Recycling centers are rather near, typically a few kilometers, for inhabitants in urban areas. In contrast, they can be tens of kilometers away from inhabitants in rural areas. It should, therefore, in principle, be easier for city dwellers to sort these waste fractions. However, as these fractions are typically large and heavy, inhabitants often need a means to transport this waste to the recycling centers. Those inhabitants who do not have easy access to the proper transport (often a car or car with trailer) have an extra obstacle to correctly sort these waste fractions. Some municipalities have services that can assist inhabitants with this transport. However, design and implementation of these types of services are still under development.

Recycling Stations

Fig. 8.3 shows a typical recycling station. Recycling stations are smaller, in land area, than recycling centers. There are also more recycling stations and they are closer to households. As opposed to recycling centers, recycling stations are for household waste that it

FIG. 8.2 A recycling center in the city of Borås, Sweden.

FIG. 8.3 Part of a recycling station with containers to collect different packaging waste fractions in the city of Borås, Sweden.

generated on a daily basis. This includes newsprint and packaging waste. There are usually five containers for the packaging waste: colored glass, transparent glass, plastic packaging, paper packaging, and metal packaging. Recycling stations are usually within walking distance from households. This is an important convenience factor that encourages household participation [13]. One reason for this is that households do not typically use the recycling stations every time waste is generated. The waste is usually temporarily stored in the home, before being transferred to the recycling station. As many homes have limited space for temporary storage, it must be easy to transfer the waste to recycling stations. This may not be done if sorting these waste fractions into the residual (mixed) fraction is perceived to be far easier. As discussed later, this residual fraction is collected in property close infrastructures, which may be perceived as being more convenient than carrying the waste to recycling stations.

Property Close Infrastructure

As mentioned earlier, the proximity of sorting and collection infrastructure to the households is a very important convenience factor that motivates participation in the waste management system. It is therefore desirable to bring this infrastructure as close to the household property as possible, and even on the property. Although this is preferred in terms of household convenience, it creates new challenges. The first challenge is the space that is needed on the household property. If a large number of waste fractions are to be sorted, then one needs either many small containers (e.g., for single-household dwellings) or a large space for many large containers (e.g., for multihousehold apartment buildings). These alternatives also increase the complexity and the costs of the logistics for waste collection, as vehicles (that should preferably collect several fractions at the same time) need to operate near the households.

There are several options for optimizing property close waste sorting and collection. These are dynamic and are being improved over time. Some systems that exist in Sweden are described in the following sections.

Two-bin system

The two-bin system is usually used in combination with the recycling centers and stations discussed earlier. In principle, the only waste fractions that cannot be collected for treatment via these centers and stations are food and residual wastes. These are the two fractions that are collected in each of the two bins. Food waste is collected from the households and used for the production of biofuel or compost (depending on where the household is located in Sweden). The residual waste is also collected from the households. As it is not practicable to material-recycle the residual waste, it is incinerated at combined heat and power plants.

Note that residual waste is collected at the household properties. Hence, people do not need to walk with this waste to the recycling stations. It may, therefore, be perceived by households as more convenient to (incorrectly) sort waste fractions such as newsprint and packaging into the residual waste. This underscores the importance of motivating households to participate correctly in the waste separation system, and not simply to follow the options that they perceive as most convenient (or, alternatively, make correct sorting more convenient than missorting).

The second two-bin system for households is to separate wet and dry fractions into different bins [5]. These bins are then sent to an MRF for further sorting. This may be convenient for households, but the fractions are of low quality compared with those sorted at other types of sorting infrastructures. This system is not implemented in Sweden.

Separate bin for each fraction

Fig. 8.4 illustrates a property close collection system where there is a separate bin for almost all waste fractions. This combines many of the nonbulky fractions that are otherwise sorted at recycling centers, the fractions sorted at recycling stations, and the food and residual fractions. Moreover, this facility is on or adjacent to the property where the households are situated. This infrastructure is clearly not suitable for single households but is very suitable for densely located multihousehold apartments. It is an infrastructure that is very convenient for the households. The fractions that can be separately collected include textile waste, light bulbs, electronic waste, newsprint, all types of packaging (collected in their separate fractions), food waste, and residual waste [5]. Hence, it is convenient not only to sort these different waste fractions but also to sort the fractions that can be material-recycled as it is to sort the residual waste.

This type of infrastructure is increasing in many regions in Sweden. In some areas, approximately 90% of multiple-household apartments have access to this type of sorting and collection system [5].

FIG. 8.4 A property close infrastructure for collection of waste fractions. The big bins are used for collection of newsprints, paper packaging, plastic packaging, metal packaging, and glass packaging; food waste; and residual waste. The small red bins are used for collection of batteries, small electronic wastes, and light bulbs.

Multicompartment bins

In contrast to the infrastructure discussed in the previous section, multicompartment bins are more suitable for single households. Fig. 8.5 shows examples of two multicompartment bins with four compartments in each bin. The compartments in the left bin are for (from top left moving clockwise) residual, plastic packaging, food, and colored glass packaging waste. The compartments in the right bin are for paper packaging, newsprints, transparent glass packaging, and metal packaging wastes. Typically, each household has its own set of multicompartment bins on its property, and these bins are emptied once every second week.

Extra small compartments, also shown in the figure, can be attached to the bins. These are for light bulbs and batteries (red small bins in Fig. 8.5). It is uncommon that these small compartments are filled every 2 weeks (or even every 2 months) and households can therefore remove them and reattach them when they are full. The authorities empty them at the same time as they empty the corresponding bin.

This type of collection infrastructure is popular among households and its use is expected to increase in Sweden [5]. As pointed out by Rousta and Dahlén [5], a more complex logistics system is needed than when using a two-bin system combined with a nearby recycling station. Vehicles that can collect all fractions need to collect from individual households, instead of a single recycling station, and these vehicles need to have four compartments for the four different waste fractions of each bin, as well as a compartment for the extra compartment that can be attached to the bin.

Sorting Infrastructure in Gathering Areas

The infrastructures discussed earlier are for waste generated at citizens' homes. However, households also generate waste in other areas. It is therefore important to have convenient infrastructure outside homes, especially in areas where a large amount of waste is generated. This usually occurs in areas where large crowds gather, especially when this occurs on a regular basis. These areas can be diverse and can be indoors or outdoors. Examples are restaurants, canteens (e.g., schools), entrances to public places such as universities and shopping centers, airports, city centers, parks, and beaches.

The type of infrastructure used in these cases must be relevant, suited to the available space, and easy to use. For example, at restaurants (in Sweden), there may be two bins: one for food waste (for biofuel or compost) and one for residual waste (incineration). Fig. 8.6 shows an example of infrastructure that can be found at entrances to public buildings. It contains compartments for residual waste, food waste, colored glass, transparent glass, metal, batteries, plastic packaging, paper packaging, and newsprints. At outdoor gathering places it is common to have a single container for residual waste, or up to four containers for gathering food waste, plastic and paper packaging waste, and residual waste. Some of these areas in Sweden also have a separate container for single-use barbeques.

THE RECYCLING BEHAVIOR TRANSITION PROCEDURE

A sustainable waste management system must promote a circular economy. This means that waste fractions must be reused or recycled without significantly losing

FIG. 8.5 Four-compartment bins. (Reprinted with permission from PWS Nordic AB.)

FIG. 8.6 An example of sorting facilities at the University of Borås, Sweden.

the inherent value of the waste. To do this, sorting the waste fractions into different materials is crucial. This is the reason that waste separation at the source is key to success in a waste management system. In source separation systems, citizens do the waste sorting in their homes. For implementing such a system, understanding and investigating the interaction between households and the technical part of the waste management system is vital. The household-technical system interface can be studied using the RBT procedure developed by Rousta et al. [6]. Fig. 8.7 is based on the RBT procedure and consists of five steps. The first step is to identify which actor, or target group, will be in focus. There are many actors in municipalities that generate waste, including households, schools, and supermarkets. Once the actor has been identified, one quantifies and characterizes the waste generated by this actor. This is done in the second step. That is, one identifies the type of waste (food, plastic packaging, etc.) as well as how much of each fraction is wasted. This is done using a pick analysis, i.e., a waste composition study [6]. Based on the results of the pick analysis, one can calculate other data such as environmental impacts and economic costs of the waste. This is, however, not always done.

The information gained in the second step is used to engage with the target group in the third step. This is typically done using interviews, focus groups, and sometimes in combination with questionnaires. This step aims to identify what is needed to improve the interface between the target group and the technical system. This is a crucial step in the procedure, as it places the target group in focus when identifying an intervention to improve the interface. Interventions may, for example, be changes in the technical system to improve the interface (e.g., changes that can improve sorting of waste) or incentives (economic incentives, laws) that can motivate the target group to interact with the technical system. Once the intervention is identified, it is designed and implemented in the fourth step. A second pick analysis is done in the fifth step and, by comparing with the first pick analysis, can be used to analyze the effect of the intervention. The last three steps can be repeated to nudge the target group to continually improve the interaction between them and the technical system.

The RBT procedure discussed earlier with reference to Fig. 8.7 merges the importance of the factors that govern participation in the waste management scheme (step 3 in Fig. 8.7) with the technical system. This procedure can be used in any waste management system, as the engagement with the target group takes into account the internal, external, and sociodemographic factors of that group. For example, a study done in a pilot area in the city of Borås [13] revealed that inhabitants in this area perceived that they would improve their participation in the waste management scheme—in this case, sorting of the food, newsprint, packaging, and residual waste—if a recycling station was located nearer to their homes. In this case, only the results of the initial pick analysis and not environmental impacts or economic costs were used when engaging with the households. The municipality and property owner therefore built a new recycling station on the property grounds. Comparison between the results of the pick analysis done before and after implementing the intervention showed statistically significant improvement of the sorting due to this intervention. For example, the amount of paper packaging, glass packaging, and newsprint that was

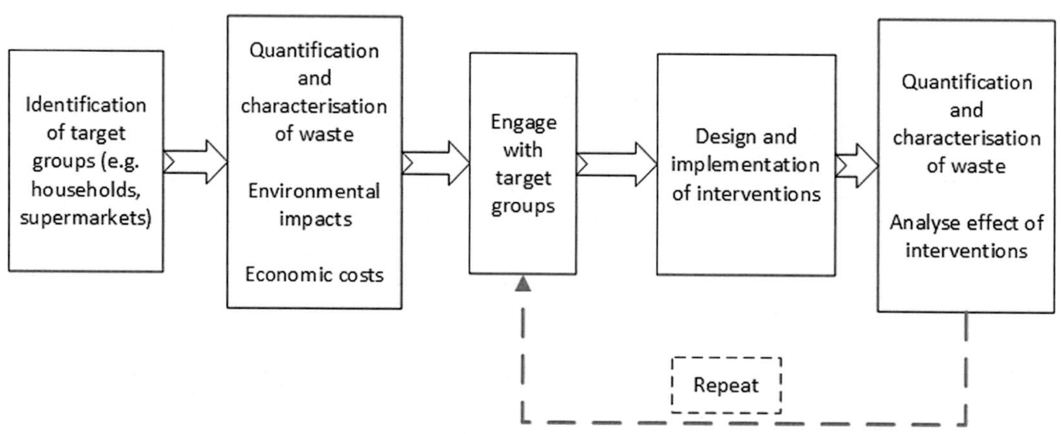

FIG. 8.7 A procedure to improve user interaction with the technical part of the waste management system. (Based on the RBT procedure by Rousta et al. [6].)

incorrectly placed in the fraction for residual waste decreased by 23.3%, 48.1%, and 49.5%, respectively, due to the intervention [13].

The RBT procedure is one of the keys for success when developing sustainable waste management systems that include waste separation at the source. By involving the target group as a vital factor in the system, it considers the local situation and the context of the place where the waste management system is implemented. This is the reason that the solutions and interventions for improvement that are identified in this procedure are potentially effective. When one investigates the needs of households to facilitate their participation in the system, their input is key for success.

Moreover, the results of pick analyses in this procedure can also be used as a guideline when interacting with the households and as raw data for other analyses such as environmental impacts and economic cost analyses. These data can also be used in the first step when establishing a new waste management system in a city. As the RBT procedure applies both qualitative and quantitative methods for data collection, and is a multidisciplinary study, the results give a holistic perspective of the interface between the social and technical aspects of the waste management system. This is necessary for improving any waste management system that includes waste separation at the source.

THE OTHER SIDE OF SOURCE SEPARATION

Source separation is the part of the waste management system where households are expected to separate their waste into different fractions. The separation of hazardous waste from other waste streams is crucial to protect the environment. Separating the recyclable materials for recycling is important for material recirculation in the society. Similarly, separating the food waste contributes to biogas and/or biofertilizer production. These are some examples that illustrate the importance of source separation in a waste management system. The goal of source separation is to decrease the environmental impacts of waste and conserve energy and material by material recirculation.

Establishing source separation as part of a waste management system is difficult, as one needs to change the attitudes and behavior of citizens toward recycling. It takes time and needs a constant effort to make it a part of a culture and a society. A survey shows that 80% of people living in Sweden participate in waste separation schemes [14]. When the majority of people separate waste, it can be considered as a norm for this country. Research also indicates that most citizens separate at the source because it is perceived as a positive contribution to environmental protection and to maintain a robust environment for coming generations [6]. This is perceived as a task that contributes to sustainable development and is a necessary complement to the benefits, which the technical part of the waste management system provide.

Source separation also changes the attitudes of people toward other environmental issues. After many years of source separation as a collective task in the Swedish society, it can be one of the reasons that the entire society thinks about and understands a wide range of environmental challenges and pays attention to, and assists in, their solutions. Hence, establishing a source separation system may, in the long term, create a wider perspective toward sustainable development. This can be a hidden and positive aspect of establishing a waste management system that includes source separation, and decision-makers should take this into consideration when establishing a waste management system.

CONCLUSIONS AND PERSPECTIVES

Generation of MSW is expected to increase due to the growth in global population and urbanization. MSW contains many different types of wastes, including food, plastic, metal, and glass wastes. In order to facilitate further treatment, these different types of wastes need to be sorted into separate fractions, with as little contamination as possible. For example, food waste needs to be sorted from the other types of wastes, with as little contamination from plastic, metal, and glass as possible. This is necessary for the efficient biological treatment of the food waste and products, such as compost, that have a high quality.

Sorting of the waste into the various fractions must be done before it is treated in the technical waste management system. This sorting can be done at the place where the waste is generated, i.e., at the source. Households generate a large quantity of MSW and should therefore also sort their waste in a system where waste is separated at the source. This requires both actions from the households and the availability of a suitable waste management system. As described in this chapter, the factors that determine how the households interact with the technical part of the waste management system are intrinsic, extrinsic, and sociodemographic factors. However, it is emphasized in this chapter that the importance of these factors, as well as the type of technical system that motivates proper sorting, depends on the local context such as culture. Hence, strategies for developing waste management systems must take the local context into account. This can be achieved by following the RBT procedure described in this chapter.

REFERENCES

[1] Kaza S, Yao L, Bhada-Tata P, Van Woerden F. What a waste 2.0: a global snapshot of solid waste management to 2050. Urban development series knowledge papers. Washington DC: The World Bank; 2018. p. 1–272.

[2] Rousta K, Richards T, Taherzadeh MJ. An overview of solid waste management toward zero landfill: a Swedish model. In: Taherzadeh MJ, Richards T, editors. Resource recovery to approach zero municipal waste. Boca Raton, USA: CRC Press; 2016. p. 1–22.

[3] Rousta R, Ordoñez I, Bolton K, Dahlén L. Support for designing waste sorting system: a mini-review. Waste Management & Research 2017;35(11):1099–111.

[4] Tchobanoglous G, Theisen H, Vigil SA. Integrated solid waste management: engineering principles and management issues. New York, USA: McGraw-Hill; 1993.

[5] Rousta K, Dahlén L. Source separation of household waste; technology and social aspects. In: Taherzadeh MJ, Richards T, editors. Resource recovery to approach zero municipal waste. Boca Raton, USA: CRC Press; 2016. p. 61–77.

[6] Rousta K, Bolton K, Dahlén L. A procedure to transform recycling behavior for source separation of household waste. Recycling 2016;1(1):147–65.

[7] Ajzen I, Fishbein M. Understanding attitudes and predicting social behaviour. Englewood Cliffs, NJ: Prentice Hall; 1980.

[8] Ajzen I. From intentions to actions: a theory of planned behaviour. Berlin, Germany: Springer; 1985.

[9] Ölander F, Thøgersen J. Understanding of consumer behaviour as a prerequisite for environmental protection. Journal of Consumer Policy 1995;18(4):345–85.

[10] Ando AW, Gosselin AY. Recycling in multifamily dwellings: does convenience matter? Economic Inquiry 2005; 43(2):426–38.

[11] Beigl P, Lebersorger S, Salhofer S. Modelling municipal solid waste generation: a review. Waste Management 2008;28(1):200–14.

[12] Miafodzyeva S, Brandt N. Recycling behaviour among householders: synthesizing determinants via a meta-analysis. Waste and Biomass Valorization 2013;4(2): 221–35.

[13] Rousta K, Bolton K, Lundin M, Dahlén L. Quantitative assessment of distance to collection point and information effects in source separation of household waste. Waste Management 2015;40:22–30.

[14] FTI. Recycling barometer. 2018. Available: http://www.ftiab.se/2461.html; (cited August 21, 2018).

CHAPTER 9

Sustainable Composting and Its Environmental Implications

QUAN WANG, PHD • MUKESH KUMAR AWASTHI, PHD • ZENGQIANG ZHANG, PHD • JONATHAN W.C. WONG, PHD

INTRODUCTION

With the continuously increasing population growth, agriculture development, and urbanization level, there is a significant increase in the production of organic waste such as livestock manure, municipal solid waste, sewage sludge, wheat straw, and other wastes all over the world [1,2]. In 2012 the global production of municipal solid waste was about 1.3 billion metric tons, and this figure is expected to reach to nearly 4.2 billion tons per year by 2050. The demand for milk and meat products is increasing, which has led to the rapid development of the global livestock and poultry breeding industry [1]. The farming methods have also been transformed from traditional decentralization and extensive to large-scale, intensive, and specialized farming system. According to statistics, the world pig production has increased from about 82 million heads in 1986 to approximately one billion heads in 2016 (Table 9.1) [2,3]. The rapid development of intensive and large-scale livestock and poultry breeding, with increasing numbers of farms, will inevitably produce a large amount of livestock and poultry manure. According to a survey by the Ministry of Environmental Protection in China, the manure production of livestock and poultry farms in China was about 3.8 billion tons in 2017. Besides, with the rapid upgrading and expansion of wastewater treatment plants throughout the developing countries, the production of the sewage sludge has been increased obviously. In China the annual output of sewage sludge can be reached to 40 million tons. For developed countries such as the European Union, the annual production of sewage sludge has reached to about 10 million tons (dry matter).

The huge quantities of organic waste generated every year have endangered human life and ecological environments. Organic wastes such as livestock manure, sewage sludge, agriculture waste, and municipal solid waste are rich in organic matter and nutrients, which can be used as fertilizers or soil amendments [4]. However, most of the organic wastes also contain organic micropollutants, heavy metals, antibiotics, pathogenic microorganisms, and other hazardous elements [1]. Improper management of these organic wastes will aggravate environment pollution, resulting in potential public health risk. The disposal of organic waste has become an urgent issue all over the world. In the past decades, various practical waste treatment technologies had been put forward to deal with and recycle organic waste. Among all the management approaches, landfilling, incineration, anaerobic digestion, and composting are the most widely used methods throughout the world. Considering the decreasing land resource and the secondary pollution from landfill sites, the use of landfills for organic waste disposal is getting more inappropriate. Anaerobic digestion is a potential biological process for converting organic waste into energy. However, the high capital investment and maintenance charge, as well as the less efficient production of methane, limit the application of anaerobic digestion technology. Compared with the other approaches, composting technology is considered a more favorable option in many developing countries because of the lower investment and operation costs, scientific expertise requirement, and technical complexity [4,5]. Composting is effective in reducing the volume and weight of the organic waste as well as in producing a stable and nutrient-rich final product (compost). The utilization of composting technology to dispose of and recycle organic waste has been widely recognized [5,6]. This chapter will introduce the basic concept of composting process, composting optimization method, potential environmental problems of composting, and application of mature compost.

Sustainable Resource Recovery and Zero Waste Approaches. https://doi.org/10.1016/B978-0-444-64200-4.00009-8

TABLE 9.1
Pig Populations of Different Countries of the World [2,3].

Region	1986	1996	2006	2016
America	52,314,000	56,123,800	61,448,900	71,500,400
Germany	37,227,600	23,736,564	26,521,300	27,376,056
France	11,842,000	14,334,813	14,837,023	12,709,379
India	10,500,000	13,200,000	11,686,000	9,084,612
Netherlands	13,481,358	13,900,000	11,200,000	12,479,000
Philippines	7,274,830	9,025,950	13,046,680	12,199,442
Poland	18,948,528	17,963,912	18,880,558	10,865,318
Spain	15,780,000	18,731,000	26,218,706	29,231,595
Brazil	32,539,344	29,202,182	35,173,824	39,950,320
Denmark	9,104,000	10,841,553	13,361,099	12,383,000
Australia	2,553,494	2,526,412	2,733,000	2,294,245
Canada	9,967,000	11,588,000	15,110,000	12,770,461
Japan	11,061,000	9,900,000	9,620,000	9,313,000
China	338,442,079	452,309,000	440,549,544	456,773,355
Global	825,577,345	910,322,598	925,337,887	981,797,339

BASIC PRINCIPLES OF COMPOSTING

Composting is a natural organic decomposition process controlled by a number of environmental conditions including pH, moisture content, porosity for air passage, soil microorganisms, carbon-to-nitrogen ratio, etc., which determine the success of a composting process [5]. By effectively controlling these composting conditions, high-quality mature compost can be obtained. But if the conditions are not good, the compost quality will be affected and also may result in environmental pollution. However, the control of these influencing factors will also vary depending on the different composting methods.

Raw materials

Livestock manure, sewage sludge, crop straw, municipal solid waste, and other organic wastes can be the potential substances for composting, while the decomposition rate of the composting materials and the quality of the final product are related to the initial materials of composting. Consequently, it is essential to know the characteristics of feedstock and the controlling parameters of composting. For livestock manure, the nitrogen and phosphorus contents in intensive farms are as high as 3.3%–14% and 0.4%–5.8%, respectively. Some researchers showed that the content of nutrients such as nitrogen, phosphorus, and potassium in livestock manure was equivalent to 70.9%, 89.8%, and 25.1% of nutrients in the same period. Meanwhile, the content of protein in air-dried chicken manure, pig manure, cow dung, and sheep manure is 24%–30%, 3.5%–4.10%, 1.7%–2.3%, and 4.10%–4.70%, respectively. Normally, the composts produced from livestock manure are always rich in organic matter and nutrients. Sewage sludge, a by-product generated from the wastewater treatment plant, has a high amount of organic carbon (57%–66%) and nitrogen (3.7%–10.6%) required for plant growth. However, sewage sludge also contains high moisture content, which should be adjusted to appropriate levels before composting. Municipal solid waste includes approximately 30%–50% of biodegradable materials, which is primarily composed of food waste. According to the report of the Food and Agriculture Organization [2], the annual production of food waste could reach about 1.3 billion tons. Except the organic matter and nutrients, the food waste also contains a high amount of lipid and salts, which increase the difficulty of food waste management. Composting technology is a low-cost and highly efficient technology for the disposal of organic waste, but due to the different properties of feedstock, the initial parameters of composting should be adjusted to insure successful composting [5].

Temperature

Temperature of the composting mass is a critical factor to indicate the microbial activities of the composting system. Microbial respiration generates heat, resulting in an increase in temperature, as well as causing the transition of bacteria from one ecological regime to another, creating the optimum temperature for the breeding of microorganisms, which can quickly and efficiently decompose organic matters. The lethal temperature for pathogenic bacteria, parasitic eggs, and maggots is generally around 50–55°C. If the compost temperature lasts for 5–7 days in this temperature range, the hygienic index of compost can be guaranteed.

During the composting process, the temperature will go through four stages of temperature changes. After the start of composting, microorganisms at their mesophilic stage will quickly decompose the easily decomposable organic substrates, which causes the increase in temperature. As the temperature rises, the compost will then enter the high-temperature period, i.e., the thermophilic phase. The high-temperature period will not immediately decline but will stabilize for some days until the readily available nutrients in the composting mass are used up. Then the temperature will enter the cooling phase that lasts for a short period and the temperature of the compost will decline to a temperature slightly higher than the ambient temperature. Then the compost temperature will remain more or less the same at the maturation phase, which may last for a long period depending on the maturity of the composting mass (Fig. 9.1).

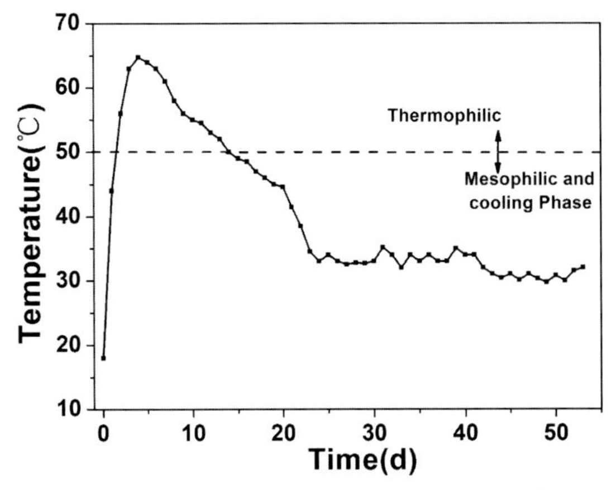

FIG. 9.1 Typical composting temperature profile.

Moisture content

An appropriate moisture content is important in creating the aerobic condition for the composting process. The amount of moisture in the pore space determines the amount of available oxygen for microbial degradation of organic matter. A moisture content of around 55%–65% is commonly used for most composting material. Lei et al. [7] reported that when the water content was 65%, it was most beneficial for cow manure composting, whereas the recommended moisture content for pig manure composting is 66%. Too high moisture content will create anaerobic condition resulting in insufficient oxygen for biodegradation, whereas too low moisture content means there is not enough moisture in the system to support microbial growth. Because of the different physical and chemical properties of compost materials, a good co-composting material may be needed to achieve the optimum moisture content for maximum composting efficiency. Food wastes, manure, and sewage sludge, which have high moisture contents of over 70%, may require the addition of dry organic material such as sawdust or dry hay to reduce the overall moisture content down to 55% to facilitate the degradation process. In fact, the addition of dry structural materials also creates better structure for the composting mass and this results in better air transfer efficiency for microbial degradation. Besides, different types of composting systems may result in different levels of moisture loss. Hence, it is important to monitor the composting process and check whether addition of water is necessary to create the optimum condition for microbial degradation process [8].

Oxygen supply

Oxygen is essential to support the microbial degradation process, and its availability depends on both the structure of the composting materials and the design of the composting reactors. Different composting systems have different optimal oxygen supply rates due to their material types, physicochemical properties, and reactor systems. Mathur [9] indicated that the oxygen supply during the high-temperature composting period should be 0.42–1.25 L/min. Lau et al. [10] considered that the optimal oxygen supply for cow manure composting was 0.04–0.08 L/min. Chowdhury et al. [11] showed that oxygen supply with 0.21 L/min was conducive to greenhouse gas (GHG) reduction in cow manure composting. However, some researchers suggested that oxygen supply during aerobic composting should between 12% and 20% [5,12].

C/N ratio

Carbon and nitrogen sources are indispensable nutrients for microbial growth and reproduction in soil. A balanced C/N ratio is a very important factor during the composting process. Extremely high and low C/N ratios can slow down the microbial activities resulting in a long composting process. In general, the initial C/N ratio of composting materials is preferably about 25–35 [5,13,14]. Composting raw materials with C/N ratio outside this recommended range should be adjusted with materials rich in either carbon, such as sawdust, wood chips, and cardboard, or nitrogen, such as soya bean residues and manure [6,8,14,15].

pH

Proper pH can provide comfortable physiologic conditions for the growth of microorganisms. The change in pH can affect the normal process of composting; therefore, effective control of pH is critical to composting. Generally, the pH value between 7.5 and 8.5 is the best for composting. Raw material such as food waste, which may have low initial pH, will have an inhibitory effect on the microbial degradation process creating the so-called acidothermophilic composting condition, resulting in a slow decomposition process [16,17]. Addition of alkaline materials such as lime, coal fly ash might be necessary to rectify the initial pH condition and facilitate the building up of the microbial population for later decomposition process [16–18].

Microorganisms Involved in Composting

There are many different types of microbes involved in the composting process. During the composting process, the population and quantity of composting microorganisms change with the gradual degradation of organic matter. With the composting process proceeding, the dominant microbial populations will significantly vary. The sources are mainly in two aspects. First, the organic waste itself contains a large number of microorganisms, including various bacteria, fungi, archaea, and protozoa [1,5]. These strains have strong decomposition ability for organic waste under certain conditions and have the characteristics of strong activity, rapid reproduction, and rapid decomposition of organic matter, which can accelerate the progress of composting and shorten the composting time [4].

On the other hand, in order to promote the composting process, improve the organic matter degradation and humification, many exogenous microbial agents such cellulose degrading bacteria, white rot fungi and Bacillus subtilis have been used during the last decades [7,9].

Bacteria, actinomycetes, fungi, and some protozoa play an important role at the various stages of composting process, which could break down most of the organic matter and produce heat [8]. Actinomycetes can decompose complex organic substances such as cellulose, lignin, and protein, which are the dominant bacteria in the decomposition of cellulose in the high-temperature stage. Fungi can use most of the lignin in the compost substrate and play an important role in the later stage of composting as moisture gradually decreases. In addition, higher animals such as rotifers, nematodes, worms, worms, beetles, and cockroaches not only partially absorb organic waste but also increase the surface area of organic waste and promote microbial growth through their movement in and swallowing of the compost [5,12].

POTENTIAL ENVIRONMENTAL RISK AND IMPROVEMENT OF CONVENTIONAL COMPOSTING

Greenhouse Gas Emissions and Mitigation

GHGs, including CH_4, N_2O, and CO_2, are produced by microorganisms consuming oxygen to decompose organic matters, which results in a partially anaerobic condition during the composting process [19,20]. According to the Fifth Assessment Report of the Intergovernmental Panel on Climate Change, CH_4 and N_2O have 25 and 296 times higher global warming potential than CO_2. Therefore their release into the atmosphere during composting is a major concern.

Mechanism of greenhouse gas emissions and environmental risks

Global climate change is a direct consequence of the greenhouse effect, which will lead to a wide influence and serious consequences on our environment [21,22]. As a major environmental problem faced by human beings, greenhouse effect has attracted the attention of governments and researchers around the globe to focus on its impact on the environment [23]. The influences of greenhouse effect on the ecological environment mainly include the following aspects: (1) It could cause sea level rise; it is estimated that the average global sea level will rise about 6 cm every 10 years and the sea level will rise by 20 cm until 2030 [24], which may result in disasters to the coastal region. (2) It has a negative effect on human health; global warming resulted in the rise of ozone concentration that in turn drastically triggers asthma and associated lung diseases in human beings. (3) It has a negative effect on the global climate change; as GHGs (carbon dioxide, methane, and water vapor) strengthen the atmospheric counter-radiation, the global temperature will generally rise, making it hard for people to settle down in the tropics. Furthermore, climate anomalies can lead to droughts, floods, lightning, hurricanes, storm surges, sandstorms, frosts, snowstorms, and a series of severe weather conditions that will eventually lead to geologic disasters such as mudslides and land subsidence.

Microorganisms can produce carbon dioxide by decomposing organic matter under aerobic conditions. The change of carbon dioxide emission not only reflects the rate of degradation of organic matter and the progress of aerobic composting but also helps evaluate the activity of aerobic microorganisms, and the degree of maturity of aerobic compost too. Most of the carbon dioxide is released during the thermophilic phase. Following the consumption of organic matter, the amount of carbon dioxide is gradually reduced. The rate of carbon dioxide production is related to the amount of oxygen, aeration, and the availability of organic carbon. In the early stages of composting, oxygen concentration is negatively correlated with carbon dioxide concentration, oxygen concentration is gradually reduced, and carbon dioxide concentration is continuously increased until there is a decrease in the availability of carbon source.

Theoretically, anaerobic digestion can be divided into three steps including the hydrolysis and fermentation stages, the hydrogen-producing acetogenesis stage, and the methane production stage. In the hydrolysis and fermentation stage, organic matter is broken down into fatty acids and alcohols by the action of fermenting bacteria. In the second stage, hydrogen-producing acetogens convert propionic acid, butyric acid, and other fatty acids and ethanol into acetic acid, CO_2, and H_2. Methane is formed by converting acetic acid through acetoclastic methanogens and by combining CO_2 and H_2 through hydrogentrophic methanogens. Methanogens are strictly anaerobic, so as long as methane is produced, it indicates poor ventilation and the presence of localized anaerobic hot spots in the composting mass. Methanogens are sensitive to temperature; when the temperature is too high, the production of methane is inhibited. The production of methane is closely related to water content, aeration, pH, redox environment, and the activity of methanogens. In aerobic composting, methane usually occurs in the early stages of composting. A large amount of methane will also be produced when the water content is high and the high-temperature period is long in the later stage. Andersen et al. [24] have reported that high water content has a great influence on the production of methane because high water content will reduce the porosity of the composting mass causing poor air diffusion, resulting in anaerobic conditions. This indicates that poor composting conditions with low thermophilic temperature and high moisture content will favor for anaerobic decomposition for methane generation.

Aerobic composting is an important source of nitrous oxide. Nitrification and denitrification are two main processes to produce N_2O. Under aerobic conditions, NH_4^+ is converted to NH_2OH by the action of bacteria and archaea, then NH_2OH is oxidized to NO_2^- by the action of hydroxylamine oxidoreductase, followed by the transformation of NO_2^- to NO_3^- through nitrite oxidoreductase [25]. The transformation of NH_4^+ to NO_3^- is considered as nitrification. N_2O is a by-product of nitrification, and nitrification is one of the main ways of nitrogen conversion in the composting process. Therefore N_2O emission is unavoidable in the composting process. NO_3^- is often converted into N_2 by denitrification or NO_3^- by alienation reduction through respiration of microorganisms under hypoxic conditions [26]. Consequently, N_2O is a by-product of denitrification, which is inhibited by the NO_3^- concentration and oxygen concentration. The composting system provides the conditions for the production of nitrous oxide because of its uneven oxygen distribution, which formed a diverse environment of internal anoxic and external aerobic conditions, and challenges the study of the source of nitrous oxide in aerobic compost. Therefore the main way of producing nitrous oxide in aerobic composting has been controversial because its production is influenced by many factors.

Mitigating the greenhouse gas emissions during composting

GHGs have gained a lot of attention because of their negative effects on the global climate change. With the purpose of reducing GHG emission during composting, many researches have been done in the recent years, developing different measures to control GHG emission. The major mitigation strategies adopted to reduce GHG emissions include addition of bulking agent, improving aeration systems, and using chemical agents or mature compost as cover materials. The addition of bulking agents is to improve the inside airflow of compost and decrease the emission of CH_4. The types of bulking agent used really depend on the local availability, especially agricultural organic wastes such as sawdust [27], wood chips [28], corn stover [19]. In the process of composting of kitchen waste, sawdust had the highest CH_4 and N_2O contents of >97.7% and 72.9%, respectively, followed by mushrooms (97% and 28.8%, respectively) and corn stalks (93% and 46.6%, respectively) [27], but it failed to show obvious reduction in NH_3 emission. In another compost fertilizer study, the combined use of precomposting and vermicomposting significantly reduced CH_4 and N_2O emissions by 84.2% and 80.9%, respectively [19].

Jiang et al. [29] performed a composting trial to show the effect of different aeration methods (continuous and intermittent) and aeration rates (0, 0.18, 0.36, and 0.54 L/[kg dry matter/min]) on GHG emissions during pig manure composting. The results indicated that GHG emissions could be significantly affected by the aeration method and aeration rate. Higher aeration rates could reduce CH_4 emission, but obviously increased NH_3 and N_2O volatilization. When compared with continuous aeration, intermittent aeration reduced NH_3 and CH_4 emissions, but increased N_2O production. Mineral additives, such as medical stone (Maifan stone) [1], bentonite, biochar [30], and phosphorus gypsum, could reduce GHG emission. Phosphorus gypsum, as the main by-product of the production of phosphoric acid, can effectively reduce CH_4 during composting of fresh pig manure [31] and kitchen waste [32] by increasing SO_4^{2-} concentration. Through the formation of struvite ($MgNH_4PO_4 \cdot 6H_2O$), magnesium and phosphate decreased NH_3 emission. Besides, adding a nitrification inhibitor dicyandiamide ($C_2H_4N_4$) can reduce the emission of N_2O [31]. Luo et al. [33] reported that mixing mature compost to the composting mass would result in higher level of reduction in GHG emission.

Nitrogen Loss and Its Controlling Technology

High-temperature composting is an effective means to reduce the volume of organic waste, to produce ecological nonhazardous composting product, and to facilitate resource utilization of livestock and poultry manure. However, in the high-temperature composting process of livestock and poultry manure, the nitrogen loss mainly caused by ammonia volatilization is serious, which not only reduces nitrogen in the final compost but also pollutes the atmosphere, endangers human and animal health, corrodes equipment, and brings acid rain damage and causes water eutrophication [1,16]. In view of this, researches have been carried out around the world to develop an appropriate technology not only to improve the composting process by reducing the loss of nitrogen but also to achieve good composting results for livestock waste.

Causes and control of ammonia release

The change in nitrogen forms during composting is mainly the result of microbial action. Nitrogen exists in different forms during the aerobic composting process, including organic nitrogen, ammonia nitrogen, nitrate nitrogen, nitrite nitrogen, ammonia gas, nitrous oxide, and nitrogen gas. There are four main ways of nitrogen conversion: ammonification, nitrification, denitrification, and bioabsorption. The main nitrogen conversion and loss pathways in composting process are shown in Fig. 9.2 [15].

Microorganisms in the composting mass convert organic nitrogen in the compost into ammonia gas through ammonification, and the ammonia gas then dissolves in the pore water to form ammonium nitrogen. Owing to the influence of temperature and pH in the composting mass, the change of ammonium nitrogen may follow four different pathways [15,16]. First, it

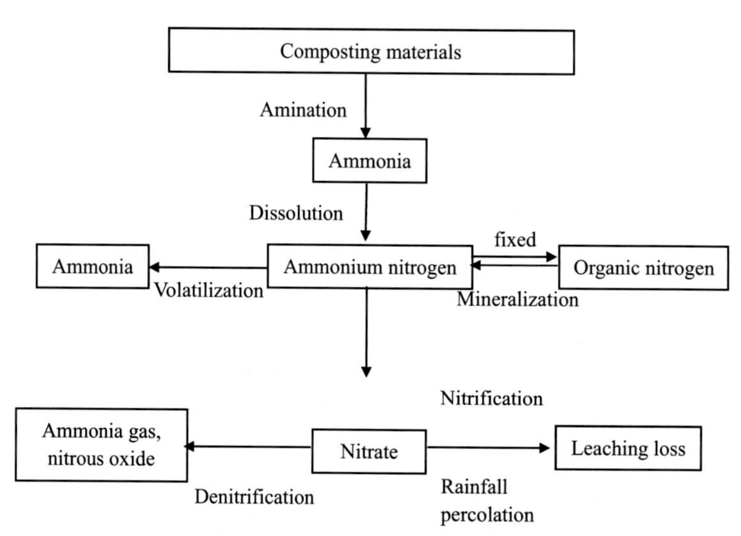

FIG. 9.2 Nitrogen transformation during aerobic composting [15].

is discharged into the atmosphere in the form of ammonia by volatilization; second, organic nitrogen can be produced by the immobilization of microorganisms; third, it is converted into nitrogen and nitrous oxide by denitrification; and fourth, through rainfall, it may start to leach out. Inorganic nitrogen (ammonium nitrogen and nitrate nitrogen) and organic nitrogen are the forms of nitrogen in the compost, and organic nitrogen accounts for 90% of nitrogen form, which is the main form of composting nitrogen. Nitrogen-containing gases (ammonia, nitrogen, and nitrous oxide) are the major form of nitrogen loss in the compost. Studies have shown that, in the process of composting, ammonia gas that gets volatilized to the atmosphere accounts for 9.6%–46% of the total organic nitrogen [27]. The decomposition of nitrogenous organic matter under ammonification is the main source of ammonium nitrogen. The organic matter in the early stage of the composting process is decomposed into ammonium nitrogen and reaches the maximum value. Then ammonium nitrogen decreases and stabilizes in the late composting stage. Nitrate nitrogen is formed by nitrification of ammonium nitrogen, and the nitrate content increases in the later stage of composting. The total amount of organic nitrogen first decreased and then rose and stabilized. Some studies have shown that at the end of composting, the nitrate nitrogen content increased significantly, whereas the content of ammonium nitrogen, organic nitrogen, and total nitrogen decreased to different extents, and nitrogen loss reached 4.3% [17,18,34].

The main factors that affect the nitrogen loss in compost are as follows:

1. Properties of materials: As mentioned in previous section, the initial parameters of the compost materials could affect the efficiency of the composting process. The suitable initial C/N ratio, moisture content, aeration rate, pH, and the particle size of the initial composting materials could improve the composting process and reduce nitrogen loss. On the other hand, improper control of composting parameters would prolong the composting period and decrease the quality of the final product.
2. Environmental parameters: The environmental parameters such as temperature, pH, and ventilation during composting greatly affect the growth and metabolism of microorganisms, which in turn affects the loss of nitrogen.

Temperature affects the growth and reproduction of microorganisms and the duration of composting. High temperature is also an important indicator for evaluating the maturity of compost. In the early stage of composting, the temperature of the reactor is not high and the ammonia volatilization is less. When the temperature rises and continues to a high level, the amount of ammonia volatilization rises sharply. Therefore controlling the high-temperature period is the way to reduce ammonia volatilization, which is beneficial to reduce the loss of nitrogen in the heap. At the same time, the high-temperature period of the compost mass will inhibit the progress of nitrification. pH can also affect the growth of microorganisms, and there are many different varieties of microorganisms in the composting process and most of the microorganisms prefer a weakly acidic or neutral environment. In the initial stage of composting, the organic matter is degraded to produce organic acid, thus decreasing the pH value and creating difficulties in the volatilization of ammonia gas. But when the temperature begins to rise, the organic acid is gradually decomposed, thus raising the pH, which in turn stimulates the volatilization of ammonia. Thus the higher pH values can reduce microbial growth and metabolism and affect microbial assimilation of nitrogen.

Ventilation is one of the important factors that affect composting. Ventilation can supply oxygen, remove moisture, generate carbon dioxide, and regulate temperature, but it will cause loss of ammonia volatilization. Reducing ventilation during the high-temperature period can reduce the volatilization of ammonia. Intermittent ventilation can significantly reduce nitrogen loss in compost than continuous ventilation [11]. Huang et al. [15] found that a turning frequency of every 2 days shortened the time required for maturity better than that with daily turning for aeration of the composting pile.

Controlling nitrogen loss during composting

In general, the nitrogen loss during composting could be reduced by adjusting the C/N ratio, ventilation, and moisture content and by adding additives.

C/N ratio. Organic materials such as livestock manure and sewage sludge having unfavorable low C/N ratio for composting can be ameliorated with materials with high carbon content, such as sawdust, straw, wheat bran, and rice hull, to increase the C/N ratio to 25–30. The immobilization of nitrogen by microorganisms can be promoted to reduce the loss of ammonia through volatilization. At the later stage of the composting process, after ammonification at the initial stage of composting, the C/N ratio of the material can be controlled to a low level to reduce both the emission of nitrous oxide to a certain extent and the loss of nitrogen [14,16,34].

Ventilation. Ventilation is one of the effective ways to provide oxygen content to interfere with the loss of nitrogen. Under the condition of ensuring sufficient oxygen concentration in the reactor, increasing the frequency of turning over could reduce the emission of nitrous oxide. The increase in the frequency of turning can destroy the anaerobic zone by providing sufficient oxygen concentration, which can possibly slow down the denitrification process and thus reduce nitrogen loss [29]. Reducing the volume and increasing the porosity of the composting mass could also promote nitrogen retention to some extent.

Additives. Composting additives are used to accelerate the composting process and improve the quality of compost products, as well as reduce the nitrogen losses. Microorganisms and organic or inorganic substances were added to enhance the composting process.

Chemical additives. The chemical additives added during the composting process are mainly acidic compounds such as superphosphate, alum, copper sulfate, magnesium chloride, bamboo vinegar, phosphoric acid, and sulfuric acid. Table 9.2 summarizes the efficiency of different chemical additives for ammonia reduction in composting processes.

Although chemical additives can effectively reduce the release of ammonia during composting, it can lead to higher levels of ammonium concentration increasing the conductivity of the compost, and also the availability of heavy metals. At the same time, the addition of sulfuric acid and phosphoric acid is corrosive to the compost materials, which could reduce the porosity of the materials and lead to the formation of anaerobic zones.

Mineral additives. The nitrogen retention of mineral additives in compost mainly depends on the adsorption performance and ion exchange capacity of the additives. Mineral additives such as zeolite, bentonite, biochar, vermiculite, and medical stone, such as Maifan stone for medicinal use, are porous and/or rich in humus and high specific surface area. Table 9.3 shows the adsorption effect on ammonium nitrogen and ammonia produced during composting process, thereby greatly reducing the loss of nitrogen.

Mineral additives are relatively safe to reduce nitrogen loss in the compost, but they do have the disadvantages in the requirement of large dosage levels, and their usage along with some microbial agents and chemical additives will achieve better results [35,36].

TABLE 9.2
The Effect of Different Chemical Additives on Nitrogen Conservation During Livestock Manure.

Composting Materials	Chemical Additives	Effect on Nitrogen Conservation	References
Poultry manure	Alum	Reduces 26% ammonia volatilization	[65]
Turkey manure	Alum	Addition of 7% alum helps reduce ammonia volatilization by 76%	[66]
Dairy manure	Alum	2.5% Alum addition reduces ammonia volatilization by 60%	[67]
Pig manure	Bamboo vinegar	Reduces 26% nitrogen loss	[68]
Pig manure	Superphosphate mixed dicyandiamide	Reduces 48.5%–52.8% ammonia volatilization	[69]
Chicken manure	Superphosphate	Addition of 10% calcium superphosphate can reduce 74% ammonia volatilization	[68]
Chicken manure	Copper sulfate, manganese sulfate	Addition of 0.5% copper sulfate and 2% manganese sulfate can reduce 39.32% nitrogen loss	[68]
Pig manure	Struvite crystallization	Reduces 45%–53% ammonia volatilization	[35]
Pig manure	Magnesium chloride	Reduces 58% ammonia volatilization	[68]

TABLE 9.3
Effects of Mineral Additives on Nitrogen Conservation During Organic Waste Composting.

Composting Materials	Additives	Effect on Nitrogen Conservation	References
Dairy manure	Zeolite	Addition of 6.25% zeolite reduces 50% ammonia release	[68]
Chicken manure	Zeolite	Addition of 15% zeolite reduces 49.13% ammonia release	[68]
Food waste	Activated carbon, zeolite	2% Activated carbon and zeolite addition reduces release of ammonia by 79%	[16,70]
Pig manure	Medical stone	Addition of 2.5%–10% medical stone reduces 27.9%–48.8% ammonia release	[1]
Chicken manure	Pumice stone	Reduces 63.89% ammonia release	[36]
Sewage sludge	Biochar and zeolite	Adding biochar and zeolite reduces the nitrogen loss by 58.03%–63.08%	[71]

Heavy Metal Pollution and Immobilization

Organic wastes such as livestock and poultry manure, as well as sewage sludge derived from agricultural and wastewater treatment process, may contain excessive levels of heavy metals (Cu, Zn, Pb, Cd, As, Hg, and Cr) [1,37]. After composting of these organic wastes, the total concentrations of the heavy metals in the compost will significantly increase depending on the characters of the composting material and condition. The bioavailability of heavy metals in the compost could be decreased or increased. After the application of the compost polluted with heavy metals to soil, plants would uptake an overdose of toxic metals that are enriched in the plant tissue and eventually flow through the food chain, affecting the entire ecosystem [38], as heavy metals cannot be removed during the microbial decomposition process. The control of the source of composting raw materials is the major control of the final heavy metal content in the final compost. It is recommended that before deciding the use of composting for treatment of organic wastes, the raw material must be collected for heavy metal determination to decide whether it is feasible for producing acceptable compost quality. If there are excessive amounts of heavy metals, composting is not advised to be performed, except that sufficient dilution can be achieved through co-composting with other organic materials to bring the final heavy metal concentrations fall within the allowable levels.

Heavy metal content in organic waste

The source of heavy metals in compost is closely associated with the types of raw materials, which are mostly attributed to the high content of heavy metals in livestock manure, sludge, and municipal solid waste. The main source of heavy metals in livestock manure is the intake of contaminated soil or animal feed. Studies had shown that the major amount of lead in cow manure was from the ingestion of lead-polluted soil by dairy cows. However, the metals currently added to feed were the main source of heavy metals in livestock manure [39]. In order to promote the growth of pigs, heavy metals were often added to the feed, but most of the heavy metals cannot be absorbed by animals, which were consequently excreted with pig manure and urine, resulting in a higher concentration of heavy metals in the animal manure. The common metals found in pig manure are Cu and Zn [40].

For achieving maximum economic benefit, the concentration of Cu in the feed can be as high as 125–250 mg/kg, and even 350–450 mg/kg; the addition of Zn to animal feed can be 2000–3000 mg/kg. The absorption rate of these trace elements in the digestive tract of animals is very low. The absorption rate of Cu is reported to be 5%–20% in piglets and 5%–10% in adult pigs. The absorption rate of Zn is reported to be about 5%–10% for piglets. Consequently, most of the Cu and Zn in the feeds will be excreted through the excreta, which is the main cause for the high content of Cu, Zn, and other metals in manure [41]. The common heavy metals present in animal manures are listed in Table 9.4. As reported by Wang et al. [42], Cu and Zn concentrations were higher in pig manure than in other animal manures, and these two elements were also the common metals found in animal manures. Composting of these manures is the main source for heavy

TABLE 9.4
Heavy Metal Contents in Animal Manures [42].

Manure Type	Cu (mg/Kg)	Zn (mg/Kg)	Hg (μg/Kg)	As (mg/Kg)	Pb (mg/Kg)	Cd (mg/Kg)	Cr (mg/Kg)
Pig	299.6 ± 382.3	599.1 ± 1194.4	28.9 ± 66.2	5.2 ± 13.8	7.3 ± 19.8	1.3 ± 1.87	31.0 ± 108.5
Dairy	56.1 ± 51.7	212.6 ± 103.3	29.1 ± 20.1	1.5 ± 1.8	8.1 ± 3.0	0.92 ± 0.86	40.1 ± 109.6
Poultry	141.7 ± 265.1	432.3 ± 287.1	17.3 ± 19.3	4.8 ± 8.7	6.1 ± 6.8	1.48 ± 1.32	67.99 ± 216.6
Sow	136.4 ± 193.3	483.5 ± 522.7	17.5 ± 18.3	5.1 ± 13.6	5.2 ± 4.3	1.11 ± 1.32	29.4 ± 110.1
Broiler	144.3 ± 208.5	351.6 ± 198.2	14.8 ± 17.6	7.9 ± 10.8	6.6 ± 10.5	1.28 ± 1.52	24.9 ± 34.8

TABLE 9.5
Heavy Metal Limits in Compost Issued by Different Countries (mg/Kg) [72,73].

Heavy Metal	Austria	Belgium	Switzerland	Denmark	France	Germany	Italy	Spain	Canada (Class A)
Cu	400	100	150			100	300	1750	100
Zn	1000	1000	500			400	500	4000	500
As				25			10		13
Cr	150	150	150			100	100	750	210
Ni	100	50	50	45	200	50	50	400	62
Pb	500	600	150	120	800	150	140	1200	150
Hg	4	5	3	1.2	8	1.0	1.5	25	0.8

metals in compost. The limits of heavy metal concentrations in compost in different countries are listed in Table 9.5. Spain allowed the highest metal concentrations in compost and more stringent standards are executed in Canada and Switzerland.

Sludge usually contains numerous kinds of heavy metals such as Cu, Zn, Cd, Ni, Cr, Hg, As, and Pb, which are concentrated in sewage sludge during the sewage treatment process by various methods including microbial absorption and adsorption. Sludge includes four components, such as particulate, biofloc, colloid, and water-soluble factions [43]. Table 9.6 shows the heavy metal contents of sludge coming from various countries, which illustrates the differences among these countries and regions, and also reflects the higher concentrations of various heavy metals in the sludge. An investigation reported that the average contents of Zn, Mn, Cu, Ni, Pb, Cd, and Cr in sludge from a sewage treatment plant in Bangkok, Thailand, were 2061, 471, 218, 25.0, 12.2, 2.1, and 19.6 mg/kg, respectively [44]. A possible reason for the higher concentration of Zn in sludge might be the use of galvanized pipes leading to higher Zn content in the sewage. Heavy metals could be transferred from the sludge to the farmland through composting, which increases the heavy metal content in plants and poses a potential pollution risk to the groundwater and surface water, and consequently limits the agricultural utilization of sludge [45].

Current approaches for heavy metal immobilization during composting

Chemical additives. Chemical additives primarily reduce heavy metals' bioavailability by reducing their active forms through complexation, precipitation, and ion exchange [37]. For example, the addition of alkaline materials such as calcium magnesium phosphate salts, phosphate rock powder, and fly ash to convert heavy metal ions into less available chemical forms through precipitation by increasing the pH of the composting mass. The addition of chemical additives during

TABLE 9.6
Heavy Metal Contents in Sewage Sludge (Mean, mg/Kg).

Countries/Regions	Cu	Zn	As	Cd	Pb	Cr	Ni	Hg	References
Spain	109	259	<0.2	<0.2	15.5	17.1	22.3	0.68	[74]
Malaysia	80	200	—	8	0.5	—	—	—	[75]
Iran	330	1908		4.1	169	213	110		[76]
Portugal	141	757		1.0	<5.6	<5.6	22.6	<1.3	[77]
Brazil	202	690		1.6	26.3	260	54.6		[78]
Egypt	538	1204		4.0	750		81.0		[79]
Japan	415	751		73.0	122	150	639		[80]
France	149	548		0.6	19.7	27.6	26.4		[81]
Portland	216	1477		3.5	167.8	44.5	23.5	0.8	[82]
Turkey	198	860		0.55	—	30.6	38.5		[83]
Shenyang	170	290		5.0	255				[84]
Xiamen	157	397	1.05	2.75	22.2	42.9	83.8		[85]
Yangzhou	252	1178		2.41	137	1312	79.7		[86]
Beijing	7.73	79.1		242	9.11				[87]
Guangzhou	4567	785		5.99	81.2	121	148		[88]
Zhaoqing	93	509		—	17.4	15.5	51		[88]
Xi'an	217	1101	17.0	10.5	166	772	46.5	3.42	[89]
Mean values in China	499	2088	25.2	3.88	112	259	167	3.18	[90]

composting process can effectively decrease the bioavailability of heavy metals, but it is easy to cause secondary pollution to the environment.

Microbiologic inoculum. The addition of a microbiologic inoculum could remove the heavy metals by adsorption of heavy metals through microorganisms and their derivatives. The microbes that can absorb heavy metals mainly include bacteria, fungi, and algae. *Bacillus* has a strong ability to adsorb metals and Pb^{2+} is adsorbed by *Bacillus licheniformis*, and the adsorption capacity can reach 224.8 mg/g in 45 min [46,47]. The adsorption capacity is many folds higher in the cell wall surface of algae, which is the main reason for their large specific surface area, providing a large number of functional groups bonded to metal ions, such as carboxyl group, hydroxyl group, amide group, amino group, and aldehyde group. These functional groups react with metal ions for a very short reaction time without any metabolic processes and energy supply. There are many advantages in the biological adsorption method, such as small investment, high efficiency, and no secondary pollution, which has good technical advantages and economic benefits. However, the current research on biosorption of heavy metals is still in the laboratory stage, which needs more research on the adsorption mechanism and the simultaneous adsorption process of multicomponent heavy metal ions [48].

Mineral additives. Mineral additives contain silicate materials that exhibit a high adsorption capacity for heavy metals to reduce their bioavailability, which is attributed to their large electrostatic force, ion exchange performance, and large cavity surface. For example, bentonite and zeolite have strong adsorption and ion exchange properties owing to their large surface area and surface energy. The physical adsorption of heavy metals during the composting of livestock and poultry has the advantages of easy acquisition of adsorbents, simple principle, and operation. But there is also the disadvantage that the adsorbent might dilute the nutrient concentration. Therefore it is necessary to further broaden the screening range of high-efficiency passivation agents.

Elimination ofAntibiotics and Antibiotic Resistance Genes

For improving economic efficiency, accelerating the growth of animals, and enhancing disease resistance, some kinds of feeds containing considerable concentrations of antibiotics and heavy metals are overused in farms. In China the amount of antibiotics used annually for animal growth and disease prevention and treatment is about 8000 tons. The main routes of antibiotics entering animals are through feeds and injection. The excessive intake of antibiotic additives (such as quinolones, peptides, tetracyclines, macrolides, sulfonamides, aminoglycosides) cannot be fully absorbed and utilized by animals, and most of them are discharged by livestock and poultry in the form of original drugs. These toxic and harmful pollutants, such as antibiotics, not only seriously threaten the safe consumption of livestock products but also cause serious potential damage to soil, waterbodies, animals, and plants. The residual antibiotics in compost and other organic wastes have the environmental risk of inducing antibiotic resistance genes (ARGs), which will lead to the emergence of drug-resistant "super bacteria" through horizontal transfer mechanism, causing serious environmental pollution and ecotoxicity. Therefore an effective treatment technology such as composting should be employed to control and reduce their environmental risk before discharged into the environment.

Aerobic compost has the efficiency to remove antibiotics, and the final compost products are proved to be effective to remove most kinds of residual antibiotics. But the removal rate of ARGs is different, which is mainly attributed to the effect of different composting substances and microbial community. Selvam et al. [49,50] reported that the levels of chlortetracycline, sulfadiazine, and ciprofloxacin were significantly reduced during pig manure composting, and chlortetracycline and sulfadiazine were significantly degraded early at 21 and 3 days, respectively. Ho et al. [51] had determined the degradation characteristics of nine antibiotics and hormones in broiler manure composting process by liquid chromatography-tandem mass spectrometry. Veterinary antibiotics and hormones are significantly degraded after 40 days of composting, and the degradation rate of antibiotics is above 99% and the half-life is about 1.3–3.8 days. The degradation of antibiotics is related to the basic physical and chemical properties of the compost, which indicates the high efficiency of composting on removal of antibiotics. Composting is also proved to be effective to recycle the nutrients in antibiotic fermentation waste and to remove the residual antibiotics. Zhang et al. [52] reported that the removal rate of penicillin was above 99% during composting penicillin fermentation fungi residue composting, and the high content of antibiotics did not affect the composting process and even promoted the microbial activity because of the nutrient elements in the antibiotic fermentation waste.

The fate of ARGs during composting is shown in Table 9.7. Selvam et al. [53] reported that the tetracycline resistance genes *tetQ*, *tetW*, *tetC*, *tetG*, *tetZ*, and *tetY*; the sulfonamide resistance genes *sul1*, *sul2*, *dfrA1*, and *dfrA7*; and the fluoroquinolone resistance gene *gyrA* can be effectively degraded during composting. Zhang et al. [54] stated that the natural zeolite and nitrification inhibitor were effective to reduce the total abundance of ARGs in compost products, which may be due to the porous structure and reduced selective pressure of heavy metals. Additionally, biochar

TABLE 9.7
Fate of Antibiotic Resistance Genes During Composting.

Genes	Fate of Genes	Substance	References
sul1, *sul2*, *dfrA1*, *dfrA7*, *tetQ*, *tetW*, *tetC*, *tetG*, *tetZ*, *gyrA*, *parC*	Below the detection line after 42 days	Pig manure	[53]
tetA, *tetC* *tetG*	Decreased by 4 logs Unchanged	Pig manure	[91]
ermA, *ermB*, *ermC*, *ermF*, *ermT*, *ermX*	Decreased by 2.3–7.3 logs	Pig manure	[92]
bla_{CTX-M}, bla_{TEM}, *ermB*, *ereA*, *tetW* *ermF*, *sul1*, *sul2*, *tetG*, *tetX*, *mefA*、*aac(6′)-Ib-cr*	Decreased by 0.3–2.0 logs Increased by 0.3–1.3 logs	Sewage sludge	[54]

prepared from mushroom residues can also effectively reduce the abundance of ARGs during the composting process and the final products.

COMPOST APPLICATION AND ITS BENEFITS

Agriculture and Economic Benefits of Compost

The application of chemical fertilizers can increase the crop yields quickly, but they also could cause soil hardening and decrease soil organic matter and pH after a long period of application, resulting in loss of soil productivity [55]. However, most proportion of the chemical fertilizers will be ran off or leached due to rain and heavy irrigation, consequently leading to environmental pollution and lower fertilizer effect. Compost is produced from organic waste, which not only contains organic matter but also is rich in micro- and macronutrients. The utilization of compost as a soil fertilizer or amendment could restore the soil quality and improve soil structure and fertility, which not only serves an important role in agricultural production but also is of great significance for improving the ecological environment [56]. Nowadays, using compost as a substitute to chemical fertilizer has become a global consensus. The application of compost could promote soil productivity and improve the crop quantity and quality, as well as increase the income of the farmers.

Effect of Compost on Soil Nutrient Availability and Microbial Activity

In general, compost has a positive influence on the nutrients and physicochemical properties of the soil, which can satisfy the nutrient demand for plant growth. When compared with the chemical fertilizer, compost has more benefits, such as nutrient comprehensive and longer fertilizer efficiency, as well as lower cost. The addition of compost could not only increase nitrogen, potassium, and phosphorus contents of soil but also incorporate some other micro- and macronutrients such as calcium, magnesium, sulfur, copper, and zinc. Many researches had indicated that the addition of compost could increase the soil nutrients and improve plant growth [57,58]. Calleja-Cervantes et al. [59] reported that the long-term application of sheep manure compost could obviously increase the total nitrogen (0.1%), available phosphorus (80.7 mg/kg), and potassium (473.8 mg/kg) contents of soil. Agegnehu et al. [60] indicated that the addition of compost could increase 24% yield of peanut. Apart from the nutrients, the character of compost is the high organic matter content and low bulk density. Soil organic carbon can be replenished by the application of compost during agricultural fertilizer management. Compost also consists of humic acid, which could increase the buffering capacity of soil and thus improve the soil quality and retain the nutrients of the soil [61]. Generally, the microbiologic properties of soil could reflect the quality of soil. Incorporating the nutrients and organic matter of compost into the soil could significantly influence the microbial activity of soil. Adding the municipal solid waste compost increased the microbial biomass, carbon, nitrogen, and sulfur content and promoted the soil respiration [61]. Abujabhah et al. [62] found that the application of compost could obviously increase the microbial abundance of soil. Increasing the total abundance of microorganisms in soil, especially some beneficial microorganisms, such as nitrogen-fixing bacteria, ammoniated bacteria, cellulose-decomposing bacteria, nitrifying bacteria, improves the metabolic intensity of the soil.

Additionally, microorganisms could influence the nutrient availability of soil by releasing enzymes such as urease, alkaline phosphatase, β-glucosidase, and protease [61]. Most of the soil processes rely on the soil enzyme activity, such as nutrient release to plants and organic matter degradation [60,62]. Compost amendment can increase the soil microbial community and promote the soil enzyme activity, and consequently improve the available nutrient content of soil [62,63]. Perucci [64] showed that the addition of municipal solid waste compost could increase the soil enzyme activities of urease, protease, alkaline phosphomonoesterase, and deaminase. The increase in soil enzyme activity may stimulate the transformation of the organic fraction of nutrients (N, P, and C) to inorganic and available forms. There are various other components in compost, such as cellulose, hemicellulose, various sugars, proteins, amino acids, amides, phospholipids, and other soluble organic compounds. These compounds, as mentioned earlier, are effectively decomposed by various microorganisms and enzymatic reactions to produce simple compounds that can be directly absorbed and utilized by crops [59].

Improvement of Crop Yield and Quality

The excessive application of chemical fertilizers and insufficient application of organic fertilizers in agricultural production processes are of great concerns for our modern agriculture. Increasing the utilization of chemical fertilizers alone cannot achieve the goal of increasing crop yields, which unfortunately increases the agricultural input and the loss of available nutrients and soil organic carbon. Additionally, the application of chemical fertilizers will also bring environmental risks such as soil compaction and soil salinization, thereby reducing the quality and yield of agricultural production. Compost and its further processing products such as organic fertilizer, bioorganic fertilizer, and organic-inorganic compound fertilizer are gradually recognized and valued by the sector. Clean compost products have higher effective nutrient content, long-lasting nutrient supply, and the positive effect on disease resistance of soil; the microorganisms from compost could also help in the nutrient uptake by plants. Compost can increase root biomass, growth range, and root-crown ratio; widen the horizontal distribution of roots; and increase the proportion of roots in deep soil layers [18,30].

Compost Standards

Appropriate application of compost not only reduces the environmental problems caused by long-term application of chemical fertilizers but also improves the economic benefits of crops. The application of compost can increase the soil organic matter and improve the quantity and quality of the crops. However, due to the different properties of the composting feedstocks, the quality of the compost should be critically assessed before applying to the soil. In order to assess the quality of the compost and ensure rational application of compost, many countries have set up compost standards. As shown in Table 9.8, the compost standards in different countries and regions are different because of the different local requirements and constraints such as soil properties, farmers' practices, and the availability of farming inputs. Prior to the application nutrient contents of soil as well as the nutrient demand of the crops. The application of excessive compost will not only inhibit plant growth but also result in economic losses. The heavy metal content in compost is also a significant factor limiting the application rate. To avoid excessive input of heavy metals derived from the addition of compost, the user needs to carry out a detailed loading calculation to estimate the long-term impact due to the application of compost.

CONCLUSIONS AND PERSPECTIVES

Composting is a promising environment-friendly technology for organic wastes that realizes the residual resource in waste and reduces the environmental risk of these organic wastes. Composting can achieve further economic value about the organic waste, replace part of chemical fertilizers, and improve agricultural production and soil structure and fertility, which can also reflect the concept of green sustainable development. However, there are some environmental problems that are difficult to avoid during the composting process, such as GHG emissions, ammonia volatilization, and high bioavailability of heavy metals, antibiotics, and ARGs, which limits the development of the

TABLE 9.8
Compost Standard of Different Countries and Regions.

	COMPOST STANDARD			
Parameters	China [93]	Europe [94]	Japan [95] (Manure Compost)	Hong Kong [96] (General Use)
pH	5.5–8.5	—	7–8.5	5.5–8.5
Moisture (%)	≤30	<45	≤70	25–35
Organic matter	≥45	>15	≥60	>20
Total nutrients (N + P + K) (%)	≥5.0	—	≥3	≥4
Cu (mg/kg)	—	≤100	—	≤700
Zn (mg/kg)	—	≤300	—	≤1300
As (mg/kg)	≤15	≤10	≤50	≤13
Hg (mg/kg)	≤2	≤1	≤2	≤1
Pb (mg/kg)	≤50	≤100	≤100	≤150
Cd (mg/kg)	≤3	≤1	≤5	≤3
Cr (mg/kg)	≤150	≤100	≤500	≤210

composting industry and the rapid elimination of organic solid waste. Therefore clean compost should be vigorously produced with the mitigation technology followed during the composting process. In addition, the quality inspection of the final products and scientific fertilization management are also important in organic fertilizer production and application. The quality control of the collected raw materials should be paid attention to screen out the heavily polluted materials, such as those polluted with higher residual antibiotics and heavy metals. The technologic development for mitigating GHG emission, heavy metal immobility, nitrogen conservation, and the rapid degradation of organic matter should be further studied. Furthermore, the mechanism of reducing the availability of ARGs as well as their internal relationship with the microbial community and their physicochemical properties also needs further exploration.

REFERENCES

[1] Wang Q, Wang Z, Awasthi MK, Jiang YH, Li RH, Ren XN, Zhao JC, Shen F, Wang MJ, Zhang ZQ. Evaluation of medical stone amendment for the reduction of nitrogen loss and bioavailability of heavy metals during pig manure composting. Bioresource Technology 2016;220: 297–304.

[2] FAO. FAO statistical yearbook 2018. Latin America and the Caribbean food and agriculture. Rome: FAO; 2018.

[3] NBSC (National Bureau of Statistics of China). China statistical yearbook. Beijing: China Statistics Press; 2018.

[4] Onwosi CO, Igbokwe VC, Odimba JN, Eke IE, Nwankwoala MO, Iroh IN, Ezeogu LI. Composting technology in waste stabilization: on the methods, challenges and future prospects. Journal of Environmental Management 2017;190:140–57.

[5] Bernal MP, Alburquerque JA, Moral R. Composting of animal manures and chemical criteria for compost maturity assessment. A review. Bioresource Technology 2009;100: 5444–53.

[6] Qian X, Shen G, Wang Z, Guo C, Liu Y, Lei Z, Zhang Z. Co-composting of livestock manure with rice straw, Characterization and establishment of maturity evaluation system. Wastes Management 2014;34:530–5.

[7] Lei DP, Huang WY, Wang XH. Effect of moisture content of substance on fermentation and heat production of cattle manure in aerobic composting. Journal of Ecology and Rural Environment 2011;27:54–7.

[8] Gajalakshmi S, Abbasi SA. Solid waste management by composting: state of the art. Critical Reviews in Environmental Science and Technology 2008;38:311–400.

[9] Mathur SP. Composting process. In: Bioconversion of waste material to industrial products. New York: Elsevier, Applied Science; 1991.

[10] Lau AK, Lo KV, Liao PH, Yu JC. Aeration experiments for swine manure composting. Bioresource Technology 1992;41:145–52.

[11] Chowdhury MA, De NA, Jensen LS. Potential of aeration flow rate and bio-char addition to reduce greenhouse gas and ammonia emissions during manure composting. Chemosphere 2014;97:16–25.

[12] Harper ER, Miller F,C, Macauley BJ. Physical management and interpretation of environmentally controlled composting ecosystem. Australian Journal of Experimental Agriculture 1992;32:657–67.

[13] Huang GF, Wu QT, Wong JWC, Nagar BB. Transformation of organic matter during co-composting of pig manure with sawdust. Bioresource Technology 2006;97:1834–42.

[14] Huang GF, Wong JWC, Wu QT, Nagar BB. Effect of C/N on composting of pig manure with sawdust. Wastes Management 2004;24:805–13.

[15] Huang GF, Wu QT, Li FB, Wong JWC. Nitrogen transformations during pig manure composting. Jounral of Environmental Science 2001;13(4):401–5.

[16] Chan MT, Selvam A, Wong JWC. Reducing nitrogen loss and salinity during 'struvite" food waste composting by zeolite amendment. Bioresource Technology 2016 [Jan]; 200:838–44.

[17] Wang X, Selvam A, Wong JWC. Influence of lime on struvite formation and nitrogen conservation during food waste composting. Bioresource Technology 2016;217: 227–32.

[18] Zhou Y, Selvam A, Wong JWC. Effect of Chinese medicinal herbal residues on microbial community succession and anti-pathogenic properties during co-composting with food waste. Bioresource Technology 2016 [Oct]; 217:190–9.

[19] Wang J, Hu Z, Xu X, Jiang X, Zheng B, Liu X, Pan X, Kardol P. Emissions of ammonia and greenhouse gases during combined pre-composting and vermicomposting of duck manure. Wastes Management 2014;34:1546–52.

[20] Fillingham MA, Vanderzaag AC, Burtt S, Baldé H, Ngwabie NM, Smith W, Hakami A, Wagner-Riddle C, Bittman S, MacDonald D. Greenhouse gas and ammonia emissions from production of compost bedding on a dairy farm. Wastes Management 2017;70:45–52.

[21] Arriaga H, Viguria M, López DM, Merino P. Ammonia and greenhouse gases losses from mechanically turned cattle manure windrows: a regional composting network. Journal of Environmental Management 2017:557–63.

[22] Christensen TH, Gentil E, Boldrin A, Larsen AW, Weidema BP, Hauschild M. C balance, carbon dioxide emissions and global warming potentials in TCA-modelling of waste management systems. Wastes Management 2009;27:707–15.

[23] Pereira RF, Cardoso EJBN, Oliveira FC, Estrada-Bonilla GA, Cerri CEP. A novel way of assessing c dynamics during urban organic waste composting and greenhouse gas emissions in tropical region. Bioresource Technology Report 2018;3:35–42. https://doi.org/10.1016/j.biteb.2018.02.002.

[24] Andersen JK, Boldrin A, Samuelsson J, Christensen TH, Scheutz C. Quantification of greenhouse gas emissions from windrow composting of garden waste. Journal of Environmental Quality 2010;39:713–24.

[25] He Y, Inamori Y, Mizuochi M. Measurements of N2O and CH4 from the aerated composting of food waste. The Science of the Total Environment 2001;254:65–74.

[26] Hao X, Chang C, Larney FJ. Carbon, nitrogen balances and greenhouse gas emission during cattle feedlot manure composting. Journal of Environmental Quality 2004;33:37–44.

[27] Yang F, Li GX, Yang QY, Luo WH. Effect of bulking agents on maturity and gaseous emissions during kitchen waste composting. Chemosphere 2013;93:1393–9.

[28] Maulini-Duran C, Artola A, Font X, Sanchez A. Gaseous emissions in municipal wastes composting: effect of the bulking agent. Bioresource Technology 2014;172:260–8.

[29] Jiang T, Li G, Tang Q, Ma X, Wang G, Schuchardt F. Effects of aeration method and aeration rate on greenhouse gas emissions during composting of pig faeces in pilot scale. Journal of Environmental Science 2015;31:124–32.

[30] Agegnehu G, Bass AM, Nelson PN, Bird MI. Benefits of biochar, compost and biochar-compost for soil quality, maize yield and greenhouse gas emissions in a tropical agricultural soil. The Science of the Total Environment 2016;543:295–306.

[31] Luo Y, Li G, Luo W, Schuchardt F, Jiang T, Xu D. Effect of phosphogypsum and dicyandiamide as additives on NH3, N2O and CH4 emissions during composting. Journal of Environmental Science 2013;25:1338–45.

[32] Yang F, Li G, Shim H, Wang Y. Effects of phosphogypsum and superphosphate on compost maturity and gaseous emissions during kitchen waste composting. Wastes Management 2015;36:70–6.

[33] Luo WH, Yuan J, Luo YM, Li GX, Nghiem LD, Price WE. Effects of mixing and covering with mature compost on gaseous emissions during composting. Chemosphere 2014;117:14–9.

[34] Wong JWC, Selvam A, Zhao ZY, Karthikeyan OP, Yu SM, Law A, Chung P. In-vessel composting of horse stable bedding waste and blood meal at different C/N ratios: process evaluation. Environmental Technology 2012; 33(22):2561–7.

[35] Jiang T, Ma X, Tang Q, Yang J, Li G, Schuchardt F. Combined use of nitrification inhibitor and struvite crystallization to reduce the NH3 and N2O emissions during composting. Bioresource Technology 2016;217:210–8.

[36] Turan NG. Nitrogen availability in composted poultry litter using natural amendments. Waste Management Resource 2009;27(1):19–24.

[37] Su DC, Wong JWC. Chemical speciation and phytoavailability of Zn, Cu, Ni and Cd in soil amended with fly ash stabilized sewage sludge. Environment International 2003;29:895–900.

[38] Wong JWC, Fang M. Effects of lime addition on sewage sludge composting process. Water Research 2000; 34(15):3691–8.

[39] Genevini PL, Adani F, Borio D, Tambone F. Heavy metal content in selected European commercial composts. Compost Science and Utilization 1997;5:31–9.

[40] Zhang FS, Li YX, Yang M, Li W. Content of heavy metals in animal feeds and manures from farms of different scales in northeast China. International Journal of Environmental Research and Public Health 2012;9:2658–68.

[41] Mantovi P, Bonazzi G, Maestri E, Marmiroli N. Accumulation of copper and zinc from liquid manure in agricultural soils and crop plants. Plant Soil 2003; 250:249–57.

[42] Wang H, Dong Y, Yang Y, Toor GS, Zhang X. Changes in heavy metal contents in animal feeds and manures in an intensive animal production region of China. Journal of Environmental Science 2013;25:2435–42.

[43] Macnicol RD, Beckett PH. The distribution of heavy metal between the principle components of digested sewage sludge. Water Research 1989;23:199–206.

[44] Parkpian P, Leong ST, Laortanakul P, Torotoro JL. Influence of salinity and acidity on bioavailability of sludge-borne heavy metals. A case study of Bangkok municipal sludge. Water, Air, & Soil Pollution 2002; 139:43–60.

[45] Mcbride MB. Mobility and solubility of toxic metals and nutrients in soil fifteen years after sludge application. Soil Sci 1997;162:487–500.

[46] Asuncion L, Nuria L, Susana M, Marqués AM. Nickel biosorption by free and immobilized cells of Pseudomonas fluorescens 4F39: a comparative study. Water, Air, & Soil Pollution 2002;135:157–72.

[47] El-Helow ER, Sabry SA, Amer RM. Cadmium biosorption by a cadmium resistant strain of Bacillus thuringiensis: regulation and optimization of cell surface affinity for metal cations. Biometals 2000;13:273–80.

[48] Gadd GM. Biosorption: critical review of scientific rationale, environmental importance and significance for pollution treatment. Journal of Chemical Technology and Biotechnology 2009;84:13–28.

[49] Selvam A, Zhao Z, Li Y, Chen Y, Leung KSY, Wong JWC. Degradation of tetracycline and sulfadiazine during continuous thermophilic composting of pig manure and sawdust. Environmental Technology 2013;34(16): 2433–41.

[50] Selvam A, Zhao Z, Wong JW. Composting of swine manure spiked with sulfadiazine, chlortetracycline and ciprofloxacin. Bioresource Technology 2012b;126: 412–7.

[51] Ho YB, Zakaria MP, Latif PA, Saari N. Degradation of veterinary antibiotics and hormone during broiler manure composting. Bioresource Technology 2013;131: 476–84.

[52] Zhang Z, Zhao J, Yu C, Dong S, Zhang D, Yu R, Wang C, Liu Y. Evaluation of aerobic co-composting of penicillin fermentation fungi residue with pig manure on penicillin degradation, microbial population dynamics and composting maturity. Bioresource Technology 2015;198: 403–9.

[53] Selvam A, Xu D, Zhao Z, Wong JWC. Fate of tetracycline, sulfonamide and fluoroquinolone resistance genes and the changes in bacterial diversity during composting of swine manure. Bioresource Technology 2012a;126: 383–90.
[54] Zhang J, Chen M, Sui Q, Tong J, Jiang C, Lu X, Wei Y. Impacts of addition of natural zeolite or a nitrification inhibitor on antibiotic resistance genes during sludge composting. Water Research 2016;91:339–49.
[55] Wong JWC, Lai KM, Su DC, Fang M, Zhou LX. Effect of applying Hong Kong biosolids and lime on nutrient availability and plant growth in acidic loamy soil. Environmental Technology 2001;22:1487–95.
[56] Wong JWC, Karthikeyan OP, Selvam A. Biological nutrient transformation during composting of pig manure and paper waste. Environmental Technology 2017;38(6):754–61.
[57] Yazdanpanah N, Mahmoodabadi M, Cerdà A. The impact of organic amendments on soil hydrology, structure and microbial respiration in semiarid lands. Geoderma 2016; 266:58–65.
[58] Agegnehu G, Bass AM, Nelson PN, Bird MI. Benefits of biochar, compost and biochar-compost for soil quality, maize yield and greenhouse gas emissions in a tropical agricultural soil. The Science of the Total Environment 2015;543:295–306.
[59] Calleja-Cervantes ME, Irigoyen I, Fernández-López M, Aparicio-Tejo PM, Menéndez S. Thirteen years of continued application of composted organic wastes in a vineyard modify soil quality characteristics. Soil Biology and Biochemistry 2015;90:241–54.
[60] Agegnehu G, Srivastava AK, Bird MI, Agegnehu G, Srivastava AK, Bird MI. The role of biochar and biochar-compost in improving soil quality and crop performance: a review. Applied Soil Ecology 2017;119: 156–70.
[61] Hargreaves JC, Adl MS, Warman PR. A review of the use of composted municipal solid waste in agriculture. Agriculture, Ecosystems & Environment 2008;123:1–14.
[62] Abujabhah IS, Bound SA, Doyle R, Bowman JP. Effects of biochar and compost amendments on soil physico-chemical properties and the total community within a temperate agricultural soil. Applied Soil Ecology 2016; 98:243–53.
[63] Crecchio C, Curci M, Mininni R, Ricciuti P, Ruggiero P. Short-term effects of municipal solid waste compost amendments on soil carbon and nitrogen content, some enzyme activities and genetic diversity. Biolfertilsoils 2001;34(5):311–8.
[64] Perucci P. Effect of the addition of municipal solid-waste compost on microbial biomass and enzyme activities in soil. Biology and Fertility of Soils 1990;10(3):221–6.
[65] Kithome M, Paul JW, Bomke AA. Reducing nitrogen losses during simulated composting of poultry manure using adsorbents or chemical amendments. Journal of Environmental Quality 1999;28(1):194–201.
[66] Delaune PB, M Jr P, Daniel TC, Lemunyon JL. Effect of chemical and microbial amendments on ammonia volatilization from composting poultry litter. Journal of Environmental Quality 2004;33(2):728–34.
[67] Lefcourt AM, Meisinger JJ. Effect of adding alum or zeolite to dairy slurry on ammonia volatilization and chemical composition. Journal of Dairy Science 2001;84(8): 1814–21.
[68] Huang X, Han Z, Shi D, Huang X, Wu W, Liu Y. Nitrogen loss and its control during livestock manure composting. Chinese Journal of Applied Ecology 2010;21:247–54.
[69] Chen S, Yuan J, Li G, He S, Zhang B. Combination of superphosphate and dicyandiamide decreasing greenhouse gas and NH3 emissions during sludge composting. Transactions of the Chinese Society of Agricultural Engineering 2017;33:199.
[70] Du L, Li G, Yuan J, Yang J. Effect of additives on NH_3 and H_2S emissions during kitchen waste composting. Transactions of the Chinese Society of Agricultural Engineering 2015;31:195.
[71] Awasthi MK, Wang M, Pandey A, Chen H, Awasthi SK, Wang Q, Ren X, Lahori AH, Li DS, Li R, Zhang Z. Heterogeneity of zeolite combined with biochar properties as a function of sewage sludge composting and production of nutrient-rich compost. Waste Management 2017;68: 760–73.
[72] Li ST, Liu R. Establishment and evaluation for maximum permissible concentration of heavy metals in biosolid wastes as organic manure. Journal of Agro-Environment Science 2006;06:777–82.
[73] Chen X, Wang D, Qiao Y, Zhang X, Li J. Analyze on the heavy metals content in China commodity organic fertilizer. Environmental Pollution and Control 2012; 34(2):72–6.
[74] Castro E, Mañas P, Heras JDL. A comparison of the application of different waste products to a lettuce crop: effects on plant and soil properties. Scientific Horticulture 2009; 123:148–55.
[75] Haroun M, Idri A, Omar S. Analysis of heavy metals during composting of the tannery sludge using physicochemical and spectroscopic techniques. J Hazard Mater 2009; 165:111–9.
[76] Nafez AH, Nikaeen M, Kadkhodaie S, Hatamzadeh M, Moghim S. Sewage sludge composting: quality assessment for agricultural application. Environ Monit Assess 2015;187:709–17.
[77] Akvarenga P, Mourinha C, Farto M, Santos T, Palma P, Sengo J, Morais MC, Cunha-Queda C. Sewage sludge, compost and other representative organic wastes as agricultural soil amendments: benefits versus limiting factors. Wastes Management 2015;40:44–52.
[78] Moretti SML, Bertoncini EI, Vitti AC, Alleoni LRF, Abreu-Junior CH. Concentration of Cu, Zn, Cr, Ni, Cd, and Pb in soil, sugarcane leaf and juice: residual effect of sewage sludge and organic compost application. Environmental Monitoring and Assessment 2016;188:163–74.

[79] Ashmawy AM, Ibrahim HS, Moniem SMA, Saleh TS. Immobilization of some metals in contaminated sludge by zeolite prepared from local materials. Toxicology and Environmental Chemistry 2012;94:1657–69.

[80] Shi WS, Liu CG, Ding DJ, Lei ZF, Yang YN, Feng CP, Zhang ZY. Immobilization of heavy metals in sewage sludge by using subcritical water technology. Bioresource Technology 2013;137:18–24.

[81] Tella M, Doelsch E, Letourmy P, Chataing S, Cuoq F, Bravin MN, Saint MH. Investigation of potentially toxic heavy metals in different organic wastes used to fertilize market garden crops. Wastes Management 2013;33: 184–92.

[82] Tüfenkçi S, Türkmen Ö, Sönmez F, Erdinç C, Sensoy S. Effects of humic acid doses and application times on the plant growth, nutrient and heavy metal contents of lettuce grown on sewage sludge-applied soils. Fresenius Environmental Bulletin 2006;15:295–300.

[83] Latare AM, Kumar O, Singh SK, Gupta A. Direct and residual effect of sewage sludge on yield, heavy metals content and soil fertility under rice-wheat system. Ecological Engineering 2014;69:17–24.

[84] Wang X, Chen T, Ge YH, Jia YF. Studies on land application of sewage sludge and its limiting factors. Journal of Hazardous Materials 2008;160:554–8.

[85] Xu JQ, Yu RL, Dong XY, Hu GR, Shang XS, Wang Q, Li HW. Effects of municipal sewage sludge stabilized by fly ash on the growth of Manilagrass and transfer of heavy metals. Journal of Hazardous Materials 2012;217–218: 58–66.

[86] Xu YQ, Xiong HX, Zhao HT, Xu J, Feng K. Structure and characteristics of dissolved organic matter derived from sewage sludge after treating by earthworm. Environment and Chemistry 2010;29:1101–5.

[87] Cheng HF, Xu WP, Liu JL, Zhao QJ, He YQ, Chen G. Application of composted sewage sludge (CSS) as a soil amendment for turf grass growth. Ecological Engineering 2007;29:96–104.

[88] Liu JY, Sun SY. Total concentrations and different fractions of heavy metals in sewage sludge from Guangzhou, China. Transactions of Nonferrous Metals Society of China 2013;23:2397–407.

[89] Wang SP, Liu XA, Zheng Q, Yang ZL, Zhang RX, Yin BH. Analysis on sewage sludge characteristics and its feasibility for landscaping in Xi'an city. China Water & Wastewater 2012;28:134–7.

[90] Yang J, Lei M, Chen TB, Gao D, Zheng GD, Guo GH, Le DJ. Current status and developing trends of the contents of heavy metals in sewage sludge in China. Frontiers of Environmental Science & Engineering 2014;8:719–28.

[91] Yu Z, Michel FC, Hansen G, Wittum T, Morrison M. Development and application of real-time PCR assays for quantification of genes encoding tetracycline resistance. Applied and Environmental Microbiology 2005;71:6926–33.

[92] Chen J, Yu Z, Michel Jr FC, Wittum T, Morrison M. Development and application of real-time PCR assays for quantification of erm genes conferring resistance to macrolides-lincosamides-streptogramin B in livestock manure and manure management systems. Applied and Environmental Microbiology 2007;73:4407–16.

[93] NY525-2012. Ministry of Agriculture of PRC Limits values organic fertilizer of China. 2012 [China].

[94] CAS2. Standards for composts. Briefing Note, UK Composting Association; 2000.

[95] Beirouti Z, Pourzamani HR, Samani MS, Vahdatpoor AR, Mohammad RJ. A review of compost quality standards and guidelines. Diet Evaluation 2011;9(4):2–3.

[96] HKORC. Compost and soil conditioner quality standards. Hong Kong Organic Resource Centre; 2005. http://www.hkorc-cert.org/download/COMPOST-SD-080124-A-Eng.pdf.

CHAPTER 10

Vermicomposting of Waste: A Zero-Waste Approach for Waste Management

KAVITA SHARMA, PHD • V.K. GARG, PHD

INTRODUCTION

Increasing quantities of waste have created several environmental problems including air pollution, water pollution, soil pollution, biodiversity loss, etc. Various greenhouse gases emitted from solid waste dumps also contribute to global warming. Currently, open dumping and landfilling are the most common solid waste disposal methods. Some other methods including pyrolysis, gasification, anaerobic digestion, composting, and biomethanation are used for resource recovery from solid wastes. In some reports, different wastes have been used to synthesize nanomaterials, but practical applications for environmental benefits are still under development [1]. Scientists are always in search of sustainable alternatives to manage solid waste without compromising the environment. Wastes produced from different sectors contain considerable quantities of biodegradable waste. This biodegradable waste can be a potential raw material for several biological methods and it can be used in composting, biomethanation, and vermicomposting processes.

Vermicomposting has gained global attention in the past few decades because of its technical simplicity and effectiveness [2]. A wide array of wastes such as animal waste [3], agricultural waste [4], industrial waste [5], and municipal waste [6] can be utilized as raw material for vermicomposting.

Vermicomposting is a bio-oxidative, mesophilic natural decomposition process in which earthworms and microorganisms synergistically mineralize organic waste substrates and convert them into nutrient-rich organic manure [7]. In this process, two useful products are obtained, namely, vermicompost and earthworm (Fig 10.1).

Vermicompost can be used as soil conditioner and earthworms can be used in medicine, feed in fish ponds. Vermicompost has several advantages over chemical fertilizers and is useful to crops. Vermicompost contains nutrients (in plant available form), humic acid, and growth hormones and hence is used extensively as an organic fertilizer at a large scale in organic farming [8]. Research conducted by various authors showed that the use of vermicompost may increase seed germination, vegetative growth of crops, and yield without compromising the soil health. Intensive agriculture has badly affected the soil health and fertility. Vermicompost application can be helpful in maintaining the soil health and fertility by improving its physical, chemical, and biological properties [2,4]. Various reports have indicated that vermicomposting can be useful in other fields also, such as wastewater treatment, soil remediation, and energy production [9,10].

Vermicomposting forms an important link in the circular economy of a country by producing energy from the waste and promoting the three *R*'s approach. China is the first country in the world that embraced the circular economy concept in 2008 [11]. Understanding the need for waste recycling, the Government of India also promoted adoption of biological techniques such as composting and vermicomposting for waste recycling.

In spite of the various benefits associated with vermicomposting, it is still not commercially very popular because of certain technologic constraints and lack of awareness. Hence, proper extension and training activities are required to explore the full potential of vermicomposting. Research is also needed

Sustainable Resource Recovery and Zero Waste Approaches. https://doi.org/10.1016/B978-0-444-64200-4.00010-4

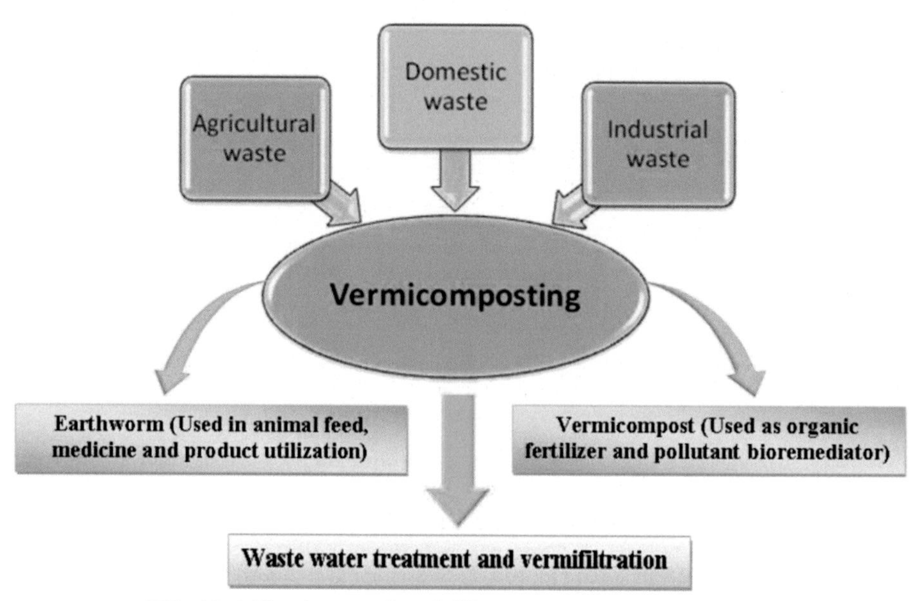

FIG. 10.1 Vermicomposting of different wastes and its application.

for a better understanding of the metabolism involved in the process, along with the proper use of vermicompost.

VERMICOMPOSTING TECHNIQUE

Brief History

The vermicomposting process was developed in the middle of 20th century. Vermicomposting has been adopted by several countries due to its importance in waste management and farming practices. The American Earthworm Technology Company produced about 500 tons vermicompost per month and Japan produces 2000–3000 tons of vermicompost per month using various organic wastes [12]. A vermicomposting plant has been set up in Sydney Waters (New South Wales), with 40 million worms and having a capacity of 200 tons waste/week. In India also, vermicomposting plants with different capacities are operating in different cities such as Bangalore and Pune [13]. The largest institute in India involved in vermiculture is the Bhawalkar Earthworm Research Institute (BERI) situated in Pune. Earthworms act as a natural decomposer and degrade various types of organic wastes.

Xiang et al. [14] collected research citation data from the Science Citation Index and conclude that earthworm research continues to increase from 2000, with the highest research output from the developed nations. However, in the Indian context, the importance of earthworm in soil management and plant growth was recognized several decades ago.

Vermicomposting

Vermicomposting is a biotechnique in which waste is converted into manure by employing earthworms and microorganisms. Gomez Brandon and Dominguez [15] defined vermicomposting as a bio-oxidative process in which earthworms interact with the decomposer community (bacteria, fungi, actinomycetes) to stabilize organic waste by altering its physical and biochemical properties. Most of the organic components are degraded and the residuals are transformed into stabilized vermicompost, which is rich in nutrients, hormones, and humic substances [16].

Earthworms

More than 4000 species of earthworms are known, which are broadly classified into three groups (epigeic, anecic, or endogeic). The vermicomposting potential of earthworms from all the three groups has been investigated by several researchers (Table 10.1). Although all earthworm species cannot be utilized for vermicomposting, several researchers have investigated the effectiveness of the three groups of soil-inhabiting earthworms [17–20]. It is evident from the available literature that epigeic earthworms comprising *Eisenia fetida, Eisenia*

TABLE 10.1
Vermicomposting Using Different Species of Earthworms.

Waste	Earthworm Species	References
MONOCULTURE (SINGLE EARTHWORM SPECIES)		
Vegetable waste, biogas slurry	*Allolobophora parva*	Suthar [17]
Bagasse, sugarcane trash, press-mud	*Drawida willsi*	Kumar et al. [101]
Water hyacinth	*Perionyx sansibaricus*	Zirbes et al. [102]
Urban sewage sludge	*Lumbricus rubellus*	Azizi et al. [18]
Dewatered sludge	*Bimastus parvus*	Fu et al. [25]
Jute mill waste	*Metaphire posthuma*	Das et al. [103]
Medicinal herbal residue	*Eisenia fetida*	Chen et al. [104]
Ruminant excreta	*E. fetida*	Sharma and Garg [3]
Paper cup waste	*Eudrilus eugeinea*	Arumugam et al. [21]
Urban waste and plant litter	*E. fetida*	Wu et al. [105]
Municipal solid waste	*E. eugeniae*	Soobhany et al. [6]
Aquaculture sludge	*Eisenia andrei*	Kouba et al. [106]
POLYCULTURE (MULTIPLE EARTHWORM SPECIES)		
Vegetable waste	*Pheretima* sp., *E. fetida, Perionyx excavatus*	Shanthi et al. [107]
Soil	*Lumbricus rubellus, Aporrectodea caliginosa, Lumbricus terrestris*	Ernst et al. [108]
Sewage sludge	*E. fetida* and *E. andrei*	Rorat et al. [109]
Vegetable market waste and rice straw	*E. fetida* and *P. excavatus*	Hussain et al. [20]
Municipal solid waste	*M. posthuma* and *E. fetida*	Sahariah et al. [44]
Kitchen vegetable waste and paddy straw	*E. fetida, E. eugeniae*, and *P. excavatus*	Hussain et al. [16]
Paper waste, cattle manure, and lawn clippings	*P. excavatus, E. eugeniae*, and *Dichogaster annae*	Martin and Eudoxie [19]

andrei, Eudrilus eugeniae, and *Perionyx excavatus* are most effective for vermicomposting, owing to the following characteristics:

- high reproduction rates,
- tolerance to a wide range of environmental conditions,
- rapid rate of vermiconversion,
- ability to feed on a wide variety of organic wastes.

Earthworm species have been studied extensively to explore their waste conversion and bioremediation potential. Based on the available data, it can be inferred that different species have different composting potentials depending on the environmental conditions and feed quality. Several authors have recommended the use of polyculture instead of monoculture in vermicomposting to hasten the process, as different species of worms may do composting in different layers of the waste. Hussain et al. [16] used earthworm consortia containing three species of earthworms (*Eisenia, Perionyx, Eudrilus*) for the degradation of vegetable waste and rice straw. It was reported that vermicompost was prepared more quickly in multispecies-mediated vermicomposting than single-species-mediated vermicomposting. The multispecies vermicompost contained higher microbial biomass and proliferation than the monospecies vermicompost. Economic analysis of multispecies- and monospecies-mediated vermicomposting revealed that earthworm consortia is more beneficial. Use of bacterial consortia with earthworms for accelerating the vermicomposting process has been reported. Arumugam et al. [21] reported that the use of bacterial consortia during paper cup waste vermicomposting produced better quality vermicompost with increased NPK (nitrogen, phosphorus, and potassium) content than the control and vermicomposting without bacterial consortia.

Precomposting and the use of bulking agent

Different arrays of wastes are utilized as feedstock for earthworms during vermicomposting. Some of the required characteristics of waste to be used as substrate for vermicomposting are pH in the range from 5 to 8, moisture content between 60% and 80%, C/N ratio around 30, and biodegradable nature [22]. Wastes are utilized for vermicomposting either directly or after mixing with some mixing agent known as bulking substrate. Use of bulking substrate during vermicomposting helps in making the waste more palatable to earthworms; increases earthworms efficiency to vermicompost; reduces the harmful effects of toxic compounds, if present in waste; and helps in maintaining the moisture content of waste. Most commonly used bulking substrates include various types of animal excreta, fruit and vegetable waste, crop residues, and paper waste. Sometimes compost and vermicompost are also utilized as bulking substrate [23]. Some authors have reported the use of dry leaves and agricultural waste as bulking substrate [24].

Various attempts have been made to compare the vermicomposting performance of different types of wastes with and without bulking substrate. Fresh sludge without bulking substrate caused earthworm mortality due to anaerobic conditions and emission of toxic gases [25]. Yadav et al. [26] also reported that earthworms could not survive in the effluent treatment plant (ETP) sludge of a food process industry. ETP sludge requires a bulking substrate to overcome the negative effects of various toxic substances present in the sludge [27]. Lignocellulosic waste has high carbon content so it needs to be mixed with a bulking substrate having low carbon content [28].

After mixing the waste with bulking substrate the next step is precomposting. Precomposting is usually preferred to eliminate anaerobic conditions and to remove any volatile gases potentially toxic to the earthworms. Arevalo et al. [29] reported that precomposting of coffee husks is necessary because they contain antinutritional factors (tannins and caffeine) that are toxic to earthworms. Fu et al. [25] reported that during the vermicomposting of sludge, small pellets of 4.5 mm stabilized fast and increased the specific surface area, which in turn provided better growing conditions for the earthworms.

Process

Vermicomposting takes place in an aerobic environment by symbiotic interactions between earthworms and microorganisms. The process commences with the ingestion of waste by earthworm, where waste enters into its stomach by mouth. After that, waste passes through the earthworm gizzard where it is grounded by ingested stones and then enters into the intestine [8]. Earthworm intestine works as a bioreactor, where various physical and biochemical activities take place. Enzymes present in the gut of earthworms (proteases, lipases, amylases, cellulases, chitinases) degrade various cellulosic and proteinaceous materials present in waste substrate [30]. The vermicomposting process is presented stepwise in Fig 10.4.

Arevalo et al. [29] investigated the activities of 19 hydrolytic enzymes within three different sections of the intestine of *E. fetida*. It was found that during vermicomposting, enzyme activity changed in a coordinated manner within each gut section and was probably influenced by selective microbial enzyme enrichment and by the availability of nutrients. All these metabolic activities convert waste into vermicast and then finally into a humus-like nutrient-enriched product called vermicompost. Apart from vermicompost, some of the portion of organic wastes is also converted to earthworm biomass as earthworms multiply by reproduction.

The whole process of vermicomposting takes place in two phases: the active phase or direct process and the maturation phase or indirect phase (Fig 10.2). During the active phase, earthworms modify the physical, chemical, and biological characteristics of waste. Earthworms ingest the waste and add various beneficial enzymes and microbes during digestion [31]. As most of the activities take place in the gut of earthworms, the active phase is also known as the gut-associated process. In the maturation phase, earthworms process the waste and alter it biochemically [8]. In this process, microbes act on vermicast and convert it into vermicompost, hence it is also known as the cast-associated process.

Changes in physico-chemical characteristics of waste

Various physicochemical (color, odor, organic matter, C/N ratio, pH, electric conductivity [EC], NPK, humification index, heavy metals, etc.) and biological (bacteria, fungi, enzyme, hormone) changes take place in the characteristics of feedstock during vermicomposting. During vermicomposting the pH and EC of waste are either increased or decreased from the initial day. Lv et al. [32] reported reduction in pH during vermicomposting of cattle dung and pig manure. However, Malinska et al. [33] reported increase in pH in the vermicompost prepared from decanter cake and rice straw. Mineralization of N and P and formation of

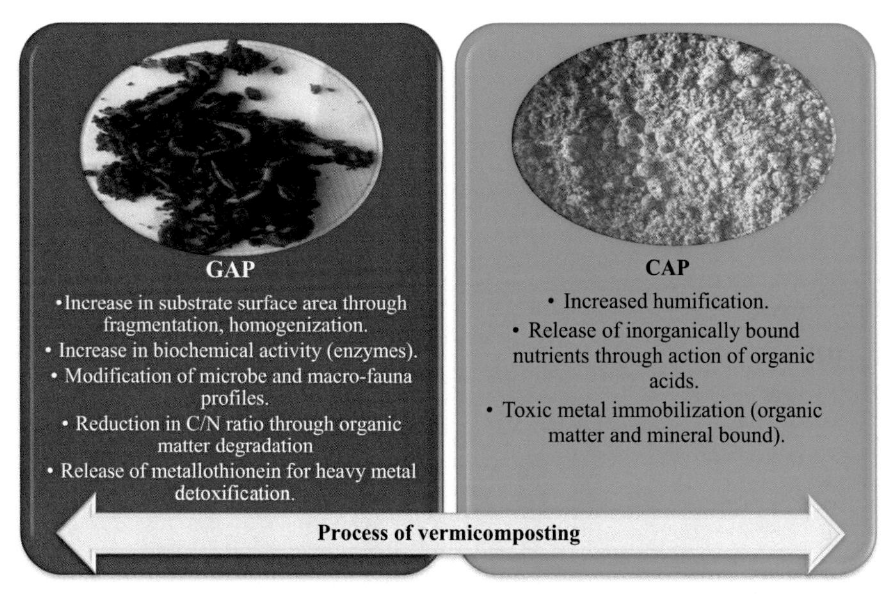

FIG. 10.2 Gut-associated process (GAP) and cast-associated process (CAP).

ammonium ions, carboxylic and phenolic groups, and humic acids are the regulatory factors for pH. Change in EC is due to the release of soluble salts and the availability of mineral salts. Various micro- and macronutrient (NPK) contents are also increased after vermicomposting. Sharma and Garg [3] reported that the NPK content increased in the different feedstock prepared with various livestock excreta and their combinations (Fig. 10.3). Similarly, Sharma and Garg [4] reported 1.2 to 2.9-fold increase in the nitrogen content and 1.1 to 2.6-fold increase in the potassium content after vermicomposting of rice straw and paper waste. Vermicomposting also enhanced the amounts of humic acid and plant growth hormones such as auxins, gibberellins, and cytokinins [34]. Orlov and Biryukova [35] reported that vermicompost contained 17%–36% of humic acid and 13%–30% fulvic acid of the total concentration of organic matter. Vermicomposting also resulted in higher concentrations of heavy metals, which may be due to weight loss, release of carbon dioxide, and mineralization [36]. Instrumental analysis proved that vermicomposting resulted in change in the surface morphology also. Kumar et al. [37] reported that vermicompost has a larger surface area that provides microsites for microbial action and retention of nutrients. According to Lim and Wu [38] and Sharma and Garg [4], vermicompost is more scattered and fragmented than the initial waste as revealed by scanning electron microscopy (SEM).

Process governing factors

A number of biotic and abiotic factors such as temperature, moisture, aeration, pH, feedstock quality, and C/N ratio affect the vermicomposting process. Earthworm species, stocking density, and worm feeding rate also significantly influence the vermicomposting process [39]. Therefore it is necessary to optimize these factors to ensure efficient conversion of organic wastes into vermicompost, along with better growth and reproduction of earthworms. Moisture content in the range of 60%–80% is most suitable for vermicomposting process. Earthworm growth and reproductive performance also affect the rate of vermicomposting.

Most of the studies have revealed that physicochemical properties of feedstock directly influence the growth and reproduction of earthworms. Mupambwa and Mnkeni [40] found that C/N ratio, feeding rate, and stocking density cumulatively affect the vermicomposting process. Martin and Eudoxie [19] investigated the comparative assessment of different earthworm species (*Perionyx excavatus, E. eugeniae, Dichogaster annae*) at a C/N ratio of 28, 36, and 53, respectively; feed rate of 1, 1.25, and 2 g feed (dry wt.)/g worm/day; and similar stocking density. The results showed that the vermicast production was affected by C/N ratio and feed rate and was variable for different species. Earthworm population and vermicast quality were highly affected by feed rate.

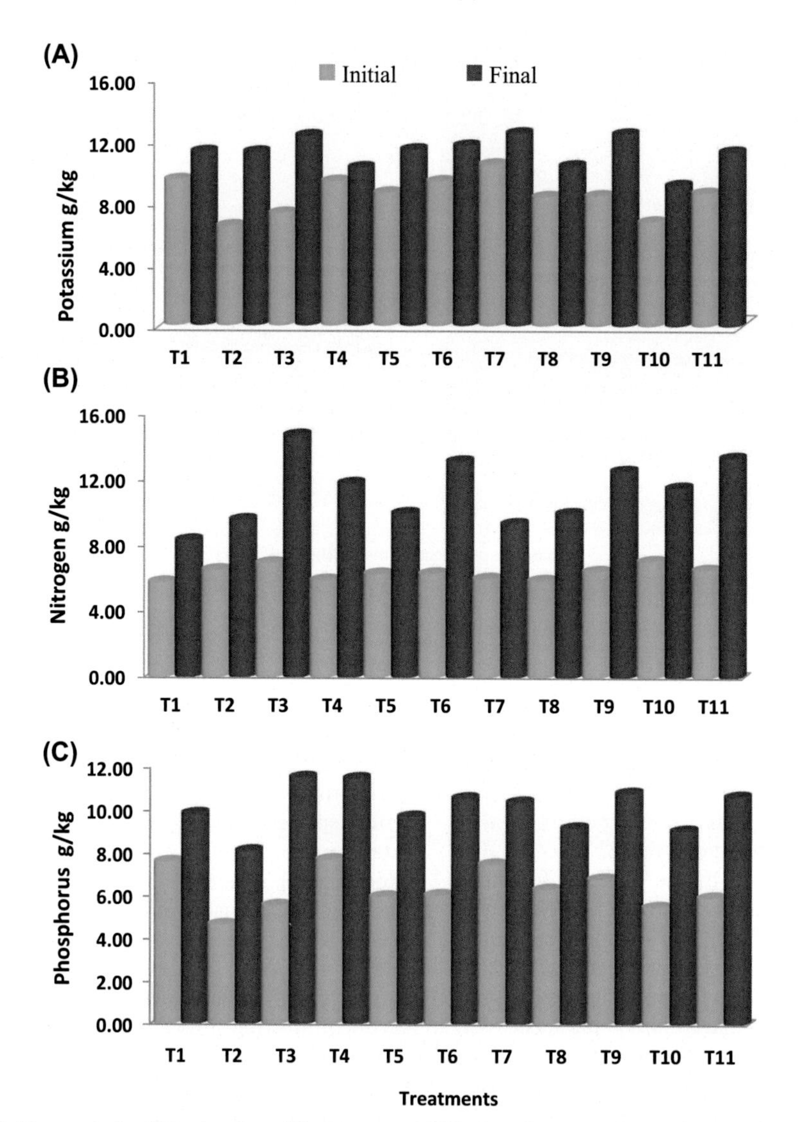

FIG. 10.3 Change in the **(A)** potassium, **(B)** nitrogen, and **(C)** phosphorus contents after vermicomposting of various treatment of livestock excreta (Sharma and Garg [3]).

Advantages of vermicomposting

The advantages of vermicomposting are given below:

- It is an eco-friendly and zero-waste technology for waste management.
- Almost all types of nontoxic organic wastes may be subjected to vermicomposting after some preprocessing.
- It consumes less energy and releases less greenhouse gases.
- Its cost is lower than that of traditional composting.
- Two useful end products obtained from vermicomposting are vermicompost and earthworms.
- It also reduces the pathogenic level in the waste and also treatment media.
- It is economically beneficial.
- It is a multiapplication technique that can be applied in the field of wastewater treatment and energy production.

VERMICOMPOSTING OF WASTE ORIGINATED FROM DIFFERENT SECTORS

Vermicomposting helps in stabilizing nontoxic organic wastes and converts them into valuable manure for agricultural applications. Studies are available on the effective vermicomposting of organic wastes generated from different sectors such as domestic, agricultural, and industrial sectors. A brief review of such studies is provided in the following sections.

Vermicomposting of Waste From Domestic (Rural/Urban) Sector

Wastes generated from the domestic sector mainly contain kitchen waste, paper waste, and food waste, which finally end up as municipal solid waste (MSW). Household wastes can effectively be vermicomposted, as they are readily biodegradable, often nontoxic, and rich in nutrients [41]. An overview of different investigations undertaken on domestic waste vermicomposting is given in Table 10.2. Huang et al. [42] investigated fruit and vegetable waste vermicomposting and reported higher bacterial and fungal densities in the vermicompost. Actinobacteria and ammonia-oxidizing bacteria were more in vermicompost than the control group. Huang and Xia [43] used the mucus of the earthworm, *E. fetida*, for the vermicomposting of food and vegetable wastes. Results revealed that earthworm mucus significantly accelerated the mineralization and humification process as well as microbial activity and bacterial population in different feedstocks. Hussain et al. [20] reported vermicomposting of vegetable market waste, rice straw, and cow dung using the earthworms *E. fetida* and *P. excavatus*. The vermicompost so produced contained a higher microbial population and bacterial strains isolated from the vermicompost were capable of improving soil health and crop growth.

MSW generation is increasing @ 5.3%/annum but a significant fraction of MSW is organic in nature (40%), which offers its suitability for vermicomposting [44] explored the possibility of MSW as feedstock using two species of earthworms, namely, *Metaphire posthuma* and *E. fetida*. Vermicomposting enhanced NPK content and decreased the toxicity by converting MSW into useful fertilizer. Physicochemical analysis of vermicompost prepared from sewage sludge revealed that vermicompost contained highly significant nitrate and humic substances, compared with control, but lower pH and water-extractable carbon [45]. In addition, MSW can be converted into stable compost and vermicompost using composting, followed by vermicomposting [27]. Composting followed by vermicomposting enhanced waste degradation and reduced the phosphorus and carbon cycling enzymes [46]. Various instrumental techniques also revealed that vermicompost prepared from MSW is more stable and homogeneous. Polysaccharide groups and aliphatic methylene groups also increased after vermicomposting [6].

Vermicomposting of Waste From Agricultural Sector

Agricultural wastes include crop residues, weeds, leaf litter, sawdust, forest waste, and livestock waste. Table 10.3 presents various vermicomposting studies on the use of agricultural waste as feedstock. Among the various agricultural wastes, livestock waste is always a preferred choice for researchers as feedstock for earthworms and as bulking substrate for vermicomposting. Livestock waste is considered as the suitable organic amendment to enhance the process of vermicomposting because of its low cost, easy availability, sufficient nutrient content, and ideal C/N ratio [3].

However, the chemical composition of livestock waste depends on the type of feed given to the animal, bedding material, and fresh or dried including the manner how excreta is collected, stored, and handled prior to vermicomposting [12,32]. Hence, differences in physicochemical characteristics of livestock waste have effects on the life cycle of earthworm species [3]. Vodounnou et al. [47] studied the vermicomposting of different animal wastes employing *E. fetida* and reported that the earthworm growth rate is in the following order: cow > pig > rabbit > poultry > sheep > compost. Similarly, Sharma and Garg [3] also utilized the excreta of different ruminants (sheep, goat, cow, and buffalo) and successfully converted it into nutrient-rich vermicompost. The study found that the highest biomass gain by earthworms was in buffalo excreta and the lowest biomass gain was in sheep excreta. Sheep bedding (rice husk) mixed with cattle manure in different ratios of 0%, 25%, 50%, 75%, and 100% was subjected to vermicomposting by Cestonaro et al. [48]. Results of the study revealed that addition of cattle manure up to 25% required 148 days to prepare vermicompost. Furthermore, increase in cattle manure proportion to sheep bedding decreased vermicomposting time. Najjari and Ghasemi [49] carried out vermicomposting of sawdust and blood powder using the earthworm, *E. fetida*. Vermicomposting enhanced the nutrient content in the waste and decreased the organic carbon and C/N ratio. Application of this vermicompost also improved the nutrient content in vegetables.

TABLE 10.2
Vermicomposting of Various Domestic Wastes.

S. No.	Type of Waste (Bulking Material)	Earthworm	Duration	Results	References
1	Food and vegetable waste	*Eisenia fetida*	90 days	Three different combinations of vegetable waste and buffalo waste were prepared. Highest growth and reproduction of earthworms was achieved in 100% buffalo waste. Vermicomposting enhanced nitrogen, total available phosphate, and total potassium contents.	Sharma and Garg [41]
2	Green waste (fallen leaves, grass clippings, branch cuttings, and biochar)		60 days	Addition of 6% biochar was found to be best as earthworm growth and vermicompost quality was considered. Biochar addition increased earthworm biomass and the juvenile and cocoon numbers of *E. fetida*. Biochar addition improved the physicochemical properties of vermicompost as well as increased the different enzymes' activities and lignin degradation rate (14%).	Gong et al. [76]
3	Fruit and vegetable waste, sludge, cow dung	*E. fetida*	45 days	Mucus of earthworm was used for food and vegetable waste vermicomposting. Mucus was extracted and inoculated into different feedstock and findings revealed that earthworm mucus also significantly accelerated the waste mineralization and humification process. Microbial activity and bacterial population were also enhanced.	Huang and Xia [43]
4	Shredded paper, cattle manure, and lawn clippings	*Perionyx excavatus, Eudrilus eugeniae*, and *Dichogaster annae*	56 days	Different wastes are combined to make C/N ratios of 28, 36, and 53. C/N ratio, feed rate, and earthworm species significantly affect vermicast quality and production. Vermiconversion increased with increasing C/N ratio but decreased with increasing the feed rate. Different parameters vary with the earthworm species.	Martin and Eudoxie [19]
5	Coffee husk and market waste	*E. fetida*	90 days	19 Hydrolytic enzymes' activities from the intestine of *E. fetida* were investigated. A study found that enzyme activity varied in a coordinated manner within each section of the earthworm gut. Microbial enzyme enrichment and nutrient availability are important factors.	Arevalo et al. [29]
6	Food waste	*E. fetida*	70 days	Use of zeolite as an amendment for food waste vermicomposting increased the decomposition process and also ammonia emission. Vermicomposting enhanced the nutrient content of food waste.	Zarrabi et al. [110]

7	Household wastes, biological and chemical sludge	*E. fetida*	70 days	C/N ratio of vermicompost was 14.5–16; N and P content increased; however, TOC decreased after vermicomposting. Earthworm accumulates heavy metals, resulting in the reduction of metal.	Amouei et al. [111]
8	Vegetable market waste (rice straw, cow dung)	*E. fetida* and *P. excavatus*	60 days	TOC decreased, pH shifted toward neutrality, and NPK content increased. A total of 45 nitrogen-fixing and 34 phosphate-solubilizing bacterial strains were isolated from earthworm gut. Two novel strains of the nitrogen-fixing bacteria *Kluyvera ascorbata* were proved to be efficient biofertilizer candidates.	Hussain et al. [20]
9	Human feces (cow dung, soil)	*E. fetida*	120 days	Total coliform levels totally reduced during vermicomposting; vermicompost was produced at the rate of 0.30 kg-cast/kg-worm/day.	Yadav et al. [112]
10	MSW (cow dung)	*Metaphire posthuma* and *E. fetida*	60 days	Vermicomposting resulted in higher bioavailability of N, P, K, and Fe. *M. posthuma* successfully converted toxic wastes into vermicompost.	Sahariah et al. [44]
11	Municipal green waste (cattle manure)	*E. fetida*	18 weeks	Comparatively, vermicomposting was found superior to composting. Vermicomposts had a lower bulk density, greater total porosity, and higher microbial biomass C than composts. Basal respiration and metabolic quotient were also higher in vermicompost.	Haynes and Zhou [113]
12	Sewage sludge (wheat straw + biochar)	*E. fetida*	14 days	Highest production of cocoons was observed after 4 weeks of vermicomposting. Trace element levels, pH, and C/N ratio decreased after vermicomposting.	Malinska et al. [33]
13	Fresh fruit and vegetable wastes (vermicompost + soil)	*E. fetida*	48 h	Among the studied wastes, banana peels proved harmful for the survival of *E. fetida*. Vermicomposting increased the bacterial diversity with significant populations of actinobacteria and ammonia-oxidizing bacteria. Vermicomposting efficiency differs with the type of waste and loading in the vermireactor.	Huang et al. [42]
14	Sewage sludge (soil)	*E. fetida* and *E. andrei*	1 month	Cd, Cu, Ni, and Pb levels were found to be reduced, whereas Zn concentration increased. *E. andrei* has higher capabilities to accumulate some metals. During the first 6 weeks of vermicomposting, riboflavin content decreased to some extent and after that was restored till the end of the 9-week experiment.	Rorat et al. [109]

Continued

TABLE 10.2
Vermicomposting of Various Domestic Wastes.—cont'd

S. No.	Type of Waste (Bulking Material)	Earthworm	Duration	Results	References
15	Digestate, kitchen waste, sewage sludge (wheat straw + paper waste + garden biowaste)	*E. andrei*	5 months	Composting and vermicomposting effects were studied on three size fractions: more than 12 mm, between 12 and 5 mm, and less than 5 mm. Finer and homogeneous final product was obtained when compared with that of composting. Nutrients were also higher in vermicompost than in compost.	Hanc and Dreslova [114]
16	Sludge (cow dung and swine manure)	*E. fetida*		Sludge + 40% swine manure proved a great medium for the growth and fecundity of *E. fetida*. Vermicompost has lower pH value, TOC, ammonium N, and C/N ratio and higher total available phosphorous content.	Xie et al. [115]
17	Sewage sludge (wood chips	*E. andrei*		Combined approach of composting-vermicomposing was proved most appropriate. Earthworm grew well in sludge.	Villar et al. [46]
18	Waste paper, chicken manure	*E. fetida*	20 days	Optimum C/N ratio for the waste mixture was 40. Total NPK concentrations increased; however, total carbon content, C/N ratio, electric conductivity, and heavy metal content gradually decreased with time	Ravindran and Mnkeni [116]

MSW, municipal solid waste; *NPK*, nitrogen, phosphorus, and potassium; *TOC*, total organic carbon.

TABLE 10.3
Vermicomposting of Various Agricultural Wastes.

S. No.	Type of waste (Bulking Material)	Earthworm	Duration	Results	References
1	Buffalo waste, sheep waste, goat waste, cow waste	*Eisenia fetida*	90 days	Maximum earthworm growth rate was achieved in the various combinations of buffalo dung and minimum growth rate in sheep waste. TOC content and C/N ratio decreased during vermicomposting, whereas total nutrient content increased.	Sharma and Garg [3]
2	Rice straw + paper waste + cow dung	*E. fetida*	105 days	Paper waste and rice straw effectively convert into nutrient-rich vermicompost. Vermicompost is more fragmented than parent feedstocks. Use of rice straw in higher ratio was not recommended.	Sharma and Garg [4]
3	*Salvinia molesta*	*E. fetida*	45 days	Chemical compounds responsible for weed allelopathic effects destroyed completely. The C/N ratio of *Salvinia* was reduced sharply from 53.9 to 9.35.	Hussain et al. [56]
4	Sewage sludge (cattle dung)	*E. fetida*	80 days	Vermicomposting modifies the structure of bacterial community in the waste and reduces the pathogenic human bacteria population.	Lv et al. [32]
5	Pig manure and rice straw	*E. fetida*	40 days	Vermicompost has higher pH, P, K, Zn, and CEC but lower available N and Cu than the parent substrate. Increment in aromatic compounds indicated high humification during vermicomposting. Earthworm tissues accumulated ^{13}C.	Zhu et al. [117]
6	Crop/tree residues	*Eudrilus* sp.		Earthworm growth and conversion efficiency vary with waste. In all the crop residues, pH, EC, and N and P levels increased, whereas C/N and C/P ratios decreased.	Thomas et al. [118]
7	Horse manure, apple pomace, grape pomace, and digestate (manure slurry, corn silage, haylage)	*Eisenia andrei*	240 days	Study evaluated vermicompost characteristics based on 120-day-old layer and 240-day-old layer in vermireactor. Maximum biomass of earthworms was in 120-day-old layer. After 240 days, microbial biomass activity decreased due to decrease in the earthworm activity, indicating a high degree of stabilization. Enzyme activities differ according to the age of the layers and the type of waste. Germination index increased after vermicomposting and was higher with apple pomace and digestate than that with horse manure and grape pomace.	Sanchez et al. [119]
8	Cow manure and wheat residues	*E. fetida*	60 days	Urease activity is a suitable indicator of vermicompost maturity and waste stabilization during the process of vermicomposting. Urease activity was highly correlated with the time of vermicomposting resulting in $r = 0.97$ for cattle manure and $r = 0.99$ for wheat waste. Urease activity showed significant correlations with the C/N ratio.	Sudkolai and Nourbakhsh [120]

Continued

TABLE 10.3
Vermicomposting of Various Agricultural Wastes.—cont'd

S. No.	Type of waste (Bulking Material)	Earthworm	Duration	Results	References
9	Wheat straw, pig dung, poultry dung, rabbit dung, cattle dung, sheep dung, and vegetal compost	*E. fetida*	90 days	Highest worm production and growth rate were obtained with cow dung followed by pig dung; however, earthworm growth decreased in vegetable compost. Maximum earthworm growth rate was found on the 90th day. Growth and worm production depend on the biochemical quality of the feedstocks.	Vodounnou et al. [47]
10	Sawdust, boxwood leaves, and cardboard compost (MSW)	*E. fetida*	100 days	MSW and carbonaceous materials in different proportions, viz., 50:50, 70:30, 85:15, and 100:0, were vermicomposted for 100 days. Vermicomposting for 75 days is sufficient for vermicompost maturity in terms of EC, WSC, DEH, and C/N ratio. Phosphorus, nitrogen, and pH levels were higher in the vermicompost.	Alidadi et al. [23]
11	*Salvinia natans* (cattle manure and sawdust)	*E. fetida*	45 days	Total concentration of heavy metals (Zn, Cu, Mn, Fe, Cr, Pb, Cd, Ni) increased; however, concentration of water-soluble and plant-available heavy metals was reduced in the final vermicompost. TCLP tests confirmed the suitability of vermicompost for agriculture.	Singh and Kalamdhad [121]
12	Leaf litter (horse dung, sheep dung)	*Perionyx excavatus*	60 days	Cashew leaf litter mixed with cow dung at 2:2 ratio was found to best in terms of vermicompost properties. The vermicompost produced had lower pH, organic carbon content, C/N ratio, C/P ratio, and lignin, cellulose, hemicellulose, and phenol content but higher NPK, DEH, and HA content than the waste and compost. Reduction in the lignocellulose and phenol content is due to the combined action of the gut lignocellulolytic microflora and earthworms during the vermicomposting process.	Parthasarathi et al. [122]
13	*Ipomoea*	*E. fetida*	30 days	Total carbon contents decreased from 527.3 to 282.8 g/kg and total nitrogen contents increased from 20.2 to 28.5 g/kg. C/N ratio of *Ipomoea* vermicompost was 9.9. Spectroscopic analysis revealed transformation of weed into potent organic fertilizer.	Hussain et al. [52]
14	Coconut husk poultry manure, pig slurry	*Eudrilus eugeniae*	21 days	Highest recovery of relative N (1.6) and K (1.3) was in 20% feedstock substitution by pig slurry, and highest P recovery (2.4) was with poultry manure substitution. Vermicompost contains higher pH, microbial biomass carbon, and macro- and micronutrients than the initial waste.	Swarnam et al. [123]

15	Cow dung, poultry manure	*E. fetida*		After vermicomposting, pH, TOC content, and C/N ratio were reduced but EC and HA were increased. Heavy metals stabilized.	Lv et al. [32]
16	Decanter cake + rice straw	*E. eugeniae*	2 weeks	Four treatments with different ratios of decanter cake and rice straw (2:1, 1:1, 1:2, 1:3) were prepared. Two parts decanter cake and one part rice straw (w/w) was found to best among all the treatments.	Lim et al. [124]
17	Crop residue (rice, wheat, corn, sugarcane)	*E. eugeniae*	90 days	Highest earthworm weight and vermicomposted matter were achieved in wheat and lowest with corn residue.	Aynehband et al. [125]
18	*Lantana*	*E. fetida*	-	C/N ratio reduced from 22.7 to 8.1; humification index from 8.38 to 2.03. FTIR spectra revealed complete degradation of phenols and sesquiterpene lactones and formation of simple compounds. GC-MS analysis revealed transformation of 24–86 constituents.	Hussain et al. [55]
19	Parthenium	*E. fetida*	—	Chemicals responsible for the allelopathic effect of parthenium weed are destroyed. Scanning electron microscopy shows marked disaggregation of the material in the vermicompost as compared with the well-formed matrix of *Salvinia* leaves.	Hussain et al. [54]
20	Tomato plant debris + paper mill sludge	*E. fetida*	6 months	Characterize HA isolated from different waste mixtures before and after vermicomposting. HA content increased by 15.9%–16.2%. Vermicompost produced from tomato debris/paper mill sludge (2:1) recorded higher C content and C/N ratio. HA from tomato debris/paper mill sludge (1:1) vermicompost showed a higher O content and O/C ratio.	Fernandez Gomez et al. [62]
21	Filter cake (cattle manure)	*E. fetida*	30 days	Positive correlation of phosphatase activities with TOC content, pH, and WSP but negative correlation with HA content. Nanopore volume found to be negatively correlated with phosphatase activities for filter cake but not for cattle manure. HA content of filter cake vermicompost was higher than that of cattle manure vermicompost	Busato et al. [126]

CEC, cation exchange capacity; *DEH*, dehydrogenase; *EC*, electric conductivity; *FTIR*, Fourier transform infrared; *GC-MS*, gas chromatography-mass spectrometry; *HA*, humic acid; *MSW*, municipal solid waste; *TCLP*, toxicity characteristic leaching procedure; *TOC*, total organic carbon; *WSC*, water-soluble carbon; *WSP*, water-soluble phosphorus.

Weeds are well known to create hurdles in the agriculture fields causing ecological and economic losses. They compete with crop plants for space, water, and nutrients and also harm the surrounding soil. All weeds cannot be used as animal fodder because of the toxic compounds present in them, which otherwise may cause health problems to animals [50]. Different weeds have different allelopathic effects that negatively affect soil biodiversity and the surrounding environment. For example, parthenium contains phenolic acids and a toxic compound parthenin, which is a sesquiterpene lactone [51]. *Ipomoea* is also a widespread weed with several bioactive compounds, which are toxic and may cause physiologic problem in human beings and animals [52]. Gotardo et al. [53] found that when goats ate *Ipomoea* during their pregnancy, their offspring are short of maternal-infant bonding and have various other disorders. Different species of weeds, such as parthenium, water hyacinth, *Salvinia*, *Ipomoea*, and *Lantana*, have successfully been used as feedstock for vermicomposting with and without a suitable bulking agent [54–57]. All studies conducted on the vermicomposting of weeds have concluded that the vermicompost prepared from weeds is rich in various macro- and micronutrients that are essential for plant growth. Vermicomposting of citronella waste and cow dung decreased the C/N ratio up to 87.7% [58].

Vermicomposting of Waste From Industrial Sector

Industrial wastes generated from sugar mills, textile, food, pulp and paper, tannery, palm mill, and milk processing units have been vermicomposted by several researchers. All their studies conclude that vermicomposts produced from different industrial wastes are pathogen free, are rich in plant nutrients, and can be applied as manure [59]. Vermicomposting also has the potential to eliminate heavy metals from the industrial waste [5,60]. Table 10.4 lists various studies that explore the vermicomposting feasibility of different industrial wastes.

Due to the possible hazardous nature of industrial wastes, they are usually mixed with different bulking agents and precomposted for a few weeks before vermicomposting. Vermicomposting of industrial waste offers a sustainable alternative to reduce the negative impact of the waste on the environment. But toxic compounds such as heavy metals (chromium, cadmium, arsenic, nickel) and polycyclic aromatic hydrocarbons present in industrial waste are considered to be carcinogenic and can be harmful to earthworms also [5,59]. Hence, they require precomposting and addition of some bulking agents, which can reduce their toxicity and make them palatable to worms. Industrial waste having appropriate moisture content and significant proportions of organic wastes can be excellent feedstock for the earthworms. Tannery waste has been reported to be 100% toxic for earthworms; therefore, cow dung as an organic substrate was added by Vig et al. [61] to enhance the nutrient contents for better survival of the worms in the system. Results showed that vermicompost produced contained higher nutrient contents and lower C/N ratio and EC so that it could further be used as a valuable manure. Precomposting removes foul odor and harmful gases from the industrial wastes [5]. Further addition of bulking substrate reduces their toxicity, provides sufficient amount of nutrients and moisture, and acts as a source of microorganisms. This makes industrial waste a feasible feedstock for vermicomposting and also speeds up the process. Elvira et al. reported that the highest growth rate of worms was obtained in paper mill sludge mixed with bulking agent. Gomez et al. [62] reported that paper mill sludge and tomato plant debris were suitable feed for the optimum growth and reproduction of *E. fetida*. It has also been reported that a higher proportion of sludge in feedstock has lethal effects and causes worm mortality [63]. Pulp and paper mill waste spiked with cow dung was vermicomposted. In the final vermicompost, the total phosphorus content increased by 76% and the total nitrogen content increased by 58.7%, with 74% reduction in the total organic carbon content [64].

Waste carbide sludge (by-product of acetylene gas production) spiked with vegetable waste, cow dung, sawdust, and dried leaves in different proportions has been vermicomposted by Varma et al. [24]. Earthworm population and biomass increased continuously during the process and 1.5%–2% sludge concentration was found most effective. It is inferred from the study that a very little amount of this sludge can be added to feedstock, which makes it doubtful to use vermicomposting technology for the management of waste carbide sludge. Ghosh et al. [65] investigated the feasibility of vermicomposting of tea factory waste, generated during tea manufacturing, after mixing with cattle manure in different ratios. The study mainly focused on the role of phytase and acid phosphatase enzymes in the phosphorus mineralization. Acid phosphatase activity was more prominent during the initial 50 days; however, phytase contributed to the latter part of vermicomposting. Similarly, Mahaly et al. [66] investigated the vermicomposting potential to biodegrade distillery sludge after mixing it with tea leaf waste in different ratios. The study reported that after 45 days of

TABLE 10.4
Vermicomposting of Various Industrial Wastes.

S. No.	Organic Waste	Earthworm Species	Duration	Salient Features	References
1	Distillery sludge waste with tea leaf residues	*Eisenia fetida*	45 days	Study reported that after 45 days of vermicomposting, nutrient content of the waste increased, although TOC and C/N ratio decreased.	Mahaly et al. [66]
2	Tea factory waste and cattle manure	*Eudrilus euginae*	80 days	Tea waste was mixed with cattle manure in different proportions. Worms survive up to 1:1 (tea waste/cattle manure) ratio and vermicompost prepared from 25% tea factory waste and 75% cattle manure was found to be best. Phytase enzyme in vermicompost was positively correlated with P mineralization. Acid phosphatase enzyme contributed to the P mineralization in the first 50 days, whereas phytase enzyme contributed to the latter part of vermicomposting.	Ghosh et al. [65]
3	Aquaculture sludge	*Eisenia andrei*	126 days	Vermicomposting of three types of aquaculture sludge was carried out at four inclusion levels. Vermicomposts produced were found suitable for use in agriculture fields and earthworms, in terms of heavy metals, were also found safe to be used as fish.	Kouba et al. [106]
4	Milk industry sludge	*E. fetida*	90 days	Nitrogen content increased between 23% and 46% and total available phosphate content increased between 39% and 47% in the vermicompost. While pH, electric conductivity, TOC content, C/N ratio, total Na and K content, and heavy metal content decreased after vermicomposting.	Singh et al. [127]
5	Agro-industrial waste	*E. fetida*	135 days	Higher nutrients were found after vermicomposting. Positive correlation was found between Mg and K and P and Ca. Successful conversion of filter cake and orange peel into vermicompost.	Pigatin et al. [128]

Continued

TABLE 10.4
Vermicomposting of Various Industrial Wastes.—cont'd

S. No.	Organic Waste	Earthworm Species	Duration	Salient Features	References
6	Fly ash	*E. fetida*	1 week	Study experimented on the effect of inoculation of EM with earthworms during vermicomposting. EM + *E. fetida* treatment resulted in higher weekly Olsen P release than *E. fetida* and EM alone and control.	Mupambwa et al. [27]
7	Tannery waste (cow dung, leaf litter)	*E. eugeniae*	25	In vermicompost, maximum concentration of phytohormones (mg/kg) was recorded as indole-3-acetic acid, 7.37; kinetin, 2.8; and gibberellic acid, 35.7. C/N ratio of vermicompost was in the range of 17.3–10.3. Microbial count was highest at 21 days of vermicomposting.	Ravindran et al. [129]
8	Jute mill waste (cow dung + vegetable waste)	*Metaphire posthuma*		TOC content, pH, and heavy metal content were reduced; NPK, humic acid and fulvic acid, and microbial carbon contents increased after vermicomposting.	Das et al. [103]
9	Leather industry waste (cow dung)	*E. fetida*	135 days	Vermicomposting process affected humic acid's biostimulant properties.	Scagalia et al. [130]
10	Press-mud sludge (cow dung)	*E. fetida*	135 days	Minimum mortality and highest earthworm weight was achieved in 25% press-mud and 75% cow dung combination. Further increasing the percentage of press-mud significantly affected the number and weight of worms.	Bhat et al. [131]
11	Palm oil mill sludge	*E. eugeniae*	—	FTIR spectra indicates increased mineralization of polysaccharides, carbohydrates, and aliphatic methylene compounds in the vermicompost. SEM analysis revealed that vermicompost was more fragmented than initial wastes and control. Larger surface area was proved by BET analysis and lower mass loss by TG analysis.	Lim and Wu [38]

TABLE 10.4
Vermicomposting of Various Industrial Wastes.—cont'd

S. No.	Organic Waste	Earthworm Species	Duration	Salient Features	References
12	Pulp and paper mill waste (cow dung + food processing waste)	*Perionyx excavatus*		It was discovered that the total phosphorus and total nitrogen contents increased by 76.1% and 58.7%, respectively, but the TOC content decreased by 74.5%. It was also concluded that vermicomposting was a better option to treat and dispose of the sludge produced by pulp and paper mills.	Sonowal et al. [64]
13	Pharmaceutical industry waste	*E. fetida*		Five different mixtures were taken, using cow dung as the mixing substrate. The vermicompost so produced was rich in nutrients.	Singh and Suthar [132]
14	Food industry waste (biogas plant slurry)	*E. fetida*	90 days	Earthworms were unable to survive in 100% food industry sludge. Vermicompost contains higher nutrient. Bulking material (60%–70%) is necessary for vermicomposting.	Yadav and Garg [26]

BET, Brunauer-Emmett-Teller; *EM*, effective microorganisms; *FTIR*, Fourier transform infrared; *NPK*, nitrogen, phosphorus, and potassium; *SEM*, scanning electron microscopy; *TG*, thermal gravimetry; *TOC*, total organic carbon.

vermicomposting, the nutrient content of the waste increased but the total organic carbon content and C/N ratio decreased. Direct use of aquacultural sludge as a fertilizer is problematic because of its pathogenic nature. Kouba et al. [106] suggested that the aquaculture sludge obtained from fish culture is also a suitable feedstock for vermicomposting and this process converts it into a safe product and earthworm biomass.

APPLICATIONS OF VERMICOMPOST

Vermicompost for Organic Farming

Soil health and sustainability is at risk due to the excessive application of agrochemicals including fertilizers and pesticides. Long-term use of chemicals not only leads to soil deprivation but also reduces crop quality and disrupts the natural balance and health of the ecosystem. Situations become more problematic due to the increasing demand of food for the increasing population. Organic manures and other organic amendments have been promoted as a replacement for chemical fertilizer to produce quality food and maintain soil health.

Organic farming is continuously gaining attention at the global level and vermicompost is one of the organic amendments in demand as organic fertilizer. Use of vermicompost as organic fertilizer helps in augmentation of organic farming. Vermicompost, added to plant-growing media, synergistically influences the plant growth directly or indirectly due to its chemical, physical, and biological properties. Compared with conventional composts, vermicomposts have better physical properties, nutrient availability, and plant-growth-promoting characteristics. Properties that make vermicompost an excellent amendment and superior fertilizer over other fertilizers are presented in Table 10.5. Vermicompost has lower C/N ratio and higher porosity, water-holding capacity, and available nutrients than compost [39]. Vermicompost also possesses finer texture and lower heavy metal content than compost [67]. Emperor and Kumar [68] found that vermicompost of tea waste contains higher number of species of

TABLE 10.5
Various Physical, Chemical, and Biological Properties of Vermicompost.

Vermicompost Properties	References
Vermicompost is a homogeneous, odor free, and peatlike compound. It is easy to apply, handle, and store.	Edwards and Bohlen [22]
Physical properties such as bulk density, porosity, moisture, and water holding capacity of the vermicompost are better than those of other manures.	Moradi et al. [133]
Vermicompost is rich in all the essential plant nutrients. Micronutrients and macronutrients are also higher in vermicompost than in the other commercially available organic manures.	Kiyasudeen et al. [100]
Vermicompost contains a higher amount of essential nutrients such as N, P, and K and it slowly releases the nutrients, which makes it an effective plant growth promoter.	Atiyeh et al. [34]
Vermicompost has the ideal C/N ratio between 15 and 20.	Hait and Tare [36]
Vermicompost is rich in beneficial microbes such as nitrogen fixers, phosphate solubilizers, and cellulose decomposers.	Parthasarathi et al. [122]
Vermicompost contains many plant growth regulator hormones such as auxins, gibberellins, cytokinins, and indole-3-acetic acid.	Scagalia et al. [130]
Vermicomposts are superior than other organic and inorganic amendments due to the high population of bacteria, actinomycetes, and fungi.	Huang et al. [42,113]
Vermicompost is free from pathogens and toxic elements.	Soobhany et al. [6]
Vermicompost improves soil structure, texture, aeration, and water holding capacity and prevents soil erosion.	Lazcano and Dominguez [69]
Vermicomposts are extremely high in various humic substances such as humines, humic acids, and fulvic acids. Humic substances bind with plant growth regulators and significantly affect crop productivity.	Canellas et al. [78]
Humic compounds present in vermicompost increase the humification process and maintain organic matter in optimum range.	Arancon et al. [77]
Vermicompost has a high concentration of hydrophilic groups and high surface area, which offer more sites to hold microbes for easy breakdown and strong absorption for nutrient grip.	Lazcano et al. [70]
Vermicompost use significantly increased the microbial biomass nitrogen, carbon content, respiration rate, and dehydrogenase activity in soil.	Manivannan et al. [134]
Various enzyme activities (dehydrogenase, urease, β-glucosidase, phosphatase, arylsulfatase) are also higher in vermicompost-amended soils.	Yang et al. [73]

bacteria, fungi, and actinomycetes than traditional compost. Soobhany et al. [6] reported that SEM images of vermicompost were more scattered and fragmented than those of compost, which confirms its maturity as manure. Thermogravimetric analysis also revealed that vermicompost has lesser mass loss than compost, which means vermicompost is more stable. Despite the differences, both compost and vermicompost modified soil properties, leading to higher nutrient content than chemical fertilizer. Biologically also, vermicompost promotes a diverse range of useful enzymes, hormones, and microbes [69].

Globally, organic food market is also growing day by day and this offers a great possibility for research in the field of vermicomposting. Significant literature is available on the agronomic influence of vermicompost on different plants including vegetables, cereal crops, ornamental and medicinal plants, trees, etc. [70]. These studies have been conducted at laboratory as well as field scales. Some of such studies that reported the potential of vermicompost as organic fertilizer are shown in Table 10.6. These studies revealed that the application of vermicompost can improve seed germination, dry biomass, and growth and development of plants including plant productivity.

A scientometric analysis of 252 research papers revealed that vermicompost improved not only the plant growth and productivity but also the physicochemical properties of soil.

Gajalakshmi and Abbasi [71] studied the effect of neem vermicompost on the growth and yield of brinjal (*Solanum melongena*). The results showed that the use of vermicompost improved plant height, root length, and fruit yield and reduced the time of flowering. Vermicompost when applied to basil plants resulted in better plant growth than with urea [72]. Similarly, Yang et al. [73] made a comparison of vermicompost, compost, and chemical fertilizer in different irrigation levels on the yield and quality of tomato plants. Results revealed that the highest yield was achieved in vermicompost applied plants. Vermicompost also increased the vitamin C content and enzymes in soil as compared with compost and chemical fertilizer [73].

Kizilkaya et al. [74] carried out an experiment to study the effect of vermicompost (prepared from sewage sludge, husk, and cattle manure) on wheat yield, and the authors reported positive effects on the wheat yield and nutrient content as compared with the wheat grown in nonvermicomposted fields. Najjari and Ghasemi [49] investigated the effect of blood powder vermicompost on the growth and nutrition of cucumber and reported more nutrient content in cucumber. Vermicompost application with ZnO nanoparticles had positive effects on the fresh and dry weight, shoot and root length, chlorophyll content, total free phenols, and sugar content in *Arachis hypogaea* L. [75]. This study gives the possibility of growing crops in zinc-deficient soils. Gong et al. [76] compared the effects of compost and vermicompost on geranium (*Pelargonium zonale*) and calendula (*Calendula officinalis*) plants. Results point out that the use of vermicompost was superior to compost in terms of plant growth and flowering. Vermicompost prepared from weeds is also safe, as it reduces the allelopathic compounds present in the weeds. Hussain et al. [57] used the *Ipomoea* weed vermicompost at 0, 2.5, 3.75, and 5 tons/ha for growing okra. Vermicompost significantly affected germination and growth in okra, and best results were obtained when vermicompost was applied at 5 tons/ha. The fruit quality was also enhanced in terms of mineral, protein, and carbohydrate contents and there were reduced incidences of disease and pest attacks.

It is evident from various studies that vermicompost application improved the plant growth due to increased nutrient content of the soil or experimental media. Enhanced plant growth has also been attributed to physical and chemical alteration, plant growth hormones, and most effectively humic acid content of vermicompost [77]. Humic substances extracted from vermicompost are capable of inducing lateral root growth in plants by stimulation of the plasma membrane proton ATPase activity, thus producing an effect similar to the one achieved by exogenous application of indole-3-acetic acid [78]. Vermicompost plays an important role in the promotion of cell division and proliferation, and hence, it is good for plant health and crop productivity.

Vermicompost in Bioremediation

Vermicomposting has been used by several authors as bioremediation technique for the removal of toxic chemicals present in soil and waste. Vermicomposting has the potential to ameliorate various pollutants in soil, such as polycyclic hydrocarbons, polychlorinated biphenyl, pesticides, and heavy metals. Vermicomposting significantly reduced the mobile arsenic content in sludge [79]. Martín-Gil et al. [80] reported that microorganisms present in the earthworm intestine and vermicompost possibly degrade and remove the toxic chemicals. Lin et al. [81] reported that use of vermicompost helps in the degradation of pentachlorophenol, chlorinated phenols, and other aromatics from the soil. Vermicompost application also helps in mitigation of Diuron, a herbicide, by promoting the development of Diuron-degrading microbial community [28]. Lin et al. [82] reported atrazine degradation in soil by using earthworms. Datta et al. [83] compared the genotoxicity of pesticide- and vermicompost-treated soil by the *Allium cepa* test. The results indicated that the pesticide-treated soil was more cytotoxic and genotoxic with higher cell aberrations.

Use of vermicompost as fertilizer also helps in suppression of plant diseases and pest attacks in plants because it provides better nutrient availability and provides greater strength, immunity, and resistance to plants against infection [84]. Xiao et al. [84] reported that compost and vermicompost are effective in eliminating root-knot nematodes (*Meloidogyne incognita*) in tomato plants. The antinematicidal efficiency of vermicompost was better than compost.

TABLE 10.6
Use of Vermicompost as Fertilizer for the Plant Growth and Development.

S. No.	Type of Waste Used for Vermicompost	Experimental Plant	Impact on Plant Growth and Development	References
1	Sawdust and blood powder	Cucumber	Use of 5% blood powder VC increased the concentration of N, P, Fe, Zn, and Cu in the leaves of cucumber.	Najjari and Ghasemi [49]
2	Leaf litter and sewage sludge	Maize	VC use exhibited significantly higher seed germination, seedling biomass, and root activity in maize.	Wu et al. [105]
3	Parthenium	*Arachis hypogaea* L.	The study investigated the synergistic effect of parthenium-based VC and zinc oxide nanoparticles on the growth of *Arachis* in zinc-deficient soil. Various growth- and yield-related parameters were significantly improved.	Rajiv and Vanathi [75]
4	*Ipomoea* weed	Lady's-finger (*Abelmoschus esculentus*)	VC applied at the rate of 0, 2.5, 3.75, and 5tons/ha enhanced the germination and growth in lady's-finger. Fruit quality was also enhanced (high in minerals, proteins, and carbohydrates). Disease incidence and pest attacks were found to be reduced with VC application. Best results were obtained with 5 tons/ha vermicompost.	Hussain et al. [57]
5	Not known	Strawberry	VC application on strawberry increases various growth parameters, fruit yield, soluble sugar levels, and vitamin C content. VC significantly improved photosynthesis rate and various microbial and enzymatic activities in soil.	Zuo et al. [135]
6	Green waste	Geranium (*Pelargonium zonale* L.) and calendula (*Calendula officinalis* L.)	Compost and VC increased bulk density, pH, EC, air space, and nutrient content and decreased porosity of growing media. Heavy metal contents are in safe limits. Soil fertility and plant growth were superior in the VC-based media.	Gong et al. [76]
7	*Salvinia molesta*	Lady's-finger (*A. esculentus*), cucumber (*Cucumis sativus*), and green gram (*Vigna radiata*)	VC enhanced the germination success and promoted the morphologic growth and biochemical content of the different plant species studied. It also bestowed plant-friendly physicochemical and biological attributes to the soil.	Hussain et al. [56]

TABLE 10.6
Use of Vermicompost as Fertilizer for the Plant Growth and Development.—cont'd

S. No.	Type of Waste Used for Vermicompost	Experimental Plant	Impact on Plant Growth and Development	References
8	Cow manure	Pepperleaf (*Piper auritum*)	Treatments included VC at the rate of 10, 20, and 30 g/plant; vermiwash at the rate of 5, 10, and 15 mL/plant; and rock phosphate at the rate of 1, 2, and 3 g/plant. VC, vermiwash, and rock phosphate had no statistically significant effect on plant growth.	Hidalgo et al. [136]
9	Distillation waste of aromatic crops	Basil (*Ocimum basilicum*)	VC and tannery sludge were added to a sodic soil at the extent of 5 tons/ha. In addition, two bacterial strains isolated from the same sodic soil were also mixed in the soil. Unamended soil served as control. VC, tannery sludge, and microbial inoculants enhanced the height, number of branches, biomass, root expansion, yield, and oil quality.	Trivedi et al. [137]
10	Food, paper, and yard waste	Green bean (*Phaseolus vulgaris*)	In plastic pots, mineral brown earth soil was fortified with VC and compost to the extent of 0% (control), 10%, 20%, 30%, 40%, 50%, and 100% (v/v). Compared with the compost and controls, VC at the rate of 40% application led to higher growth and yield. With further increase in the compost and VC concentration, the growth and yield declined.	Soobhany et al. [138]
11	Vegetable waste, agriculture crop residue, weeds, and cow dung	Tomato and cabbage	In field experiment, VC, compost, farmyard manure, and inorganic fertilizers were supplemented to the soil to make the NPK 75-60-60 t/ha for tomato and 120-60-60 t/ha for cabbage soil. Combined treatment of VC, compost, and inorganic fertilizers stimulated the growth, yield, quality, and storage longevity of both tomato and cabbage.	Goswami et al. [139]
12	Not stated	Basil (*O. basilicum* L.)	Mixed organic fertilizer gives maximum yield. Organic fertilizer improved Ca, Mg, and vitamin C contents and antioxidant activity in the basil leaf. Major volatile constituents were methyl chavicol and linalool.	Pandey et al. [140]

continued

TABLE 10.6
Use of Vermicompost as Fertilizer for the Plant Growth and Development.—cont'd

S. No.	Type of Waste Used for Vermicompost	Experimental Plant	Impact on Plant Growth and Development	References
13	Not stated	Japanese mint (*Mentha arvensis* L.)	Increased growth is attributed to VC. Integration of FYM, VC, and poultry manure resulted in maximum herbage yield, oil recovery, and oil quality	Bajeli et al. [141]
14	Cattle manure	Sweet corn (*Zea mays*)	Treatments include main plots consisting of five VC application rates (5, 10, 15, 20, and 25 Mg/ha) and subplots consisted of five liquid organic fertilizer application rates (0%, 25%, 50%, 75%, and 100%, initial concentration). VC noticeably increased the NPK uptake and plant height, plant leaf area, shoot weight, weight of the husked and unhusked ears, diameters of the ears, and weight of the husked ear per plot. The higher the VC concentration, the greater the impact.	Muktamar et al. [142]
15	Rubber plant leaves	Pineapple (*Ananas comosus*)	VC improved pineapple yield and fruit weight and increased average length, width, and the number of leaves per plant	Chaudhuri et al. [143]
16	Cow dung and *Leucaena leucocephala* leaves	Sunflower (*Helianthus annuus*)	In a randomized complete block design, plants were grown in three sets, which comprise set 1, recommended dose of NPK; set 2, 1 kg/plant VC; and set 3, 1 L/plant biogas slurry. In addition, five pots from each set were irrigated with 0.5, 4.8, and 8.6 dS/m EC levels of saline water. VC and biogas slurry improved the growth, yield, nitrate, and protein content and decreased sodium-induced inhibitory effects on sunflower. An increase in the nitrogen-assimilating enzymes was also recorded.	Jabeen and Ahmad [144]
17	Pig manure and wheat straw	Radish (*Raphanus sativus*), onion (*Allium cepa*), marigolds (*Tagetes* spp.), and grass (*Trifolium* spp.)	Plants were grown in soil mixed with VC in concentrations of 0%, 2%, 4%, 6%, 10%, and 12.5%. In another set of greenhouse outdoor	Kovshov and Iconnicov [145]

TABLE 10.6
Use of Vermicompost as Fertilizer for the Plant Growth and Development.—cont'd

S. No.	Type of Waste Used for Vermicompost	Experimental Plant	Impact on Plant Growth and Development	References
			experiment, VC was added to the soil to the extent of 0, 1, 3, 5, and 7 kg/m^2. In both the pot and the plot experiments, VC improved the growth of all the plants studied.	
18	Buffalo manure	Maize (*Z. mays*)	VC was applied at the rate of 20 t/ha to soil pretreated with chemical fertilizers and it was observed that growth (plant biomass) and yield of maize was improved.	Doan et al. [146]
19	Warm cast, chicken manure, horse manure, maize straw, and rice chaff	Tomato	VC 91.43 g/kg soil (0.64 g N/kg soil) chicken compost, horse compost, and chemical fertilizer. VC increases the sugar/acid ratio; vitamin C, lycopene, and soluble protein levels; and the activities of the enzymes acid phosphatase, catalase, and urease.	Yang et al. [73]
20	Food waste, paper waste	Peppers	VC significantly increased plant shoot biomass and marketable fruit weights and decreased yields of nonmarketable fruit of peppers.	Arancon et al. [77]

EC, electric conductivity; *FYM*, farmyard manure; *NPK*, nitrogen, phosphorus, and potassium; *VC*, vermicompost.

Possible reasons for bioremediation by vermicomposting are

- strong interaction of humic substances with toxic organic compounds accelerates their biodegradation [28],
- higher microbial diversity significantly stimulates degradation of toxic substances [42],
- the higher porosity and large surface area of vermicompost provide more adsorption sites [10],
- earthworm mucus, urine, and cast mineral bacterial interaction also promote soil decontamination [85],
- earthworms accumulate heavy metals via skin absorption or through their intestine [86].

Vermicompost for Wastewater Treatment

Vermicomposting has also been also used as a cost-effective technique for wastewater treatment. Vermifiltration is the process in which vermicompost prepared from different wastes is utilized for wastewater treatment. Taylor et al. [87] used a vermicompost filter bed for the domestic wastewater treatment. Li et al. [88] used swine manure vermicompost, and Lourenco and Nunes [89] compared different stage vermifiltration and reported that four-stage vermifiltration significantly reduced various pollutants, such as biological oxygen demand, chemical oxygen demand, and total suspended solid, present in wastewater. Large surface area and high cation exchange capacity make vermicompost a suitable low-cost organic adsorbent [10,37]. Hence, vermicompost is used for the removal of heavy metals and dyes from wastewater [9]. Zhu et al. [90] reported that vermicompost prepared from cattle manure removed lead and cadmium from wastewater. Vermicompost adsorbs heavy metals more effectively than cattle manure. Vermicompost successfully adsorbs cadmium, copper, nickel, zinc, and chromium from the synthetic effluents. Jordao et al. [91] reported 100% efficiency of vermicompost in removing heavy metals from (Cu^{2+}, Ni^{2+}, Al^{3+}, Fe^{2+}, and Zn^{+}) galvanoplastic effluents. This may be attributed to the increased pH during the experiment and the presence of phosphates in vermicompost. Not only heavy metals but also different cationic dyes such as crystal violet and methylene blue were removed using vermicompost as the adsorbent [92]. Barbosa et al. [93] studied the effect of dried vermicompost as the adsorbent for remediation of cadmium, copper, manganese, nickel, lead, and zinc from laboratory wastewater. The study reported that vermicompost increased the pH of the solution and efficiently adsorbed all the heavy metals. Manyuchi et al. [94] used vermifiltration for decontamination of distillery wastewater and found more than 90% reduction in various pollutants, along with nutrient-rich vermicompost.

VERMICOMPOSTING IN A CIRCULAR ECONOMY

The concept of circular economy is an approach to achieve economic growth, giving importance to sustainability and environmental development. Circular economy emphasized the utilization of different by-products and recycling of materials via different methods. From the perspective of sustainable development, Korhonen et al. [11] suggested that circular economy should concentrate on the use of nature's cycles for preserving materials, energy, and nutrients for economic use. Circular economy mainly focuses on the less utilization of resources to increase the efficiency of materials and energy use, minimum waste production, by-product utilization, and recycling and reuse of waste.

China is a country with a developing economy and it has embraced the "circular economy" concept. Using 2003 as a baseline, China achieved various goals in 2010 of increasing the reuse of industrial solid waste by a rate of around 60% over 7 years [95]. During vermicomposting, waste (which has negative environmental and economic value) is converted into useful products, namely, vermicompost and earthworms, which have positive environmental and economic value (Fig 10.5). Vermicomposting is considered a viable waste management method and it is an important link within the circular economy. Hence, the vermicomposting technology must be centralized to maintain global sustainability and circular economy without any harm to the environment.

Economic analysis of vermicomposting revealed that except a few cases, vermicomposting is a profitable activity. A case study in Maharashtra (India) reported that vermicomposting of a temple's solid wastes generates profit [96]. However, profit from vermicomposting is variable and depends on the market location, vermicompost cost and quality, transportation cost, and waste availability. Bogdanov [97] reported that vermicompost costs between 13,625 and 68,125 INR/ton. However, Moledor et al. [98] reported that vermicompost costs Rs. 155,325 to Rs. 171,057.63 per ton. In Uganda, the selling cost of vermicompost was Rs. 5.45/kg [99]. Not only vermicompost but also earthworm biomass is a valuable by-product obtained during vermicomposting and it can be utilized as animal feed, in medicine, and for preparing other products.

CONSTRAINTS IN THE POPULARIZATION OF VERMICOMPOSTING

Vermicomposting is regarded as a clean, sustainable, and zero-waste approach to manage organic wastes but there are still some constraints in the popularization of vermicomposting. Instead of increasing research in the field of vermicomposting, practical application of vermicomposting needs more attention. Vermicomposting on a large scale is required to solve the problem of waste disposal effectively and on a global level. One of the major constraints is the lack of awareness and proper knowledge regarding vermicomposting and the use of vermicompost. It is necessary to guide farmers about vermicomposting and the appropriate use of vermicompost by organizing various training and extension activities. Innovative and effective agricultural activities must be developed to educate and assist farmers for organic farming. They should be educated about the process of vermicomposting and quantities of vermicompost that should be applied to achieve the best results in agricultural fields [9]. Higher cost of organic fertilizer than synthetic fertilizer is also an obstacle for farmers to adopt organic farming on a large scale. Mass application of vermicompost is not achieved due to the failure of policy implementation related to vermicompost technology.

Maintaining a continuous supply of organic waste, water, temperature, and moisture are major hurdles that complicate the process of vermicomposting [100]. Most of the wastes require a bulking substrate (most extensively, animal manure) and precomposting before vermicomposting. Transportation cost also makes the process more expensive. The combined effect of all these problems has hindered the commercial scale popularity of vermicomposting. However, a few studies are available that reported direct vermicomposting of organic wastes (phytomass) without any precomposting or manure supplementation [98].

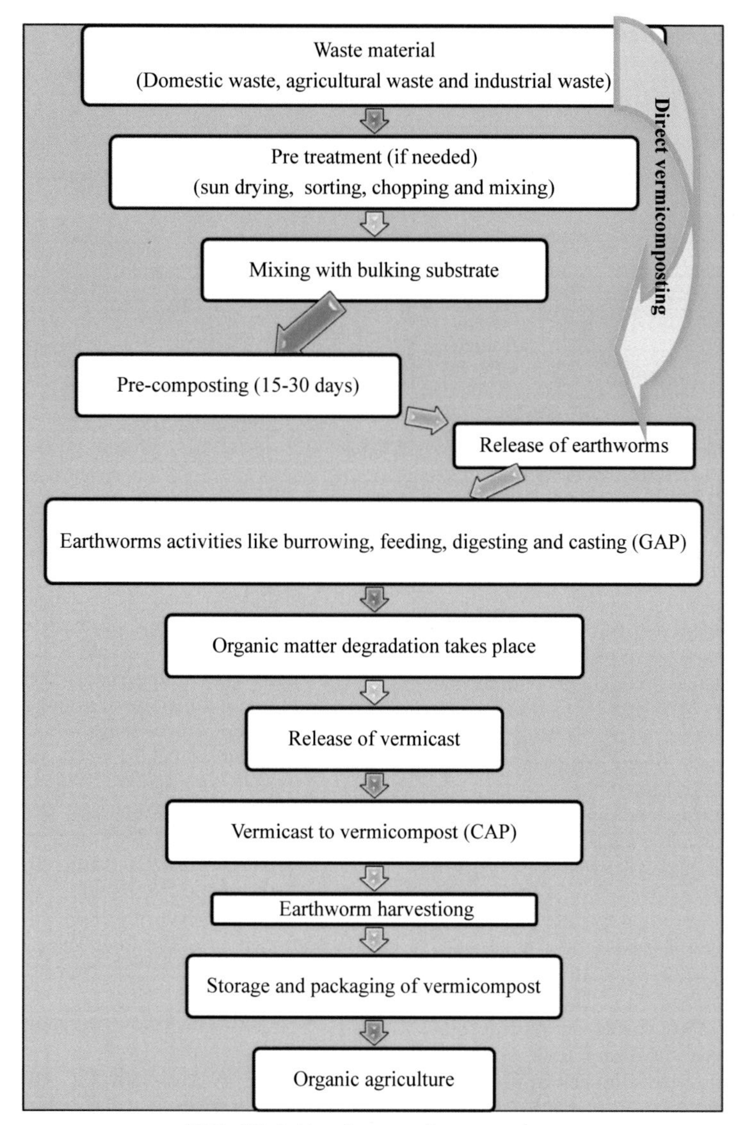

FIG. 10.4 Vermicomposting procedure.

Furthermore, some problems are also associated with the use of vermicompost as fertilizer, such as phytotoxic substances, high salt concentrations, and heavy metal content, which negatively affect plant growth and development [39]. In some cases, use of immature vermicompost also prevented seed germination and plant growth [2]. This may be due to the difference in soil type and meteorologic conditions of that particular area. Future research must be focused on the development of high rates of vermicomposting directly without any supplementation and on proficiency in the technology. In-depth research is required to study the composition of immature vermicompost and its application failure and to determine vermicompost concentrations under typical soil-water plant-micrometeorologic regimes. This will help in the augmentation of organic farming and popularization of vermicomposting for environmental sustainability.

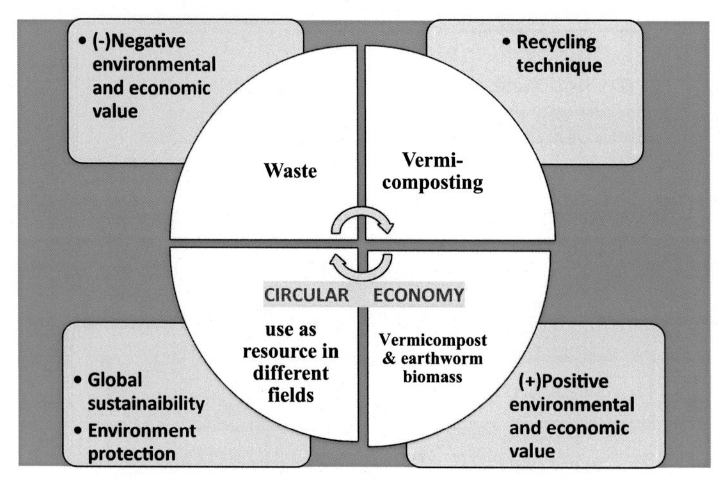

FIG. 10.5 Vermicomposting versus circular economy.

REFERENCES

[1] Griffin S, Sarfraz M, Farida V, Jawad M. No time to waste organic waste: nanosizing converts remains of food processing into refined materials. Journal of Environmental Management 2018;210:114–21.

[2] Hussain N, Abbasi SA. Efficacy of the vermicomposts of different organic wastes as "Clean" fertilizers: state-of-the-Art. Sustainability 2018;10(4):1205.

[3] Sharma K, Garg VK. Vermi-modificaion of ruminant excreta using *Eisenia fetida*. Environmental Science and Pollution Research 2017a;24(24):19938–45.

[4] Sharma K, Garg VK. Comparative analysis of vermicompost quality produced from rice straw and paper waste employing earthworm *Eisenia fetida* (Sav.). Bioresource Technology 2018;24(8):7829–36. 2018.

[5] Lee LH, YeongWu T, Shak KPY, Lim SL, Ng KY, Nguyen MN, Teoh WH. Sustainable approach to biotransform industrial sludge into organic fertilizer via vermicomposting: a mini-review. Journal of Chemical Technology and Biotechnology 2017;93:925–35.

[6] Soobhany N, Gunasee S, Rago YP, Joyram H, Raghoo P, Mohee R, Garg VK. Spectroscopic, thermogravimetric and structural characterization analyses for comparing Municipal Solid Waste composts and vermicomposts stability and maturity. Bioresource Technology 2017a;263: 11–9. https://doi.org/10.1016/j.biortech.2017.03.161.

[7] Pramanik P, Chung YR. Changes in fungal population of fly ash and vinasse mixture during vermicomposting by *Eudrilus eugeniae* and *Eisenia fetida*: documentation of cellulase isozymes in vermicompost. Waste Management 2011;31:1169–75.

[8] Dominguez J, Edwards CA. Biology and ecology of earthworm species used for vermicomposting. In: Vermiculture technology-earthworms, organic wastes, and environmental management. Boca Raton, FL: CRC Press; 2011. p. 27–40.

[9] Pereira MG, Neta LCS, Fontes MPF, Souza AN, Matos TC, Sachdev EL, Santos AV, Souza MOG, Andrade MAVS, Paulo GMM, Ribeiro JN, Ribeiro AVFN. An overview of the environmental applicability of vermicompost: from wastewater treatment to the development of sensitive analytical methods, vol. 2014. Hindawi Publishing Corporation; 2014. Article ID 917348, 14 pages.

[10] Dey MD, Das S, Kumara R, Doleya R, Bhattacharya, Mukhopadhyaya R. Vermiremoval of methylene blue using *Eisenia fetida*: a potential strategy for bioremediation of synthetic dye-containing effluents. Ecological Engineering 2017;106:200–8.

[11] Korhonen J, Honkasalo A, Seppala J. Circular economy: the concept and its limitations. Ecological Economics 2018;143:37–46.

[12] Edwards CA. Breakdown of animal, vegetable and industrial organic wastes by earthworms. In: Edwards CA, Neuhauser EP, editors. Earthworms in waste and environmental management. The Hague, the Netherlands: SPB Academic Publishing; 1988. p. 21. 31.

[13] Sinha RK. Vermiculture biotechnology for waste management and sustainable agriculture. In: Sinha RK, editor. Environmental crisis and human's at risk. India: INA Shree Publication; 1996. p. 233–40.

[14] Xiang H, Zhang J, Zhu Q. Worldwide earthworm research: a scientometric analysis, 2000–2015. Scientometrics 2015;105:1195–207.

[15] Gomez Brandon M, Dominguez J. Recycling of solid organic wastes through vermicomposting: microbial community changes throughout the process and use of vermicompost as a soil amendment. Critical Reviews in Environmental Science and Technology 2014; 44(12):1289–312.

[16] Hussain N, Das S, Goswami L, Das P, Sahariah B, Bhattacharya SS. Intensification of vermitechnology for kitchen vegetable waste and paddy straw employing earthworm consortium: assessment of maturity time, microbial community structure, and economic benefit. Journal of Cleaner Production 2018;182:414–26.
[17] Suthar S. Potential of *Allolobophora parva* (Oligochaeta) in vermicomposting. Bioresource Technology 2009; 100(24):6422–7.
[18] Azizi AB, Lim MPM, Noor ZM, Abdullah N. Vermiremoval of heavy metal in sewage sludge by utilising *Lumbricus rubellus*. Ecotoxicology and Environmental Safety 2013;90:13–20.
[19] Martin M, Eudoxie G. Feedstock composition influences vermicomposting performance of *Dichogaster annae* relative to *Eudrilus eugeniae* and *Perionyx excavatus*. Environmental Science and Pollution Research 2018:1–10.
[20] Hussain N, Singh A, Saha S, Kumar MVS, Bhattacharya P, Bhattacharya SS. Excellent N-fixing and P-solubilizing traits in earthworm gut-isolated bacteria: a vermicompost based assessment with vegetable market waste and rice straw feed mixtures. Bioresource Technology 2016a;222:165–74.
[21] Arumugam K, Renganathan S, Oluranti BO, Muthunarayanan V. Investigation on paper cup waste degradation by bacterial consortium and *Eudrilus eugeinea* through vermicomposting. Waste Management 2017;74:185–93.
[22] Edwards CA, Bohlen PJ. Biology and ecology of earthworms. London: Chapman and Hall; 1996.
[23] Alidadi H, Hosseinzadeh A, Najafpoor AA, Esmaili H, Zanganeh J, Takabi MD, Piranloo FG. Waste recycling by vermicomposting: maturity and quality assessment via dehydrogenase enzyme activity, lignin, water soluble carbon, nitrogen, phosphorous and other indicators. Journal of Environmental Management 2016;182: 134–40.
[24] Varma VS, Yadav J, Das S, Kalamdhad AS. Potential of waste carbide sludge addition on earthworm growth and organic matter degradation during vermicomposting of agricultural wastes. Ecological Engineering 2015; 83:90–5.
[25] Fu X, Huang K, Chen X, Li F, Guangyu C. Feasibility of vermistabilization for fresh pelletized dewatered sludge with earthworms *Bimastus parvus*. Bioresource Technology 2015;175:646–50.
[26] Yadav A, Suthar S, Garg VK. Dynamics of microbiological parameters, enzymatic activities and worm biomass production during vermicomposting of effluent treatment plant sludge of bakery industry. Environmental Science and Pollution Research 2015;22:14702–9.
[27] Mupambwa HA, Ravindran B, Mnkeni PNS. Potential of effective micro-organisms and *Eisenia fetida* in enhancing vermi-degradation and nutrient release of fly ash incorporated into cow dung–paper waste mixture. Waste Management 2016;48:165–73.
[28] Castillo JM, Beguet J, Martin-Laurent F, Romero E. Multidisciplinary assessment of pesticide mitigation in soil amended with vermicomposted agroindustrial wastes. Journal of Hazardous Materials 2016;304:379–87.
[29] Arevalo OB, Guillen-Navarro K, Huerta E, Cuevas R, Calixto-Romo MA. Enzymatic dynamics into the *Eisenia fetida* (Savigny, 1826) gut during vermicomposting of coffee husk and market waste in a tropical environment. Environmental Science and Pollution Research 2018;25(2):1576–86.
[30] Gupta R, Garg VK. Vermiremediation and nutrient recovery of non-recyclable paper waste employing *Eisenia fetida*. Journal of Hazardous Materials 2009;162:430–9.
[31] Lores M, Gomez Brandon M, Perez-Diaz D, Dominguez J. Using FAME profiles for the characterization of animal wastes and vermicomposts. Soil Biology and Biochemistry 2006;38:2993–6.
[32] Lv B, Xing M, Yang J. Speciation and transformation of heavy metals during vermicomposting of animal manure. Bioresource Technology 2016;209:397–401.
[33] Malinska K, Zabochnicka Swiateka M, Caceresb R, Marfa O. The effect of precomposted sewage sludge mixture amended with biochar on the growth and reproduction of *Eisenia fetida* during laboratory vermicomposting. Ecological Engineering 2016;90: 35–41.
[34] Atiyeh RM, Subler S, Edwards CA, Bachman G, Metzger JD, Shuster W. Effects of vermicomposts and composts on plant growth in horticultural container media and soil. Pedobiologia 2000;44:579–90.
[35] Orlov DS, Biryukova ON. Humic substances of vermicomposts. Agrokhimiya 1996;12:60–7.
[36] Hait S, Tare V. Transformation and availability of nutrients and heavy metals during integrated composting–vermicomposting of sewage sludges. Ecotoxicology and Environmental Safety 2012;79:214–24.
[37] Kumar MS, Rajiv P, Rajeshwari S, Venckatesh R. Spectroscopic analysis of vermicompost for determination of nutritional quality. Spectrochimica Acta Part A: Molecular and Biomolecular Spectroscopy 2015;135:252–5.
[38] Lim SL, Wu TY. Characterization of matured vermicompost derived from valorization of palm oil mill by product. Journal of Agricultural and Food Chemistry 2016;64:1761–9.
[39] Lim SL, Wu TY, Lim PN, Shak KPY. The use of vermicompost in organic farming: overview, effects on soil and economics. Journal of the Science of Food and Agriculture 2015;95(6):1143–56.
[40] Mupambwa HA, Mnkeni PNS. Optimizing the vermicomposting of organic wastes amended with inorganic materials for production of nutrient-rich organic fertilizers: a review. Environmental Science and Pollution Research 2018:1–19.
[41] Sharma K, Garg VK. Management of food and vegetable processing waste spiked with buffalo waste using earthworms (*Eisenia fetida*). Environmental Science & Pollution Research 2017b;24(8):7829–36.
[42] Huang K, Xia H, Li F, Wei Y, Cui G, Fu X, Chen X. Optimal growth condition of earthworms and their vermicompost features during recycling of five different

fresh fruit and vegetable wastes. Environmental Science and Pollution Research 2016;23:13569–75.
[43] Huang K, Xia H. Role of earthworms' mucus in vermicomposting system: biodegradation tests based on humification and microbial activity. The Science of the Total Environment 2018;610:703–8.
[44] Sahariah B, Goswami L, Kim KH, Bhattachatyya P, Bhattacharya SS. Metal remediation and biodegradation potential of earthworm species on municipal solid waste a parallel analysis between *Metaphire posthuma* and *Eisenia fetida*. Bioresource Technology 2015;180:230–6.
[45] Yang J, Lv B, Zhang J, Xing M. Insight into the roles of earthworm in vermicomposting of sewage sludge by determining the water-extracts through chemical and spectroscopic methods. Bioresource Technology 2014; 154:94–100.
[46] Villar I, Alves D, Perez-Diaz D, Mato S. Changes in microbial dynamics during vermicomposting of fresh and composted sewage sludge. Waste Management 2016; 48:409–17.
[47] Vodounnou DSJV, Kpogue DNS, Tossavi CE, Mennsah GA, Fiogbe ED. Effect of animal waste and vegetable compost on production and growth of earthworm (*Eisenia fetida*) during vermiculture. International Journal of Recycling of Organic Waste in Agriculture 2016;5:87–92.
[48] Cestonaro T, de Mendonça Costa MSS, de Mendonça Costa LA, Pereira DC, Rozatti MA, Martins MFL. Addition of cattle manure to sheep bedding allows vermicomposting process and improves vermicompost quality. Waste Management 2017;61:165–70.
[49] Najjari F, Ghasemi S. Changes in chemical properties of sawdust and blood powder mixture during vermicomposting and the effects on the growth and chemical composition of cucumber. Scientia Horticulturae 2018; 232:250–5.
[50] Sharma K, Garg VK. Vermicomposting of lignocellulosic waste: a biotechnological tool for waste management. Phytoremediation of environmental pollutants, vol. 1. Taylor and Francis; 2017c. p. 26.
[51] Picman J, Picman AK. Autotoxicity in *Parthenium hysterophorus* and its possible role in control of germination. Biochemical Systematics and Ecology 1984;12:287–92.
[52] Hussain N, Abbasi T, Abbasi SA. Vermicomposting-mediated conversion of the toxic and allelopathic weed ipomoea into a potent fertilizer. Process Safety and Environmental Protection 2016b;103:97–106.
[53] Gotardo AT, Pfister JA, Raspantini PC, Górniak SL. Maternal ingestion of *Ipomoea carnea*: effects on goat-kid bonding and behavior. Toxins 2016;8(3):74.
[54] Hussain N, Abbasi T, Abbasi SA. Vermicomposting transforms allelopathic parthenium into a benign organic fertilizer. Journal of Environmental Management 2016c;180:180–9.
[55] Hussain N, Abbasi T, Abbasi SA. Transformation of toxic and allelopathic lantana into a benign organic fertilizer through vermicomposting. Spectrochimica Acta Part A: Molecular and Biomolecular Spectroscopy 2016d;163:162–9.
[56] Hussain N, Abbasi T, Abbasi SA. Generation of highly potent organic fertilizer from pernicious aquatic weed *Salvinia molesta*. Environmental Science and Pollution Research 2017;25:4989–5002.
[57] Hussain N, Abbasi T, Abbasi SA. Evaluating the fertilizer and pesticidal value of vermicompost generated from a toxic and allelopathic weed ipomoea. Journal of the Saudi Society of Agricultural Sciences 2018.
[58] Deka H, Deka S, Baruah C, Das J, Hoque S, Sarma H, Sarma N. Vermicomposting potentiality of *Perionyx excavatus* for recycling of waste biomass of java citronella – an aromatic oil yielding plant. Bioresource Technology 2011;102:11212–7.
[59] Bhat SA, Singh S, Singh J, Kumar S, Vig AP. Bioremediation and detoxification of industrial wastes by earthworms: vermicompost as powerful crop nutrient in sustainable agriculture. Bioresource Technology 2018; 252:172–9.
[60] Mupondi LT, Mnkeni PNS, Muchaonyerwa P, Mupambwa HA. Vermicomposting manure-paper mixture with igneous rock phosphate enhances biodegradation, phosphorus bioavailability and reduces heavy metal concentrations. Heliyon 2018;4(8):e00749.
[61] Vig AP, Singh J, Wani SH, Dhaliwal SS. Vermicomposting of tannery sludge mixed with cattle dung into valuable manure using earthworm *Eisenia fetida* (Savigny). Bioresource Technology 2011;102(17):7941–5.
[62] Gomez FMJ, Nogales R, Plante A, Plaza C, Fernandez MJ. Application of a set of complementary techniques to understand how varying the proportion of two wastes affects humic acids produced by vermicomposting. Waste Management 2015;35:81–8.
[63] Yadav A, Garg VK. Feasibility of nutrient recovery from industrial sludge by vermicomposting technology. Journal of Hazardous Materials 2009;168(1):262–8.
[64] Sonowal PKD, Khwairkpam M, Kalamdhad AS. Feasibility of vermicomposting dewatered sludge from paper-mills using *Perionyx excavatus*. European Journal of Environmental Science 2013;3:17–26.
[65] Ghosh S, Goswami AJ, Ghosh GK, Pramanik P. Quantifying the relative role of phytase and phosphatase enzymes in phosphorus mineralization during vermicomposting of fibrous tea factory waste. Ecological Engineering 2018;116:97–103.
[66] Mahaly M, Senthilkumar AK, Arumugam S, Kaliyaperumal C, Karupannan N. Vermicomposting of distillery sludge waste with tea leaf residues. Sustainable Environment Research 2018;28(5):223–7.
[67] Wu TY, Lim SL, Lim PN, Shak KPY. Biotransformation of biodegradable solid wastes into organic fertilizers using composting or/and vermicomposting. Chemical Engineering Transactions 2014;39:1579–84.
[68] Emperor GN, Kumar K. Microbial population and activity on vermicompost of *Eudrilus eugeniae* and *Eisenia fetida* in different concentrations of tea waste with cow dung and kitchen waste mixture. International Journal of Current Microbiology and Applied Sciences 2015; 4(10):496–507.

[69] Lazcano C, Dominguez J. The use of vermicompost in sustainable agriculture: impact on plant growth and soil fertility. Soil Nutrition 2011;10:1–23.
[70] Lazcano C, Revilla P, Malvar RA, Dominguez J. Yield and fruit quality of four sweet corn hybrids (Zea mays) under conventional and integrated fertilization with vermicompost. Journal of the Science of Food and Agriculture 2011;91(7):1244–53.
[71] Gajalakshmi S, Abbasi SA. Neem leaves as a source of fertilizer cum pesticide vermicompost. Bioresource Technology 2004;92:291–6.
[72] Cabanillas C, Stobbia D, Ledesma A. Production and income of basil in and out of season with vermicomposts from rabbit manure and bovine ruminal contents alternatives to urea. Journal of Cleaner Production 2013;47: 77–84.
[73] Yang L, Zhao F, Chang Q, Li T, Li F. Effects of vermicomposts on tomato yield and quality and soil fertility in greenhouse under different soil water regimes. Agricultural Water Management 2015;160:98–105.
[74] Kizilkaya RF, Turkay SH, Turkmen C, Durmus M. Vermicompost effects on wheat yield and nutrient contents in soil and plant. Archives of Agronomy and Soil Science 2012;58:696–777.
[75] Rajiv P, Vanathi P. Effect of Parthenium based vermicompost and zinc oxide nanoparticles on growth and yield of *Arachis hypogaea* L. in zinc deficient soil. Biocatalysis and agricultural biotechnology 2018;13: 251–7.
[76] Gong X, Li S, Sun X, Wang L, Cai L, Zhang J, Wei L. Green waste compost and vermicompost as peat substitutes in growing media for geranium (*Pelargonium zonale* L.) and calendula (*Calendula officinalis* L.). Scientia Horticulturae 2018;236:186–91.
[77] Arancon NQ, Edwards CA, Bierman P, Metzger JD, Lucht C. Effects of vermicomposts produced from cattle manure, food waste and paper waste on the growth and yield of peppers in the field. Pedobiologia 2005;49: 297–306.
[78] Canellas LP, Olivares FL. Physiological responses to humic substances as plant growth promoter. Chemical and Biological Technologies in Agriculture 2014;1(1):3.
[79] Vasickova J, Manakova B, Sudoma M, Hofman J. Ecotoxicity of arsenic contaminated sludge after mixing with soils and addition into composting and vermicomposting processes. Journal of Hazardous Materials 2016;317: 585–92.
[80] Martín-Gil J, Gómez-Sobrino E, Correa-Guimaraes A, Hernández-Navarro S, Sánchez-Báscones M, del Carmen Ramos-Sánchez M. Composting and vermicomposting experiences in the treatment and bioconversion of asphaltens from the Prestige oil spill. Bioresource Technology 2008;99(6):1821–9.
[81] Lin Z, Baia J, Zhen Z, Laoa S, Li W, Wua Z, Li Y, Spirod B, Zhang D. Enhancing pentachlorophenol degradation by vermicomposting associated bioremediation. Ecological Engineering 2016;87:288–94.
[82] Lin Z, Zhen Z, Ren L, Yang J, Luo C, Zhong L, Hu H, Liang Y, Li Y, Zhang D. Effects of two ecological earthworm species on atrazine degradation performance and bacterial community structure in red soil. Chemosphere 2018;196:467–75.
[83] Datta S, Singh J, Singh J, Singh S, Singh S. Assessment of genotoxic effects of pesticide and vermicompost treated soil with *Allium cepa* test. Sustainable Environment Research 2018;28(4):171–8.
[84] Xiao Z, Liu M, Jiang L, Chen X, Griffiths BS, Li H, Hu F. Vermicompost increases defense against root-knot nematode (*Meloidogyne incognita*) in tomato plants. Applied Soil Ecology 2016;105:177–86.
[85] Sinha R, Hahn K, Singh GPK, Suhane RK, Reddy A. Organic farming by vermiculture: producing safe, nutritive and protective foods by earthworms (Charles Darwin's friends of farmers). American Journal of Experimental Agriculture 2011;1:363–99.
[86] Swati A, Hait S. Fate and bioavailability of heavy metals during vermicomposting of various organic waste—a review. Process Safety and Environmental Protection 2017;109:30–45. https://doi.org/10.1016/j.psep.2017.03.031.
[87] Taylor M, Clarke W, Greenfield P. The treatment of domestic wastewater using small-scale vermicompost filter beds. Ecological Engineering 2003;21:197–203.
[88] Li YS, Xiao YQ, Qiu JP, Dai YQ, Robin P. Continuous village sewage treatment by vermifiltration and activated sludge process. Water Science and Technology 2009; 60(11):3001–10.
[89] Lourenco N, Nunes LM. Optimization of a vermifiltration process for treating urban wastewater. Ecological Engineering 2017;100:138–46.
[90] Zhu W, Du W, Shen X, Zhang H, Ding Y. Comparative adsorption of $Pb^{2+}$and Cd^{2+} by cow manure and its vermicompost. Environmental Pollution 2017;227: 89–97.
[91] Jordao CP, Fernandes RBA, de Lima Ribeiro K, de Barros PM, Fontes MPF, de Paula Souza FM. A study on Al (III) and Fe (II) ions sorption by cattle manure vermicompost. Water, Air, and Soil Pollution 2010; 210(1–4):51–61.
[92] Pereira MG, Korn M, Santos BB, Ramos MG. Vermicompost for tinted organic cationic dyes retention. Water, Air, and Soil Pollution 2009;200(1–4):227–35.
[93] Barbosa LPG, de Freitas TOP, Pereira MG. Use of vermicompost for the removal of toxic metal ions of synthetic aqueous solutions and real wastewater. Ecletica Quimica 2018;43:35–43.
[94] Manyuchi MM, Mbohwa C, Muzenda E. Biological treatment of distillery wastewater by application of the vermifiltration technology. South African Journal of Chemical Engineering 2018;25:74–8.
[95] Singh A, Singh GS. Vermicomposting: a sustainable tool for environmental equilibria. Environmental Quality Management 2017;27:23–40.
[96] Gurav MV, Pathade GR. Production of vermicompost from temple waste (Nirmalya): a case study. Universal

Journal of Environmental Research and Technology 2011;1(2):182–92.

[97] Bogdanov P. Vexatious vermicomposting: problems leading to failure in large-scale projects. In: 3rd International Scientific and Practical Conference "Vermicomposting and Vermiculture as basis of ecological landownership in XXI century–problems, outlooks, achievements."; 2013.

[98] Moledor S, Chalak A, Fabian M, Talhouk SN. Socioeconomic dynamics of vermicomposting systems in Lebanon. Journal of Agriculture, Food Systems, and Community Development 2016;6(4):145–68.

[99] Lalander CH, Komakech AJ, Vinneras B. Vermicomposting as manure management strategy for urban smallholder animal farms – kampala case study. Waste Management 2015;39:96–103.

[100] Kiyasudeen KS, Ibrahim H, Quaik S, Ismail SA. Vermicompost, its applications and derivatives. In: Jegatheesun JV, editor. Prospects of organic waste management and the significance of earthworms. Applied environmental science and engineering for a sustainable future. Zurich, Switzerland: Springer International Publishing; 2015. p. 201–30.

[101] Kumar R, Singh BL, Kumar U, Verma D, Shweta. *Drawida willsi* Michalsen activates cellulolysis in pressmud vermireactor. Bioresource Technology 2010;101(23):9086–91.

[102] Zirbes VL, Renard Q, Dufey J, Khanh Tu P, Duyet HN. Volarisation of water hyacinth in vermicomposting using an epigeics earthworm *P. Exacavatus* Biotechnology, Agronomy, Society and Environment 2011;15(1):85–93.

[103] Das S, Deka P, Goswami L, Sahariah B, Hussain N, Bhattacharya SS. Vermiremediation of toxic jute mill waste employing *Metaphire posthuma*. Environmental Science and Pollution Research 2016;23(15):15418–31.

[104] Chen Y, Chang SK, Chen J, Zhang Q, Yu H. Characterization of microbial community succession during vermicomposting of medicinal herbal residues. Bioresource Technology 2018;249:542–9.

[105] Wu D, Yu X, Chu S, Jacobs DF, Wei X, Wang C, Long F, Chen X, Zeng S. Alleviation of heavy metal phytotoxicity in sewage sludge by vermicomposting with additive urban plant litter. The Science of the Total Environment 2018;633:71–80.

[106] Kouba A, Lunda R, Hlavac D, Kuklina I, Hamackova J, Randak T, Kozak P, Koubova A, Buric M. Vermicomposting of sludge from recirculating aquaculture system using *Eisenia andrei*: technological feasibility and quality assessment of end-products. Journal of Cleaner Production 2018;177:665–73.

[107] Shanthi NR, Bhoyar RV, Bhide AD. Vermicomposting of vegetable waste. Compost Science & Utilization 1993;1(4):27–30.

[108] Ernst G, Felten D, Vohland M, Emmerling C. Impact of ecologically different earthworm species on soil water characteristics. European Journal of Soil Biology 2009;45(3):207–13.

[109] Rorat A, Suleiman H, Grobelak A, Grosser A, Kacprzak M, Płytycz B, Vandenbulcke F. Interactions between sewage sludge-amended soil and earthworms—comparison between *Eisenia fetida* and *Eisenia andrei* composting species. Environmental Science and Pollution Research 2016;23:3026–35.

[110] Zarrabi M, Mohammadi A, Al-musawi T, Saleh HN. Using natural clinoptilolite zeolite as an amendment in vermicomposting of food waste. Environmental Science and Pollution Research 2018:1–10.

[111] Amouei AI, Yousefi Z, Khosravi T. Comparison of vermicompost characteristics produced from sewage sludge of wood and paper industry and household solid wastes. Journal of Environmental Health Science and Engineering 2017;15:5. https://doi.org/10.1186/s40201-017-0269-z.

[112] Yadav KD, Tare V, Ahammed MM. Vermicomposting of source-separated human faeces for nutrient recycling. Waste Management 2010;30:50–6.

[113] Haynes RJ, Zhou YF. Comparison of the chemical, physical and microbial properties of composts produced by conventional composting or vermicomposting using the same feedstocks. Environmental Science and Pollution Research 2016;23:10763–72.

[114] Hanc A, Dreslova M. Effect of composting and vermicomposting on properties of particle size fractions. Bioresource Technology 2016;27:186–9.

[115] Xie D, Wu W, Hao X, Jiang D, Li X, Bai L. Vermicomposting of sludge from animal wastewater treatment plant mixed with cow dung or swine manure using *Eisenia fetida*. Environmental Science and Pollution Research 2016. https://doi.org/10.1007/s11356-015-5928-y.

[116] Ravindran B, Mnkeni PNS. Bio-optimization of the carbon-to-nitrogen ratio for efficient vermicomposting of chicken manure and waste paper using *Eisenia fetida*. Environmental Science and Pollution Research 2016;23:16965–76.

[117] Zhu W, Yao W, Shen X, Zhang W, Xu H. Heavy metal and δ 13 C value variations and characterization of dissolved organic matter (DOM) during vermicomposting of pig manure amended with 13C-labeled rice straw. Environmental Science and Pollution Research 2018:1–10.

[118] Thomas GV, Mathew AE, Baby G, Mukundan MK. Bioconversion of residue biomass from a tropical homestead agro-ecosystem to value added vermicompost by *Eudrilus* species of earthworm. Waste Biomass Valorization 2018. https://doi.org/10.1007/s12649-018-0203-3.

[119] Sanchez GM, Tausnerova H, Hanc A, Tlustos P. Stabilization of different starting materials through vermicomposting in a continuous-feeding system: changes in chemical and biological parameters. Ecological Engineering 2017;106:200–8.

[120] Sudkolai ST, Nourbakhsh F. Urease activity as an index for assessing the maturity of cow manure and wheat residue vermicomposts. Waste Management 2017;64:63–6.

[121] Singh WR, Kalamdhad AS. Transformation of nutrients and heavy metals during vermicomposting of the invasive green weed *Salvinia natans* using *Eisenia fetida*. International Journal of Recycling of Organic Waste in Agriculture 2016;5:205–20.

[122] Parthasarathi K, Balamurugan M, Prashija KV, Jayanthi L, Basha SA. Potential of *Perionyx excavatus* (Perrier) in lignocellulosic solid waste management and quality vermifertilizer production for soil health. International Journal of Recycling of Organic Waste in Agriculture 2016;5:65–86.

[123] Swarnam TP, Velmurugan A, Pandey SK, Roy SD. Enhancing nutrient recovery and compost maturity of coconut husk by vermicomposting technology. Bioresource Technology 2016;207:76–84.

[124] Lim SL, Wu TY, Sim EYS, Lim PN, Clarke C. Biotransformation of rice husk into organic fertilizer through vermicomposting. Ecological Engineering 2012;41:60–4.

[125] Aynehband A, Gorooei A, Moezzi AA. Vermicompost: an eco-friendly technology for crop residue management in organic agriculture. Energy Procedia 2017;141:667–71.

[126] Busato JG, Papa G, Canellas LP, Adani F, Oliveira AL, Leao TP. Phosphatase activity and its relationship with physical and chemical parameters during vermicomposting of filter cake and cattle manure. Journal of the Science of Food and Agriculture 2016;96:1223–30.

[127] Singh S, Bhat SA, Singh J, Kaur R, Vig AP. Earthworms converting milk processing industry sludge into biomanure. The Open Waste Management Journal 2017;10(1).

[128] Pigatin LBF, Atoloye IO, Obikoya OA, Borsato AV, Rezende MOO. Chemical study of vermicomposted agroindustrial wastes. International Journal of Recycling Organic Waste in Agriculture 2016;5:55–63.

[129] Ravindran B, Wong JW, Selvam A, Sekaran G. Influence of microbial diversity and plant growth hormones in compost and vermicompost from fermented tannery waste. Bioresource Technology 2016;217:200–4.

[130] Scagalia B, Nunes RR, Rezende MOO, Tambone F, Adani F. Investigating organic molecules responsible of auxin-like activity of humic acid fraction extracted from vermicompost. The Science of the Total Environment 2016;562:289–95.

[131] Bhat SA, Singh J, Vig AP. Effect on growth of earthworm and chemical parameters during vermicomposting of pressmud sludge mixed with cattle dung mixture. Procedia Environmental Sciences 2016;35:425–34.

[132] Singh S, Suthar S. Vermicomposting of herbal pharmaceutical industry solid wastes. Ecological Engineering 2012;39:1–6.

[133] Moradi H, Fahramand M, Sobhkhizi A, Adibian M, Noori M, Shila abdollahi Rigi K. Effect of vermicompost on plant growth and its relationship with soil properties. IJFAS Journal 2014:333–8. 2014-3-3.

[134] Manivannan S, Balamurugan M, Parthasarathi K, Gunasekaran G, Ranganathan LS. Effect of vermicompost on soil fertility and crop productivity beans (*Phaseolus vulgaris*). Journal of Environmental Biology 2009;30:275–81.

[135] Zuo Y, Zhang J, Zhao R, Dai H, Zhang Z. Application of vermicompost improves strawberry growth and quality through increased photosynthesis rate, free radical scavenging and soil enzymatic activity. Scientia Horticulturae 2018;233:132–40.

[136] Hidalgo LMC, Gomez-Hernandez DE, Villalobos-Maldonado JJ, Abud-Archila M, Montes-Molina JA, Enciso-Saenz S, Ruiz-Valdiviezo VM, Gutierrez-Miceli FA. Effects of vermicompost and vermiwash on plant, phenolic content, and anti-oxidant activity of Mexican Pepperleaf (*Piper auritum* Kunth) cultivated in phosphate rock potting media. Compost Science & Utilization 2017;25(2):95–101.

[137] Trivedi P, Singh K, Pankaj U, Verma SK, Verma RK, Patra DD. Effect of organic amendments and microbial application on sodic soil properties and growth of an aromatic crop. Ecological Engineering 2017;102:127–36.

[138] Soobhany N, Mohee R, Garg VK. A comparative analysis of composts and vermicomposts derived from municipal solid waste for the growth and yield of green bean (*Phaseolus vulgaris*). Environmental Science and Pollution Research 2017;24(12):11228–39.

[139] Goswami L, Nath A, Sutradhar S, Bhattacharya SS, Kalamdhad A, Vellingiri K, Kim KH. Application of drum compost and vermicompost to improve soil health, growth, and yield parameters for tomato and cabbage plants. Journal of Environmental Management 2017;200:243–52.

[140] Pandey V, Patel A, Patra DD. Integrated nutrient regimes ameliorate crop productivity, nutritive value, antioxidant activity and volatiles in basil (*Ocimum basilicum* L.). Industrial Crops and Products 2016;87:124–31.

[141] Bajeli J, Tripathia S, Kumara A, Tripathia A, Upadhyay RK. Organic manures a convincing source for quality production of Japanese mint (*Mentha arvensis* L.). Industrial Crops and Products 2016;83:603–6.

[142] Muktamar Z, Sudjatmiko S, Chozin M, Setyowati N, Fahrurrozi F. Sweet corn performance and its major nutrient uptake following application of vermicompost supplemented with liquid organic fertilizer. International Journal on Advanced Science, Engineering and Information Technology 2017;7:602–8.

[143] Chaudhuri PS, Paul TK, Dey A, Datta M, Dey SK. Effects of rubber leaf litter vermicompost on earthworm population and yield of pineapple (*Ananas comosus*) in West Tripura, India. International Journal of Recycling of Organic Waste in Agriculture 2016;5:93–103.

[144] Jabeen N, Ahmad R. Growth response and nitrogen metabolism of sunflower (*Helianthus annuus* L.) to vermicompost and biogas slurry under salinity stress. Journal of Plant Nutrition 2017;40(1):104–14.

[145] Kovshov SV, Iconnicov DA. Growing of grass, radish, onion and marigolds in vermicompost made from pig manure and wheat straw. Indian Journal of Agriculture Research 2017;51:327–32.

[146] Doan TT, Henry-des-Tureaux T, Rumpel C, Janeau J, Jouquet P. Impact of compost, vermicompost and biochar on soil fertility, maize yield and soil erosion in Northern Vietnam: a three year mesocosm experiment. The Science of the Total Environment 2015;514:147–54.

FURTHER READING

[1] Elvira C, Sampedro L, Dominguez J, Mato S. Vermicomposting of wastewater sludge from paper pulp industry with nitrogen rich materials. Soil Biology and Biochemistry 1997;29:759–62.

[2] Abbasi SA, Nayeem Shah M, Abbasi T. Vermicomposting of phytomass: limitations of the past approaches and the emerging directions. Journal of Cleaner Production 2015;93:103–14.

CHAPTER 11

Biogas From Wastes: Processes and Applications

SAMIR KUMAR KHANAL, PHD • TJOKORDA GDE TIRTA NINDHIA, PHD • SAOHARIT NITAYAVARDHANA, PHD

INTRODUCTION

Biogas consists of primarily methane (50%–75% by volume) and carbon dioxide (25%–40% by volume), with small amounts of hydrogen sulfide, nitrogen, and oxygen, and is produced via anaerobic digestion (AD) of diverse feedstocks [1]. AD offers several merits such as generation of renewable energy for different applications ranging from stationary power to transportation fuel, remediation of diverse organic wastes, mitigation of greenhouse gas (GHG) emission, and generation of nutrient-rich digestate (residuals) for agricultural applications. Thus AD process is probably one of the very few technologies that address the three grand challenges of 2050, namely, energy and food insecurity and environment. There are thousands of large-scale anaerobic digesters currently in operation globally utilizing diverse organic feedstocks such as agri-residues/wastes, energy crops, sewage sludge, animal manures, organic fraction of municipal solid waste (OFMSW), and food wastes. In addition, there are several million small-scale digesters (also known as household digester) currently in operation in the developing countries, providing biogas for cooking and lighting along with producing digestate as organic fertilizer for agricultural applications.

Globally, 2017 million metric tons of waste was generated in 2018 and is expected to increase to 2586 million metric tons and 3401 million metric tons in 2030 and 2050, respectively [2]. With stringent regulation on disposal of organic wastes in landfills coupled with several environmental concerns such as GHG emissions, surface water and groundwater contamination, odor emanation, and transmission of vectors via birds and insects, there has been significant effort to eliminate or reduce the disposal of organic wastes into landfill. In this respect, biogas technology offers a unique opportunity for both waste remediation and recovery of resources. Importantly, a significant amount of revenue could be generated via a tipping fee as part of waste management. This chapter provides an overview of waste-to-biogas technology, some opportunities and challenges, and future perspectives.

GLOBAL WASTE GENERATION

The global population is expected to increase from 7.7 billion in 2018 to 9.7 billion in 2050 [3]. Such rapid increase in population will lead to an increased demand for resources such as food, feed, and fuel, while, at the same time, generating large amount of wastes. The global solid waste generation is expected to increase from 2017 million metric tons in 2018 to 3401 million metric tons in 2050, with a global average waste generation of 0.74 kg/capita/day [2]. Food waste generation occurs throughout the supply chain—from agricultural production to final household consumption. Approximately, one-third of all food produced for human consumption worldwide is discarded as waste, representing about 1.3 billion tons waste per year [4]. In Europe and North America, an average of 95–115 kg of food waste is generated per person per year [4]. There are several reasons for such a large generation of food wastes. For example, at the production level, farmer-buyer sales agreements result in farm crops being wasted, due to quality standards lacking perfect shape or appearance, and at the consumer level, insufficient purchase planning and best-before-use; residual, uneaten, food components; and, last but not the least, the careless attitude of the consumers.

Sustainable Resource Recovery and Zero Waste Approaches. https://doi.org/10.1016/B978-0-444-64200-4.00011-6

ANAEROBIC DIGESTION PATHWAY

AD of complex organic matters is carried out by diverse microbial communities in a series of a multi-step biochemical processes as elucidated in Fig. 11.1.

Complex organic matters such as carbohydrates, proteins, and lipids are hydrolyzed into simple soluble compounds such as monomeric and oligomeric sugars, amino acids, and long-chain fatty acids, respectively, through the actions of extracellular enzymes excreted by the hydrolytic microorganisms. The process is known as hydrolysis and is a rate-limiting step in AD of high solid feedstocks such as organic wastes. The hydrolysis rate and subsequent methane production are primarily governed by the particle size. Thus often various pretreatment methods are adopted to improve the hydrolysis rate and overall digestibility. These soluble substrates generated after hydrolysis are then broken into intermediates such as short-chain volatile fatty acids (VFAs), simply known as VFAs (e.g., propionic, butyric, caproic, and valeric acids); ethanol; and lactic acid by the fermenting bacteria. The process is known as acidogenesis, which often occurs rapidly and leads to accumulation of VFAs. The digester often fails due to accumulation of VFAs and thus precise process control and monitoring is critical for stable operation of the AD process. The intermediates (VFAs and ethanol) are then converted into acetic acid, hydrogen, and carbon dioxide by the syntrophic hydrogen-producing acetogenic bacteria. The process is known as acetogenesis and is thermodynamically unfavorable unless the produced hydrogen is removed/consumed by H_2-consuming microbes (e.g., hydrogenotrophs, homoacetogens, and/or sulfate reducers). Thus there exists a symbiotic relationship between H_2-producing acetogenic bacteria and H_2-consuming microbes such as hydrogenotrophic methanogens, homoacetogens, and/or sulfate-reducing bacteria (if sulfate is present in the wastes). Finally, the acetic acid and hydrogen + carbon dioxide are converted into methane via acetoclastic or acetotrophic methanogenesis and hydrogenotrophic methanogenesis, respectively.

FACTORS AFFECTING ANAEROBIC DIGESTION

For the successful bioconversion of organic wastes into biogas, several factors including environmental conditions should be maintained within the comfort range of anaerobic microbes. Some of these factors include temperature, pH, hydraulic retention time (HRT), solids retention time (SRT), organic loading rate, mixing, concentration of solids, alkalinity, nutrients, trace elements, toxic compounds, and reactor configuration.

Organic Loading Rate

Loading rate, also known as the volumetric organic loading rate, is the feeding rate of organic matter into the digester. Loading rate determines the amount of

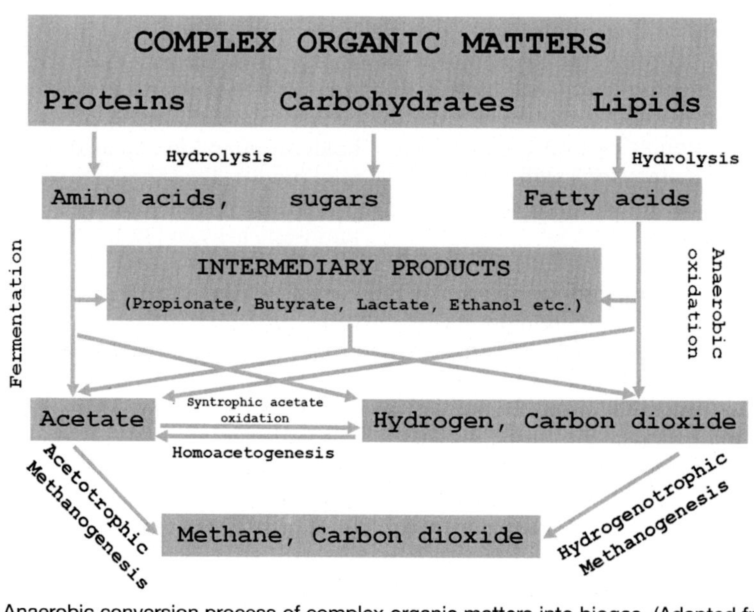

FIG. 11.1 Anaerobic conversion process of complex organic matters into biogas. (Adapted from Ref. [5].)

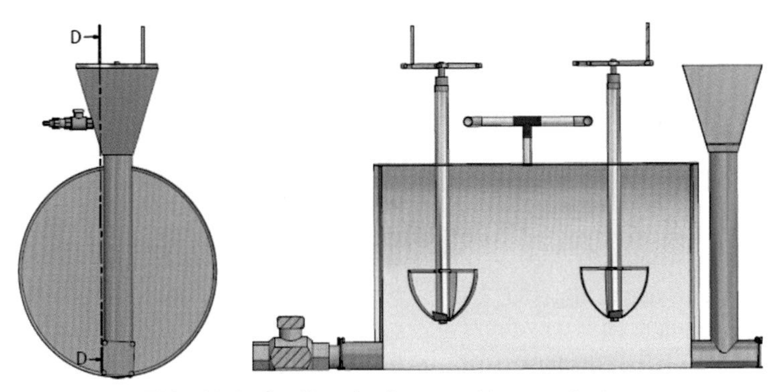

FIG. 11.2 Small-scale digester with manual mixer.

substrate that the microorganisms are capable of stabilizing in a given time per unit reactor volume (kg volatile solids [VS]/m^3-day). The possible maximum organic loading rate, however, depends on the amount of biomass retained in the bioreactor, substrate types (concentration, biodegradability), environmental conditions (temperature, pH), and reactor configuration, among others. For a stable digester performance, overloading must be avoided. Some major indicators for reactor overloading are decline in the methane content and biogas generation rate, decrease in reactor pH, and rapid increase in VFAs and VFA/alkalinity ratio [6]. As the optimum loading rate depends on many factors, the loading rate should be such that the pH is within a neutral range, the total VFAs are between 1500 and 4500 mg as acetic acid equivalent (HAc)/L, and the ammonium nitrogen concentration is below 4500 mg/L [6]. For high solid wastes such as OFMSW, vegetable waste, and fruit waste, the optimum organic loading rate varies from 0.3 to 2.5 kg VS/m^3/day [7–10].

Mixing

A certain level of mixing is necessary for maintaining homogeneity within the digester. Proper mixing assures effective digester volume utilization and prevents temperature stratification. Some natural mixing occurs in an anaerobic digester. However, such mixing alone is not sufficient, especially at high solids loading rates. Because propeller and screw mixing are costly to install and energy intensive in operation, these types of mixing are mainly used in commercial-scale digesters. However, in the case of a small-scale digester such as household digester, recirculation of slurry and gas could satisfactorily mix the digester contents. Another alternative could be the modification of an inlet design to create sufficient head for feedstocks such that sufficient disturbance is created in the digester's contents during feeding. In regions with considerable wind velocity, harvesting wind energy by using simple (wooden or metallic) blades to rotate the propeller could create sufficient periodic mixing of the digester contents. In some instances, the reactor is designed with a manual mixing system, as shown in Fig. 11.2.

Temperature

Temperature has two major effects in AD. First, it affects the microbial activity and second, it affects the viscosity of the bioreactor contents. There are three temperature regimes for anaerobic microbes, namely, psychrophilic, mesophilic, and thermophilic. As the anaerobic conversion of organic matter has its highest activity at 5–15°C (psychrophilic), 35–40°C (mesophilic), or at about 55°C (thermophilic) [11], most of the digesters are operated at either mesophilic or thermophilic condition. As active microbial communities at two temperature regimes are quite different, the shifting of temperature from mesophilic to thermophilic and vice versa can result in a sharp decline in biogas yield until the predominant microbial communities are established [12]. Thermophilic digestion has certain merits over mesophilic digestion, such as shorter retention time, better digestion efficiency, and higher rate of destruction of pathogens. However, thermophilic microbes are more sensitive than mesophilic microbes to temperature fluctuations, inhibitory compounds, and loading variations, among other factors. Moreover, thermophilic digestion needs to be well buffered, as it has a higher VFA/alkalinity ratio, and also requires higher energy input for maintaining the higher digester temperature. Hence, increase in biogas production from thermophilic process must be balanced against the increased energy requirement for operating and maintaining the digester.

In the countries with tropical climate (annual mean temperature between 24 and 28°C), ambient temperature is quite favorable for biogas production. However, in cold mountainous regions of some developing countries such as China, northern India, and Nepal or in regions with warm summer and cold winter, lower biogas yield is a major issue for the installation of household digesters. In such regions, the effects of lower temperature can be compensated to some extent by increasing the retention time in the digester. But longer retention time requires a larger digester volume, which ultimately increases installation costs. For example, in northern India with an annual average temperature of 15°C, a normal functioning digester must have a retention time of 50–55 days compared with 30 days in the tropical south. Note that increasing the retention time from 30 to 50–55 days requires a 1.8-times larger digester volume. Coating the digester with an insulating material could help to maintain digester temperature within the desired range [13]. Another option could be site selection for the biogas plant such that the plant faces toward the sun and is well protected from cold winds. Additionally, simple practices of covering the digester with locally available insulating materials such as crop residues, compost piles, and sand (sand walls avoid rapid cooling of the system during the night by keeping the digester at 10°C compared with the 0°C ambient temperature) could be helpful in preventing rapid heat loss from the digester [14].

pH

pH is one of the most critical factors for AD. The interactions of the produced VFAs and carbon dioxide with ammonia and bicarbonate present in the system govern the pH of the digester [15]. The optimum pH for hydrolysis and acidogenesis is between 5.5 and 6.5, whereas methanogens prefer a slightly alkaline pH in the range of 7.8–8.2 [16–18]. Below pH 6.0, unionized VFAs become inhibitory to methanogens. The same issue occurs with unionized (free) ammonia at a pH greater than 8.0. As methanogens are more sensitive to pH fluctuations, and methanogenesis is often a rate-limiting step for highly soluble waste streams, a pH close to neutral (6.8–7.4) should be maintained [5]. The digester can be maintained at a desired pH range (6.8–7.4) by manipulating the organic loading rate and via codigestion of different feedstocks. Ammonia toxicity at higher pH resulting from the digestion of nitrogen-rich feedstocks such as chicken litter and pig and dairy cattle manure and urine can be overcome by codigesting such feedstocks with feedstocks high in carbon content such as agri-residues/wastes. Similarly, lower pH resulting from the rapid hydrolysis of readily degrading feedstocks such as food wastes can be addressed by digesting such feedstocks with animal manure, which has good buffering capability. Moreover, animal manure is the most readily available and commonly used feedstock for biogas production in the developing countries.

Nutrients and Trace Elements

For proper operation of an anaerobic digester, a balanced supply of nutrients and trace elements is necessary, as these components are the building blocks for microbial cells. Nutrients and trace elements also provide a suitable physico-chemical condition for optimum growth of microorganisms.

The amount of nitrogen (N) and phosphorus (P) required for normal operation of a digester depends on the organic content measured as the chemical oxygen demand (COD) of the substrate (low solids content). Nutrient, especially nitrogen (N) and phosphorus (P), supplementation is generally based on COD/N/P ratios. For low-strength waste streams, N and P are supplemented as per the COD/N/P ratio of 1000:7:1, and for high-strength wastes, N and P are supplemented based on the COD/N/P ratio of 350:7:1 [19]. These ratios provide a C/N ratio of at least 25:1, which is optimal for biogas production [20]. One option to make sure that N and P are not limited in the digester, is to measure the residual values of ammonium nitrogen (NH_4^+-N) and orthophosphate-phosphorus (HPO_4^--P) in the digester effluent or digestate. For non-nutrient-limiting conditions, the residual values of 5 mg/L of NH_4^+-N and 1–2 mg/L of HPO_4^--P are commonly recommended [20]. As methanogens have unique enzyme systems, their micro-nutrient requirements are different from other microbes. They especially need trace elements such as cobalt (Co), iron (Fe), and nickel (Ni). Other trace elements essential for methanogens include selenium, tungsten, molybdenum, barium, calcium, magnesium, and sodium. In a household digester, a simple way to achieve a balance of nutrients in the digester is to codigest different feedstocks, such as animal manure, crop residues, and food wastes, that complement each other with respect to both macro- and micro-nutrients.

Toxicity and Inhibition

Anaerobic microorganisms are sensitive to certain compounds present in the feedstocks (e.g., heavy metals, halogenated compounds, cyanides, antibiotics) as well as some metabolic by-products of microorganisms (e.g., ammonia, sulfide, and VFAs). The inhibitory concentration of a substance, however, depends on the ionic strength of the toxic material, pH, organic loading rate, temperature, the presence of other materials, and the ratio of the concentration of toxic substances to microbial biomass [21]. Although all microorganisms require cations for their metabolism, cations at certain concentration are toxic to microbes. Methanogens are more sensitive to cations than acidogens. In general, relative toxicity of cations increases with valence and with atomic weight [22].

In AD, the toxicity of heavy metals can be reduced either by adding precipitating anions such as sulfide or by operating the digester at a maximum allowable pH, as most of the heavy metal hydroxides are less soluble at higher pH [23]. However, it may create problems in disposal or land application of the digestate, as heavy metals associated with the biomass may become more available to the environment, particularly in acidic environment. To some extent, the heavy metal toxicity of the slurry can be reduced by adding dry organic residues, which dilute the concentration of heavy metals.

For normal AD, the concentration of VFAs should not exceed 2000–3000 mg HAc/L [24]. Moreover, *n*-butyrate, *n*-valerate, and *n*-caproate are more toxic to methanogens than the isoforms of these acids [25]. Also, ammonia toxicity is often a problem in the AD of protein-rich feedstocks such as slaughterhouse wastes and urine and swine, dairy cattle, and poultry manures due to the rapid deamination of protein constituents. Free ammonia has been found to be much more toxic than ammonium ions. In some cases, however, methanogens have shown acclimation to some toxicants, which revealed the possibility of AD of industrial wastes containing toxic compounds [26]. VFA toxicity can be overcome by reducing or stopping substrate feeding to the digester until VFA levels drop to an acceptable limit. Similarly, toxicity caused by ammonia can be corrected by diluting the digester contents with water. Recirculation of air-dried digestate could be another option that increases the total solids (TS) content of the digester and dilutes the concentration of the ammonia nitrogen. Also, maintenance of a proper C/N ratio by feeding carbon-rich feedstocks such as crop residues/agriwastes can overcome the ammonium nitrogen toxicity.

Carbon-to-Nitrogen (C/N) Ratio

The C/N ratio represents the amount of carbon and nitrogen in the feedstock and is an important process parameter for high solids AD. Both carbon and nitrogen are vital for the microbial cell growth and functioning. Nitrogen present in the feedstock facilitates the synthesis of amino acids, proteins, and nucleic acids, while carbon acts as the structural unit as well as the energy source for microbes. Part of the organic nitrogen present in feedstock is converted into ammonia. Ammonia produced in the digester also helps neutralize the volatile acids produced by fermentative bacteria and helps maintain the pH in the neutral range. In general, anaerobic microbes utilize carbon 25–30 times faster than nitrogen. Thus for efficient biogas production, the C/N ratio in the feedstocks should be maintained at 20–30:1 [27]. Excess carbon, as in the case of crop residues, results in the accumulation of CO_2 in the biogas, whereas feedstocks rich in nitrogen such as urine, slaughterhouse wastes, and swine and poultry manures can lead to ammonia accumulation in the digester. Proper C/N ratios in the digester can be achieved by co-digesting feedstocks rich in carbon, such as crop residues, with nitrogen-rich feedstocks such as animal manure, urine, and slaughterhouse wastes. In addition to alleviating the problem associated with the digestion of these feedstocks alone, co-digestion also improves biogas yields [28]. For instance, co-digestion of potato waste (C/N ratio of 35) and beet leaf (C/N ratio of 14) increased the methane yield by 60% compared with the digestion of potato waste alone [29].

BIOREACTOR SYSTEM

Owing to the high solids nature of the organic wastes, there are very limited bioreactor systems that could be applied for AD. Based on the TS content in digestion process, the AD process is divided into wet digestion (TS $<15\%$) and dry digestion or solid-state AD (TS of 20%–40%) [30]. The bioreactors employed for wet and dry digestion are discussed in the following sections.

Wet Digestion

The most common bioreactor systems for wet digestion of organic wastes are discussed in the following sections.

Continuous-stirred tank reactor

Continuous-stirred tank reactor (CSTR) is by far the most widely adopted AD bioreactor system for organic waste digestion. The substrates are fed either continuously or semi-continuously and the reactor contents are mixed using a mechanical mixer, and in some cases via biogas recirculation. The TS content may vary from 1% to 10%. In CSTR the reactor contents are well mixed and the concentration of the constituents is ideally the same throughout the reactor. In CSTR, SRT = HRT and is usually maintained around 20–50 days. Due to rapid dilution of the substrate in the reactor, CSTR can withstand higher shock loading. CSTR is easier to build and operate, and the generated biogas is often stored on top of the digester using flexible ethylene propylene diene monomer storage membrane, Biolene, which gives the characteristic dome shape to the digester, as shown in Fig. 11.3.

Plug-flow digester

Plug-flow digester is a long rectangular concrete tank or polyethylene tube covered with hard concrete or polypropylene in which mixing mainly takes place in the lateral direction and there is no mixing in the longitudinal direction (direction of substrate flow) (Fig. 11.4). Plug-flow digester is ideally suited for high solids feedstock such as cattle manure with TS content 12%–15% and operated at HRT (SRT) of 15–30 days. The substrate is degraded as it moves through the reactor, with daily feeding of the substrate in a flow pattern similar to piston. There is no mixing in the digester; however, heating could be provided by recirculating hot water through pipes. Plug-flow digester is low cost and simple in design.

Covered anaerobic lagoon

Covered anaerobic lagoon is a low-cost bioreactor for digesting waste streams with TS content of 0.3%–5% and retention times could vary from 3 to 6 months (Fig. 11.5). Lagoon system is ideal for animal waste (dairy waste/swine manure) digestion. Anaerobic lagoon is an earthen pit in which an impermeable liner such as clay or plastic is placed at the bottom and sides to prevent liquid seepage. Often the lagoon is operated without mixing and temperature control. The biogas produced is collected in the headspace of the covered lagoon.

Dry Digestion

Dry AD has several merits over wet AD, such as no requirement of water addition; easy handling and broader applications of digestate; less nutrient runoff/loss during storage; absence of problems related to foaming, sedimentation, and surface crust; and savings on energy and cost associated with mixing [31,32]. Furthermore, dry digestion technology does not require preprocessing (e.g., size reduction) of feedstocks such as agri-residues/wastes, which could otherwise accumulate in the digesters [33].

Dry digestion could be carried out in batch or continuous mode. In batch mode digestion, the digestate is mixed with the wastes (1:1 ratio) and is placed in an airtight reactor for digestion. Sometimes leachate is spread as a source of inoculum as well as to maintain certain moisture content. Usually no mixing is provided. Continuous dry digestion systems employ several propriety processes such as Dranco, Kompogas, and Valorga [1]. Dranco digestion system has a vertical design without a mechanical mixing unit. Mixing is achieved via recirculation of digestate drawn from the bottom and mixed with fresh wastes (6:1 ratio) [1]. Kompogas digestion system is similar to a plug-flow

FIG. 11.3 Full-scale continuous-stirred tank reactor with membrane top for gas storage. (Courtesy of Samir Khanal (AD plant operated by the University of Hohenheim, Germany).)

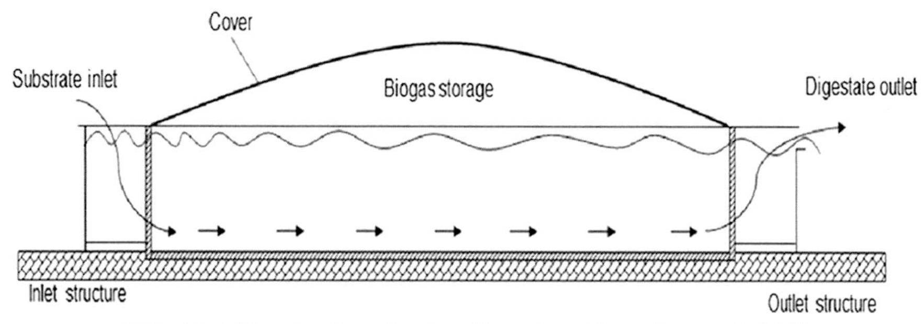

FIG. 11.4 The plug-flow digester. (Reproduced from, Source: Ref. [1].)

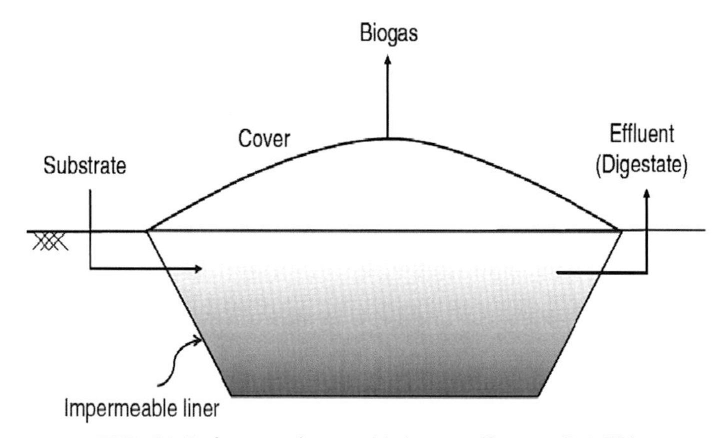

FIG. 11.5 A covered anaerobic lagoon. (Source: Ref. [1].)

digester with a horizontally placed tank assisted by slow-rotating horizontal paddle mixers, which facilitate homogenization. Valorga digestion system consists of a vertical cylindrical steel tank with a central baffle [1,34,35].

BIOGAS UTILIZATION

Biogas has many different applications, as summarized in Fig. 11.6. Biogas can be directly used for cooking. This is the most common use of biogas in the developing countries where millions of household digesters are currently in operation. Researchers at the University of Udayana, Bali, Indonesia, are testing small-scale biogas-based electricity generation system (1−3 kW) for household/farm applications in the rural communities of the developing countries. Biogas (after cleaning for H_2S removal) could be used as boiler fuel. For large-scale digester, the produced biogas is often used for electricity and heat generation using a combined heat and power (CHP) unit with capacity ranging from several 100 kW to megawatts. There are several large-scale facilities for biogas upgrading (with methane content >95% v/v) to obtain a quality equivalent to natural gas to be injected into gas grid or use as a transportation fuel (e.g., compressed natural gas).

ANAEROBIC DIGESTION BIOREFINERY

As AD of many complex wastes such as agri-wastes/residues, cattle manures, pulp and paper mill wastes, and wood residues merely converts 20%−30% of the organic matter into biogas, 70%−80% of the organic matter is still present in the digestate [36]. Digestate is currently being used as organic fertilizer, which is a low-value product. Moreover, such digestates do not have high nutrient value. Thus there is a need to convert the digestate into higher value products. AD biorefinery provides an opportunity for conversion of such digestate into a plethora of biobased products and biochemicals, as elucidated in Fig. 11.7.

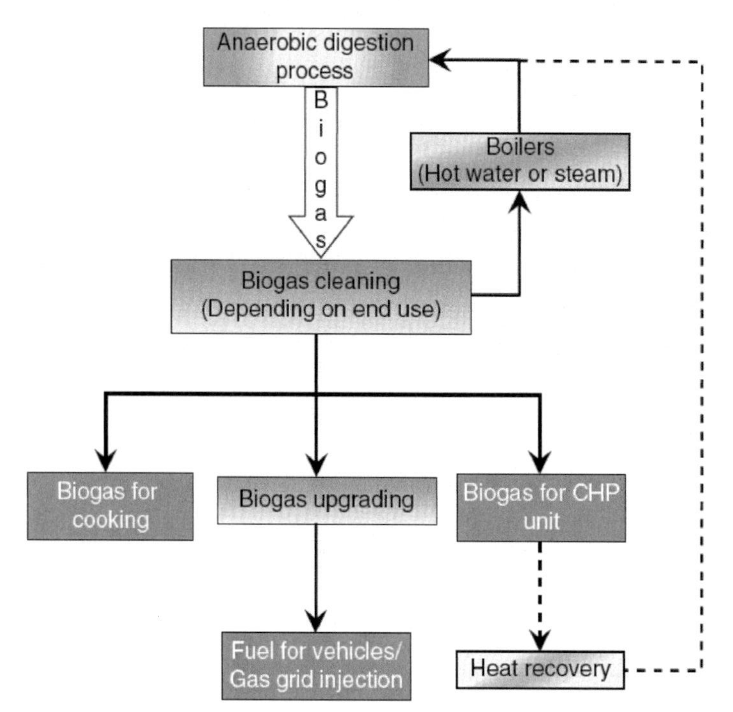

FIG. 11.6 Various applications of biogas. (Source: Ref. [1].)

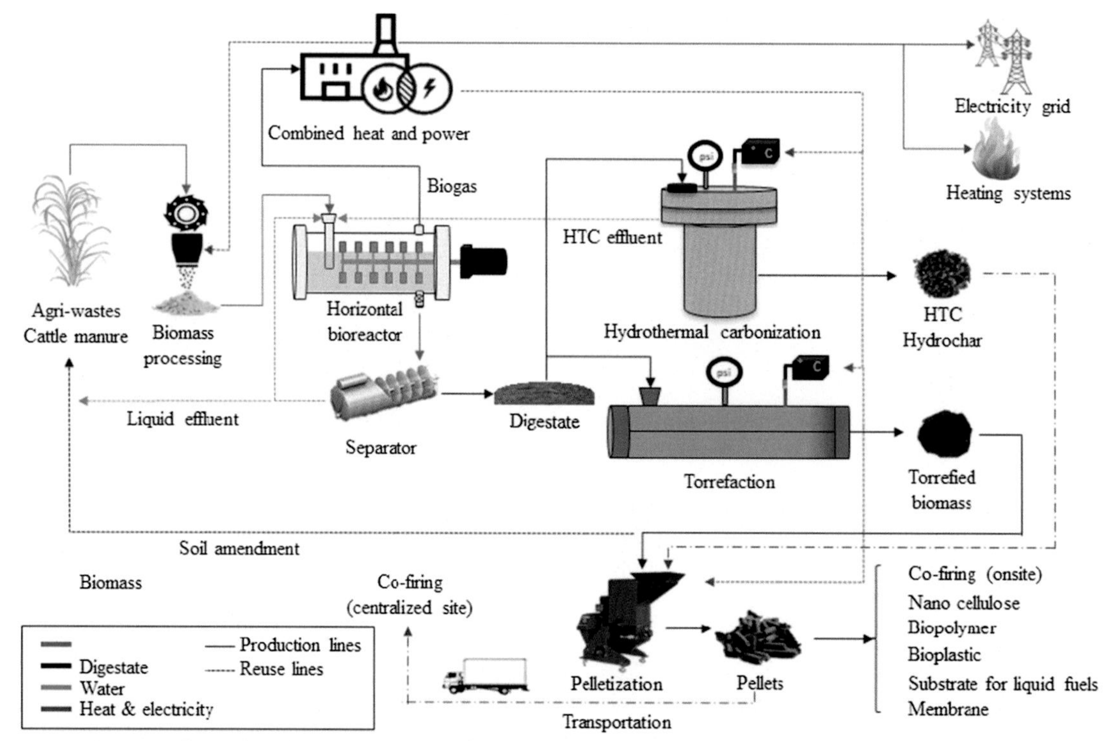

FIG. 11.7 Anaerobic digestion biorefinery for complete utilization of organic wastes. *HTC*, hydrothermal carbonization. (Source: Ref. [36].)

CONCLUSION AND FUTURE PERSPECTIVES

With significant increase in organic waste production, there is a critical need to effectively manage the wastes. Biogas production via AD is one of the ideal platforms for valorization of wastes with concomitant recovery of resources. The process also curtails GHG emission by avoiding methane emission from landfills and supplementing renewable energy. However, biogas as a sole product does not always provide much economic incentive. Thus in addition to producing biogas, there is a critical need to utilize the AD technology for generating high-value products. Adoption of the anaerobic biorefinery concept even at the farm level could provide additional revenue beyond biogas. In case of the developing countries, the capital cost of installing a small-scale digester (household digester) is still beyond the reach of poor households. Thus there is a need for governmental/nongovernmental supports/incentives.

ACKNOWLEDGMENTS

The authors would like to thank the USDA supplemental fund, College of Tropical Agriculture and Human Resources, University of Hawaii at Manoa (UHM), USA. We would also like to thank Renisha Karki (M.S. student at UHM) for helping with organizing the references and Dr. Surendra K.C. (Junior Researcher, UHM) for reviewing the first draft.

REFERENCES

[1] Li Y, Khanal SK. Bioenergy: principles and applications. John Wiley & Sons; 2017.
[2] Kaza S, et al. What a waste 2.0: a global snapshot of solid waste management to 2050. The World Bank; 2018.
[3] Bureau PR. 2018 world population data sheet. 2018. p. 1–18.
[4] FAO. Global food losses and food waste–Extent, causes and prevention. 2011.
[5] Khanal SK. Anaerobic biotechnology for bioenergy production: principles and applications. John Wiley & Sons; 2008.
[6] Braun R. Anaerobic digestion: a multi-faceted process for energy, environmental management and rural development. In: Ranalli P, editor. Improvement of crop plants for industrial end uses. Springer; 2007. p. 335–416.
[7] Alvarez R, Liden G. Semi-continuous co-digestion of solid slaughterhouse waste, manure, and fruit and vegetable waste. Renewable Energy 2008;33(4):726–34.
[8] de Laclos HF, Desbois S, Saint-Joly C. Anaerobic digestion of municipal solid organic waste: Valorga full-scale plant in Tilburg, The Netherlands. Water Science and Technology 1997;36(6–7):457–62.
[9] Mata-Alvarez J, et al. Anaerobic digestion of the Barcelona central food market organic wastes: experimental study. Bioresource Technology 1992;39(1):39–48.
[10] Nguyen P, Kuruparan P, Visvanathan C. Anaerobic digestion of municipal solid waste as a treatment prior to landfill. Bioresource Technology 2007;98(2):380–7.
[11] Van Haandel AC, Lettinga G. Anaerobic sewage treatment: a practical guide for regions with a hot climate. John Wiley & Sons; 1994.
[12] Ward AJ, et al. Optimisation of the anaerobic digestion of agricultural resources. Bioresource Technology 2008; 99(17):7928–40.
[13] Molnar L, Bartha I. High solids anaerobic fermentation for biogas and compost production. Biomass 1988; 16(3):173–82.
[14] Herrero JM. Transfer of low-cost plastic biodigester technology at household level in Bolivia. Livestock Research for. Rural Development 2007;19(12).
[15] Chawla O. Advances in biogas technology. New Delhi: Indian Council of Agricultural Research; 1986.
[16] Kim J, et al. Effects of various pretreatments for enhanced anaerobic digestion with waste activated sludge. Journal of Bioscience and Bioengineering 2003; 95(3):271–5.
[17] Kim M, et al. Hydrolysis and acidogenesis of particulate organic material in mesophilic and thermophilic anaerobic digestion. Environmental Technology 2003;24(9): 1183–90.
[18] Yu H-Q, Fang HHP. Acidogenesis of dairy wastewater at various pH levels. Water Science and Technology 2002; 45(10):201–6.
[19] Speece RE. Anaerobic biotechnology for industrial wastewater treatment. Environmental Science Technology 1983;17(9):416A–27A.
[20] Gerardi MH. The microbiology of anaerobic digesters. John Wiley & Sons; 2003.
[21] Kugelman IJ, McCarty PL. Cation toxicity and stimulation in anaerobic waste treatment. Water Pollution Control Federation 1965;37(1):97–116.
[22] Hayes TD, Theis TL. The distribution of heavy metals in anaerobic digestion. Water Pollution Control Federation 1978;50(1):61–72.
[23] Price EC, Cheremisinoff PN. Biogas: production and utilization. Ann Arbor, MI: Ann Arbor Science Publishers, Inc.; 1981.
[24] Taiganides E. Biogas: energy recovery from animal wastes. Part 1. World Animal Review 1980;(35):2–12.
[25] Hajarnis S, Ranade D. Inhibition of methanogens by n- and iso-volatile fatty acids. World Journal of Microbiology and Biotechnology 1994;10(3):350–1.
[26] Parkin G, Speece R. Attached versus suspended growth anaerobic reactors: response to toxic substances. Water Science and Technology 1983;15(8–9):261–89.
[27] Haque MS, Haque NN. Studies on the effect of urine on biogas production. Bangladesh Journal of Scientific & Industrial Research 2006;41(1):23–32.
[28] Capela I, et al. Impact of industrial sludge and cattle manure on anaerobic digestion of the OFMSW under

mesophilic conditions. Biomass and Bioenergy 2008;32(3):245–51.

[29] Parawira W, et al. Anaerobic batch digestion of solid potato waste alone and in combination with sugar beet leaves. Renewable Energy 2004;29(11):1811–23.

[30] Tchobanoglous G, et al. Integrated solid waste management: engineering principles and management issues, vol. 949. New York: McGraw-Hill; 1993.

[31] Angelonidi E, Smith SR. A comparison of wet and dry anaerobic digestion processes for the treatment of municipal solid waste and food waste. Water and Environment Journal 2015;29(4):549–57.

[32] Kothari R, et al. Different aspects of dry anaerobic digestion for bio-energy: an overview. Renewable and Sustainable Energy Reviews 2014;39:174–95.

[33] Chiumenti A, da Borso F, Limina S. Dry anaerobic digestion of cow manure and agricultural products in a full-scale plant: efficiency and comparison with wet fermentation. Waste Management 2018;71:704–10.

[34] Walter A, et al. Biotic and abiotic dynamics of a high solid-state anaerobic digestion box-type container system. Waste Management 2016;49:26–35.

[35] Weiland P. Biogas production: current state and perspectives. Applied Microbiology and Biotechnology 2010;85(4):849–60.

[36] Sawatdeenarunat C, et al. Decentralized biorefinery for lignocellulosic biomass: integrating anaerobic digestion with thermochemical conversion. Bioresource Technology 2018;250:140–7.

CHAPTER 12

Dry Anaerobic Digestion of Wastes: Processes and Applications

REGINA J. PATINVOH, PHD

INTRODUCTION

Biogas production from organic wastes has technologically advanced in the recent years; this cost-effective technology has been productive in managing energy demands, carbon emissions, and environmental pressures. The anaerobic digestion (AD) process to produce biogas involves many stages with specific optimum process conditions as explained in the literature [1–3]. The digestion process requires a consortium of microorganisms working synergistically and sometimes the feedstock used are difficult to handle; this makes the process a bit complex. During the digestion process, a negative effect on any of the stages involved (hydrolysis, fermentation, acetogenesis, and methanogenesis) has a direct effect on the other and the process can become unsteady or can fail completely [4]. There are a diversity of papers dealing with the pretreatment of feedstock [5–8], codigestion of various feedstocks [9–13], bioaugmentation with anaerobic enzymes [14–16], and acclimatization of microorganisms to feedstock [17,18] to stabilize the digestion process, to reduce the inhibitory effect of harmful compounds, and to increase the digestion rate, as well as the biogas yield and the digestate quality.

Conventional digestion processes (wet AD) to produce biogas involves the use of a large volume of water, over 90% of the constituents of the bioreactor is water; the total solids (TS) content of the feedstock is usually between 0.5% and 15% [19]. This is a major challenge for countries with shortage of water; it is obvious that there will be competition for water as population increases and industrial development progresses. Additionally, dewatering of the residue after biogas production requires high energy consumption and results in loss of nutrients. So the use of digestate as biofertilizer with less dewatering systems or without dewatering is inevitable. The use of forest and crop residues with high solid content (usually between 15% and 50%) is also becoming very attractive due to their abundance in nature and high energy value. In view of these issues, the dry AD technology is gaining momentum in both research and industries.

This chapter focuses on dry AD processes for biogas production. It discusses the importance of the start-up phase and its correlation with the microbial communities, as well as the major processes involved in solid-based AD systems. It also addresses current applications in the industry to enhance the performance of the process and for effective industrial feasibility.

DRY ANAEROBIC DIGESTION TECHNOLOGY

This new technology is designed to process more organic wastes per reactor volume with TS content greater than 20% [20]. This approach provides better economic feasibility because reactor volume is minimized, lower energy is required for heating, and cost of handling the residue after biogas production is reduced [21,22]. Dry AD processes are flexible regarding the acceptance of debris (nonbiodegradable materials); these materials will not contribute to the biogas yield but does not affect the degradation process. More so, the microbial morphology in the bioreactor changes as the TS content changes [23,24]. The microbial communities that are dominant in dry digestion processes may differ from those in wet digestion processes as the TS content increases [18,25]. Yi et al. [24] investigated microbial communities using pyrosequencing technology while treating food waste with TS content from 5% to 20% under the mesophilic condition. The results showed better volatile solids (VS) reduction and methane yield in reactors with higher TS content and that *Methanosarcina* was dominant in three reactors showing an increasing trend with increasing TS content.

Sustainable Resource Recovery and Zero Waste Approaches. https://doi.org/10.1016/B978-0-444-64200-4.00012-8

Li et al. [18] reported that *Anaerococcus* species are abundant in solid-state AD (dry AD) and that these species are responsible for improved degradation efficiency and methane yield. For dry AD processes to accommodate high loading rates, it is essential that the microorganisms are adapted/acclimatized to the feedstock and there is sufficient contact between viable microorganisms and the feedstock. It is also necessary to have favorable environmental conditions for the microbial communities in the bioreactor under the operating conditions [26]. In view of these vital conditions, dry AD process requires a start-up phase before the digestion process and appropriate process conditions for an improved performance.

Start-Up Phase and Generation of Adapted Microbial Cultures

In dry AD processes, it is important that the inoculum is collected from an active anaerobic bioreactor, preferably from a dry AD plant because inoculum with high solid content is needed. Obtaining an inoculum with high solid content is vital because centrifuging the inoculum at high speed to obtain as many bacteria as possible (solid inoculum) may result in significant reductions in bacterial viability [27], which could affect the experimental outcome. Manure can as well be used for plants that does not have access to inoculum from an active AD plant. It is also important to have an appropriate inoculum-to-substrate ratio; the amount of solid inoculum needed depends on the substrate and the feedstock of the previous bioreactor from which the inoculum is taken. Patinvoh et al. [17] reported an initial inoculum-to-substrate ($VS_{inoculum}$-to-$VS_{substrate}$) ratio of 1 in a textile bioreactor treating manure with straw, which was reduced to 0.5 after an acclimatization period of 232 days.

The start-up phase is important for the generation of microbial cultures adapted to the feedstock; the period can be very long due to the low growth rates of the anaerobic microorganisms. If inoculum adapted to feedstock that is similar to the one of interest is available then the period can be shorter; it can take just a few days or it can take a longer period, up to 1 or 2 months or longer, if the right microorganisms are not acclimatized. A start-up period of 40 days was reported in the literature before dry digestion of manure with straw in a continuous plug flow bioreactor [28], whereas just about 28 days was adequate when cosubstrates (chicken feathers, citrus wastes, wheat straw, and manure with straw) were used [13] under mesophilic conditions.

Optimum Conditions for Dry Anaerobic Digestion Process Stability

Organic loading rate

The organic loading rate (OLR) should be moderate at the initial stage and then increased gradually until the bioreactor's optimum threshold is reached. At higher OLRs, the organic compounds are not fully degraded, which results in accumulation of surface active agents and by-products that promotes foaming [29]. Foaming is not a common phenomenon with dry AD processes; however, if the OLR is too high, there can be volatile fatty acid (VFA) accumulation leading to inhibition of methanogens resulting in process failure [30]. Overloading and fluctuations of bioreactor loading should be avoided by daily monitoring of organic loading and the bioreactor's performance. Early indicators of process instability are VFA concentration, alkalinity ratio (VFA/alkalinity ratio), hydrogen concentration (should be typically less than 100 ppm), and redox potential (should be lower than −300 mV) [4]. The VFA/alkalinity ratio lower than 0.3 is generally considered to indicate stable processes [4]. A continuous loading rate is vital for the stability of the process [31] because variations in the loading rate often result in variations in biogas production rates, which can affect the methane yield. Alterations in the composition or energy content of two different batches of the same feedstock can also result in biogas production instability, even though the loading rate remains the same [4]. Patinvoh et al. [28] reported an OLR of 6 gVS/L/d with a corresponding retention time of 28 days favoring process instability while treating manure with straw at 22% TS content in a continuous plug flow bioreactor.

Solid content

The optimum TS content usually varies between 20% and 35% depending on the feedstock being treated. Fernández et al. [20] studied the effect of TS content on dry AD of organic fraction of municipal solid wastes under mesophilic conditions. The results showed reduction in methane production from 7.01 to 5.53 L as the TS content increased from 20% to 30%. Patinvoh et al. [13] also reported 13.5% reduction in methane yield as the TS content increased from 21% to 32% during co-digestion of citrus wastes with chicken feathers and wheat straw. The experimental results also showed a blockage in the inlet and outlet parts of the continuous plug flow bioreactors as the TS content increased to 32% with an increased OLR from 2 to 3.8 gVS/L/d. High ammonia nitrogen concentration has been reported by Chen et al. [32] as the TS concentration of swine manure was increased from 20% to 35% in continuous dry AD process. Ammonia is toxic to the microbes at higher concentrations, especially to the methanogens, and as such reduction in biogas yield was observed as the TS content increased from 20% to 35%.

At very high TS contents (≥40%), digestion can come to a complete halt, as there is insufficient water available for the growth of the microorganisms [23,33]. Therefore pretreatment of the feedstock is necessary when digesting at higher solid contents, especially for lignocellulosic and keratin-rich wastes, because it helps increase the availability of the feedstock to the microorganisms and thereby increases the biogas yield [6,34,35]. Ammonia inhibition from protein-rich wastes can be resolved by low loading or by dilution [36]. Membrane applications have also been reported to be effective in protecting the microorganisms from the toxic effects of feedstock and/or product inhibition [37,38]. Additionally, a nutrient medium is necessary to supply the macronutrients, trace elements, and vitamins needed for microbial growth and activity [39].

Temperature

Dry AD under mesophilic conditions may require a long start-up phase for better performance, whereas the rate of digestion is faster under thermophilic conditions [40]. Additionally, there is increased rate of pathogen reduction during thermophilic digestion processes [19]. However, thermophilic microbes are more sensitive to temperature changes as well as feedstock and/or product inhibition. Thus most dry AD biogas plants are operated under mesophilic conditions and the process is more stable.

DIGESTION PROCESSES

Dry digestion of wastes to produce biogas includes a series of biochemical transformations under specific conditions as explained in the sections Introduction and Dry Anaerobic Digestion Technology; it is commonly carried out either in batch or continuous processes. This section explains the batch and continuous processes in relation to dry fermentation and their industrial applications.

Batch Processes

Batch processes are common and simple to operate. In this process, the undiluted feedstock is mixed with an appropriate amount of inoculum (this is needed only at the initial stage) and thereafter the mixture is fed into an airtight bioreactor under strict anaerobic conditions. The bioreactor remains closed until decomposition of the organic fraction of the feedstocks into biogas (methane, carbon dioxide, and other gases) is completed; the biogas produced is continuously collected in a gas holder [17]. When there is no more gas production, the bioreactor is opened for the next run; the process can be repeated by replacing a portion of the digestate with fresh undiluted feedstock. The remaining portion of the digestate can be stored as a biofertilizer or posttreated using an aerobic composting process.

Industrially, several bioreactors are run in parallel; this ensures continuous collection and treatment of the feedstock while the biogas is continuously produced. However, the process can be inhibited and the digestion process can become inefficient because of the supply of a large volume of feedstock to the microbial communities at once and inadequate mixing [30]. In addition, steady biogas production is not achievable because the condition in the process changes continuously due to cell metabolism [41] and the limited opportunity to control the process.

Continuous Processes

The continuous process allows a periodic feedstock supply without alteration to the microbes, which minimizes substrate or product inhibition. Moderate amount of the feedstock is fed at the initial stage and then increased gradually to reach the target; balance between feed and discharge is maintained for a sufficient retention time to achieve steady-state condition [42]. The retention time is dependent on the feed rate (OLR), dilution ratio, and digestion temperature. There can be VFA accumulation if the OLR is too high, resulting in instability in the system and subsequent process failure [30].

Plug flow bioreactors are commonly used for continuous dry digestion processes; feedstock added at the inlet of the reactor are transported slowly toward the outlet taking several rotations [28]. The process requires mixing of the incoming (fresh) feedstock with a portion of the digestate to avoid acidification and blockage of the reactor at the inlet.

SUITABLE DESIGN TECHNOLOGIES FOR DRY DIGESTION PROCESSES

The major challenges of the dry digestion processes are handling, pumping, and mixing of the highly solid and viscous waste streams. Therefore the process requires pumps specifically designed for handling highly solid slurries, special mixing techniques, and appropriate technology for operation. This section describes applicable technologies in industries to circumvent some of these difficulties.

These technologies employ transportation of wastes using conveyor belts, screws, and pumps especially designed for highly viscous streams; these pumps are more expensive than centrifugal pumps used in wet fermentation [43].

BEKON Technology

The BEKON Technology is a single-step batch digestion process that allows digestion of solid wastes with TS content up to 50% [44]; several reactors are run simultaneously to allow continuous production of biogas. The fermentation process runs at the mesophilic condition (34–37°C) and the temperature is regulated through heated floor and walls. The inventive part of this technology is that it allows continuous inoculation to provide moisture for the microorganisms during the digestion, as shown in Fig. 12.1; the percolation liquid is collected and recirculated back continuously into the reactor. This is a good practice because recirculation of the liquid can facilitate the growth of the microorganisms [45] and thereby improves methane yield while also reducing degradation time. However, liquid recirculation sometimes can result in a high risk of acidification, thereby leading to instability in the process.

Textile Bioreactor Technology

Dry AD of solid wastes using textile bioreactors is a novel technology designed to enhance global advancement of biogas technology, especially in the developing countries; the technology is robust and simple to operate [17]. The bioreactor is made from textile materials; this makes it less combustible, resistant to tearing, lightweight, and environmentally benign. It is also UV treated to prevent easy degradation due to exposure to sunlight and is weather resistant, which increases its life span. More so, it is composed of coated polymers as a protection against corrosive gases such as H_2S [30]. Textile biogas bioreactors range in size from 1 to 1000 m^3 and are suitable for both small- and large-scale biogas processes.

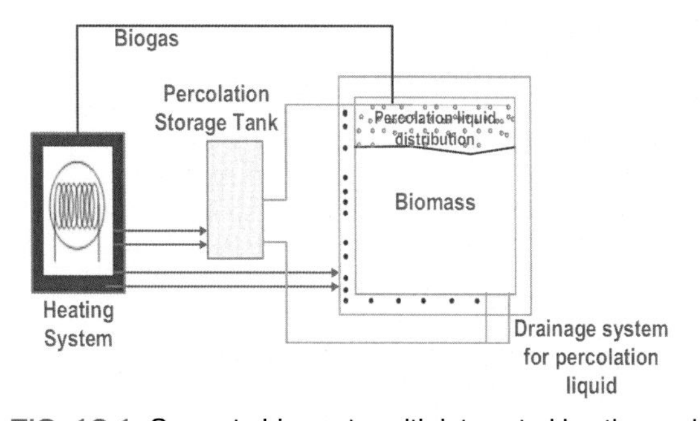

FIG. 12.1 Concrete bioreactor with integrated heating and percolation liquid systems. (Adapted from Innovative Solutions for Cities and Agriculture. BEKON Energy Technologies GmbH & Co. KG., in Bioenergy via Dry Fermentation. http://www.cityofpaloalto.org/civicax/filebank/documents/19875 Accessed 2014/03/03.)

Additionally, high investment cost impedes biogas production from agricultural wastes from reaching its full potential; the use of conventional bioreactors contributes to about 25%–35% of the fermentation capital investment cost [46]. In order to reduce the technical and economic barriers of biogas production in the agricultural sector, the use of robust, simple, and cost-effective textile bioreactors for dry digestion of solid wastes is a promising approach.

The reactor worked successfully for the dry digestion of manure with straw at 22%–30% TS (solid content of the feedstock), as reported by Patinvoh et al. [17]. A long-term acclimatization period of 232 days resulted in a 58% increase in methane yield. However, the dry digestion process can be optimized for maximum production through mild pretreatment and biogas recirculation, as shown in Fig. 12.2. This will enhance mixing in the textile bioreactor, thereby reducing the solid retention time.

Continuous Dry Anaerobic Digestion Technologies

For continuous dry fermentation of solid wastes, plug flow bioreactors with horizontal and vertical orientations are commonly used for industrial purposes. Technologies with horizontal orientations are the Västblekinge Miljö AB (VMAB), STRABAG, and KOMPOGAS technologies, whereas the VALORGA and DRANCO technologies are based on plug flow principles with vertical orientations. This section describes the operating conditions, outstanding features, and mixing techniques of the various technologies.

Eisenmann technology

Eisenmann provides a strategic technology for dry solid-based AD systems. A typical example of this technology is the VMAB technology in Sweden. The VMAB technology is designed for dry digestion of household wastes (about 75% sorted household wastes and 25% garden wastes) at *Mörrum* in Sweden. At present, there are many biogas plants in Sweden, of which only one located at Mörrum operates on dry digestion principle. This is a low level of such installations compared to available agricultural residues such as straw, grass, and

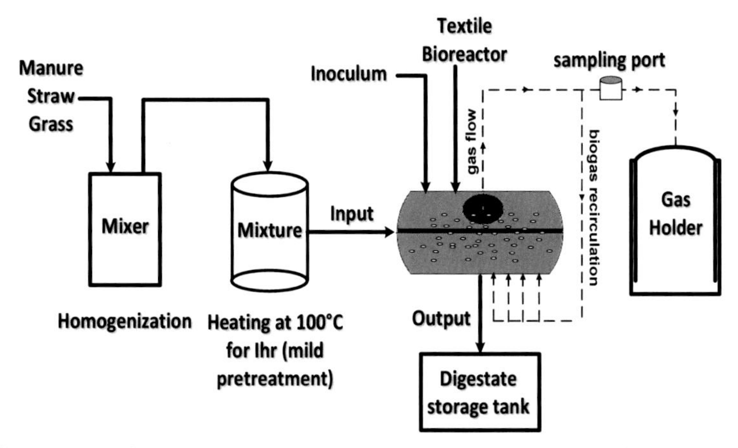

FIG. 12.2 Optimization of horizontal textile bioreactor with biogas recirculation for mixing enhancement.

manure; these types of wastes usually have a low moisture content, which makes them suitable for dry digestion processes. This new technology is based on a plug flow bioreactor principle that has a horizontal orientation. The digestion process was started initially under thermophilic conditions but presently operates on both thermophilic and mesophilic conditions depending on the bioreactor process performance.

Pretreatment: The collected solid wastes from the municipalities are stored and thereafter transported to a mill to reduce the particle size below 80 mm. The particles are then separated using the star screen technology, where particles are separated based on the rotational speed of the stars. Some undigested materials pass through the screen; such materials do not affect the degradation process as in the case of wet digestion. Roughly 90% of the separated wastes are transported to the bioreactor for the digestion process, while the remaining debris are incinerated.

Digestion: The separated wastes are stored in a storage tank and thereafter transported through a vertical pump into a horizontal plug flow bioreactor. Manure with inoculum from a wet digestion plant is used as a start-up, and the bioreactors are fed continuously with a feedstock TS content of 35%. The solid content inside the bioreactor is about 11%–14% and the digestate TS content varies between 17% and 24%. The most outstanding feature of this technology is that there are horizontal impellers constructed inside the bioreactor, which have a rotational speed of about 2–4 rpm; they allow flow of material and minimal mixing, thereby making mixing easier and reducing the solid retention time. Additionally, the bioreactor has an online sensor for measuring the concentration of hydrogen sulfide and the flow rates of carbon dioxide, methane, oxygen, and biogas. Oxygen is sometimes sparged from the top of the bioreactor for reduction of hydrogen sulfide concentration during the digestion process.

Digestate: The residue obtained after biogas production is pumped to a hydrocyclone and separated into two fractions (solid and liquid fractions). The solid fractions are composted, while a portion of the liquid fraction is sometimes sent back to the bioreactor and the remaining fraction is stored as biofertilizer.

STRABAG technology

The STRABAG technology is similar to the Eisenmann technology, with difference in the mixing inside the bioreactor. This technology is designed for the treatment of solid waste with TS content between 15% and 45% using horizontal plug flow reactors, as shown in Fig. 12.3. The bioreactors are equipped with agitators arranged transverse to the flow direction to prevent formation of floating scum and settlement of materials [47].

KOMPOGAS technology

This technology is similar to the STRABAG technology with cylindrical reactors, as shown in Fig. 12.3. Mixing is enhanced by impellers inside the bioreactors, with slow rotation; these impellers also aid in homogenization, degassing, and resuspension of heavier particles [43]. The process requires careful adjustment of the TS content in the bioreactor to around 23%; the flow of material is reduced or stopped at higher TS values [43].

VALORGA technology

This technology is different from the previous ones explained in that it uses biogas recirculation as a means of enhancing mixing in the bioreactor, as shown in

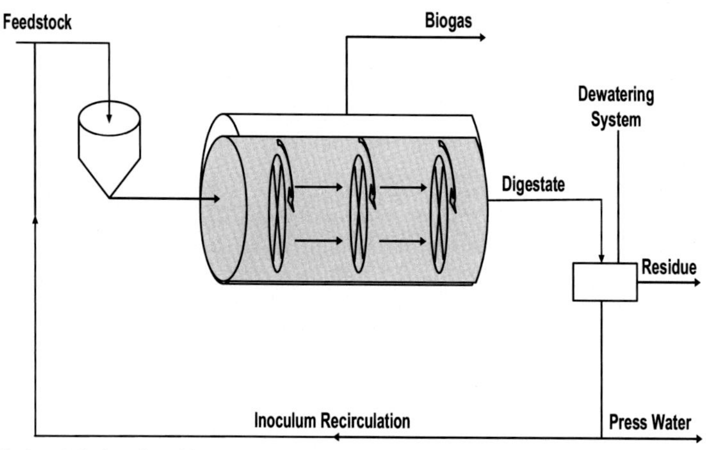

FIG. 12.3 Horizontal plug flow bioreactor design with inoculum recirculation. (Adapted from Ref. [43].)

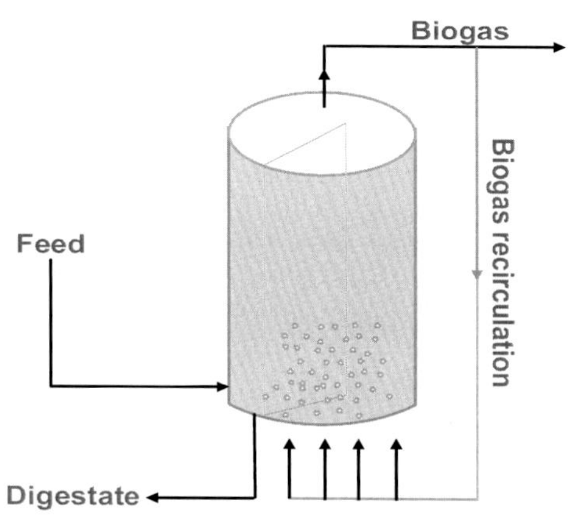

FIG. 12.4 Vertical plug flow bioreactor design with biogas recirculation. (Adapted from Ref. [43])

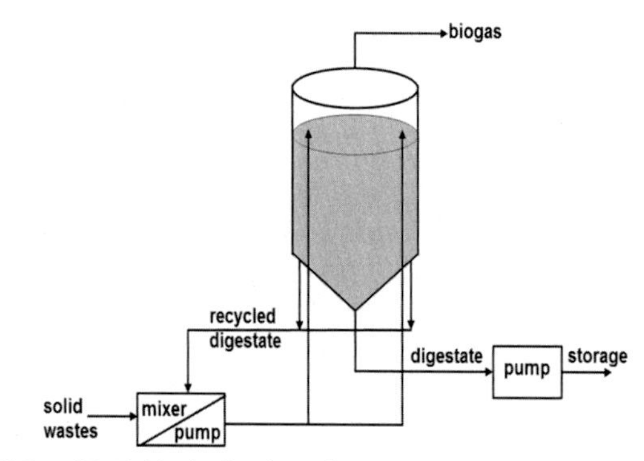

FIG. 12.5 Vertical plug flow bioreactor design with digestate recirculation for adequate mixing of solid wastes. (Adapted from Baere http://www.ows.be/wp-content/uploads/2013/02/The-DRANCO-technology-2012.pdf[49].)

Fig. 12.4. The process operates semicontinuously under mesophilic conditions and there are no mechanical parts inside the bioreactor; as such, no maintenance with mechanical device is required. The collected solid wastes are first screened in a rotating screen or sorted mechanically and crushed to a particle size below 80 mm. Thereafter, the crushed waste is mixed with a portion of the liquid fraction of the process residue, thereby adjusting the feedstock TS content to around 30%, which is pumped into bioreactor using piston pumps [48]. During the digestion process, mixing is enhanced automatically through injection of compressed biogas at the bottom of the bioreactor every 15 min [43]. The residue after biogas production is dewatered using screw press, and the liquid fraction is transported to a hydrocyclone to remove heavy particles and passed through a flocculation-filtration unit to remove suspended solids, whereas the solid fraction undergoes aerobic composting [48].

DRANCO technology

This technology is a single-stage system developed in Belgium for the treatment of municipal solid wastes, with a reactor TS content of 15%–40% [49]. It is also based on a plug flow bioreactor principle but with a vertical orientation, as shown in Fig. 12.5. The feedstock is pumped to the top of the reactor through feeding tubes and extracted through a conical outlet at the bottom. There is no mixer inside the reactor, but mixing is enhanced by recycling the digestate (one part of fresh waste mixed with six parts of the digestate) [43].

CONCLUSIONS AND PERSPECTIVES

Dry AD is an improved technology for biogas production; it addresses the problem of large water volume in conventional bioreactors and increases the acceptability of feedstocks. This technology is gaining interest as new bioreactor designs are being developed to enhance mixing in the system, thereby increasing the production efficiency. However, there is need for further research on the behavior and growth of the dominant microbial communities in dry digestion processes; this will help reduce the start-up period and solid retention time. Another important factor for future expansion of this technology is the development of methods for continuous monitoring of this process, as it is sensitive to inhibition and inhibitors that enter the reactor through feedstocks are higher in concentration due to the reduced water content. This will make it easier to control the process and get maximum production.

REFERENCES

[1] Deublein D, Steinhauser A. Biogas from waste and renewable resources. Wiley-VCH; 2011.

[2] Schnürer A, Jarvis A. Microbiological handbook for biogas plants. Swedish Waste Management U2009:03. 2010. 2017/02/09]; Available from: http://www.eac-quality.net/fileadmin/eac_quality/user_documents/3_pdf/Microbiological_handbook_for_biogas_plants.pdf.

[3] Friehe J, Weiland P, Schattauer A. Fundamentals of anaerobic digestion. 2016/05/31]. In: Guide to biogas from production to use. 5th revised edition. Fachagentur Nachwachsende Rohstoffe e. V. (FNR); 2010. Available from: https://mediathek.fnr.de/media/downloadable/files/samples/g/u/guide_biogas_engl_2012.pdf.

[4] Drosg B. Process monitoring in biogas plants. IEA Bioenergy; 2013, ISBN 978-1-910154-03-8. p. 1–38. IEA Bioenergy 978-1-910154-03-8.

[5] Patinvoh RJ, et al. Innovative pretreatment strategies for biogas production. Bioresource Technology 2017;224:13–24.

[6] Patinvoh RJ, et al. Biological pretreatment of chicken feather and biogas production from total broth. Applied Biochemistry and Biotechnology 2016;180(7):1401–15.

[7] Bolado-Rodríguez S, et al. Effect of thermal, acid, alkaline and alkaline-peroxide pretreatments on the biochemical methane potential and kinetics of the anaerobic digestion of wheat straw and sugarcane bagasse. Bioresource Technology 2016;201:182–90.

[8] Theuretzbacher F, et al. Steam explosion pretreatment of wheat straw to improve methane yields: investigation of the degradation kinetics of structural compounds during anaerobic digestion. Bioresource Technology 2015;179:299–305.

[9] Mata-Alvarez J, et al. A critical review on anaerobic co-digestion achievements between 2010 and 2013. Renewable and Sustainable Energy Reviews 2014;36:412–27.

[10] Zeshan, Karthikeyan OP, Visvanathan C. Effect of C/N ratio and ammonia-N accumulation in a pilot-scale thermophilic dry anaerobic digester. Bioresource Technology 2012;113:294–302.

[11] Pagés Díaz J, et al. Co-digestion of different waste mixtures from agro-industrial activities: kinetic evaluation and synergetic effects. Bioresource Technology 2011;102(23):10834–40.

[12] Yenigün O, Demirel B. Ammonia inhibition in anaerobic digestion: a review. Process Biochemistry 2013;48(5–6):901–11.

[13] Patinvoh RJ, et al. Dry anaerobic Co-digestion of citrus wastes with keratin and lignocellulosic wastes: batch and continuous processes. In: Waste and biomass valorization; 2018. p. 1–12.

[14] Peng X, et al. Impact of bioaugmentation on biochemical methane potential for wheat straw with addition of Clostridium cellulolyticum. Bioresource Technology 2014;152:567–71.

[15] Zhang J, et al. Bioaugmentation with an acetate-type fermentation bacterium Acetobacteroides hydrogenigenes improves methane production from corn straw. Bioresource Technology 2015;179:306–13.

[16] Forgács G, et al. Pretreatment of chicken feather waste for improved biogas production. Applied Biochemistry and Biotechnology 2013;169(7):2016–28.

[17] Patinvoh RJ, et al. Cost effective dry anaerobic digestion in textile bioreactors: experimental and economic evaluation. Bioresource Technology 2017;245(Part A):549–59.

[18] Li A, et al. A pyrosequencing-based metagenomic study of methane-producing microbial community in solid-state biogas reactor. Biotechnology for Biofuels 2013;6(1):3.

[19] Li Y, Park SY, Zhu J. Solid-state anaerobic digestion for methane production from organic waste. Renewable and Sustainable Energy Reviews 2011;15(1):821–6.

[20] Fernández J, Pérez M, Romero LI. Effect of substrate concentration on dry mesophilic anaerobic digestion of organic fraction of municipal solid waste (OFMSW). Bioresource Technology 2008;99(14):6075–80.

[21] Karthikeyan OP, Visvanathan C. Bio-energy recovery from high-solid organic substrates by dry anaerobic bio-conversion processes: a review. Reviews in Environmental Science and Bio/Technology 2013;12(3):257–84.

[22] Jha AK, et al. Research advances in dry anaerobic digestion process of solid organic wastes. African Journal of Biotechnology 2011;10(64):14242–53.

[23] Motte J-C, et al. Total solids content: a key parameter of metabolic pathways in dry anaerobic digestion. Biotechnology for Biofuels 2013;6:164.

[24] Yi J, et al. Effect of increasing total solids contents on anaerobic digestion of food waste under mesophilic conditions: performance and microbial characteristics analysis. PLoS One 2014;9(7):e102548.

[25] Di Maria F, et al. Solid anaerobic digestion batch with liquid digestate recirculation and wet anaerobic digestion of organic waste: comparison of system performances and identification of microbial guilds. Waste Management 2017;59:172–80.

[26] Lettinga G. Anaerobic digestion and wastewater treatment systems. Antonie Van Leeuwenhoek 1995;67(1):3–28.

[27] Pembrey RS, Marshall KC, Schneider RP. Cell surface analysis techniques: what do cell preparation protocols do to cell surface properties? Applied and Environmental Microbiology 1999;65(7):2877–94.

[28] Patinvoh RJ, et al. Dry fermentation of manure with straw in continuous plug flow reactor: reactor development and process stability at different loading rates. Bioresource Technology 2017;224:197–205.

[29] Moen G. Anaerobic digester foaming: causes and solutions. In: Water environment and technology. Alexandria: Water Environment Federation; 2003. p. 70–3.

[30] Patinvoh R. Biological pretreatment and dry digestion processes for biogas production. PhD Thesis. Sweden: University of Boras; 2017.

[31] Gallert C, Winter J. Propionic acid accumulation and degradation during restart of a full-scale anaerobic biowaste digester. Bioresource Technology 2008;99(1):170–8.

[32] Chen C, et al. Continuous dry fermentation of swine manure for biogas production. Waste Management 2015;38:436–42.

[33] Friehe J, Weiland P, Schattauer A. Guide to Biogas from production to use. c.r.e. 5th 2010. Gülzow, 2010.

[34] Taherzadeh MJ, Karimi K. Pretreatment of lignocellulosic wastes to improve ethanol and biogas production: a review. International Journal of Molecular Sciences 2008;9(9):1621–51.

[35] Johnson DK, Elander RT. Pretreatments for enhanced digestibility of feedstocks. In: Biomass recalcitrance. Blackwell Publishing Ltd; 2009. p. 436–53.

[36] DaSilva, E.J., Biogas generation: developments, problems, and tasks - an overview, in Division of scientific research and higher education,. Unesco, Paris, France.

[37] Dasa KT, et al. Inhibitory effect of long-chain fatty acids on biogas production and the protective effect of membrane bioreactor. BioMed Research International 2016;2016:9.

[38] Youngsukkasem S, et al. Biogas production by encased bacteria in synthetic membranes: protective effects in toxic media and high loading rates. Environemental Technology 2013;34(13–14):2077–84.

[39] Angelidaki I, et al. Defining the biomethane potential (BMP) of solid organic wastes and energy crops: a proposed protocol for batch assays. Water Science and Technology 2009;59(5):927–34.

[40] Hartmann H, Ahring BK. Strategies for the anaerobic digestion of the organic fraction of municipal solid waste: an overview. Water Science and Technology 2006;53(8):7–22.

[41] Smith JE. Bioprocess/fermentation technology. In: Biotechnology. Cambridge: Cambridge University Press; 2009. p. 49–72.

[42] Brethauer S, Wyman CE. Review: continuous hydrolysis and fermentation for cellulosic ethanol production. Bioresource Technology 2010;101(13):4862–74.

[43] Vandevivere P, De Baere L, Verstraete W. Types of anaerobic digester for solid wastes. In: Biomethanization of the organic fraction of municipal solid wastes. Iwa Publishing; 2003. p. 111–40.

[44] Liu H, et al. Enhancement of sludge dewaterability with filamentous fungi Talaromyces flavus S1 by depletion of extracellular polymeric substances or mycelium entrapment. Bioresource Technology 2017;245:977–83.

[45] Kusch S, Oechsner H, Jungbluth T. Effect of various leachate recirculation strategies on batch anaerobic digestion of solid substrates. International Journal of Environment and Waste Management 2012;9(1–2):69–88.

[46] Maiorella BL, et al. Economic evaluation of alternative ethanol fermentation processes. Biotechnology and Bioengineering 2009;104(3):419–43.

[47] GmbH SU. Dry digestion. STRABAG Umweltanlagen GmbH; 2017/09/16. Available from: http://www.strabag-umwelttechnik.com/databases/internet/_public/files.nsf/SearchView/A9C0D56F88B4274EC1257746005 04FB5/$File/3_4%20Trockenvergaerung_e%20d.pdf.

[48] Fruteau de Laclos H, Desbois S, Saint-Joly C. Anaerobic digestion of municipal solid organic waste: valorga full-scale plant in Tilburg, The Netherlands. Water Science and Technology 1997;36(6):457–62.

[49] Baere, L.D. The DRANCO technology: a unique digestion technology for solid organic wastes. http://www.ows.be/wp-content/uploads/2013/02/The-DRANCO-technology-2012.pdf. Accessed 2017/09/16 2012.

CHAPTER 13

Combustion of Waste in Combined Heat and Power Plants

TOBIAS RICHARDS, PHD

INTRODUCTION

Combustion was first a method used for minimizing solid waste. The idea was that combustion reduced the amount of solids (on both a volumetric and a mass basis), leaving room for more waste. It was thereby possible to reduce the volume by up to 90% and the mass by 80%. However, it did not take long before new features as well as risks were discovered. New features included the ability to destroy toxic materials and thus avoid their release into the environment and to utilize the heat that was released to produce steam. The risks that were discovered were associated with the materials produced, both solid and gaseous: the solid material formed (i.e., ash) could contain material that was readily soluble when it came into contact with water (such as rain) and therefore be released into, and transported by, water systems. This release could contain not only relatively harmless materials (such as sodium and potassium ions) but also harmful heavy metals. It is possible that unburned material in the gases formed may result in emissions of chlorinated compounds (such as dioxins and furans) together with catalysts (such as copper). Uncontrolled combustion, with areas of relatively low temperatures and a low concentration of oxygen, leads to the emission of dioxin and other toxic materials—a common problem is emission of vaporized mercury, which was not properly addressed initially. However, much has happened since the early days of waste combustion: new, modern boilers have high degrees of efficiency and they simultaneously radically reduce the amount of unwanted emissions produced. It has also become obvious that the alternative of using landfill sites, irrespective of the fact that land is scarce, is not a sustainable option; it is, in fact, prohibited in many countries (e.g., in the European Union) to send organic material to these sites. One of the main reasons for this ban is the uncontrolled formation of methane gases that results from landfills; methane is a highly potent greenhouse gas that should be avoided, as it contributes to raising the global temperature.

WASTE MATERIALS FOR COMBUSTION

Waste that is combusted generally comes from two very different sources: municipal solid waste and industrial waste. Both these fractions have a rather wide span in composition. Industrial waste that is to be combusted could, for example, contain wood from demolition sites, waste from the pet food industry, fiber sludge, and various kinds of plastics. Such waste generally has a high heating value and low moisture content. Municipal solid waste differs in composition depending on the location and varies between countries as well as within each country. This leads to differences in heating values, and one major impact is the content of food and garden wastes in the waste collected, as these wastes have a high moisture content. A country that implements presorting of food waste will thus have waste with a higher heating value, which is easier to handle in most boilers and could, in most cases, also increase the overall efficiency of the boiler, as the heat loss in the flue gas will be reduced.

A difference between waste and other fuels is the relatively high concentration of alkali (sodium in particular), sulfur, and chlorine. These chemical species undergo many chemical reactions that could eventually lead to deposits and corrosion on the heat transfer surfaces in the boiler. The formation of a deposit will reduce the effective heat transfer and require a more frequent cleaning schedule. Corrosion will eventually reduce the strength of the piping, which could lead to the formation of holes and the subsequent shut down of the operation. In addition to these issues, others are present. Of special importance is the presence of heavy metals such as zinc, lead, and mercury because

Sustainable Resource Recovery and Zero Waste Approaches. https://doi.org/10.1016/B978-0-444-64200-4.00013-X

they have a low melting point and may be vaporized; if they are vaporized, they might react with the available chlorine and increase both corrosion and the formation of deposits. Iron and copper act as catalysts in the formation of dioxin in the temperature range of 400–250°C.

COMBUSTION

The purpose of combustion is to convert all the material in the fuel into a fully oxidized product that is generally gaseous. In the case of waste combustion, this means that the incoming carbon content should produce carbon dioxide and any hydrogen should yield water (i.e., steam). The basic reaction of organic material with no inorganic content is shown in Reaction (R1).

$$C_xH_yO_z + (x + y/4 - z/2)O_2 \rightarrow x\,CO_2 + y/2H_2O \quad \textbf{(13.R1)}$$

However, the mechanism is a little more complex and occurs in different steps. The first step is drying, whereby the water in the fuel is released as steam; this occurs practically without any other chemical reaction and is a part of the heating process. Drying is an endothermic process, meaning that heat is required. Drying is followed by pyrolysis, which is a heat-induced cracking reaction that leads to the release of gaseous products from the solid, such as carbon oxides and various hydrocarbons of different molecular weights. No oxygen is need, therefore, to be present during pyrolysis and, depending on the chemical composition, pyrolysis could be either slightly endothermic or slightly exothermic. The solid that remains after pyrolysis is composed of inorganic material, along with the char formed. The char has a high content of carbon, as well as some hydrogen and oxygen. The higher the temperature of pyrolysis, the higher the fraction of carbon present. The final step is reaction with oxygen, which is exothermic and provides the high temperatures in the boiler; it could occur with both the gaseous species that forms and the remaining solid material.

It is important that the heat and mass transfer inside a boiler is good to provide not only efficient and rapid heating during the initial phases but also good mass transfer, which will ensure good contact between the reacting species and oxygen. All gases must have passed through a zone with a temperature of at least 850°C for a minimum period of 2 s [1].

BOILER TYPES

Two main types of boilers are used in waste-to-energy applications: stoker (grate) and fluidized bed. Several other options are, however, used for more specific fractions [2]. Rotary boilers, for example, are commonly used for combusting hazardous materials, but they are not included here.

A stoker boiler has a bed (often on a moving grate to ensure continuous transport) through which the primary air passes; the species being formed is thereby initially swept away and, in the final section, the remaining solids are burned. Secondary air is used above the grate to provide good mixing and thus good contact between oxygen and the gases formed. The stoker boiler is the most common type of boiler used, mainly based on its reliable operation and the large number of reference cases from around the world. Although it has minimum requirements for the fuel being pretreated, it is more difficult to operate properly if the fuel input is nonuniform. This could lead to higher emissions of carbon monoxide, which is an indication that the conditions within the boiler vary and that mixing is nonuniform.

A fluidized bed boiler has an inert bed of fine particles (often with quartz sand). The fuel is fed into the boiler, which is fluidized with air and recirculated flue gases. Once the fuel is in the fluidized bed, the bed itself ensures efficient mixing in terms of both mass and heat transfer, creating a uniform concentration and temperature. Secondary air is introduced above the bed in the "freeboard" for the final combustion and to ensure a sufficiently high temperature. This type of fluidized bed is known as the bubbling fluidized bed and is used for moderate-sized boilers in the order of 20 MW_{th}. Larger fluidized bed boilers generally use a circulating fluidized bed that has a higher power production per unit of cross-sectional area unit. The fuel used in a fluidized bed boiler requires a higher degree of pretreatment than that used in a stoker boiler to maintain the high mixing properties, which is done mainly by size reduction. In contrast to the stoker boiler, a fluidized bed does not contain any moving parts, which reduces the cost of the plant. On the other hand, the running costs are greater due to the fuel requiring extended pretreatment, the addition of sand, and more material being collected in the ash-handling system.

PARTS OF A BOILER

After collection, the solid waste is transported to a temporary storage facility where there is the possibility of subjecting it to pretreatment. This is especially important for fluidized bed boilers because it is preferable that the size of the material be reduced. There are several different options for size reduction, with the most common being the hammer mill. Although this mill is efficient, it requires substantial maintenance, which increases the cost of operation. However, when pretreatment has been carried out, the fuel is much more mixed than the initial waste that was collected, and operation will therefore be easier. After pretreatment, the waste is transported to the fuel bin; from here, the operator can control the feed input so that it matches the requirements of the production and also achieves as even combustion properties as possible. A typical result of rapidly changing the fuel's properties can be seen as spikes of carbon monoxide, indicating the need for more oxygen to achieve complete combustion. Another result would be a rapid change in temperature, either an increase or a decrease, indicating an input of fuel with a different heating value.

Combustion Section

Inside the boiler, the material is combusted. A high combustion temperature will lead to a high formation of NO_x, which is undesirable. Regulations nevertheless demand a minimum temperature of 850°C [1], which provides the operating window that can be controlled in several ways: by decreasing the amount of fuel input, by changing the amount of primary and secondary air, or by changing the amount of recirculated flue gases. In a stoker boiler, the walls are constructed of boiler tubes, where water at high pressure is evaporated. This increases the heat transfer from the flue gases, which occurs mainly by radiation from the combustion flame. In a fluidized bed boiler, it is common that at least the fluidization zone is made of ceramic to prevent possible erosion of the water tubes. This induces higher recirculation of flue gases with respect to the heat load in order to control the temperature.

During combustion, it is essential that the level of oxygen present is controlled. If all oxygen is present at the same position, a very rapid reaction would occur, with high temperatures and a great risk of uneven mixing; some parts would be left very hot and others rather cold, moreover with no access to oxygen. A way of preventing this would be to run the boiler with a high amount of excess air, but this will result in low energy efficiency because the flue gases leave the boiler at an elevated temperature. Several levels of air "intake" are used instead, ensuring better and more homogeneous combustion as well as distribution of the temperature profile. For both the fluidized bed boiler and the stoker boiler, the definite minimum amount of primary air is controlled by the content of organic carbon in the ash.

The flue gases formed after combustion contain some inorganic material because of either evaporation (as fumes) or entrainment (as small particles). This could cause several different issues in the following equipment, with the two most severe being corrosion and deposits. It should also be noted that the higher the temperature of these inorganic particles, the greater the risk of deposits forming, as they melt at elevated temperatures. One possible solution for reducing the magnitude of these problems is to include an "empty draft" after combustion—a passage with a large open space and planar heat exchangers with the purpose of decreasing the temperature of the flue gas sufficiently so that the particles become solid. Combining this with a return shaft (in practice, this is a U-bend) allows a substantial amount of particles to be removed from the flue gas.

Heat Exchangers

Proceeding further in the boiler is the majority of the heat exchangers. Traditionally, in power boilers, the first heat exchanger section is known as the "superheater section" because it has the biggest driving force for transferring heat; the difference in temperature between the hot and cold sides in the heat exchanger is greatest in this section. This gives a heat exchanger with the smallest area possible, and thus the lowest cost. However, this is only possible when the risks of corrosion and erosion are low enough without jeopardizing operation and this, unfortunately, is not the situation in the case of waste combustion. In particular, the presence of alkali (sodium and potassium) together with chlorine is highly corrosive, and other species are also present that could enhance corrosion. Therefore the maximum temperature of superheated steam in waste combustion is often limited to slightly above 400°C, whereas it is at least 100°C higher for ordinary biomass. In addition, heat exchanger sections could also be arranged in such a way that the highest temperatures are prevented from occurring simultaneously at the hot and cold sides. Fig. 13.1A represents the traditional setup with a minimum heat exchanger area, whereas Fig. 13.1B represents the modified setup for a corrosive environment. Here, the penalty is a heat exchanger with a larger area, although this will prolong the operating time before replacement is necessary.

Another aspect of the combustion of municipal solid waste and the amounts of inorganic materials contained in the flue gases is the large amount of deposits that occur, some of which end up on the surface of the heat exchangers. An increase in pressure drop results if the deposits are not removed. A high pressure drop leads to either the fan before the chimney requiring

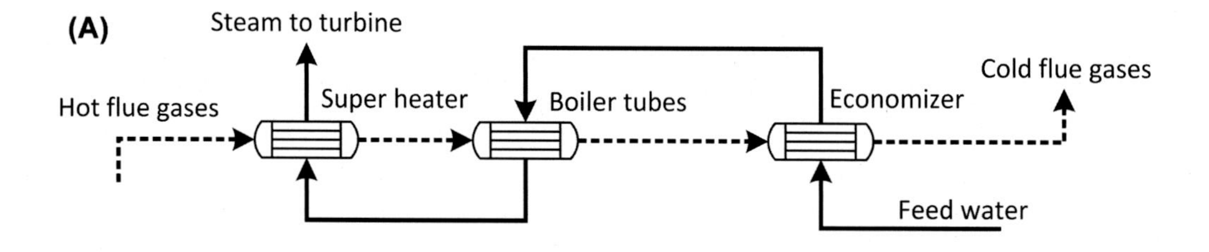

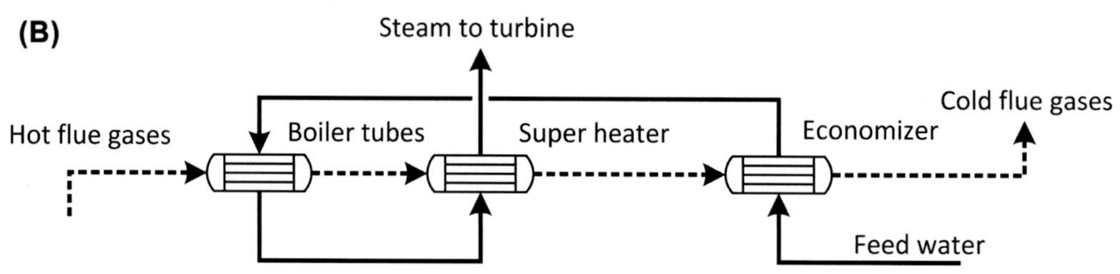

FIG. 13.1 Heat exchangers in the Rankine cycle. **(A)** An arrangement to minimize the surface area of the heat exchanger. **(B)** A setup often used in waste combustion.

more power or the combustion zone having a higher pressure. The former leads to higher internal use of electricity, whereas the latter demands a greater control, and more costly construction, of the boiler because, for safety reasons, it operates at a slightly decreased pressure (subatmospheric). Where deposits occur depends on several factors, of which the three most important are composition, the temperature of the tube walls, and the temperature of the flue gas. The composition and the temperature of the flue gas determine the degree of "stickiness" of the particles (i.e., their ability to adhere to other surfaces) and is related to the amount of smelt. The temperature of the surface determines whether the particles stick to it or fall off.

Some manufacturers of waste power boilers employ heat exchangers in a horizontal alignment. Such boilers have a larger footprint than those with a vertical alignment but have the advantage that they are easier to both handle and replace, when required.

When a circulating fluidized bed boiler is used, a cyclone is added before the heat exchangers to remove the bed material; in a bubbling fluidized bed boiler, the cyclone is more often placed after the boiler tubes and superheaters. This difference in position is due to the amount of bed particles present in the flue gas. Grate boilers do not need a cyclone.

Cleaning

The final cleaning is done using an electrostatic filter in combination with some kind of textile filter or a flue gas condenser. It is also important to handle the sulfur oxides (waste contains generally 0.1%–0.4% sulfur) and the acid gas (hydrogen chlorine) formed during combustion [3]. This is most commonly done by mixing the flue gases with lime (calcium oxide) or slaked lime (calcium hydroxide), resulting in the formation of gypsum (calcium sulfate) and calcium chlorine. This could be carried out (1) in a dry system, where powdered lime hydrate reacts mainly on a textile filter; (2) in a semidry system, where a slurry is sprayed into the flue gases for immediate reaction; or (3) in a wet system, where the flue gases pass through a scrubber system. In the dry and semidry systems, activated carbon is often used in combination with lime, as this would also capture heavy metals and dioxins.

A flue gas condenser may be introduced into a waste boiler that produces both heat and power. It will also serve as a cleaning device because most of the particles (even the smallest ones) can then be removed. However, the condensed water has a low pH and must therefore be neutralized and cleaned before being released. A flue gas condenser could increase heat production by approximately 15% of the fuel heating value, depending on the fuel being used and its moisture content.

Other substances found in the flue gases include various nitrogen oxides (NO_x). During combustion, the substance found is, to a large extent, nitrogen oxide (NO), although this compound oxidizes in the atmosphere over time to form nitrogen dioxide (NO_2). As mentioned earlier, it is preferable that the concentration of oxygen is kept low during combustion, especially during the initial fuel-rich phase to minimize the formation of NO_x. Moreover, it is preferable that the temperature is kept below 1300°C to prevent the formation of thermal NO_x. These two ways are referred to as primary measures. Secondary measures include two different reduction methods, namely, selective noncatalytic reduction (SNCR) and selective catalytic reduction (SCR). In the SNCR method, either ammonia or urea is injected into the upper part of the combustion chamber, where the targeted temperature of the flue gases is 850–950°C [4]. This method makes it possible to reduce the content of NO_x by 60% without allowing any significant amount of ammonia or urea to pass through unreacted. In the SCR method, more than 90% of the NO_x can be removed but, as it requires a catalyst, it increases costs. On the other hand, the reaction is highly efficient, as almost no unreacted ammonia or urea is present [5]. The sensitivity of the catalyst means that this reactor is placed at the end of the gas cleaning stage, so the flue gases may have to be reheated to reach the optimal reaction temperature.

The Rankine Cycle

A combined heat and power boiler is equipped with the Rankine cycle to produce electricity. This cycle is composed of four key parts: the generation of high-pressure steam, a turbine, a condenser, and a pump. Steam is generated in the boiler and then transferred to the turbine. Depending on the production setup (i.e., the preferred outcome and local restrictions, such as the cooling sources available), the back-end pressure of the turbine may vary between locations and individual plants. A lower pressure means a higher production of electricity but at the expense of less usable heat from the condenser. In systems with heat production, the back-end pressure of the turbine is around 1 bar, which would generate a water temperature close to 100°C. When the initial temperature and pressure of 420°C and 40 bar, respectively, are used, the theoretic power output for a given heat input would decrease by 28% if the condenser temperature is changed from 30 to 100°C (although, in reality, the decrease is smaller as the efficiency of the turbine drops when a humid area is encountered).

EFFICIENCY OF A COMBINED HEAT AND POWER PLANT

There are many ways of assessing and measuring efficiency. One of the most obvious method is measuring the energy efficiency, which is basically the amount of energy transferred from the original feedstock to the desired products. In the case of a combined heat and power plant using waste as the feedstock, the energy efficiency may be described as

$$n_{\text{eff}} = \frac{\text{Electricity} + \text{heat}}{\text{Heating value of waste}}$$

The commonly used term heating value is actually the lower heating value, which originates from the time before flue gas condensers existed. However, the addition of flue gas condensers allows the energy efficiency, in some cases, to exceed 100%, and it is, therefore, more relevant to use the higher heating value. In the energy balance, the internal use of electricity should also be considered, which, in a heat and power plant, would be mainly by the pump in the Rankine cycle and the fan for extracting the flue gases. These normally consume less than 1% of the power produced in the turbine and are often neglected. The remaining energy losses are found to be due to the housing (construction and pipes) and emissions via the stack of flue gases (with temperatures up to 200°C). Minimizing heat losses would obviously increase the levels of energy efficiency.

Another option is to evaluate the exergy efficiency. In this case, the amount of energy as well as its quality are evaluated. Normally this is defined as the theoretic ability to perform work. The theoretic ability to perform work possessed by a heat flow is given by the Carnot efficiency (temperatures are given in K).

$$n_{\text{Carnot}} = 1 - \frac{T_{\text{surrounding}}}{T_{\text{gas}}}$$

This means that gas at a high temperature has a greater theoretic ability to perform work, and a gas flow at the same temperature as the surrounding has no possibility of performing work because there is no driving force to be utilized. It is possible to use this concept for even more complex materials, and it is often divided into three parts. The first part is the consideration of the kinetic and potential energy, together with the temperature; the second part is the possible reaction needed to form the chemical species that are present naturally in the surrounding (requiring a definition

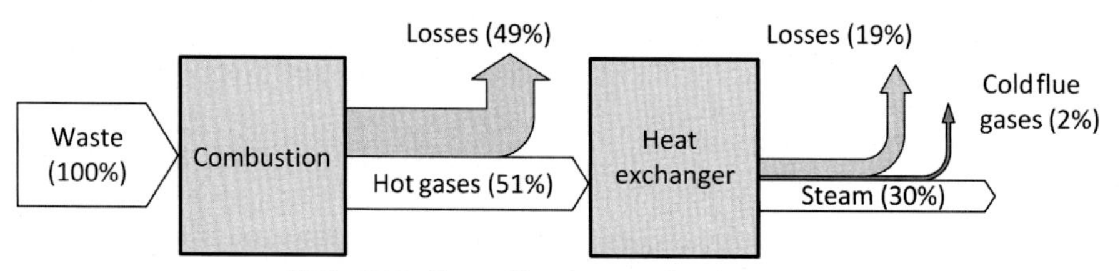

FIG. 13.2 Exergy flow in a combustion plant.

for each compound); and the third part is the mixing work related to different concentrations of the reacted species on entering the surrounding, as well as separating the species from the surrounding, which may be necessary for the chemical reaction to occur. When considering turning waste into energy, the third part is very small in comparison to the other two parts. Furthermore, the kinetic and potential energy in the combustion and flue gas sections can often be neglected, as they are much smaller than both the exergy from the chemical reactions and the temperature. Although energy efficiency does not distinguish between heat and power, exergy efficiency does. About 1 kW of electricity has the exergy value of 1 kW but 1 kW of district heating water at 100°C only has an exergy value of 0.23 kW (using a standard surrounding temperature of 15°C). Considering the loss of exergy in combustion, the majority occurs within the actual combustion process: about 49% is lost when energy stored chemically is transformed into fully oxidized flue gases at 850°C (Fig. 13.2). An additional 21% of the initial amount is lost in the heat exchanger when producing steam at 420°C and 40 bar, provided that the cold flue gases go directly to the chimney. It should be noted that this example does not consider any heat loss due to the lack of insulation to prevent other leakages from occurring, and the only energy loss is through the chimney (the cold flue gases), accounting for about 20%. The fuel composition used here is on a dry basis: 48% C, 6.3% H, 26% O, and 18% ash, with the remainder comprising N, S, and Cl. The heating value is 11,500 kJ/kg as received and the moisture content is 35%.

WASTE COMBUSTION IN A CIRCULAR ECONOMY

Waste combustion has been considered as being the last resort for the collected material; the primary reason for utilizing combustion has been to reduce mass and volume. However, it is important to understand that waste materials have substantial values in addition to their energy content that is already being utilized. These other values are found in the solid residue, i.e., ash. Bottom ash from the combustion section contains much of the iron and copper present in the waste fuel and can be separated by using magnet or eddy current separators. The remaining ash (the fraction with a low concentration of heavy metals) is often used as construction material for roads and to cover landfills. In countries with subsequent treatment (e.g., plasma melting), the remaining ash has been sintered to form an almost impermeable product with low leachability [6] that can be used in construction work close to habitation.

Along with these simple metals, other metals are also present. These are often bound chemically to the ash particles and mean that more advanced methods for separation must be used. Success has been achieved in the removal of, for example, zinc, which is enriched in the fly ash and could be removed by leaching. Many metals, including zinc, copper, lead, and cadmium, are released during a leaching cleaning system, where recirculated acid flows from a wet flue gas. The leaching stage is able to release almost all the cadmium (above 85%), over 60% of the zinc, and lesser amounts of copper and lead [7–9]. The next step is purification of the solution; although this can be performed by several methods, the most promising one appears to be electrolysis. The economics of the process depends on both the concentration of zinc in the fly ash and the current price of the recovered metal, which varies between different countries.

Another promising material suitable for recovery is phosphorus. This is a metal that is vital for plant growth and is present in fertilizers. Phosphorus enters into the system via food waste but the main source is, in fact, sewage sludge. Extensive research and trials have been undertaken to distribute the sludge, after subjecting it to some conditioning, directly on farm land, but the risk of spreading toxic matter, hormones, pesticides, and unused medicines is substantial. Thermal treatment is necessary here, but then problem arises if the phosphorus has not been recovered. Phosphorus is a natural resource of significant importance, and it is also limited and becoming rather scarce. Another problem is that new discoveries of phosphorus contain high concentrations of cadmium, which is a heavy metal that should be avoided by farmers because it is taken up by plants and then accumulates in the human body (primarily in the kidneys, where it generates reactive oxygen species and activates cell death pathways [10]). Phosphorus is a stable metal with a low vapor pressure even at elevated temperatures. Most of it is, therefore, found in the bottom ash (i.e., the ash in the combustion zone) and has thus been separated early on from many of the heavy metals. In principle, it is possible to use this ash as part of a fertilizer, provided that the concentrations of heavy metals are sufficiently low. The concentration of phosphorus depends very much on the feedstock, and dedicated campaigns where such a fuel is used could be of great benefit. It is important that the behavior in the boiler is monitored and estimated before a large amount of sewage sludge is added; both the composition and chemistry of the ash will change, which could cause severe agglomeration and large lumps in stoker boilers.

IMPROVING WASTE COMBUSTION

The combustion of waste material has the potential of increasing its efficiency as well as increasing the recovery of material. Increasing energy efficiency may be done by employing primarily two different measures. The first measure is to include flue gas condensers wherever possible. This could add an additional 10%–15% to the boiler effect. However, to utilize a flue gas condenser, the condensation temperature has to be higher than the return temperature of the district heating network. This generally implies temperatures above 70°C, and more likely higher. Not all the water is condensed at these temperatures, so a substantial amount of water remains—saturated air at 80°C contains 558 g of water per kg of dry air but is reduced to 281 g/kg when the temperature is decreased to 70°C. An extra option is to add a heat pump and reduce the temperature to about 30°C and then the content will be only 28 g/kg. This, however, requires heat to raise the temperature to the necessary levels. The second measure is basically to lower the temperature of the flue gas; e.g., lowering the temperature, as mentioned earlier, from 200 to 150°C reduces the energy loss from 22% to 19%. However, reducing the temperature may cause low-temperature corrosion because salt solutions could be formed from, for example, alkali chlorides and sulfuric compounds [11]. In addition, it is important that condensation in the chimney is avoided; low temperatures may also reduce the natural draft and result in the stack fan (should there be one) requiring more power.

The exergy efficiency can be improved in several ways. One option is to improve the efficiency of the Rankine cycle, which would increase the electricity production per mass of waste fuel. This, however, can never go beyond the exergy content of the steam. Measures that can be taken here are (1) improving the design of feed water heaters (an increase in both their number and temperatures may increase the steam flow for a given heat load), (2) reducing the temperature in the condenser (operating the district heating network at a lower temperature would give a lower back-pressure in the turbine), and (3) reheating the steam at a medium pressure from the turbine (a costly way of improving the efficiency of the turbine and avoiding condensation in it).

Another option is to enhance the properties of the steam while simultaneously avoiding severe corrosion, especially in the superheaters. A plant in Lahti (Finland) uses pregasification and then removes the alkali in a high-temperature filter before the gas is burned in a gas boiler. In this way, the steam's properties are increased to 540°C and 121 bar. There are also ways of making improvements to an existing boiler, e.g., sulfur recirculation can be utilized, yielding high concentrations of sulfur inside the boiler. The sulfate thus formed will compete with chloride to combine with alkali metals, thereby reducing corrosion. This could be used to either increase the properties of the steam or reduce costs by using less expensive material in the heat exchangers. It will, however, require a wet cleaning system and the addition of hydrogen peroxide, but it is not expected to increase emissions of sulfur from the stack [12].

CONCLUSIONS AND PERSPECTIVES

The combustion of waste is currently an excellent option for sustaining progressive work with respect to the environment. There are several reasons for this: fewer landfill areas are necessary, the amount of emitted greenhouse gases is reduced, toxic materials are destroyed, more material is recovered, and waste is converted into valuable products. The reduction in the amount of greenhouse gas emitted is primarily due to the combustion of organic material that would otherwise form methane gas, which has a much higher environmental impact than carbon dioxide. Examples of toxic materials are different chemicals, medicines, and pesticides, all of which should not be released into the environment; these substances react completely at high temperatures when oxygen is present. Furthermore, it is possible to recover metals from the ash that remains after combustion. This has been used mainly for the recovery of iron and copper, although today other options include zinc and phosphorus. The main products of combustion are power and heat, both of which originate from high-pressure steam.

There are some features that are specific to waste when it is used as a fuel. Primarily, waste has a high inorganic content, especially the contents of alkali metals and halogens. These substances form salts that are corrosive and could cause deposits. Both corrosion and deposits will result in the electric efficiency being relatively low because the properties of the steam must be reduced to ensure a sufficiently long life span of the heat exchanger tubes. In addition, the composition of waste varies, making it difficult to operate the combustion unit properly. An airflow that is too high would increase the formation of NO_x and too low an airflow would not ensure complete combustion, so the concentration of carbon monoxide will increase. Moreover, there is the requirement of keeping the temperature above a minimum value. Failing to adjust the airflow correctly would impact all these conditions. Finally, cleaning the flue gas must be done appropriately. Sulfur, for example, must be separated from the flue gas or it will be released as sulfur dioxide, which will react in the atmosphere and cause acid rain. Dioxins and furans are very toxic substances that are formed during poorly controlled combustion operations or in specific temperature intervals when catalysts such as copper are present, and they must be removed from the flue gases.

Future Perspectives

Waste combustion is currently used in many countries; with respect to its generation of organic waste, Sweden is one of the countries with the highest capacity for combustion. In 2017, more than 7.6 million tons of waste was used as fuel in combustion facilities [13]. Of this, 6.1 million tons was used in dedicated waste combustion units with energy recovery [14] and the remainder was used in industrial units. This high capacity is a result of the overall high degree of energy efficiency of the combined heat and power combustion facilities in question. In Europe, the total amount of waste used for energy recovery is 105 million tons, with 196 million tons being sent to landfill [15]. There is an obvious need to minimize the amount of waste going to landfills, from both a local and global environmental perspective. The global perspective is largely associated with landfill emissions of greenhouse gases, whereas the local perspective is associated with the requirement of space, as well as the risk of substances leaking out that could contaminate water and soil.

It is important to strive after a better economy with the resources available, be it energy or materials. Many countries are now discussing and initiating the implementation of a circular economy with a higher degree of material utilization to reduce the amount of waste that is being created. This, in turn, will alter the composition of the waste available for combustion; in particular, plastics and textiles, with their high heating values, will be separated to a larger extent. An increased circular economy, which implies increased recycling of materials, will not remove the need for combustion completely because increased recycling means that harmful substances will accumulate in the materials. A function capable of removing such substances is therefore necessary and combustion is one such function, which also takes care of materials that are unsuitable for further recycling. The life span of, for example, paper and textile fibers is shortened each time they are subjected to mechanical recycling and, eventually, they become too short to function properly. In the final stage, combustion can utilize the energy content in the production of heat and power.

One of the biggest problems of modern waste combustion is the solid that remains, i.e., the ash, which is often 15%–20% (on a weight basis) of the initial amount of waste. It is composed of different fractions: the bottom ash is often regarded as being nonhazardous and used in some construction work, mainly on landfills, whereas the fly ash contains most of the heavy metals and is currently sent to landfill sites. These landfills will, however, eventually be full to capacity and alternatives are therefore necessary. It is possible to utilize these ashes for the further recovery of material: one option is to select the fuel in such a way that the resulting ash can be used for other purposes. This would, for example, allow nutrients from sewage sludge to be made available as fertilizers without risking the spread of undesirable substances such as pathogens and heavy metals. Another option is to utilize these ashes as construction material other than in landfills, provided that the chemical composition and leaching properties are within acceptable ranges.

The combustion of waste has been used very successfully in the production of district heating, as mentioned earlier; however, it is not certain that this demand will remain at the same level in the future. It has to compete with other technologies (such as heat pumps) and, as new houses are also increasingly better insulated, the demand for district heating will decrease. Moreover, waste heat from industries is also competing within the existing district heating network. Increases in demand will be from new building constructions and from more efficient electric appliances. The introduction of district cooling as a parallel network will also increase in demand, especially when the requirement for heating is relatively low, i.e., when temperature outdoors is high. This can be of importance to other countries too.

REFERENCES

[1] European Commission. Directive 2000/76/EC on the incineration of waste. Official Journal of the European Communities 2000:L332–91.

[2] Brunner CR. Waste-to-energy combustion, incineration technologies. In: Tchobanoglous G, Kreith F, editors. Handbook of solid waste management. New York: McGraw-Hill; 2002. p. 13.13–84.

[3] Blomqvist, E. & Jones, F. reportDetermination of the fossil carbon content in combustible municipal solid waste in Sweden. Report U2012:05, Avfall Sverige, ISSN 1103-4092.

[4] Tabasová A, Kropáč J, et al. Waste-to-energy technologies: impact on environment. Energy 2012;44(1):146–55.

[5] Kitto JB, Stultz SC, editors. Steam. 41st ed. Barberton, OH: Babcock & Wilcox; 2005.

[6] Shin-ichi S, Masakatsu H. Municipal solid waste incinerator residue recycling by thermal processes. Waste Management 2000;20:249–58.

[7] Karlfeldt Fedje K, Andersson S, Modin O, Frändegård P, Pettersson A. Opportunities for Zn recovery from Swedish MSWI fly ashes. In: Proceedings of SUM 2014, May 19–21, Bergamo, Italy; 2014.

[8] Schlumberger S, Bühler J. Metal recovery in fly and filter ash in waste to energy plants. Stockholm, Sweden: Ash; 2012. 2012, January 25–27.

[9] Schlumberger S, Schuster M, Ringmann S, Koralewska R. Recovery of high purity zinc from filter ash produced during the thermal treatment of waste and inerting of residual materials. Waste Management & Research 2007;25: 547–55.

[10] Johri N, Jacquillet G, Unwin R. Heavy metal poisoning: the effects of cadmium on the kidney. Biometals October 2010;23(5):783–92. https://doi.org/10.1007/s10534-010-9328-y. Epub 2010 Mar 31.

[11] Vainio E, Kinnunena H, Laurén T, Brink A, Yrjas P, DeMartini N, Hupa M. Low-temperature corrosion in co-combustion of biomass and solid recovered fuels. Fuel November 15, 2016;184:957–65. https://doi.org/10.1016/j.fuel.2016.03.096.

[12] Andersson S, Blomqvist EW, et al. Sulphur recirculation for increased electricity production in Waste-to-Energy plants. Waste Management 2014;34(1):67–78. https://doi.org/10.1016/j.wasman.2013.09.002.

[13] Eurostat. Treatment of waste by waste category, hazardousness and waste management operations. 2018. http://appsso.eurostat.ec.europa.eu/nui/show.do?dataset=env_wastrt&lang=en.

[14] Sverige A. Svensk Avfallshantering 2018. 2018, 31. https://www.avfallsverige.se/avfallshantering/avfallsfakta/svensk-avfallshantering/.

[15] Eurostat. Management of waste excluding major mineral waste, by waste management operations. 2018. http://appsso.eurostat.ec.europa.eu/nui/show.do?dataset=env_wasoper&lang=en.

CHAPTER 14

Gasification Technologies and Their Energy Potentials

YANING ZHANG, PHD • YUNLEI CUI, MSC • PAUL CHEN, PHD • SHIYU LIU, MSC • NAN ZHOU, PHD • KUAN DING, PHD • LIANGLIANG FAN, PHD • PENG PENG, PHD • MIN MIN • YANLING CHENG, PHD • YUNPU WANG, PHD • YIQIN WAN, PHD • YUHUAN LIU, PHD • BINGXI LI, PHD • ROGER RUAN, PHD

INTRODUCTION

The energy dependence of the world is a serious economic and national security issue that we must deal with in our time. As the world population increased monotonically in the range of 6.52–7.35 billion with an annual increase rate of 1.27% during the years 2006–15, the world primary energy consumption nearly increased monotonically in the range of $11.27–13.15 \times 10^3$ Mtoe (million tonnes oil equivalent) with an annual increase rate of 1.67% during the same period [1]. This primary energy consumption included $3.19–3.83 \times 10^3$ Mtoe of coal, with an annual increase rate of 2.01%; $3.96–4.36 \times 10^3$ Mtoe of oil, with an annual increase rate of 1.01%; and $2.61–3.20 \times 10^3$ Mtoe of natural gas, with an annual increase rate of 2.26%. Among these primary energy resources, natural gas had the highest annual increase rate.

As compared with the raw solid fuels, gaseous fuels are generally more efficient, versatile, and easily controllable, and the used combustion units are simpler. Gasification technology is a useful and efficient method that can convert a wide range of low-energy-density fuels into high-value flammable gases, and it therefore attracts significant attention all over the world.

As compared with the other technologies including physical technologies, chemical technologies, and biochemical technologies, thermochemical technologies are generally more versatile and can essentially convert all the organic components of the feedstock into a range of products, which are mainly biofuels [2], and they are therefore introduced in this chapter.

THERMOCHEMICAL TECHNOLOGIES

Generally, thermochemical technologies include liquefaction, carbonization, combustion, pyrolysis, and gasification. In some publications, some other methods are also included, e.g., incineration [3], torrefaction [4], hydrogenation [5,6].

Liquefaction

Liquefaction is a thermochemical process in which biomass undergoes complicated chemical reactions in a solvent medium to form mainly liquid products (biooil or bio-oil). Hydrothermal liquefaction is a process in which water is used as the reaction medium, and the process is carried out in sub-/supercritical water (200–400°C) under sufficient pressure to liquefy biomass for biooil production [7].

The heavy oil obtained from the liquefaction process is a viscous tarry lump, which may sometimes cause handling troubles. In this case, some organic solvents (e.g., propanol, butanol, acetone, methyl ethyl ketone, ethyl acetate) are added to the reaction system. Generally, catalytic aqueous liquefaction may result in higher biooil yield than the noncatalytic aqueous liquefaction [8]. The biooil yield from the liquefaction method is generally much lower than that from the pyrolysis method and the biooil from the liquefaction method is more viscous than that from the pyrolysis method [8].

Biooils obtained from the liquefaction process generally contain high contents of volatile organic acids, alcohols, aldehydes, ethers, esters, ketones, and nonvolatile components. These oil components could be catalytically upgraded to yield an organic distillate product, which is rich in hydrocarbons and useful chemicals [8].

Sustainable Resource Recovery and Zero Waste Approaches. https://doi.org/10.1016/B978-0-444-64200-4.00014-1

On the other hand, the feedstock feeding system for liquefaction is more complex and expensive than pyrolysis, rendering a low interest in liquefaction [5,9].

Carbonization

Carbonization is a process that typically heats biomass feedstock in a kiln or retort (pyrolysis) at temperatures around 400°C (generally between 300 and 900°C) in the absence of air [10,11]. The produced biochar is also known as charcoal, which is a porous, carbon-enriched, grayish black solid [10]. It can also be produced from torrefaction, gasification, hydrothermal carbonization, etc.

As compared with the raw feedstocks, biochars have some advantages. Table 14.1 shows the main characteristics including the H/C ratio, O/C ratio, and higher heating value (HHV) of some raw feedstocks and the produced biochars. It is observed that the H/C ratio, O/C ratio, and HHV of the raw feedstocks are 1.40–1.68, 0.55–0.77, and 16.20–24.80 MJ/kg, respectively. The H/C ratio, O/C ratio, and HHV of the biochars are 0.12–1.24, 0.08–0.49, and 22.50–35.70 MJ/kg, respectively. Generally, biochars have much higher HHVs and much lower H/C and O/C ratios than the raw feedstocks. These are mainly caused by the increases in the carbon content. For instance, the fixed carbon content was increased from 16.39 to 81.03 wt% (4.94 times) when the white pine was pyrolyzed to biochar [11].

As compared with the raw feedstocks, the obtained high-quality biochars have much more usages including (1) solid fuel in boilers, (2) feedstock in further gasification process, (3) carbon sequestration, (4) production of activated carbon, (5) making carbon nanotubes, and (6) making briquets [5,11].

Combustion

Combustion is a thermochemical process where fuel is burnt in an oxygen-excess atmosphere (air or oxygen) and the chemical energy stored in the fuel is released to produce heat, which can be used for cooking, space heating, and electricity generation. The overall reactions are

$$\text{Biofuel} + O_2 \rightarrow CO_2 + H_2O + \text{heat} \quad (14.1)$$

$$\text{Biofuel} + \text{Air} \rightarrow CO_2 + H_2O + N_2 + \text{heat} \quad (14.2)$$

If the biofuel contains ash, S, or N, then ash, SO_x, or NO_x would also be included in Eqs. (14.1) and (14.2) [12,13].

TABLE 14.1
Characteristics of Different Biochars [11].

	FEEDSTOCK			BIOCHAR		
Feedstock	**H/C**	**O/C**	**HHV (MJ/Kg)**	**H/C**	**O/C**	**HHV (MJ/Kg)**
White pine	1.40	0.62	18.06	0.43	0.08	31.41
Fruit cuttings	1.61	0.77	17.27	0.12	0.09	29.02
Switchgrass	1.56	0.77	19.50	0.70	0.24	26.60
Barley straw	1.45	0.66	17.34	0.40	0.11	26.02
Safflower seeds	1.68	0.62	24.80	0.43	0.21	30.27
Eucalyptus sawdust	1.47	0.71	16.69	0.81	0.27	26.19
Barley straw	1.45	0.66	17.34	0.86	0.21	27.49
Corn stover	1.62	0.73	16.20	0.94	0.18	27.76
Spruce	1.49	0.65	19.94	1.24	0.49	22.50
Birch	1.56	0.68	20.42	1.24	0.49	22.50
Maize silage	1.58	0.55	22.30	1.13	0.09	35.70
Coconut fiber	1.41	0.71	18.40	0.66	0.15	30.60
Eucalyptus leaves	1.59	0.72	18.90	1.01	0.22	29.40
Corn stalk	1.58	0.65	17.51	0.94	0.17	29.21
Wood	1.65	0.68	17.93	0.90	0.22	28.38

HHV, higher heating value.

For theoretic combustion, the C and H contents in the fuel would be converted to CO_2 and H_2O, respectively, and the produced heat can result in a high temperature of ~2000°C (theoretically). Practically, the combustion would be incomplete, thereby leading to the production of CO, and the heat would be diffused (to the environment), thereby leading to a relatively low temperature of 800–1000°C [9]. For a better combustion, excess air or oxygen is usually used [14].

Theoretically, it is possible to (completely) combust any type of biomass or biofuel, but practically, combustion is feasible only when the fuel has a moisture content less than 50% [5,9]. To resolve this issue, a preheating step is usually used.

Although combustion is an efficient and mature technology to convert biofuel and generate heat, it is actually not a technology to produce biofuel because no biofuels would be produced after the complete combustion of the original feedstock (mainly CO_2 and H_2O are left).

Pyrolysis

Pyrolysis is typically defined as the thermochemical decomposition of biomass feedstock at medium (300–800°C) to high temperatures (800–1300°C) in an inert atmosphere [15]. Some similar definitions are reported in other publications [5,8,9,14,16–18].

The overall reaction of biomass feedstock is

$$\text{Biofuel} + \text{heat} \rightarrow \text{liquid} + \text{syngas} + \text{solid} \quad (14.3)$$

The products of liquid, syngas, and solid are all actually valuable fuels and are defined as biooil , biosyngas, and biochar, respectively.

The chemical reactions during the pyrolysis process include [19]

$$C + 2H_2O \rightarrow 2H_2 + CO_2 \quad \Delta H = +75\ \text{kJ/mol} \quad (14.4)$$

$$C + H_2O \rightarrow H_2 + CO \quad \Delta H = +131\ \text{kJ/mol} \quad (14.5)$$

$$CH_4 + 2H_2O \rightarrow 4H_2 + CO_2 \quad \Delta H = +165\ \text{kJ/mol} \quad (14.6)$$

$$CH_4 + H_2O \rightarrow CO + 3H_2 \quad \Delta H = +206\ \text{kJ/mol} \quad (14.7)$$

When compared with the other thermochemical technologies including liquefaction and combustion, pyrolysis generally has the following advantages: (1) the main product is biooil and the yield may be as high as 75%; (2) the biooil may have a high content of carbon; (3) the biooil may have low nitrogen and sulfur contents; (4) the HHV of biooil may be very high, e.g., 42 MJ/kg, which is comparable with those of fossil fuels; (5) the residence time is generally short, which decreases the operational cost; (6) the desired product (biooil, biosyngas, or biochar) could be produced by adjusting the operational parameters; (7) the biooil could be easily stored or transported; and (8) the biomass feedstock may not need to be processed [20].

According to the operating parameters such as heating rate, pyrolysis temperature, and residence time, conventional electric pyrolysis can be generally classified into three groups: (1) slow pyrolysis, (2) fast pyrolysis, and (3) flash pyrolysis. In some publications, catalytic pyrolysis, microwave pyrolysis, vacuum pyrolysis, and hydropyrolysis are also included [2,8,20]. For slow pyrolysis, the heating rate, pyrolysis temperature, and residence time are <1°C/s, 300–700°C, and >450 s, respectively. For fast pyrolysis, they are 10–300°C/s, 550–1250°C, and 0.5–20 s, respectively. For flash pyrolysis, they are >1000°C/s, 800–1300°C, and <0.5 s, respectively [15,21]. These different conditions generally result in different pyrolysis results. Table 14.2 shows the pyrolysis results obtained from different pyrolysis technologies. It is observed that flash pyrolysis generally favors biooil production, followed by fast pyrolysis and slow pyrolysis.

A new pyrolysis technology, microwave-assisted pyrolysis, has been developed and widely used [22–24], and it has drawn serious attention because of its

TABLE 14.2
Pyrolysis Results Obtained From Different Pyrolysis Technologies [58].

Pyrolysis	Operating Conditions	Results
Slow pyrolysis	Temperature: 300–700°C Vapor residence time: 10–100 min Heating rate: 0.1–1°C/s Feedstock size: 5–50 mm	Biooil: ~30 wt% Biochar: ~35 wt% Gases: ~35 wt%
Fast pyrolysis	Temperature: 400–800°C Vapor residence time: 0.5–5 s Heating rate: 10–200°C/s Feedstock size: <3 mm	Biooil: ~50 wt% Biochar: ~20 wt% Gases: ~30 wt%
Flash pyrolysis	Temperature: 800–1000°C Vapor residence time: < 0.5 s Heating rate: > 1000°C/s Feedstock size: <0.2 mm	Biooil: ~75 wt% Biochar: ~12 wt% Gases: ~13 wt%

advantages over the conventional electric pyrolysis, which are shown in Table 14.3. These advantages are due to the different heat and mass transfer mechanisms. For the conventional electric pyrolysis, heat is transferred from high-temperature gas to the fuel particle surface through the convection mechanism and it is then further transferred from the outside surface to the inside core through the conduction mechanism. A temperature gradient from the outside to the inside of the feedstock particle is formed, and the released volatile diffuses from the inside core to the outside surface through a higher temperature region. For the microwave-assisted pyrolysis, microwave penetrates the feedstock particle and the microwave energy is transformed into thermal energy, which constantly accumulates inside the biomass particle and is then transferred outward. A temperature gradient from the inside to the outside of the particle is formed, and the released volatile diffuses from the inside core to the outside surface through a lower temperature region [15].

Gasification

Gasification is also a thermochemical process in which the reactions between fuel and the gasification agent take place and syngas (also known as producer gas, product gas, synthetic gas, or synthesis gas) is produced.

TABLE 14.3
Comparison Between Microwave-Assisted Pyrolysis and Conventional Electric Pyrolysis [15,59,60].

Microwave-Assisted Pyrolysis	Conventional Electric Pyrolysis
Conversion of energy	Transfer of energy
Noncontact heating	Contact heating
Hot spot	No hot spot
Selective	Nonselective
Lower thermal inertia and faster response	Higher thermal inertia and slower response
Lower energy consumption	Higher energy consumption
Rapid heating	Slow heating
Shorter reaction times	Longer reaction times
Volumetric heating	Superficial heating
Higher level of control	Lower level of control
Improved product yields	Lower product yields

The syngas is mainly composed of CO, H_2, N_2, CO_2, and some hydrocarbons (CH_4, C_2H_4, C_2H_6, etc.). Very small amounts of H_2S, NH_3, and tars may also be included [25].

In general, biomass gasification is the thermochemical conversion of organic (waste) feedstock in a high-temperature environment, through which biomass can be converted not only to syngas for energy generation but also to chemicals; for instance, methane, ethylene, adhesives, fatty acids, surfactants, detergents, and plasticizers [26].

Based on the gasification agents used, biomass gasification processes can be divided into air gasification (using air), oxygen gasification (using oxygen), steam gasification (using steam), carbon dioxide gasification (using carbon dioxide), supercritical water gasification (using supercritical water), etc. Generally, oxygen gasification, steam gasification, carbon dioxide gasification, and supercritical water gasification result in higher HHVs of syngas than those obtained by air gasification; however, air gasification is the most widely studied and applied process because the gasification agent (air) is cheap, the reaction process is easy, the reactor structure is simple, etc.

For biomass gasification with steam, carbon dioxide, or supercritical water, the overall gasification reaction is generally endothermic, and external heating is therefore required during the whole gasification process. However, for biomass gasification with air or oxygen, the overall gasification may be endothermic or exothermic. These reactions can be controlled or changed by varying the air or oxygen content. Generally, a specific air or oxygen content corresponds to a specific gasification temperature if no external heat is provided. If higher gasification temperature is required or designed, external heat should be input or a higher air or oxygen content is needed [27].

The place where the gasification reactions take place is known as a gasifier. A gasifier is a very important factor that affects the gasification processes, reactions, and products. Generally, gasifiers can be classified into three broad groups: fixed bed gasifiers (or moving bed gasifiers), fluidized bed gasifiers, and entrained flow gasifiers [25]. The fixed bed gasifiers include updraft, downdraft, horizontal-draft, etc. The fluidized bed gasifiers include bubbling fluidized bed, circulated fluidized bed, double circulated fluidized bed, etc. A review of the 50 gasifier manufacturers in Europe, United States, and Canada showed that 75% of the designs were downdraft fixed beds, 2.5% were updraft fixed beds, 20% were fluidized beds, and 2.5% were the other designs [25,28]. Table 14.4 shows the main

TABLE 14.4
Main Characteristics of Different Gasifiers [19,25,61].

Gasifier	Characteristics
Fixed bed	1. Small capacity (0.01–10 MW) 2. Can handle large and coarse particles 3. Low product gas temperature (450–650°C) 4. High particulate content in gas product stream 5. High gasification agent consumption 6. Ash is removed as slag or dry 7. May result in high tar content (0.01–150 g/Nm^3)
Fluidized bed	1. Medium capacity (1–100 MW) 2. Uniform temperature distribution 3. Better gas-solid contact 4. High operating temperature (1000–1200°C) 5. Low particulate content in the gas stream 6. Suitable for feedstocks with low ash fusion temperature 7. Ash is removed as slag or dry
Entrained flow	1. Large capacity (60–1000 MW) 2. Needs finely divided feed material (<0.1–0.4 mm) 3. Very high operating temperatures (>1200°C) 4. Not suitable for high-ash-content feedstocks 5. Very high oxygen demand 6. Short residence time 7. Ash is removed as slag 8. May result in low tar content (negligible)

characteristics of different gasifiers. Using those characteristics, specific gasifiers can be selected and designed.

BIOMASS GASIFICATION

According to the gasification agent used, biomass gasification can be classified as air gasification, oxygen gasification, steam gasification, hydrogen gasification, carbon dioxide gasification, etc. Sometimes, a gasification technology can be a combined one, i.e., steam-O_2 gasification, air-CO_2 gasification, etc. [29,30].

Air Gasification

For the air gasification of biomass, the gasification process occurs between 800 and 1800°C and the reactions can be divided into four parts: drying, devolatilization, combustion, and reduction [31–34].

Drying or moisture release:

$$\text{Wet biofuel} \rightarrow \text{dry biofuel} + H_2O \quad (14.8)$$

Devolatilization:

$$\text{Dry biofuel} \rightarrow CO,\ CO_2,\ CH_4,\ C_2H_4,\ H_2O + \text{carbon and primary tar } (CH_xO_y) \quad (14.9)$$

$$\text{Primary tar} \rightarrow CO,\ CO_2,\ CH_4,\ C_2H_4,\ H_2 + \text{secondary tar} \quad (14.10)$$

$$\text{Secondary tars} \rightarrow C,\ CO,\ H_2 \quad (14.11)$$

Combustion:

$$CO + 0.5O_2 \rightarrow CO_2 \quad \Delta H = -283\ \text{kJ/mol} \quad (14.12)$$

$$H_2 + 0.5O_2 \rightarrow H_2O \quad \Delta H = -242\ \text{kJ/mol} \quad (14.13)$$

$$CH_4 + 0.5O_2 \rightarrow CO + 2H_2 \quad \Delta H = -110\ \text{kJ/mol} \quad (14.14)$$

$$C + O_2 \rightarrow CO_2 \quad \Delta H = -393.5\ \text{kJ/mol} \quad (14.15)$$

$$C + 0.5O_2 \rightarrow CO \quad \Delta H = -123.1\ \text{kJ/mol} \quad (14.16)$$

Reduction:

$$\text{Dry reforming reaction: } CH_4 + CO_2 \rightarrow 2CO + 2H_2 \quad \Delta H = +247\ \text{kJ/mol} \quad (14.17)$$

$$\text{Steam reforming methanization: } CH_4 + H_2O \rightarrow CO + 3H_2 \quad \Delta H = +206\ \text{kJ/mol} \quad (14.7)$$

$$\text{Water gas shift reaction: } CO + H_2O \rightarrow CO_2 + H_2 \quad \Delta H = -40.9\ \text{kJ/mol} \quad (14.18)$$

$$\text{The Boudouard equilibrium reaction: } C + CO_2 \rightarrow 2CO \quad \Delta H = +159.9\ \text{kJ/mol} \quad (14.19)$$

$$\text{Water gas reaction (steam reforming): } C + H_2O \rightarrow CO + H_2 \quad \Delta H = +118.5\ \text{kJ/mol} \quad (14.5)$$

$$\text{Methane production reaction: } C + 2H_2 \rightarrow CH_4 \quad \Delta H = -87.5\ \text{kJ/mol} \quad (14.20)$$

For air gasification, the gasification agent is air and the oxygen in air is not fully oxidized with biomass, giving off heat for gasification, and the gasification renders a gas stream with hydrogen, carbon monoxide, carbon dioxide, methane, and other hydrocarbons, as well as a large amount of nitrogen [26]. Air gasification is relatively simple and economical and is the easiest way to realize gasification because air is readily available and air gasification can provide enough heat on its own without requiring external heat sources. However, the large amount of nitrogen in the air (79%) does not participate in the gasification reaction and the nitrogen dilutes the content of syngas, so the calorific value of the generated gas is relatively low, i.e., about 3–5 MJ/Nm^3 [35].

As gasification is a sensitive and complex process, it is highly influenced by the biomass composition, particle size, gasification temperature, and equivalence ratio (ER). ER is a crucial operating variable in biomass gasification. It is the ratio between the actual amount of air required for gasification per unit mass of biomass and the theoretic amount of air required for complete combustion and it is calculated as [36]

$$ER = \frac{Air_i}{Air_j} \quad (14.21)$$

where Air_j is the stoichiometric air (mol or Nm^3) and Air_i is the actual amount of air supplied (mol or Nm^3).

Oxygen Gasification

Oxygen (O_2) gasification uses oxygen as the gasification agent, and its reaction principle is very similar to that of air gasification. The use of pure oxygen as a gasification agent provides producer gas with a higher HHV than that of its counterpart obtained from air gasification because of the absence of nitrogen in the syngas [30]. The presence of oxygen gives an exothermic contribution, so that the higher the oxygen content in the feeding stream, the higher the achieved temperature but the lower the heating value of the produced syngas [37]. Compared with air gasification, oxygen gasification has higher reaction temperature, faster reaction rate, higher thermal efficiency, and higher cost. The main components in syngas are carbon monoxide, hydrogen, methane, etc. and the calorific value of syngas is about 10–18 MJ/Nm^3 [35]. For oxygen gasification, the oxygen equivalence ratio (OER) is a crucial factor that significantly affects the reaction process and results. OER refers to the ratio of actual oxygen supplied to the stoichiometric oxygen [38]:

$$OER = \frac{Oxygen_i}{Oxygen_j} \quad (14.22)$$

where $Oxygen_j$ is the stoichiometric oxygen (mol or Nm^3) and $Oxygen_i$ is the actual amount of oxygen supplied (mol or Nm^3).

Steam Gasification

Steam (H_2O) gasification refers to the reaction of steam with biomass at high temperatures, including both water gas reaction and water gas shift reaction. Steam as an oxidizing agent is attracting more attention because it gives better results in terms of H_2/CO and hydrogen yield ($\%H_2$ in syngas) during the gasification processes. Hydrogen is a fuel with high calorific value and it is a highly efficient energy source, which has become a renewed focus of interest in the industry, thanks to its environmental advantages. The presence of H_2O gives an endothermic contribution, so that the higher the water in the feed (as fuel moisture as well as added steam), the lower the temperature in the reactor but the higher the potential heating value of the produced syngas [37]. The main reactions are listed in the following [39].

Drying:

$$\text{Wet biofuel} \rightarrow \text{dry biofuel} + H_2O \quad (14.8)$$

Pyrolysis:

$$\text{Dry biofuel} \rightarrow CO,\ CO_2,\ CH_4,\ C_2H_4,\ H_2O + \text{carbon and primary tar } (CH_xO_y) \quad (14.9)$$

$$\text{Primary tar} \rightarrow CO,\ CO_2,\ CH_4,\ C_2H_4,\ H_2 + \text{secondary tar} \quad (14.10)$$

Reduction:

$$\text{Water gas reaction}: C + H_2O \rightarrow CO + H_2 \quad \Delta H = +118.5\ \text{kJ/mol} \quad (14.5)$$

$$\text{Methanation reaction}: C + 2H_2 \rightarrow CH_4 \quad \Delta H = -87.5\ \text{kJ/mol} \quad 14.20$$

$$\text{Steam reforming methanization}: CH_4 + H_2O \rightarrow CO + 3H_2 \quad \Delta H = +206\ \text{kJ/mol} \quad (14.7)$$

$$\text{Water gas shift reaction}: CO + H_2O \rightarrow CO_2 + H_2 \quad \Delta H = -40.9\ \text{kJ/mol} \quad (14.18)$$

$$\text{The Boudouard reaction}: C + CO_2 \rightarrow 2CO \quad \Delta H = +159.9 \text{ kJ/mol} \quad (14.19)$$

Introducing steam to the gasification process is advantageous because it improves the H_2 content in syngas by raising the partial pressure of H_2O inside the gasifier. Steam/carbon ratio (SCR) is a crucial operating variable in biomass gasification, which is the ratio between steam mass flow rate and the total carbon feed mass flow rate [40]:

$$SCR = \frac{n_s}{n_c} \quad (14.23)$$

where n_s is the steam mass flow rate (kg/s) and n_c is the total carbon feed mass flow rate (kg/s).

Hydrogen Gasification

Hydrogen (H_2) gasification has been developed to increase the gas production level, especially methane (CH_4) by introducing external hydrogen (H_2) into the gasification process. During this process, hydrogen reacts with carbon and water to produce large amounts of methane. Currently, most of the research has been performed under high H_2 pressure, which facilitates yielding a higher CH_4 production rate from the hydrogen gasification process. However, the required severe operating conditions, including a high temperature and H_2 pressure, result in a high level of energy consumption and several safety concerns. Hydrogen gasification under mild conditions is more attractive but it requires the addition of a suitable catalyst to increase the otherwise unacceptably low reaction rate [41].

It is suggested that the hydrogenation of CO and water gas shift reactions primarily take place during hydrogen gasification, and the gasification process includes the reactions of char (heterogeneous reactions) and tar and light gas (homogeneous reactions). Cracking and reforming of volatile matter and tar occur under the temperature range of 600−650°C, whereas char gasification becomes active when the temperature is higher than 600°C. The main reactions are listed in the following [41,42].

Drying:

$$\text{Wet biofuel} \rightarrow \text{dry biofuel} + H_2O \quad (14.8)$$

Pyrolysis:

$$\text{Dry biofuel} \rightarrow CO,\ CO_2,\ CH_4,\ C_2H_4,\ H_2O + \text{carbon and primary tar } (CH_xO_y) \quad (14.9)$$

$$\text{Primary tar} \rightarrow CO, CO_2, CH_4, C_2H_4, H_2 + \text{secondary tar} \quad (14.24)$$

Reduction:

$$\text{CO hydrogenation reaction}: C + 3H_2 \rightarrow CH_4 + H_2O \quad (14.24)$$

$$\text{Water gas shift reaction}: CO + H_2O \rightarrow CO_2 + H_2 \quad \Delta H = -40.9 \text{ kJ/mol} \quad (14.18)$$

$$\text{Methanation reaction}: C + 2H_2 \rightarrow CH_4 \quad \Delta H = -87.5 \text{ kJ/mol} \quad (14.20)$$

$$\text{Water gas reaction}: C + H_2O \rightarrow CO + H_2 \quad \Delta H = 131 \text{ kJ/mol} \quad (14.5)$$

Carbon Dioxide Gasification

Carbon dioxide (CO_2) as an oxidizing agent has been explored for biomass gasification in only a limited number of studies. From the viewpoint of CO_2 consumption and CO yield enhancement, the utilization of CO_2 is more preferable as a gasification agent. Also, CO_2 is a major component of flue gases from many industries. It is a major cause of global warming and leads to various health issues. A major drawback with carbon dioxide gasification is that an external source of heat is constantly required to maintain the gasification temperature, as the heat supplied by partial combustion of biomass in air or oxygen is not enough.

Carbon dioxide gasification consists of three major steps: drying, devolatilization (pyrolysis), and gasification. The main reactions are listed in the following [43,44].

Drying:

$$\text{Wet biofuel} \rightarrow \text{dry biofuel} + H_2O \quad (14.8)$$

Pyrolysis:

$$\text{Dry biofuel} \rightarrow CO,\ CO_2,\ CH_4,\ C_2H_4,\ H_2O + \text{carbon and primary tar } (CH_xO_y) \quad (14.9)$$

$$\text{Primary tar} \rightarrow CO,\ CO_2,\ CH_4,\ C_2H_4,\ H_2 + \text{secondary tar} \quad (14.10)$$

Reduction:

$$\text{The Boudouard reaction}: C + CO_2 \rightarrow 2CO \quad \Delta H = +159.9 \text{ kJ/mol} \quad (14.19)$$

$$\text{Dry reforming reaction}: CO_2 + CH_4 \rightarrow 2H_2 + 2CO \quad \Delta H = +246.9\ \text{kJ/mol} \quad (14.17)$$

$$\text{Reverse water gas shift reaction}: CO_2 + H_2 \rightarrow H_2O + CO \quad \Delta H = +41.2\ \text{kJ/mol} \tag{14.25}$$

$$\text{Water gas reaction}: C + H_2O \rightarrow CO + H_2 \quad \Delta H = 131\ \text{kJ/mol} \quad (14.5)$$

For carbon dioxide gasification, CO_2/C is a crucial operating variable during the process and is defined as the ratio of CO_2 flow rate to the total biomass flow rate [44]:

$$\frac{CO_2}{C} = \frac{m_c}{m_{bio}} \tag{14.26}$$

where m_c is the CO_2 fed into the reactor by gasification agent (mol) and m_{bio} is the carbon in the feedstock (mol).

Combined Gasification

Gasification performances of biomass are significantly varied or improved by the combination of gasification technologies.

Rupesh et al. [45] studied the air-steam gasification of rice husk and the results showed that the H_2 yield was increased from 10.5 to 14.0 vol% (33.3% increase) when steam was involved in the air gasification, and the mechanisms involved were mainly the water gas shift reaction ($CO + H_2O \rightarrow CO_2 + H_2$) and steam reforming reaction ($CH_4 + H_2O \rightarrow CO + 3H_2$). However, the CO yield was decreased from 14 to 4 vol% (71.4% decrease) when steam was involved in the air gasification, and the mechanism involved was mainly the water gas shift reaction ($CO + H_2O \rightarrow CO_2 + H_2$).

Yu et al. [46] studied air-CO_2 gasification of pine sawdust and the results showed that the CO yield was increased from 0.48 to 0.6 Nm^3/kg (25% increase) when CO_2 was involved in the air gasification, and the mechanism involved was mainly the Boudouard reaction ($C + CO_2 \rightarrow 2CO$).

ENERGY POTENTIALS OF GASIFICATION TECHNOLOGIES

Mostly, the energy potential of a gasification technology can be assessed or evaluated by cold gasification efficiency (CGE), gasification system efficiency [29,45,47], energy efficiency [25], exergy efficiency [43,48], etc. Sometimes, syngas HHV [42,49], syngas yield [39,50], CH_4 yield [23,51], and H_2 yield [39,50] can also be used to evaluate the energy potential of a gasification technology. Among these evaluating methods, CGE is the most frequently used one and is defined as [32,52]

$$CGE = \frac{LHV_g \times F_g}{LHV_s \times F_s} \tag{14.27}$$

where LHV_g is the lower heating value (LHV) of syngas (MJ/Nm^3), F_g is the production rate of syngas (Nm^3/s), LHV_s is the LHV of biomass (MJ/kg), and F_s is the feeding rate of biomass (kg/s).

The LHV of syngas (LHV_g) can be calculated as

$$LHV = \sum v_i \times LHV_i \tag{14.28}$$

where v_i is the volumetric concentration of a combustible gas i (i.e., H_2, CO, CH_4, etc.) in the syngas (%) and LHV_i indicates the LHV of a combustible gas i (MJ/Nm^3).

In some publications, CGE is also defined as [32]

$$CGE = \frac{HHV_g \times F_g}{HHV_s \times F_s} \tag{14.29}$$

where HHV_g is the HHV of syngas (MJ/Nm^3), F_g is the production rate of syngas (Nm^3/s), HHV_s is the HHV of biomass (MJ/kg), and F_s is the feeding rate of biomass (kg/s).

Air Gasification

Because the gasification agent used is air, which contains a high volumetric content of N_2 (e.g., 79 vol%), the compositions of syngas are dominated by N_2 and the HHV of syngas is significantly diluted by N_2, yielding the result that air gasification of biomass generally has lower CGEs.

Table 14.5 shows the CGEs of air gasification of different biomass feedstocks at different gasification temperatures and ERs. It is observed that the CGEs are between 36% and 84%. Air gasification of rice husks showed the lowest CGE (36%) mainly because the heating value of syngas was very low (4 MJ/m^3). Air gasification of black pine showed the highest CGE (84.26%) because of the facts that (1) CO and H_2 were produced by char reduction reactions when the temperature was high and (2) the appropriate ER ensured a high carbon conversion efficiency (91.03%).

The CGE of an air gasification process is significantly varied by gasification temperature. Tavares et al. [53] studied air gasification of polyethylene terephthalate/vine prunings at different gasification temperatures. The results showed that the CGE was

TABLE 14.5
CGEs of Biomass Air Gasification.

Biomass	CGE (%)	Temperature, ER	References
Rice husks	36	850°C, 0.26	[54]
Rice husks	37	870°C, 0.29	[54]
Rice husks	36.5	850°C, 0.32	[54]
PET/vine prunings	49	600°C, 0.7	[53]
PET/vine prunings	72	800°C, 0.7	[53]
PET/vine prunings	75	1000°C, 0.7	[53]
Wood	80	1000°C	[48]
Torrefied wood	82	750°C, 0.34	[62]
Black pine	84.26	950°C, 0.35	[36]

CGE, cold gasification efficiency; *ER*, equivalence ratio; *PET*, polyethylene terephthalate.

TABLE 14.6
CGEs of Biomass Oxygen Gasification.

Biomass	CGE (%)	Temperature, OER	References
Rice husk	28	600°C, 0.62	[30]
Mangrove	30	600°C, 0.71	[30]
Rice straw	36	700°C, 0.2	[55]
Rice straw	38	700°C, 0.24	[37]
Rice straw	38.5	700°C, 0.24	[55]
Rice straw	41	750°C, 0.2	[55]
Rice straw	44	800°C, 0.2	[55]
Palm frond	50	600°C, 0.56	[30]
PET/vine prunings	55	800°C, 0.3	[53]
Algae	60	600°C, 0.61	[30]
Vine prunings	85	1100°C, 0.3	[37]

CGE, cold gasification efficiency; *OER*, oxygen equivalence ratio; *PET*, polyethylene terephthalate.

increased from 49% to 75% (53% increase) when the gasification temperature was increased from 600 to 1000°C (67% increase).

The CGE of an air gasification process is varied also by the ER. Zhao et al. [54] studied air gasification of rice husks at different ERs. The results showed that the CGE was increased from 36% to a maximum of 37% (2.8% increase) when ER was increased from 0.26 to 0.29 (12% increase).

Oxygen Gasification

As compared with air gasification, oxygen gasification results in a syngas with no or less N_2 and may also convert more chars (mainly contains carbon element) to combustible gases, i.e., CO, through incomplete (or partial) combustion ($C + 0.5O_2 \rightarrow CO$), thus it theoretically has a higher CGE.

Table 14.6 shows the CGEs of oxygen gasification of different biomass feedstocks at different gasification temperatures and OERs. It is observed that the CGEs are between 28% and 85%. These values (28%–85%) are generally lower than the CGEs of air gasification (36%–84%). This is mainly because the combustible gases (i.e., H_2, CO, CH_4, etc.) are combusted or consumed by the rich oxygen, and the mechanisms involved may be combustion ($H_2 + 0.5O_2 \rightarrow H_2O$, $CO + 0.5O_2 \rightarrow CO_2$, or $CH_4 + 0.5O_2 \rightarrow CO + 2H_2$), etc.

In Table 14.6, oxygen gasification of rice husks showed the lowest CGE (28%) because the gasification temperature was low and the syngas had a low hydrogen content (2.2%). Oxygen gasification of vine prunings showed the highest CGE (85%) because the high gasification temperature favored the dry reforming reaction ($CH_4 + CO_2 \rightarrow 2CO + 2H_2$), steam reforming methanization ($CH_4 + H_2O \rightarrow CO + 3H_2$), the Boudouard reaction ($C + CO_2 \rightarrow 2CO$), and water gas reaction ($C + H_2O \rightarrow CO + H_2$), which resulted in more CO and H_2 in the syngas.

The CGE of an oxygen gasification process is significantly varied by the gasification temperature. Liu et al. [55] studied oxygen gasification of rice straw at different gasification temperatures. The results showed that the CGE was increased from 41% to a maximum of 44% (7% increase) when the gasification temperature was increased from 750 to 800°C (7% increase).

The CGE of an oxygen gasification process is varied also by the OER. Liu et al. [55] studied air gasification of rice straw at different OERs. The results showed that the CGE was increased from 36% to 38.5% (7% increase) when the OER was increased from 0.20 to 0.24 (20% increase).

Steam Gasification

Similar to oxygen gasification, steam gasification also avoids the dilution of N_2. On the other hand, steam

TABLE 14.7
CGEs of Biomass Steam Gasification.

Biomass	CGE (%)	Temperature, SCR	References
Rice husk	82	800°C, 1.02	[30]
Coal/ biomass	83	1100°C, 1	[56]
Algae	87	800°C, 1.22	[30]
Mangrove	88	800°C, 1.5	[30]
Coal/ biomass	89	700°C, 1	[56]
Coal/ biomass	90	800°C, 1	[56]
Coal/ biomass	94	800°C, 2	[56]
Palm frond	98	800°C, 0.95	[30]

CGE, cold gasification efficiency; *SCR*, steam/carbon ratio.

gasification would generate H_2-rich syngas mainly through water gas reaction ($C + H_2O \rightarrow CO + H_2$), steam reforming methanization ($CH_4 + H_2O \rightarrow CO + 3H_2$), and water gas shift reaction ($CO + H_2O \rightarrow CO_2 + H_2$). Consequently, the CGE of steam gasification would be significantly increased.

Table 14.7 shows the CGEs of steam gasification of different biomass feedstocks at different gasification temperatures and SCRs. It is observed that the CGEs are between 82% and 98%. These values (82%–98%) are generally much higher than the CGEs of air gasification (36%–84%) and oxygen gasification (28%–85%). This is mainly because the combustible gases (i.e., CO, CH_4, H_2, etc.) are generated by the steam.

In Table 14.7, steam gasification of rice husks showed the lowest CGE (82%) mainly because the HHV of the producer gas was lower (14.6 MJ/kg), which was due to the lower hydrogen content (2.2%). Steam gasification of palm frond showed the highest CGE (98%) mainly because the HHV of the producer gas was high (25.34 MJ/kg), which was due to the fact that the biomass was completely converted into producer gas because of the high oxygen content in palm frond.

The CGE of a steam gasification process is significantly varied by the gasification temperature. Xiang et al. [56] studied steam gasification of coal/biomass at different temperatures. The results showed that the CGE was increased from 89% to a maximum of 90% (1% increase) when the gasification temperature was increased from 700 to 800°C (14% increase), whereas the CGE was decreased from 90% to 83% (8% decrease) when the gasification temperature was increased from 800 to 1100°C (37.5% increase).

The CGE of a steam gasification process is varied also by the SCR. Xiang et al. [56] studied the steam gasification of coal/biomass at different SCRs. The results showed that the CGE was increased from 90% to 94% (4% increase) when SCR was increased from 1 to 2 (100% increase).

Carbon Dioxide Gasification

Carbon dioxide gasification would also result in a high CGE because (1) it avoids N_2 dilution and (2) it produces rich CO and H_2 (mainly through the dry reforming reaction: $CH_4 + CO_2 \rightarrow 2CO + 2H_2$, the Boudouard reaction: $C + CO_2 \rightarrow 2CO$, steam reforming methanization: $CH_4 + H_2O \rightarrow CO + 3H_2$, and water gas reaction: $C + H_2O \rightarrow CO + H_2$). However, it would yield a low CGE if the gasification agent (carbon dioxide) is not efficiently or completely consumed (the carbon dioxide in syngas dilutes the LHV of syngas).

Generally, carbon dioxide gasification would result in a high CGE. Table 14.8 shows the CGEs of carbon dioxide gasification of different biomass feedstocks at different gasification temperatures and CO_2/C ratios. It is observed that the CGEs are between 50% and 87%. These values (50%–87%) are generally higher than the CGEs of air gasification (36%–84%) and oxygen gasification (28%–85%). This is mainly because more CO is produced during the carbon dioxide gasification process, and the mechanism involved is mainly the Boudouard reaction ($C + CO_2 \rightarrow 2CO$).

In Table 14.8, carbon dioxide gasification of rice straw showed the lowest CGE (50%) because the lower gasification temperature (700°C) resulted in a lower carbon conversion efficiency (78%). Carbon dioxide gasification of pine sawdust showed the highest CGE (87%) due to the facts that (1) the optimal CO_2/C ratio (0.25) resulted in the highest carbon conversion efficiency (98%) and (2) the high gasification temperature (1000°C) was also conducive to the carbon conversion efficiency.

The CGE of carbon dioxide gasification process is significantly varied by the gasification temperature. Yu et al. [46] studied the carbon dioxide gasification of rice straw and pine sawdust at different gasification temperatures. The results showed that the CGE of pine sawdust was increased from 58% to a maximum of 87% (50% increase) when the gasification temperature was increased from 700 to 1000°C (43% increase) and that the CGE of rice straw was increased from 50% to a

TABLE 14.8
CGEs of Biomass Carbon Dioxide Gasification.

Biomass	CGE (%)	Temperature, CO_2/C Ratio	References
Rice straw	50	700°C, 0.25	[46]
Pine sawdust	58	700°C, 0.25	[46]
Rice straw	61	1000°C, 1	[46]
Rice straw	68	1000°C, 0	[46]
Rice straw	71	1000°C, 0.5	[46]
Rice straw	75	1100°C, 0.25	[46]
Pine sawdust	75	1000°C, 1	[46]
Pine sawdust	79	900°C, 0.25	[46]
Pine sawdust	82	1000°C, 0.5	[46]
Pine sawdust	87	1000°C, 0.25	[46]

CGE, cold gasification efficiency.

maximum of 75% (50% increase) when the gasification temperature was increased from 700 to 1100°C (57% increase).

The CGE of the carbon dioxide gasification process is varied also by the CO_2/C ratio. Yu et al. [46] studied the carbon dioxide gasification of rice straw and pine sawdust at different CO_2/C ratios. The results showed that the CGE of rice straw was increased from 68% to 71% (4% increase) when the CO_2/C ratio was increased from 0 to 0.5, whereas the CGE of pine sawdust was decreased from 87% to 75% (13.8% decrease) when the CO_2/C ratio was increased from 0.25 to 1 (300% increase). These also indicate that the CGE of a carbon dioxide gasification process is varied by the biomass feedstock.

Hydrogen Gasification

Hydrogen gasification would also result in a high CGE because (1) it avoids N_2 dilution, (2) it produces rich CH_4 (through the CO hydrogenation reaction: $C + 3H_2 \rightarrow CH_4 + H_2O$ and methanation reaction: $C + 2H_2 \rightarrow CH_4$), and (3) it produces rich H_2 (through the water gas shift reaction: $CO + H_2O \rightarrow CO_2 + H_2$ and water gas reaction: $C + H_2O \rightarrow CO + H_2$).

Generally, the gas component yield is mainly used to assess hydrogen gasification of biomass. These indexes include CH_4 yield [51,57], CO yield [41,51], etc. These high–heating-value combustible gases contribute to the heating value (HHV and LHV) of syngas, and they therefore also contribute to the CGEs of biomass hydrogen gasification processes. However, the CGEs of biomass hydrogen gasification processes are seldom reported.

SUMMARY AND FUTURE OUTLOOK

This chapter gives an up-to-date knowledge of the gasification technologies and their energy potentials. Thermochemical technologies, i.e., liquefaction, carbonization, combustion, pyrolysis, and gasification, are first introduced and the biomass gasification technologies are then detailed, including air gasification, oxygen gasification, steam gasification, hydrogen gasification, carbon dioxide gasification, and combined gasification. Energy potentials of these gasification technologies are also discussed. The results show that air gasification is the most widely studied and applied gasification technology because air is cheap and the technology is mature. Oxygen gasification, steam gasification, hydrogen gasification, carbon dioxide gasification, and combined gasification generally result in better performances because of the higher quality syngas produced. However, these processes are generally more difficult to organize or control. Generally, all the gasification technologies can be used to gasify a biomass. However, the processes and results may be quite different owing to the differences in the feedstock characteristics. Studying feedstock characteristics, optimizing gasification processes, and developing new technologies are of significant importance in the future.

ACKNOWLEDGMENTS

This study was supported in part by grants from the National Natural Science Foundation of China (Grant No. 51606048), Minnesota's Environment and Natural Resources Trust Fund (ENRTF) through the processes of the Legislative-Citizen Commission on Minnesota Resources (LCCMR), and the University of Minnesota Center for Biorefining.

REFERENCES

[1] Zhang Y, Gao X, Li B, Li H, Zhao W. Assessing the potential environmental impact of woody biomass using quantitative universal exergy. Journal of Cleaner Production 2018;176:693–703.

[2] Bhaskar T, Pandey A. Advances in thermochemical conversion of biomass–Introduction. Chapter 1 in recent advances in thermo-chemical conversion of biomass. 2015. p. 3–30.

[3] Okoro OV, Sun Z, Birch J. Meat processing waste as a potential feedstock for biochemicals and biofuels e A review of possible conversion technologies. Journal of Cleaner Production 2017;142:1583–608.

[4] Williams CL, Dahiya A, Porter P. Introduction to bioenergy. Chapter 1 in bioenergy. 2015. p. 5–36.

[5] Goyal HB, Seal D, Saxena RC. Bio-fuels from thermochemical conversion of renewable resources: a review. Renewable and Sustainable Energy Reviews 2008;12: 504–17.

[6] Suali E, Sarbatly R. Conversion of microalgae to biofuel. Renewable and Sustainable Energy Reviews 2012;16: 4316–42.

[7] Chen H, Zhou D, Luo G, Zhang S, Chen J. Macroalgae for biofuels production: progress and perspectives. Renewable and Sustainable Energy Reviews 2015;47:427–37.

[8] Naik SN, Goud VV, Rout PK, Dalai AK. Production of first and second generation biofuels: a comprehensive review. Renewable and Sustainable Energy Reviews 2010;14(2): 578–97.

[9] Champagne P. Biomass. Chapter 9 in future energy. 2008. p. 151–70.

[10] Guo M, Song W, Buhain J. Bioenergy and biofuels: history, status, and perspective. Renewable and Sustainable Energy Reviews 2015;42:712–25.

[11] Kambo HS, Dutta A. A comparative review of biochar and hydrochar in terms of production, physico-chemical properties and applications. Renewable and Sustainable Energy Reviews 2015;45:359–78.

[12] Zhang Y, Chen P, Liu S, Fan L, Zhou N, Min M, Cheng Y, Peng P, Anderson E, Wang Y, Wan Y, Liu Y, Li B, Ruan R. Microwave-assisted pyrolysis of biomass for bio-oil production. Chapter 6, Pyrolysis. InTech; 2017. p. 129–66.

[13] Zhang Y, Liu S, Fan L, Zhou N, Omar MM, Peng P, Anderson E, Addy M, Cheng Y, Liu Y, Li B, Snyder J, Chen P, Ruan R. Oil production from microwave-assisted pyrolysis of a low rank American brown coal. Energy Conversion and Management 2018;159:76–84.

[14] Fokaides PA, Christoforou E. Life cycle sustainability assessment of biofuels. Chapter 3 in Handbook of Biofuels Production. 2nd ed. 2016. p. 41–60.

[15] Zhang Y, Chen P, Liu S, Peng P, Min M, Cheng Y, Anderson E, Zhou N, Fan L, Liu C, Chen G, Liu Y, Lei H, Li B, Ruan R. Effects of feedstock characteristics on microwave-assisted pyrolysis-a review. Bioresource Technology 2017;230:143–51.

[16] Du C, Zhao X, Liu D, Lin CSK, Wilson K, Luque R, Clark J. Introduction: an overview of biofuels and production technologies. Chapter 1 in Handbook of Biofuels Production. 2nd ed. 2016. p. 3–12.

[17] Ho DP, Ngo HH, Guo W. A mini review on renewable sources for biofuel. Bioresource Technology 2014;169: 742–9.

[18] Jayasinghe P, Hawboldt K. A review of bio-oils from waste biomass: focus on fish processing waste. Renewable and Sustainable Energy Reviews 2012;16:798–821.

[19] Manara P, Zabaniotou A. Towards sewage sludge based biofuels via thermochemical conversion – a review. Renewable and Sustainable Energy Reviews 2012;16: 2566–82.

[20] Azizi K, Moraveji MK, Najafabadi HA. A review on biofuel production from microalgal biomass by using pyrolysis method. Renewable and Sustainable Energy Reviews 2018;82:3046–59.

[21] Taghizadeh-Alisaraei A, Assar HA, Ghobadian B, Motevali A. Potential of biofuel production from pistachio waste in Iran. Renewable and Sustainable Energy Reviews 2017;72:510–22.

[22] Liu S, Zhang Y, Fan L, Zhou N, Tian G, Zhu X, Cheng Y, Wang Y, Liu Y, Chen P, Ruan R. Bio-oil production from sequential two-step catalytic fast microwave-assisted biomass pyrolysis. Fuel 2017;196:261–8.

[23] Fan L, Zhang Y, Liu S, Zhou N, Chen P, Cheng Y, Addy M, Lu Q, Omar MM, Liu Y, Wang Y, Dai L, Anderson E, Peng P, Lei H, Ruan R. Bio-oil from fast pyrolysis of lignin: effects of process and upgrading parameters. Bioresource Technology 2017;241:1118–26.

[24] Zhang Y, Zhao W, Li B, Xie G. Microwave-assisted pyrolysis of biomass for bio-oil production: a review of the operation parameters. Journal of Energy Resources 2018;140(4):040802.

[25] Zhang Y, Zhao Y, Gao X, Li B, Huang J. Energy and exergy analyses of syngas produced from rice husk gasification in an entrained flow reactor. Journal of Cleaner Production 2015;95:273–80.

[26] Lozano FJ, Lozano R. Assessing the potential sustainability benefits of agricultural residues; Biomass conversion to syngas for energy generation or to chemicals production. Journal of Cleaner Production 2018;172: 4162–9.

[27] Zhang Y, Li B, Li H, Liu H. Thermodynamic evaluation of biomass gasification with air in autothermal gasifiers. Thermochimica Acta 2011;519(1–2):65–71.

[28] Balat M, Balat M, Elif K, Balat H. Main routes for the thermo-conversion of biomass into fuels and chemicals. Part 2: gasification systems. Energy Conversion and Management 2009;50(12):3158–68.

[29] Parvez AM, Mujtaba IM, Wu T. Energy, exergy and environmental analyses of conventional, steam and CO_2-enhanced rice straw gasification. Energy 2016;94: 579–88.

[30] Adnan MA, Susanto H, Binous H, Muraza O, Hossain MM. Feed compositions and gasification potential of several biomasses including a microalgae: a thermodynamic modeling approach. International Journal of Hydrogen Energy 2017;42:17009–19.

[31] de Oliveira JL, da Silva JN, Martins MA, Pereira EG, da Conceição M, e Oliveira TB. Gasification of waste from coffee and eucalyptus production as an alternative source of bioenergy in Brazil. Sustainable Energy Technologies and Assessments 2018;27:159–66.

[32] Yao Z, You S, Ge T, Wang CH. Biomass gasification for syngas and biochar co-production: energy application and economic evaluation. Applied Energy 2018;209: 43–55.

[33] Gao X, Zhang Y, Li B, Yu X. Model development for biomass gasification in an entrained flow gasifier using intrinsic reaction rate submodel. Energy Conversion and Management 2016;108:120–31.

[34] IEA. Main reactions during biomass gasification. 2015. http://www.ieatask33.org/app/webroot/files/file/various/Main%20reactions%20during%20biomass%20gasification.pdf.

[35] Aydin ES, Yucel O, Sadikoglu H. Numerical and experimental investigation of hydrogen-rich syngas production via biomass gasification. International Journal of Hydrogen Energy 2018;43:1105–15.

[36] Kihedu JH, Yoshiie R, Naruse I. Performance indicators for air and air-steam auto-thermal updraft gasification of biomass in packed bed reactor. Fuel Processing Technology 2016;141:93–8.

[37] Biagini E. Study of the equilibrium of air-blown gasification of biomass to coal evolution fuels. Energy Conversion and Management 2016;128:120–33.

[38] Song YC, Ji MS, Feng J, Li WY. Product distribution from Co-gasification of coal and biomass in a fluidized-bed reactor. Energy Sources Part A 2015;37:2550–8.

[39] Parthasarathy P, Narayanan KS. Hydrogen production from steam gasification of biomass: influence of process parameters on hydrogen yield-A review. Renewable Energy 2014;66:570–9.

[40] Favas J, Monteiro E, Rouboa A. Hydrogen production using plasma gasification with steam injection. International Journal of Hydrogen Energy 2017;42: 10997–1005.

[41] Maneewan K, Krerkkaiwan S, Sunphorka S, Vitidsant T. Catalytic effect of biomass pyrolyzed char on the atmospheric pressure hydrogasification of giant leucaena (leucaena leucocephala) wood. Industrial & Engineering Chemistry Research 2014;53(30):11913–9.

[42] Perna A, Minutillo M, Jannelli E. Hydrogen from intermittent renewable energy sources as gasification medium in integrated waste gasification combined cycle power plants: a performance comparison. Energy 2016;94: 457–65.

[43] Shen Y, Ma D, Ge X. CO_2-looping in biomass pyrolysis or gasification. Sustain Energy Fuel 2017;1:1700–92.

[44] Sadhwani N, Adhikari S, Eden MR, Li P. Aspen plus simulation to predict steady state performance of biomass-CO_2 gasification in a fluidized bed gasifier. Biofuel Bioprod Bio 2018;12:379–89.

[45] Rupesh S, Muraleedharan C, Arun P. A comparative study on gaseous fuel generation capability of biomass materials by thermo-chemical gasification using stoichiometric quasi-steady-state model. International Journal of Energy Environment Engineering 2015;6:375–84.

[46] Yu H, Chen G, Xu Y, Chen D. Experimental study on the gasification characteristics of biomass with CO_2/air in an entrained-flow gasifier. Bioresources 2016;11(3): 6085–96.

[47] Al-Zareer M, Dincer I, Rosen M. Influence of selected gasification parameters on syngas composition from biomass gasification. Journal Energy Resources 2018;140:041803.

[48] Shayan E, Zare V, Mirzaee I. Hydrogen production from biomass gasification; a theoretical comparison of using different gasification agents. Energy Conversion and Management 2018;159:30–41.

[49] Sadhwani N, Adhikari S, Eden MR. Biomass gasification using carbon dioxide: effect of temperature, CO_2/C ratio, and the study of reactions influencing the process. Industrial & Engineering Chemistry Research 2016;55(10): 2883–91.

[50] Jin CL, Yang RF, Farahani MR. Hydrogen and syngas production from biomass gasification for fuel cell application. Energy Sources Part A 2018;40(5):553–7.

[51] Zheng N, Zhang J, Wang J. Parametric study of two-stage hydropyrolysis of lignocellulosic biomass for production of gaseous and light aromatic hydrocarbons. Bioresource Technology 2017;244:142–50.

[52] Sun S, Zhao Y, Ling F, Su F. Experimental research on air staged cyclone gasification of rice husk. Fuel Processing Technology 2009;90:465–71.

[53] Tavares R, Ramos A, Rouboa A. Microplastics thermal treatment by polyethylene terephthalate-biomass gasification. Energy Conversion and Management 2018; 162:118–31.

[54] Zhao Y, Feng D, Zhang Z, Sun S, Che H, Luan J. Experimental study on autothermal cyclone air gasification of biomass. Journal Energy Resources 2018;140(4):042001.

[55] Liu L, Huang Y, Cao J, Liu C, Dong L, Xu L, Zha J. Experimental study of biomass gasification with oxygen-enriched air in fluidized bed gasifier. The Science of the Total Environment 2018;626:423–33.

[56] Xiang X, Gong G, Shi Y, Cai Y, Wang C. Thermodynamic modeling and analysis of a serial composite process for biomass and coal co-gasification. Renewable and Sustainable Energy Reviews 2018;82:2768–78.

[57] Zhang J, Zheng N, Wang J. Co-hydrogasification of lignocellulosic biomass and swelling coal. Earth Environmental Science 2016;40(1):012042.

[58] Jenkins RW, Sutton AD, Robichaud DJ. Pyrolysis of biomass for aviation fuel. Chapter 8 in biofuels for aviation. 2016. p. 191–215.

[59] Zhang Y, Fan X, Li B, Li H, Gao X. Assessing the potential environmental impact of fuel using exergy-cases of wheat straw and coal. International Journal of Exergy 2017; 23(1):85–100.

[60] Bundhoo ZMA. Microwave-assisted conversion of biomass and waste materials to biofuels. Renewable and Sustainable Energy Reviews 2018;82:1149–77.

[61] Damartzis T, Zabaniotou A. Thermochemical conversion of biomass to second generation biofuels through integrated process design – a review. Renewable and Sustainable Energy Reviews 2011;15:366–78.

[62] Ramos-Carmona S, Pérez JF. Effect of torrefied wood biomass under an oxidizing environment in a downdraft gasification process. Bioresources 2017;12(3):6040–61.

CHAPTER 15

Syngas Fermentation for Bioethanol and Bioproducts

HARIS NALAKATH ABUBACKAR, PHD • MARÍA C. VEIGA, PHD • CHRISTIAN KENNES, PHD

INTRODUCTION

Over the past decades, an increased interest in biobased technologies for the production of fuels and chemicals has been observed, thereby shifting from fossil fuel–based production to a more sustainable production technology [1]. Use of fossil fuels and the associated CO_2 emission has been recognized as the major cause for global climate change. Many countries have thus put forward policies and challenges that they mandate to follow in order to achieve low carbon emissions. For instance, the European council revised its objectives for reducing greenhouse gas emissions by 80%–95% from 1990 to 2050. In addition, for promoting renewable energy usage, the European Union revised its targets, now established as 27% renewable energy consumption by 2030 (revised European Directive 2009/28/EC). Ethanol, a high-octane and oxygenated chemical, is considered an interesting alternative biofuel. Bioethanol is produced traditionally from starch- or sugar-based feedstocks, known as first-generation bioethanol production technology. This arises a dilemma of food versus fuel usage of such feedstock. Another approach consists in utilizing lignocellulosic biomass as feedstock in a second-generation technology. This process is more complex than the first-generation technology, as it involves several steps such as pretreatment (chemical, physical, enzymatic, or a combination of these), hydrolysis (chemical or enzymatic), and fermentation of sugars released from the feedstock [2–4]. However, the lignin fraction that constitutes about 15%–30% (dry weight basis) of lignocellulosic materials does not yield any fermentable sugars upon pretreatment or hydrolysis. An alternative approach to generate bioethanol is through the hybrid gasification and fermentation process. It is a feedstock-flexible process in which many carbonaceous materials can be gasified, including lignocellulosic material and municipal solid waste, to generate syngas rich in CO, CO_2, and/or H_2. Such gas mixture can later be utilized by bacteria, e.g., some acetogens, as their carbon and energy source to generate products such as bioethanol, besides acetic acid [5]. Other products can also be obtained, in a similar way, through such anaerobic fermentation technology. In case of ethanol, optimization of the fermentation process and conditions can ultimately lead to improved bioethanol production, as a biofuel, even without any accumulation of acetic acid [6]. Some companies such as LanzaTech are focusing on scaling up this gas fermentation technology for bioethanol production at the commercial scale.

SYNTHESIS GAS AND WASTE GAS

Synthesis gas, also known as syngas, is mainly composed of CO, H_2, and CO_2. In addition, it may contain other gases such as nitrogen and methane. It is produced through a thermochemical process called gasification, which converts carbonaceous materials such as biomass, municipal wastes, coal, petroleum, and tires under controlled amount of oxidant such as oxygen, air, and CO_2, inside a gasifier to obtain syngas [7]. The syngas composition varies depending on the type of gasifier and its operational conditions, as well as the feedstock properties. Low amounts of C_2H_2, C_2H_4, C_3H_8, ash, char, tars, O_2, phenol, NH_3, H_2S, HCN, NO_x, and COS can also be generated [8]. Through gasification a basically complete conversion of lignocellulosic materials to syngas is possible. Thus, as mentioned earlier, the lignin fraction, which constitutes 15%–30% (dry weight basis) of lignocellulosic materials, is also gasified and can be utilized for the generation of bioethanol or other products through the hybrid gasification-fermentation process [9]. Waste gases from certain industries such as the steel milling industry, petroleum refining industry, nonferrous metal manufacturing industry, and other industries such as

Sustainable Resource Recovery and Zero Waste Approaches. https://doi.org/10.1016/B978-0-444-64200-4.00015-3

carbon black, ammonia, methanol, and coke manufacturing industries contain CO, CO_2 or syngas-related gas mixtures as well [10,11].

MICROBIAL FERMENTATION OF SYNGAS

A promising approach for the conversion of syngas into biofuels and platform chemicals is through bioprocesses utilizing a specific group of strictly anaerobic organisms, called acetogens. These bacteria use the C1 molecules present in syngas as the carbon and/or energy source and use hydrogen as the energy source in the presence of CO/CO_2, following the reductive acetyl-CoA or the Wood-Ljungdahl (WL) pathway to incorporate these molecules into acetyl-CoA. Later these molecules are reduced further to yield different platform chemicals or fuels such as acetic acid, ethanol, butyric acid, butanol, 2,3-butanediol, hexanol, or lactic acid at different concentrations depending on the syngas composition, fermentation conditions, and acetogenic biocatalyst [12–14]. Several acetogens have proven their ability to metabolize C1 gas molecules, including *Clostridium autoethanogenum, Clostridium ljungdahlii, Clostridium ragsdalei, Clostridium coskatii,* and *Clostridium carboxidivorans. C. ljungdahlii* is the first acetogen identified for its ability to produce ethanol from syngas [15,16]. Apart from ethanol and acetic acid, *C. ljungdahlii, C. autoethanogenum,* and *C. ragsdalei* have been reported to produce 2,3-butanediol [17]. *C. carboxidivorans,* on the other hand, produces a broad range of biocommodities through gas fermentation, including acetic acid, ethanol, butyric acid, butanol, hexanoic acid, and hexanol [18,19].

BIOPROCESSING OF SYNGAS TO VALUE-ADDED PRODUCTS

Following the WL pathway, acetogens subsequently reduce acetyl-CoA to butyric acid, hexanoic acid, butanol, hexanol, 2,3-butanediol, and lactic acid, besides acetic acid and ethanol. Integrating syngas fermentation with other bioprocessing would further expand the spectrum of possible products. Octanoic acid, octanol, malic acid, and C16 and C18 lipids are some of the products that can be obtained by this way. They have a higher market value and are less soluble in aqueous phase than ethanol and are thereby potentially easier to separate from the fermentation broth [20,21]. Chain elongation of short-chain carboxylic acid, e.g., acetic acid, to medium-chain carboxylic acids has been performed by some bacterial strains obtaining energy, carbon, and electrons from ethanol or lactate and employing the reverse β-oxidation reactions. Some studies have been carried out aimed at utilizing acetate and ethanol produced during syngas fermentation to generate medium-chain carboxylic acids, with strains such as *Clostridium kluyveri* catalyzing both ethanol oxidation and reverse β-oxidation [21–23]. Generally, the conversion of ethanol generates ATP by substrate-level phosphorylation (SLP) and does also generate NADH, a reducing equivalent necessary for the subsequent reverse β-oxidation pathway [20]. Diender et al. [22] observed in batch bottle studies that coculturing *C. autoethanogenum* and *C. kluyveri* would result in the production of butyric acid, hexanoic acid, butanol, and hexanol from CO/syngas, compared with the production of only acetate and ethanol if *C. autoethanogenum* was used as single strain in the syngas fermentation process. Another study using a coculture of *C. ljungdahlii* and *C. kluyveri* in a continuous reactor produced a variety of metabolites such as ethanol, acetate, 2,3-butanediol, butyric acid, hexanoic acid, butanol, hexanol, and octanol. Chain elongation to yield medium-chain fatty acids is performed by *C. kluyveri,* while the conversion to the corresponding alcohols is supposed to be performed by *C. ljungdahlii* [23]. Another approach to increase the product spectrum from syngas was successfully demonstrated by coupling anaerobic syngas fermentation with aerobic processes such as malic acid production by the fungus *Aspergillus oryzae* [24]; C16 and C18 lipid production by an engineered yeast, *Yarrowia lipolytica* [25]; and biopolymer (polyhydroxyalkanoate [PHA]) production by using enriched PHA accumulating biomass [26]. Biomethane produced using hydrogenotrophic methanogens is another product obtained from syngas [27–29].

ADVANTAGES OF THE BIOLOGICAL ROUTE OVER THE THERMOCHEMICAL ROUTE FOR THE PRODUCTION OF CHEMICALS FROM SYNGAS

There are basically two categories of methods available for the conversion of syngas to various platform chemicals, namely, the biological (syngas fermentation) and the thermochemical routes (the Fischer-Tropsch process). Syngas fermentation is a simple and highly specific process performed by biocatalysts, which can grow and produce metabolites at near ambient temperatures and pressures. Instead, the thermochemical alternative does generally take place at temperatures around 200–300°C and pressures of 10–40 bar using catalysts based on metals such as iron or cobalt. These catalysts are relatively more expensive than biocatalysts and are less tolerant to impurities present in syngas such as sulfur and tar [30]. These impurities are mostly nontoxic to bacteria and sometimes even considered to be growth stimulators [31]. For example, the presence of sulfur in the syngas mixture stimulates the growth of anaerobes by reducing the medium [32]. In addition, byproducts generated from syngas fermentation are less harmful to the environment. One of the main advantages of syngas fermentation over the thermochemical route is the nonrequirement of a fixed H_2/CO ratio. Chemical catalysts in turn require specific H_2/CO ratios in the gas mixture, which need to be adjusted to the required ratio optimal for each metal catalyst in the water gas shift reaction. Thus, in terms of energy and infrastructure input, syngas fermentation processes offer several advantages and high flexibility over the thermochemical route [10,32].

A main limitation of syngas fermentation is the low solubility of the gaseous substrates in the fermentation medium, mainly when using suspended growth bioreactors. The solubilities of CO and H_2 in aqueous phase are, respectively, 0.028–0.019 and 0.0016–0.0013 $g\,L^{-1}$ water at 25–45°C and 1 atm [13]. The CO and H_2 solubilities are only about 60% and 3% of the oxygen solubility, respectively [13]. This will result in scarcity of substrate availability for the bacteria during fermentation. In addition, the energy yield for the bacteria when using syngas is low compared with that of sugar fermentation, resulting generally in slow growth rates and low cell concentrations, which reduce the overall metabolite productivity of the process.

Another major bottleneck in the syngas fermentation process is the low mass transfer of sparingly soluble syngas compounds to the fermentation medium [33,34]. An ideal environment for syngas fermentation should allow for high mass transfer without causing any substrate-level inhibition, thereby achieving high cell concentrations. The major diffusion resistance to mass transfer from the bulk gas phase to the bulk liquid phase is exerted by the stagnant layer at the gas-liquid (GL) interface [12]. The rate of mass transfer $\frac{dn}{dt}$ (mol/h) is calculated by

$$-\frac{1}{V}\frac{dn}{dt} = K_{L}a(C_{i} - C_{L})$$

where V is the liquid volume into which the gas is transferred (L), n is the molar gas substrate concentration in the gas phase, K_La is the overall mass transfer coefficient (h^{-1}), a is the GL interface area (m^2), C_i is the concentration of the gaseous substrate at the GL interface (mol/L), and C_L is the concentration of the gaseous substrate in the bulk liquid phase (mol/L).

THE WOOD-LJUNGDAHL PATHWAY FOR METABOLITE PRODUCTION

During growth with inorganic substrates such as syngas compounds, acetogens follow the metabolic pathway called the WL pathway, also known as acetyl-CoA biochemical pathway (Fig. 15.1) [35,36]. This reductive pathway allows the acetogens to utilize CO or CO_2/H_2 or a mixture of all three gases as the sole carbon and energy sources for growth and production of the acetyl-CoA molecule, which is further reduced to form different metabolites such as ethanol [14]. The reducing equivalents are generated by a hydrogenase from H_2 (Eq. 15.1), via biochemical water gas shift reaction (Eq. 15.2), or from redox mediators such as reduced ferredoxin (Fd^{2-}) and NAD(P)H [37].

$$H_2 \xrightarrow{\text{Hydrogenase}} 2H^+ + 2e^- \quad (15.1)$$

$$CO + H_2O \xrightarrow{\text{CODH}} CO_2 + 2H^+ + 2e^- \quad (15.2)$$

The WL pathway consists of a methyl branch (also known as eastern branch) and a carbonyl (western) branch [36]. The eastern branch provides the methyl moiety of acetyl-CoA via reduction of one molecule of CO_2 using six electrons. The first step in the eastern branch is the reduction of CO_2 to formate by the enzyme formate dehydrogenase (FDH). The next step is an ATP-consuming step in which formate is combined with tetrahydrofolate or H_4folate to form 10-formyl-H_4folate catalyzed by 10-formyl-H_4folate synthetase [38]. The 5,10-methenyl-H_4folate cyclohydrolase converts 10-formyl-H_4folate to 5,10-methenyl-H_4folate along with the removal of a water molecule [39]. Further reduction is carried out by methylene-H_4folate dehydrogenase to generate 5,10-methylene-H_4folate, which is then reduced to form (6*S*)-5-CH_3-H_4folate by methylene-H_4folate reductase [40,41].

In the western branch, the methyl group of CH_3-H_4folate is transferred to the cobalt center, Co (I), of the corrinoid iron-sulfur protein (CFeSP) by methyltransferase, resulting in the formation of an organometallic intermediate, methyl-Co(III)-CFeSP [42–44]. The bifunctional CO dehydrogenase (CODH) catalyzes two reactions: conversion of CO_2 to CO, which can serve as the carbonyl moiety of the acetyl-CoA, and formation of a complex with acetyl-CoA synthase (ACS), catalyzing the acetyl-CoA formation. For the synthesis of acetyl-CoA, the CO transfers to the A cluster of ACS and the methyl group from the methylated CFeSP transfers to the CODH/ACS complex. The methyl and carbonyl groups condense to form an acetyl metal. Finally, the acetyl metal combines

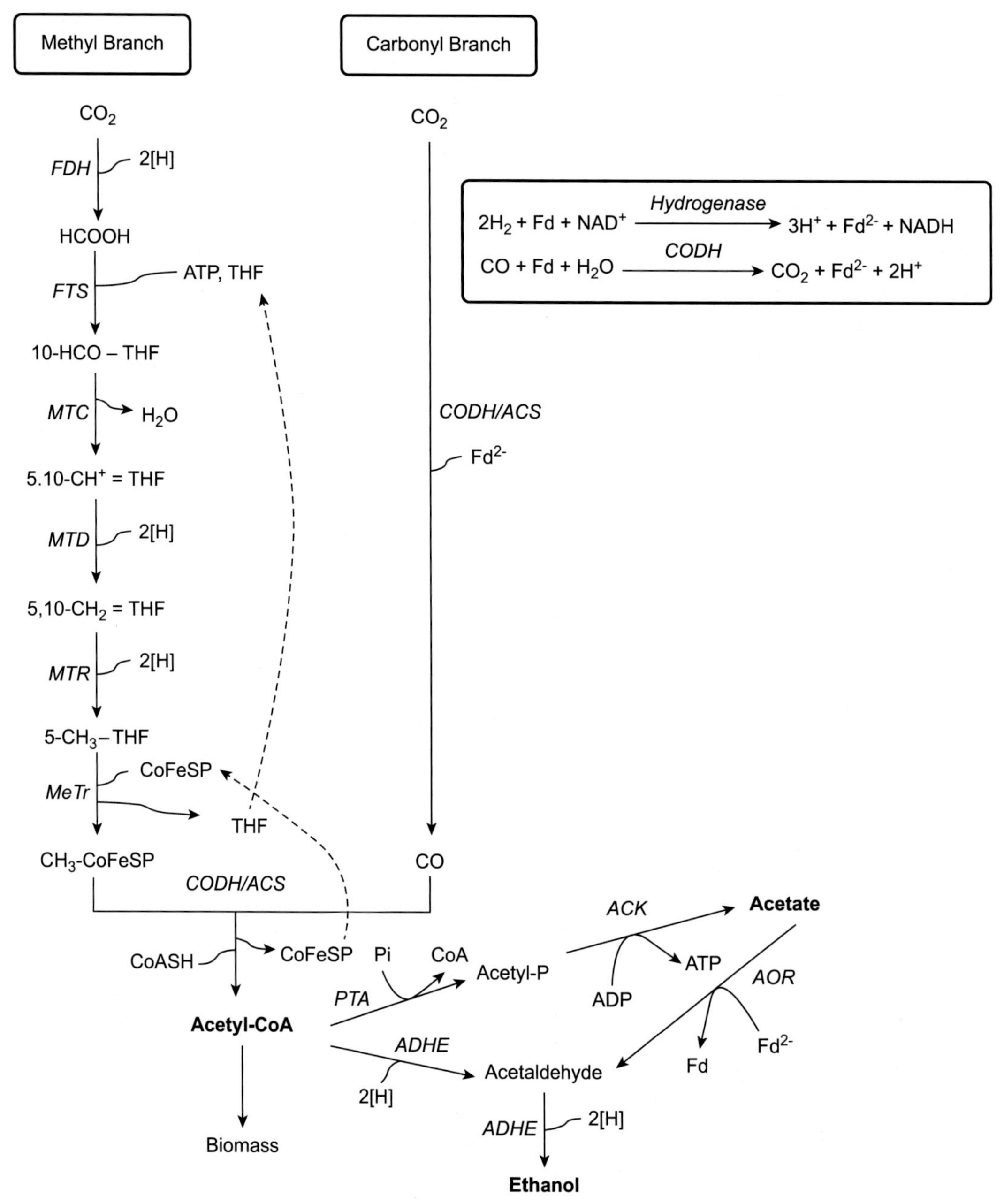

FIG. 15.1 The Wood-Ljungdahl pathway. *2[H]*, reducing equivalents (NADH or NADPH); *ACK*, acetate kinase; *ADHE*, aldehyde/alcohol dehydrogenase; *AOR*, aldehyde oxidoreductase; *CODH/ACS*, carbon monoxide dehydrogenase/acetyl-CoA synthase; *CoFeSP*, corrinoid iron-sulfur protein; *Fd^{2-}*, reduced ferredoxin; *Fd*, ferredoxin; *FDH*, formate dehydrogenase; *FTS*, 10-formyl-H_4folate synthetase; *MeTr*, methyltransferase; *MTC*, 5,10-methenyl-H_4folate cyclohydrolase; *MTD*, 5,10-methylene-H_4folate dehydrogenase; *MTR*, 5,10-methylene-H_4folate reductase; *PTA*, phosphotransacetylase; *THF*, cofactor tetrahydrofolate. (modified, with permission, from reference [19])

with the coenzyme (CoA), facilitated by ACS, to form acetyl-CoA, the key intermediate [45].

Acetyl-CoA is converted to various metabolites through the action of various enzymes. The production of acetate from acetyl-CoA is an energy conservation step in which ATP is generated by SLP. Acetyl-CoA is first converted to acetyl phosphate by phosphotransacetylase. Through the action of acetate kinase (ACK), acetyl phosphate is converted to acetate and 1 mole of ATP is generated through SLP [5,37]. On the other hand, ethanol can theoretically be produced in two ways: in the first way, acetyl-CoA is reduced to acetaldehyde and later to ethanol by the enzyme aldehyde/alcohol dehydrogenase (ADHE), whereas in the second way, ethanol is produced from the already produced acetate via aldehyde ferredoxin oxidoreductase (AFOR), which catalyzes the conversion of acetate to acetaldehyde using the electrons provided by the reduced ferredoxin (Fd^{2-}) [46,47].

The western branch is unique in anaerobes and provides the carbonyl group for acetyl-CoA either by directly using CO from the medium or by involving a (CODH) that catalyzes the conversion of CO_2 to CO. The presence of H_2 in the syngas mixture is beneficial in such a way that it can supply electrons necessary for the reduction, thereby improving CO assimilation for the production of metabolites. The stoichiometry of acetic acid and ethanol production from CO and CO/H_2 is given below [13]

$$4CO + 2H_2O \rightarrow CH_3COOH + 2CO_2 \quad (15.3)$$

$$2CO_2 + 4H_2 \rightarrow CH_3COOH + 2H_2O \quad (15.4)$$

$$2CO + 2H_2 \rightarrow CH_3COOH \quad (15.5)$$

$$6CO + 3H_2O \rightarrow C_2H_5OH + 4CO_2 \quad (15.6)$$

$$2CO_2 + 6H_2 \rightarrow C_2H_5OH + 3H_2O \quad (15.7)$$

$$3CO + 3H_2 \rightarrow C_2H_5OH + CO_2 \quad (15.8)$$

$$2CO + 4H_2 \rightarrow CH_3CH_2OH + H_2O \quad (15.9)$$

According to Eq. (15.6), only one-third of the carbon can be converted to ethanol and the remaining two-thirds will be lost as CO_2 if the sole carbon source is CO. However, if hydrogen is present in a ratio of 1:2 (CO/H_2), no carbon will be lost as CO_2. However, studies on CO or a syngas mixture have found that the amount of ethanol produced will often not agree with the theoretic stoichiometry, as there are several factors that influence the syngas fermentation process in acetogens.

ENERGY CONSERVATION IN ACETOGENS BELONGING TO THE GENUS *CLOSTRIDIUM*

In the WL pathway, 1 mole of ATP is required to combine formate with H_4folate and to synthetize 10-formyl-H_4folate, and 1 mole of ATP is conserved by SLP during the conversion of acetyl phosphate to acetate by ACK. Hence, there is no net ATP formation in the WL pathway by SLP. Therefore in order to grow autotrophically on CO, the bacteria depend on another mode of energy conservation, e.g., using chemiosmotic ion gradient–driven phosphorylation. On the basis of how the ion gradient across the membrane is established, acetogens are classified into various classes. Out of the large group of acetogens, three acetogens are considered model organisms to explain the different ways of transmembrane ion gradient: *Moorella thermoacetica*, *Acetobacterium Woodii*, and *C. ljungdahlii* [37]. *M. thermoacetica* has a membrane-bound heme group containing a protein cytochrome and lipid-soluble protein menaquinone. However, for growth and acetate formation on H_2 and CO_2, it does not require sodium ions (Na^+) [48]. Conversely, *A. Woodii* cells do not possess cytochromes and quinones. Its growth and acetate formation strongly depend on Na^+ while grown autotrophically [49]. In contrast to the two abovementioned organisms, *C. ljungdahlii* neither has cytochrome nor is independent of Na^+ [50].

The ferredoxin-dependent redox reactions are involved in the formation of metabolites from CO during the reversible oxidation of CO to CO_2 and during the reduction of acetaldehyde to acetic acid. Ferredoxin is an iron-sulfur protein having [2Fe-2S] or [4Fe-4S] clusters. The redox potential of ferredoxin is around −400 mV and even lower, e.g., −500 mV physiologically [51]. The reduction of low-potential ferredoxin by physiologic electron donors is endergonic (reaction that consumes energy, i.e., the free energy change is positive) and has to be coupled to an exergonic reaction (energy release, i.e., free energy change is negative) catalyzed by a flavin-based cytoplasmic enzyme complex. This process is called flavin-based electron bifurcation and occurs in anaerobic bacteria and archaea [52,53]. Until now, six cytoplasmic electron-bifurcating complexes have been characterized [51].

1. The butyryl-CoA dehydrogenase/electron transfer flavoprotein complex (Bcd/EtfAB), which catalyzes the endergonic reduction of ferredoxin with NADH by coupling with the exergonic reduction of crotonyl-CoA to butyryl-CoA [54].

2. The MvhADG/HdrABC complex (heterodisulfide reductase (HdrABC)-[NiFe]-hydrogenase (MvhAGD) complex) of methanogenic archaea, which couples the exergonic reduction of CoM-S-S-CoB, the heterodisulfide of coenzyme M and enzyme B, by H_2 with the endergonic reduction of ferredoxin with H_2 [55]. Another complex belonging to this group that has been discovered is the FdhAB/HdrABC complex [56].
3. The HydABC(D) complex from anaerobic bacteria couples the reversible ferredoxin reduction with H_2 to NAD^+ reduction [57]; this complex is composed of three subunits (HydA, HydB, and HydC) in some anaerobic bacteria such as *Thermotoga maritima* and in various acetogens such as *C. ljungdahlii* and *M. thermoacetica* [37,58]. However, the complex in *A. woodii* contains four subunits [57].
4. The NfnAB complex from bacteria and archaea catalyzes the coupling reaction of reduction of ferredoxin with NADPH to the reduction of NAD^+. The enzyme NADH-dependent reduced ferredoxin: NADP oxidoreductase (Nfn) is a heterodimer with two subunits NfnA and NfnB and both subunits are iron-sulfur flavoproteins. In most clostridia the Nfn complex is present, with the exception of *Clostridium acetobutylicum*. The complex is in a fused state in *C. ljungdahlii* and may be in separate units such as in *M. thermoacetica* [51,59].
5. The caffeyl-CoA reductase/electron transfer flavoprotein complex of *A. woodii* drives the endergonic ferredoxin reduction with NADH using the energy gained from the exergonic NADH-dependent reduction of caffeyl-CoA [60].
6. The recently reported electron-bifurcating [FeFe]-hydrogenase, which is NADP dependent and forms a complex with FDH (fdhA/hytA-E) in *C. autoethanogenum* grown on CO. This complex is proposed to have the following function: reversible coupled reduction of ferredoxin and $NADP^+$ with H_2 or formate as reductant and reversible formation of formate from H_2 and CO_2 [59].

Membrane-associated integral protein complexes oxidize the reduced ferredoxin and generate a chemiosmotic gradient that is used for the generation of ATP. Two such complexes are multisubunit ferredoxin-NAD^+ oxidoreductase (Rnf) complexes coupling the oxidation of reduced ferredoxin through NAD^+ reduction with either Na^+ or H^+ translocation and an energy-converting hydrogenase (Ech) complex that catalyzes the oxidation of reduced ferredoxin, coupling the electron transfer to H^+ forming H_2 and H^+ translocation across the membrane [61,62]. Both the membrane-associated enzyme complexes lead to the formation of an ion gradient across the membrane that triggers the membrane-associated F_0F_1 or A_0A_1 ATP synthase complex resulting in the formation of ATP by phosphorylation of ADP [37]. The Rnf complex was first discovered in the purple nonsulfur bacterium *Rhodobacter capsulatus* and has six subunits, RnfA–E and RnfG. This complex is present in various aerobic and anaerobic bacteria. In case of the acetogens, *A. woodii* [63] and *C. ljungdahlii* [50]), the Rnf complex translocates Na^+ and H^+ ions, respectively, whereas the Ech complex is present in various anaerobic bacteria and is used for either H_2 formation or H_2 consumption.

In summary, energy conservation in acetogens results from the use of flavin-based electron-bifurcating enzyme complexes that couple exergonic reactions to endergonic reactions to reduce ferredoxin by using HydABC(D) complexes (Reaction 15.10). Part of the reduced ferredoxins are used up in the reductive WL pathway where one among the reactions that require Fd^{2-} is the conversion of CO_2 to enzyme-bound CO. Rest of the reduced ferredoxins get oxidized at the Rnf or Ech complex creating a chemiosmotic ion gradient across the cytoplasmic membrane as a driving force for the synthesis of ATP (Reactions 15.11 and 15.12).

$$Fd_{ox} + 2H_2 + NAD^+ \rightarrow Fd^{2-} + NADH + 3H^+ \quad (\Delta G^0 = -21\ kJ/mol) \tag{15.10}$$

$$Fd^{2-} + H^+ + NAD^+ \rightarrow Fd_{ox} + NADH + \Delta\mu H^+/Na^+ (\text{Rnf complex}) \tag{15.11}$$

$$Fd^{2-} + 2H^+ \rightarrow Fd_{ox} + H_2 + \Delta\mu H^+/Na^+ (\text{Ech complex}) \tag{15.12}$$

Energy conservation in the acetogen *C. ljungdahlii* was described to involve more enzyme complexes (Fig. 15.2). Apart from the electron-bifurcating [FeFe] hydrogenase that functions to reduce ferredoxin, it contains another electron-bifurcating [FeFe] hydrogenase that forms a complex with FDH using H_2, reduced ferredoxin, or NADPH as electron donors [64]. The latter type of enzyme is also present in *C. autoethanogenum*, as mentioned earlier. Other FDHs form a complex with small ferredoxin-like proteins, which might use reduced ferredoxin as the source of electrons [37]. However, *C. ljungdahlii* also contains the electron-bifurcating Nfn complex that converts 2 moles of NADP to 2 moles of NADPH using 1 mole of NADH and 1 mol of reduced ferredoxin [64]. In addition to these two enzyme complexes, *C. ljungdahlii* contains a dimeric enzyme (MetF and MetV) that helps reduce methylene-H_4folate to methyl-H_4folate. Nevertheless, this bacterium has an Rnf complex that allows to

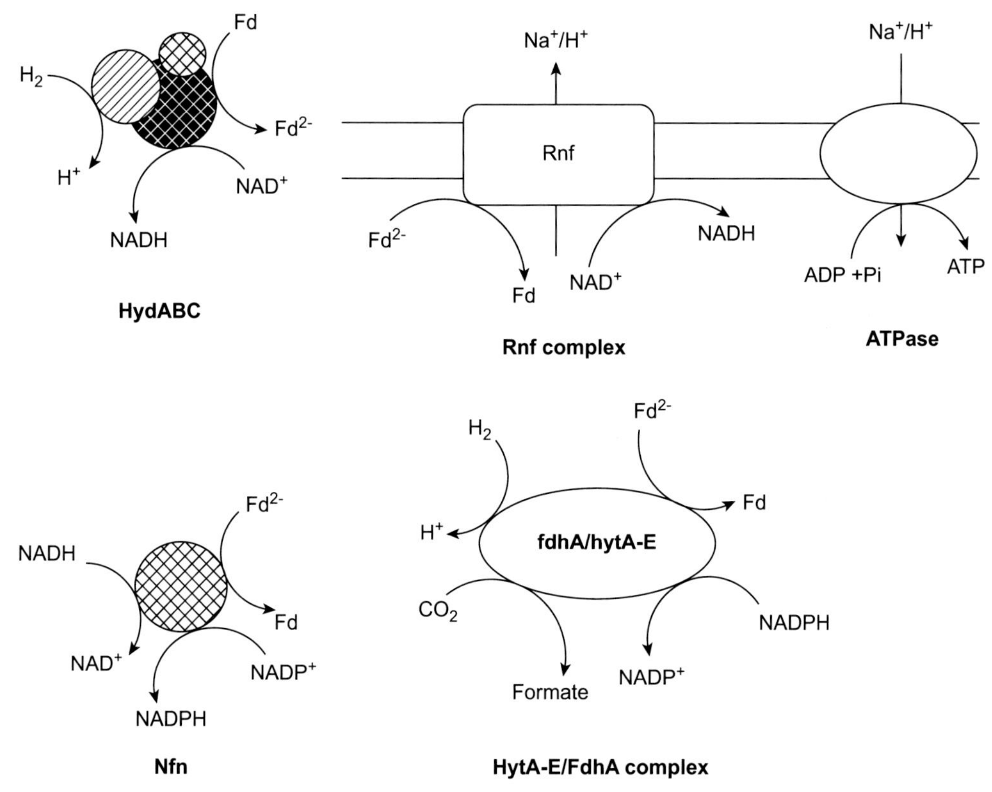

FIG. 15.2 Enzyme complexes involved in the energy conservation in *Clostridium*.

transport H^+ ions across the cytoplasmic membrane by coupling the electron transfer from reduced ferredoxin to NAD^+. This ion gradient helps synthetize ATP using ATP synthase.

METABOLIC PHASES IN *CLOSTRIDIA* PRODUCING ALCOHOLS THROUGH GAS FERMENTATION

The metabolic pathway of solventogenic acetogens is characterized by two distinctive phases: acetogenesis and solventogenesis [65,66]. During C1 fermentation or organic carbon assimilation, acetogens utilize the carbon present in the fermentation medium for their growth and convert it into carboxylic acids such as acetic acid with the generation of ATP. This acetogenic phase occurs when the bacteria are maintained at their optimal growth pH and temperature, and with ample supply of nutrients. However, when the bacteria enter the stationary growth phase, solventogenesis will start, which results in the assimilation of the carboxylic acids produced during the acetogenic phase with their conversion to the respective alcohols. This is the case, for example, of acetic acid conversion to ethanol. The solventogenic phase is characterized by increased concentrations of alcohols and occurs when the bacteria face harsh environmental conditions such as a pH lower than the optimal growth pH, limited nutrient supply, presence of sufficient amount of reducing agents, and presence of specific trace metals in the medium.

PH-INDUCED METABOLIC SHIFT FROM ACIDOGENESIS TO SOLVENTOGENESIS

pH is a key factor that influences the distribution of metabolites during CO/syngas fermentation. Its effect has been observed from various studies and is well documented [18,67]. Most strains of *Clostridium* have a near optimal growth pH ranging between 5.5 and 7.5 [15]. Studies performed by maintaining the pH in that range or at near-neutral pH reported facilitated growth with simultaneous generation of acids. However, reducing the pH below the optimal growth value, for example, near pH 5, will result in the accumulation of alcohols in the bioreactor [68,69]. The influence of pH is due to the stress environment created that disrupts the cellular acid-base homeostasis in the following ways:

- *Acid shock*: A sudden change in the culture pH results in the loss of self-buffering and proton-exchange potential of the cells.
- *Accumulation of membrane-permeant acids:* Low-molecular-weight carboxylic acids, with pK_a values ranging between 3 and 5, produced during fermentation at optimal growth pH get protonated when the fermentation pH is low or near acidic. They can easily pass across the cell membrane. However, once they are inside the cytoplasm, these metabolites get deprotonated due to the higher cytoplasmic pH. This leads to acidification of the cytoplasm, damaging the cell homeostasis.

The stress condition that leads to loss in homeostasis is overcome by bacteria by consuming and converting these membrane-permeant acids into their corresponding alcohols, thus allowing bacteria to survive under this low acidic pH conditions [70]. In addition, the fermentation pH affects the charge of the cell membrane, which interferes with the nutrient absorption of the cells and the enzyme activities. Although a low pH stimulates alcohol production in the CO/syngas fermentation, it will reduce the overall productivity, as fermentation at a pH lower than the optimal growth pH is unfavorable for the cell growth [71].

PROCESS PARAMETERS AND STRATEGIES FOR ENHANCING ALCOHOL PRODUCTION

Trace Metals

Production of ethanol by syngas fermentation is one of the most studied processes and it has reached the pre-commercial stage. Certain specific trace metals are present at the active site of the enzymes that are involved in the pathway that leads to the production of alcohols as well as the conversion of acids to their corresponding alcohols [6,72]. As explained earlier, the production of alcohols is a two-stage process characterized by the conversion of the gaseous substrates into acids in the first stage, followed by the conversion of those acids into alcohols. The presence of specific trace metals in the fermentation medium may greatly improve the production of alcohols during the solventogenic phase. One such key metal is tungsten, which is a cofactor of the enzyme AFOR [59]. This enzyme is involved in the catalysis of the conversion of acids such as acetic acid, produced during the acetogenic phase, into acetaldehyde with electrons from the reduced ferredoxin (Fd^{2-}). An increased accumulation of ethanol and a high conversion rate of acetic acid into ethanol were achieved in *C. autoethanogenum* when tungsten was added to the fermentation medium [6]. However, in the absence of tungsten, the production of ethanol decreased and only limited conversion of acetic acid to ethanol was observed [71]. Another metalloenzyme is alcohol dehydrogenase (ADHE), which is involved in the reduction of acetyl-CoA to ethanol. It was observed that the absence of Fe^{2+} in the medium resulted in a reduction in the activity of ADHE, thereby resulting in a decrease in ethanol production [72].

Yeast Extract and Its Concentration

Yeast extract is quite commonly used in bioprocesses and fermentation studies. It is a complex nutrient and a source of nitrogen, minerals, and vitamins, among others. Thus, it is a growth stimulant and has often been used in small amounts for syngas fermentation in order to promote solventogenesis, thereby improving ethanol production. The presence of vitamin B_{12} in yeast extract plays a significant role in acetyl-CoA generation from CO_2. The enzyme that catalyzes the transfer of a methyl group of methyl-H_4folate to the cobalt center of CFeSP is a vitamin B_{12}-dependent enzyme. Insufficient amount of vitamin B_{12} will lead to a reduction in methyl transfer, thereby resulting in NAD(P)H accumulation and subsequent improvement in ethanol production [73,74]. From the several syngas/CO fermentation studies performed so far, it has been well documented that reducing the concentration of yeast extract greatly enhances ethanol production [75,76]. However, in some laboratory studies, at least 0.01% yeast extract was present in the medium in order to maintain cell growth [77], although cheaper alternative sources of nutrients, with a similar effect, might be used as well [78,79]. In a study performed to understand the influence of a mixture of vitamins on CO fermentation, it was found that the addition of vitamins had no effect on ethanol production, at least in the specific case of that study done with a *C. autoethanogenum* strain [6]. However, those studies were performed in the presence of 0.5 g/L yeast extract. Although the addition of a specific vitamin solution did not show any effect on ethanol production, the presence of yeast extract in the medium may have provided a sufficient amount of vitamins, such as vitamin B_{12}, necessary for bacterial growth and metabolite production.

Sequential Acetogenic and Solventogenic Process Using Two Bioreactors in Series

Lowering the initial fermentation pH as a way to induce solventogenesis in syngas fermentation will result in a long lag phase with limited bacterial growth and will then also limit the overall productivity of metabolites.

For continuous syngas fermentation, a two-stage system with two bioreactors in series can be recommended. In such a two-stage process, the first reactor (acetogenic reactor) is focused on providing optimal growth conditions that promote acetogenesis, by maintaining an optimal growth pH and temperature and by supplying a nutrient-rich medium. This results in cell growth and the production of acids. In the second reactor (solventogenic reactor), harsh conditions, such as a pH lower than the optimal growth pH and a limited nutrient supply, will not support growth anymore but will stimulate solventogenesis. Besides, the presence of an abundant amount of reducing agent at the solventogenic stage favors the production of more reduced metabolites such as ethanol [68,80,81]. A combination of two stirred tank reactors (STRs) or a hybrid system with an STR and a bubble column has been tested for syngas fermentation [68]. A low pH in the solventogenic reactor limits growth; however, cell washout can be overcome by continuously feeding the cell biomass from the acetogenic reactor as well as by connecting a cell recycling unit to the solventogenic reactor.

Cyclic pH Shifts Using a Single Bioreactor

Instead of using two reactors in series to facilitate acetogenesis and solventogenesis separately, another strategy can be applied based on cyclically shifting the pH from a high value to a low value [82]. This allows the bacteria to generate enough cells, along with acids at the higher pH, and once the bacteria reach the stationary phase, the high pH can be decreased to a lower pH value and maintained constant until all the acids are converted to alcohols. Maintaining the pH constant at the high value and later shifting to a lower pH is necessary, respectively, for a rapid cell growth and acid production and later for a fast conversion of the produced acids into alcohols. When the complete conversion of acids has taken place in the first cycle, the pH can be shifted back to a higher value to promote a new cycle of acid production followed by its conversion to additional alcohol when later lowering the pH again. This cyclic pH shift approach thus improves the overall alcohol production by using a single bioreactor, which in turn reduces the investment costs of the process. One rational explanation for the cyclic acid buildup and subsequent conversion to alcohol during the cyclic pH shifts could be the oxidation and reduction of ferredoxin (Fd), an iron-sulfur protein that mediates electron transfer. Conversion of acetic acid to acetaldehyde is mediated by the electrons provided by the reduced ferredoxin (Fd^{2-}) [14]. This conversion is catalyzed by the enzyme AFOR resulting in the accumulation of ferredoxin (Fd), which can be useful later on in the higher pH stage for further assimilation of CO to CO_2 catalyzed by CODH. The CODH oxidizes CO with simultaneous reduction of ferredoxin (Fd). The reduced ferredoxin (Fd^{2-}) can be useful later on for the conversion of acid(s) into alcohol(s) during the lower pH stage. An experiment was performed using *C. autoethanogenum* and applying the cyclic pH shift strategy to study CO bioconversion to ethanol [82]. In that study, an ethanol concentration of 7143 mg/L was achieved by applying three cyclic pH shifts between the highest pH of 5.75 and the lowest pH of 4.75, with occasional partial medium replenishment [82].

Gas-Adsorbing Particles

The enhancement of GL mass transfer using catalytic gas-adsorbing particles has been demonstrated by some researchers [83,84]. The most important factor that improves the GL mass transfer is the attachment of these particles at the GL interface, which is determined by several parameters such as surface tension, viscosity, and density of the liquid; size, lyophobicity, and other characteristics of the catalyst particle; and process parameters. It was demonstrated that GL mass transfer enhancement through the addition of such particles is due to four different mechanisms in a stirred slurry reactor: (1) boundary layer mixing, (2) shuttling (refers to the transport of particles from the GL interface to the bulk liquid), (3) coalescence inhibition, and (4) boundary layer reaction or grazing effect [85]. Shuttling prevails when the particle size is smaller or equal to the GL interface layer, which is about 5–25 μm. However, one reported study confirmed that the shuttling effect is not significant in enhancing the GL mass transfer and its enhancement activity decreases when increasing the stirring rate [85]. The adhesion rate of smaller particles to the GL interface is higher than that for larger particles. Also the adhesion rate of smaller particles is lower than that for larger particles, thereby improving the average residence time at the GL interface. In this respect, nanoparticles have been tested in enhancing CO mass transfer [83]. The enhancement in mass transfer of CO appeared to be greatly influenced by the surface properties (hydrophobic or hydrophilic) and the size of the nanoparticles. It was found that mesoporous silica material (MCM41) nanoparticles with a size of ~250 nm and having mercaptan groups attached showed ~1.9-times enhancement in CO mass transfer when compared with those materials without any nanoparticle addition. Kim et al. [86] studied the effect of various nanoparticles along with changing their functional group from hydrophilic to hydrophobic group. Studies on syngas fermentation using 0.3% wt concentration of methyl-functionalized silica nanoparticles showed improvement in cell, ethanol, and acetic acid production by *C. ljungdahlii* by 34.5%, 166.1%, and 29.1%, respectively, when compared with studies without any nanoparticle addition. Also, dissolved concentrations of CO, CO_2, and H_2 were increased by 272.9%, 200.2%, and 156.1%, respectively, by using methyl-functionalized silica nanoparticles. The feasibility of reusability of nanoparticles functionalized with methyl group (cobalt ferrite-silica nanoparticles) was also tested [87]. Dissolved concentrations of syngas when using these nanoparticles remained the same even after reusing five times. In addition, in syngas fermentation studies performed with *C. ljungdahlii*, the production improved by 227.6%, 213.5%, and 59.6%, respectively, for cell mass, ethanol, and acetic acid accumulation [87].

BIOREACTOR CONFIGURATION

Mass transfer of less soluble gaseous substrates such as CO and H_2 to the liquid medium is a major bottleneck in the syngas fermentation process. An optimal bioreactor system should not only offer an environment that can properly control aspects such as the production pH and temperature but also facilitate the increase in mass transfer of these sparingly soluble gas substrates in the medium without inhibiting bacterial growth and thereby improving the overall efficiency of ethanol production [88,89]. The STR is the most extensively studied reactor system at research level that is operated in batch, semibatch, or continuous mode with or without cell recycling or by coupling it to other bioreactor types such as a bubble column in some two-stage studies [68,74,80]. Improvement of mass transfer in an STR is achieved by increasing the bubble breakup with the help of baffles, by increasing the impeller speed, and/or by using a microbubble sparger [90,91]. Increasing the impeller speed for gas breakup, thereby increasing the interfacial area for mass transfer, is not economically feasible at large scale, as the required power input is proportional to the cube of the impeller speed. In addition, foam formation and the shear stress created due to high mixing are harmful to microbial cells. Mass transfer can be increased in an STR by increasing the gas flow rate, but this will reduce gas utilization during syngas fermentation.

Alternative bioreactors tested for syngas fermentation are column reactors such as the bubble column reactor (BCR), the gas-lift reactor (GLR), and bioreactors with packing materials such as the trickle bed reactor (TBR) or the monolith biofilm reactor (MBR), or reactors with special membrane modules (hollow fiber membrane [HFM] reactor) (Table 15.1) [92–95]. These reactors do not have any mechanically moving part, thus lowering the power input. In BCR and GLR, the gas dispersion by using microbubble diffusors greatly enhances the mass transfer; however, problems such as cell washout and limited cell concentration are drawbacks of these systems. In the case of biofilm-attached reactors such as in MBR, HFM reactor, and TBR, the mass transfer can be significantly improved by increasing the liquid flow rate. After long-term operation, biofouling problems may occur, which is one of the drawbacks of these systems. However, this is less significant in an MBR due to the large size of the monolithic channels. Syngas fermentation using *C. carboxidivorans* P7 was compared in BCR and MBR at a syngas flow rate of 300 mL/min, liquid flow rate of 500 mL/min, and a dilution rate of 0.48 day^{-1}. A 57% enhancement of CO consumption rate (mmol/L/day), as well as an increase of 27% in H_2 consumption rate (mmol/L/day), of 53% in ethanol productivity (g/L/day), and of 17% in ethanol/acetate molar ratio, was observed in MBR when compared with BCR [93]. The same research group carried out studies in HFM at various liquid and gas flow rates and obtained a maximum ethanol concentration of 23.93 g/L with an ethanol-to-acetic acid ratio of 4.79 at a syngas flow rate of 200 mL/min, a liquid recirculation rate of 200 mL/min, and a dilution rate of 0.12 day^{-1} [95]. Devarapalli et al.[94] conducted syngas fermentation in a TBR using 6-mm soda lime glass beads as packing material. They obtained 5.7 and 12.3 g/L, respectively, of ethanol and acetic acid at a cocurrent liquid flow rate of 200 mL/min and a syngas feed of 4.6 sccm (standard cubic centimeters per minute). Production of 98% and 83% more ethanol and acetic acid was reported, respectively, during cocurrent over countercurrent flow. This was attributed to the improved GL contact and the possibility to avoid phenomena such as gas bypassing and flooding during cocurrent operation [94].

CLOSTRIDIUM AUTOETHANOGENUM, A MODEL ACETOGEN FOR ETHANOL PRODUCTION FROM SYNGAS OR CO

C. autoethanogenum is a gram-positive anaerobe, first isolated from rabbit feces using CO as the sole carbon and energy source [96]. It is a rod-shaped bacterium with a size of 0.5 × 3.2 μm. After a long period of incubation, cells cease dividing and thus change to continuous chains of encapsulated filaments of size 0.6 × 42.5 μm. It has the ability to grow chemoautotrophically by utilizing CO as the sole carbon and energy source and converting the gaseous

TABLE 15.1
Various Bioreactor Configuration Studies Performed for Syngas Fermentation to Produce Ethanol.

Reactor Configuration	Syngas Composition (%)	Microorganism	Ethanol Production (g/L)	Ethanol/ Acetic Acid	References
HFM-BR	$CO = 20, H_2 = 5$ $CO_2 = 15, N_2 = 60$	*Clostridium carboxidivorans* P7	23.93	4.79	[95]
MBR	$CO = 20, H_2 = 5$ $CO_2 = 15, N_2 = 60$	*C. carboxidivorans* P7	4.89	2.1	[93]
h-RPB	$CO = 20, H_2 = 5$ $CO_2 = 15, N_2 = 60$	*C. carboxidivorans* P7	7.0	1.6	[97]
CSTR	$CO = 55, H_2 = 20$ $CO_2 = 10, Ar = 15$	*Clostridium ljungdahlii*	6.5	1.53	[98]
STR	$CO = 100$	*Clostridium autoethanogenum*	0.867	NA	[6]
TBR	$CO = 38, H_2 = 28.5$ $CO_2 = 28.5, N_2 = 5$	*Clostridium ragsdalei* P11	5.7	0.607	[94]
STR	$CO = 20, H_2 = 5$ $CO_2 = 15, N_2 = 60$	*C. ragsdalei* P11	25.26	6.8	[99]
CSTR-BC	$CO = 60, H_2 = 35$ $CO_2 = 5$	*C. ljungdahlii* ERI-2	19.73	3	[68]
CSTR (cell recycle)	$CO = 55, H_2 = 20$ $CO_2 = 10, Ar = 15$	*C. ljungdahlii*	48	21	[77]
ICR	$CO = 13, H_2 = 14$ $CO_2 = 5, N_2 = 68$	*C. ljungdahlii* ERI-2	2.74	0.64	[77]

CSTR, continuous stirred tank reactor with continuous liquid and gas flow; *CSTR-BC*, CSTR bubble column; *h-RPB*, horizontally oriented rotating packed bed; *HFM-BR*, hollow fiber membrane biofilm reactor; *ICR*, immobilized cell reactor; *MBR*, monolith biofilm reactor; *STR*, stirred tank reactor with liquid batch; *TBR*, trickle bed reactor.
Ethanol/acetic acid ratio is represented in molar concentrations.

substrate to metabolites such as acetic acid, ethanol, 2,3-butanediol, and lactic acid [17]. In addition, it can grow heterotrophically by utilizing various organic carbon sources such as pyruvate, xylose, arabinose, fructose, rhamnose, and L-glutamate [96]. This bacterium has the optimum pH for growth between 5.8 and 6.0. The impact of various parameters, for example, pH, yeast extract, and trace metal concentrations has been examined in bioreactors and bottles to understand acetogenesis and solventogenesis in *C. autoethanogenum* [6,75]. As observed in most CO-utilizing clostridia, this strain produces acids (acetic acid) at its optimal growth pH, along with cell growth, and when the pH is decreased to a lower value, alcohol (ethanol) production will increase. Yeast extract has also shown to have a negative impact on ethanol production. In a study performed in serum bottles at two different initial pH values and yeast extract concentrations, a 200% enhancement in maximum ethanol production was observed when the initial pH and yeast extract concentrations were decreased from 5.75 to 4.75 and 1.6 to 0.6 g/L, respectively, when the strain was grown with CO as the sole carbon and energy source [75]. In a fermentation study performed in gas sampling bags, the maximum ethanol and cell concentrations were obtained with 1 g/L yeast extract, which was the highest yeast extract concentration tested. However, the molar ethanol/acetic acid ratio obtained was higher at 0.6 and 0.8 g/L of yeast extract than at 1 g/L of yeast extract [76]. Enhancement in ethanol production, by about 128%, was also observed with the addition of tungsten in the medium compared with a medium without the addition of tungsten, during CO fermentation. A complete conversion of the produced acetic acid was achieved when the fermentation pH was shifted from growth pH (higher) to production pH (lower) in the presence of tungsten [6]. The influence of tungsten on ethanol production can be attributed to its positive effect on enhancing the activity of the enzyme AFOR that catalyzes the conversion of the produced acetic acid to acetaldehyde.

CONCLUSIONS AND PERSPECTIVES

Extensive reactor studies have been carried out with wild-type bacterial strains for syngas fermentation, leading to the production of a variety of platform chemicals and biofuels. However, certain genes of these strains were inactivated to direct the carbon flow toward ethanol production [47]. Studies with these mutant strains are still necessary to be tested in a reactor for long-term syngas fermentation processes. As described in the earlier sections, acetogens grown on syngas face challenges such as a low achievable maximum biomass concentration or limitations in substrate availability due to the low GL mass transfer of poorly water-soluble syngas compounds. Different possible strategies to improve GL mass transfer have been put forward over the past few years, including the addition of gas-adsorbing particles such as nanoparticles or using various impeller designs and schemes. Although most of them were tested for their performance under abiotic conditions, research is still required to verify their applicability in a real syngas fermentation process and to evaluate their cost-efficiency, which may be questionable. The application of pressurized bioreactors for syngas fermentation would to some extent alleviate the barrier that slows down the transfer of gaseous substrates to the biocatalyst. Pressurized systems will also increase operating costs. Besides the commonly used STR at laboratory scale, some other different bioreactor configurations have been tested, often aimed at trying to improve mass transfer and thereby the productivity of the process. Those reactors have often only been tested in one or a few studies and thus further studies should be warranted. For the successful implementation of syngas fermentation processes at the commercial scale, there are still concerns about the bacterial performance that can be achieved with syngas generated from gasification of lignocellulosic materials or by using real stack gas from factories. Impurities present in these sources might either inhibit or improve the activity of key enzymes involved in the syngas fermentation pathway. In addition, the utilization of different types of wastes, such as food waste, for gasification and subsequent fermentation still needs to be explored further.

ACKNOWLEDGMENTS

The authors thank the Spanish Ministry of Economy and Competitiveness (MINECO) and European FEDER funds for their financial support of the research on biofuel production and on the bioconversion of pollutants from waste gases, wastewaters, and solid waste to bioproducts (project CTQ2017-88292-R). HNA also thanks the Xunta de Galicia (Spain) for his postdoctoral fellowship (ED481B 2016/195-0).

REFERENCES

[1] Chen H, Qiu W. Key technologies for bioethanol production from lignocellulose. Biotechnology Advances 2010; 28(5):556–62.

[2] Paulova L, Patakova P, Branska B, Rychtera M, Melzoch K. Lignocellulosic ethanol: technology design and its impact on process efficiency. Biotechnology Advances 2015; 33(6):1091–107.

[3] Kennes D, Abubackar HN, Diaz M, Veiga MC, Kennes C. Bioethanol production from biomass: carbohydrate vs syngas fermentation. Journal of Chemical Technology and Biotechnology 2016;91(2).

[4] Wu M, Liu H, Guo J, Yang C. Enhanced enzymatic hydrolysis of wheat straw by two-step pretreatment combining alkalization and adsorption. Applied Microbiology and Biotechnology 2018. https://doi.org/10.1007/s00253-018-9335-4.

[5] Bengelsdorf FR, et al. Bacterial anaerobic synthesis gas (syngas) and CO_2+H_2 fermentation. Advances in Applied Microbiology 2018;103:143–221.

[6] Abubackar HN, Veiga MC, Kennes C. Carbon monoxide fermentation to ethanol by *Clostridium autoethanogenum* in a bioreactor with no accumulation of acetic acid. Bioresource Technology 2015;186.

[7] Sikarwar VS, et al. An overview of advances in biomass gasification. Energy & Environmental Science 2016; 9(10):2939–77.

[8] Ramachandriya KD, et al. Critical factors affecting the integration of biomass gasification and syngas fermentation technology. AIMS Bioengineering 2016;3(2): 188–210.

[9] Wainaina S, Horváth IS, Taherzadeh MJ. Biochemicals from food waste and recalcitrant biomass via syngas fermentation: a review. Bioresource Technology 2018; 248:113–21.

[10] Molitor B, et al. Carbon recovery by fermentation of CO-rich off gases - turning steel mills into biorefineries. Bioresource Technology 2016;215:386–96.

[11] van Groenestijn JW, Abubackar HN, Veiga MC, Kennes C. Bioethanol. In: Kennes C, Veiga MC, editors. Air pollution prevention and control: bioreactors and bioenergy. Chichester, UK: John Wiley & Sons, Ltd; 2013. p. 431–63.

[12] Phillips JR, Huhnke RL, Atiyeh HK. Syngas fermentation: a microbial conversion process of gaseous substrates to various products. Fermentation 2017;3(2):28.

[13] Fernández-Naveira Á, Veiga MC, Kennes C. H-B-E (hexanol-butanol-ethanol) fermentation for the production of higher alcohols from syngas/waste gas. Journal of Chemical Technology and Biotechnology 2017; 92(4):712–31.
[14] Bengelsdorf FR, Straub M, Dürre P. Bacterial synthesis gas (syngas) fermentation. Environmental Technology 2013; 34(13–14):1639–51.
[15] Abubackar HN, Veiga MC, Kennes C. Biological conversion of carbon monoxide: rich syngas or waste gases to bioethanol. Biofuels, Bioproducts & Biorefining 2011;5(1).
[16] Bengelsdorf FR, et al. Industrial acetogenic biocatalysts: a comparative metabolic and genomic analysis. Frontiers in Microbiology 2016;7:1–15.
[17] Köpke M, et al. 2,3-Butanediol production by acetogenic bacteria, an alternative route to chemical synthesis, using industrial waste gas. Applied and Environmental Microbiology 2011;77(15):5467–75.
[18] Fernández-Naveira Á, Veiga MC, Kennes C. Effect of pH control on the anaerobic H-B-E fermentation of syngas in bioreactors. Journal of Chemical Technology and Biotechnology 2017;92(6):1178–85.
[19] Fernández-Naveira Á, Abubackar HN, Veiga MC, Kennes C. Production of chemicals from C1 gases (CO, CO_2) by *Clostridium carboxidivorans*. World Journal of Microbiology and Biotechnology 2017;33(3).
[20] Spirito CM, Richter H, Rabaey K, Stams AJM, Angenent LT. Chain elongation in anaerobic reactor microbiomes to recover resources from waste. Current Opinion in Biotechnology 2014;27:115–22.
[21] Gildemyn S, Molitor B, Usack JG, Nguyen M, Rabaey K, Angenent LT. Upgrading syngas fermentation effluent using *Clostridium kluyveri* in a continuous fermentation. Biotechnology for Biofuels 2017;10(1):1–15.
[22] Diender M, Stams AJM, Sousa DZ. Production of medium-chain fatty acids and higher alcohols by a synthetic co-culture grown on carbon monoxide or syngas. Biotechnology for Biofuels 2016;9(1):1–11.
[23] Richter H, Molitor B, Diender M, Sousa DZ, Angenent LT. A narrow pH range supports butanol, hexanol, and octanol production from syngas in a continuous co-culture of *Clostridium ljungdahlii* and *Clostridium kluyveri* with in-line product extraction. Frontiers in Microbiology 2016; 7(1773).
[24] Oswald F, et al. Sequential mixed cultures: from syngas to malic acid. Frontiers in Microbiology 2016;7(891).
[25] Hu P, et al. Integrated bioprocess for conversion of gaseous substrates to liquids. Proceedings of the National Academy of Sciences 2016;113(14):3773–8.
[26] Lagoa-Costa B, Abubackar HN, Fernández-Romasanta M, Kennes C, Veiga MC. Integrated bioconversion of syngas into bioethanol and biopolymers. Bioresource Technology 2017;239:244–9.
[27] Daglioglu ST, Karabey B, Ozdemir G, Azbar N. CO_2 utilization via a novel anaerobic bioprocess configuration with simulated gas mixture and real stack gas samples. Environmental Technology 2017;3330: 1–7.
[28] Zabranska J, Pokorna D. Bioconversion of carbon dioxide to methane using hydrogen and hydrogenotrophic methanogens. Biotechnology Advances 2018;36(3):707–20.
[29] Guneratnam AJ, et al. Study of the performance of a thermophilic biological methanation system. Bioresource Technology 2017;225:308–15.
[30] Ail SS, Dasappa S. Biomass to liquid transportation fuel via Fischer Tropsch synthesis - technology review and current scenario. Renewable and Sustainable Energy Reviews 2016;58:267–86.
[31] Ahmed A, Cateni BG, Huhnke RL, Lewis RS. Effects of biomass-generated producer gas constituents on cell growth, product distribution and hydrogenase activity of *Clostridium carboxidivorans* P7T. Biomass and Bioenergy 2006;30(7):665–72.
[32] Hu P, Jacobsen LT, Horton JG, Lewis RS. Sulfide assessment in bioreactors with gas replacement. Biochemical Engineering Journal 2010;49(3):429–34.
[33] Orgill JJ, Atiyeh HK, Devarapalli M, Phillips JR, Lewis RS, Huhnke RL. A comparison of mass transfer coefficients between trickle-bed, Hollow fiber membrane and stirred tank reactors. Bioresource Technology 2013;133:340–6.
[34] Ungerman AJ, Heindel TJ. Carbon monoxide mass transfer for syngas fermentation in a stirred tank reactor with dual impeller configurations. Biotechnology Progress 2007;23(3):613–20.
[35] Ragsdale SW, Pierce E. Acetogenesis and the Wood-Ljungdahl pathway of CO_2 fixation. Biochimica et Biophysica Acta (BBA) – Proteins & Proteomics 2008; 1784(12):1873–98.
[36] Ragsdale SW. The Eastern and Western branches of the Wood/Ljundahl pathway:How the east and west were won. BioFactors 1997;6:3–11.
[37] Schuchmann K, Müller V. Autotrophy at the thermodynamic limit of life: a model for energy conservation in acetogenic bacteria. Nature Reviews Microbiology 2014; 12(12):809–21.
[38] McGuire JJ, Rabinowitz JC. Studies on the mechanism of formyltetrahydrofolate synthetase. The *Peptococcus aerogenes* enzyme. Journal of Biological Chemistry 1978; 253:1079–85.
[39] Poe M, Benkovic SJ. 5-formyl- and 10-formyl-5,6,7,8-tetrahydrofolate. Conformation of the tetrahydropyrazine ring and formyl group in solution. Biochemistry 1980;19:4576–82.
[40] Moore R, Brien EO. Purification and characterization of nicotinamide methylenetetrahydrofolate dehydrogenase from *Clostridium formicoaceticum* from adenine communication. Science 1974;249(16):5250–3.
[41] Clark JE, Ljungdahl LG. Purification and properties of 5,10-methylenetetrahydrofolate reductase, an iron-sulfur flavoprotein from *Clostridium formicoaceticum*. Journal of Biological Chemistry 1984;259(17):10845–9.
[42] Svetlitchnaia T, Svetlitchnyi V, Meyer O, Dobbek H. Structural insights into methyl transfer reactions of a corrinoid iron-sulfur protein involved in acetyl-CoA synthesis. Proceedings of the National Academy of Sciences of the United States of America 2006;103(39):14331–6.

[43] Doukov T, Seravalli J, Stezowski JJ, Ragsdale SW. Crystal structure of a methyltetrahydrofolate- and corrinoid-dependent methyltransferase. Structure 2000;8(8):817–30.

[44] Ragsdale SW, Lindahl PA, Münck E. Mössbauer, EPR, and optical studies of the corrinoid/iron-sulfur protein involved in the synthesis of acetyl coenzyme A by *Clostridium thermoaceticum*. Journal of Biological Chemistry 1987;262(29):14289–97.

[45] Doukov TI, Iverson TM, Seravalli J, Ragsdale SW, Drennan CL. A Ni-Fe-Cu center in a bifunctional carbon monoxide dehydrogenase/acetyl-CoA synthase. Science 2002;298(5593):567–72.

[46] Norman ROJ, Millat T, Winzer K, Minton NP, Hodgman C. Progress towards platform chemical production using *Clostridium autoethanogenum*. Biochemical Society transactions 2018;46(3):523–35.

[47] Liew F, Henstra AM, Köpke M, Winzer K, Simpson SD, Minton NP. Metabolic engineering of *Clostridium autoethanogenum* for selective alcohol production. Metabolic Engineering 2017;40:104–14.

[48] Müller V. Energy conservation in acetogenic bacteria. Applied and Environmental Microbiology 2003;69(11): 6345–53.

[49] Heise R, Müller V, Gottschalk G. Sodium dependence of acetate formation by the acetogenic bacterium *Acetobacterium woodii*. Journal of Bacteriology 1989;171(10):5473–8.

[50] Tremblay PL, Zhang T, Dar SA, Leang C, Lovley DR. The Rnf complex of *Clostridium ljungdahlii* is a proton-translocating ferredoxin:NAD+ oxidoreductase essential for autotrophic growth. mBio 2013;4:e00406–12.

[51] Buckel W, Thauer RK. Energy conservation via electron bifurcating ferredoxin reduction and proton/Na+ translocating ferredoxin oxidation. Biochim Biophys Acta – Bioenerg 2013;1827(2):94–113.

[52] Buckel W, Thauer RK. Flavin-based electron bifurcation, a new mechanism of biological energy coupling. Chemistry Review 2018;118(7):3862–86.

[53] Herrmann G, Jayamani E, Mai G, Buckel W. Energy conservation via electron-transferring flavoprotein in anaerobic bacteria. Journal of Bacteriology 2008;190(3): 784–91.

[54] Li F, Hinderberger J, Seedorf H, Zhang J, Buckel W, Thauer RK. Coupled ferredoxin and crotonyl coenzyme A (CoA) reduction with NADH catalyzed by the butyryl-CoA dehydrogenase/Etf complex from *Clostridium kluyveri*. Journal of bacteriology 2008;190(3):843–50.

[55] Kaster A-K, Moll J, Parey K, Thauer RK. Coupling of ferredoxin and heterodisulfide reduction via electron bifurcation in hydrogenotrophic methanogenic archaea. Proceedings of the National Academy of Sciences 2011; 108(7):2981–6.

[56] Costa KC, Yoon SH, Pan M, Burn JA, Baliga NS, Leigh JA. Effects of H_2 and formate on growth yield and regulation of methanogenesis in *Methanococcus maripaludis*. Journal of Bacteriology 2013;195(7):1456–62.

[57] Schuchmann K, Müller V. A bacterial electron-bifurcating hydrogenase. Journal of Biological Chemistry 2012; 287(37):31165–71.

[58] Schut GJ, Adams MWW. The iron-hydrogenase of *Thermotoga maritima* utilizes ferredoxin and NADH synergistically: a new perspective on anaerobic hydrogen production. Journal of Bacteriology 2009;191(13):4451–7.

[59] Wang S, Huang H, Kahnt HH, Mueller AP, Köpke M, Thauer RK. NADP-Specific electron-bifurcating [FeFe]-hydrogenase in a functional complex with formate dehydrogenase in *Clostridium autoethanogenum* grown on CO. Journal of Bacteriology 2013;195(19):4373–86.

[60] Bertsch J, Parthasarathy A, Buckel W, Müller V. An electron-bifurcating caffeyl-CoA reductase. Journal of Biological Chemistry 2013;288(16):11304–11.

[61] Hedderich R, Forzi L. Energy-converting [NiFe] hydrogenases: more than just H_2 activation. Journal of Molecular Microbiology and Biotechnology 2006;10(2–4):92–104.

[62] Schmehl M, et al. Identification of a new class of nitrogen fixation genes in *Rhodobacter capsulatus*: a putative membrane complex involved in electron transport to nitrogenase. Molecular and General Genetics 1993; 241(5–6):602–15.

[63] Müller V, Imkamp F, Biegel E, Schmidt S, Dilling S. Discovery of a ferredoxin:NAD+-oxidoreductase (Rnf) in *Acetobacterium woodii:* a novel potential coupling site in acetogens. Annals of the New York Academy of Sciences 2008;1125:137–46.

[64] Nagarajan H, et al. Characterizing acetogenic metabolism using a genome-scale metabolic reconstruction of *Clostridium ljungdahlii*. Microbial Cell Factories 2013;12(1): 1–13.

[65] Richter H, Molitor B, Wei H, Chen W, Aristilde L, Angenent LT. Ethanol production in syngas-fermenting: *Clostridium ljungdahlii* is controlled by thermodynamics rather than by enzyme expression. Energy & Environmental Science 2016;9(7):2392–9.

[66] Kumar M, Gayen K, Saini S. Role of extracellular cues to trigger the metabolic phase shifting from acidogenesis to solventogenesis in *Clostridium acetobutylicum*. Bioresource Technology 2013;138:55–62.

[67] Abubackar HN, Bengelsdorf FR, Dürre P, Veiga MC, Kennes C. Improved operating strategy for continuous fermentation of carbon monoxide to fuel-ethanol by clostridia. Applied Energy 2016;169:210–7.

[68] Richter H, Martin ME, Angenent LT. A two-stage continuous fermentation system for conversion of syngas into ethanol. Energies 2013;6(8):3987–4000.

[69] Ganigué R, Sánchez-Paredes P, Bañeras L, Colprim J. Low fermentation pH is a trigger to alcohol production, but a killer to chain elongation. Frontiers in Microbiology 2016;7:1–11.

[70] Slonczewski JL, Fujisawa M, Dopson M, Krulwich TA. Cytoplasmic pH measurement and homeostasis in bacteria and archaea. Advances in Microbial Physiology 2009; 55(317):1–79.

[71] Abubackar HN, Veiga MC, Kennes C. Ethanol and acetic acid production from carbon monoxide in a *clostridium* strain in batch and continuous gas-fed bioreactors. International Journal of Environmental Research and Public Health 2015;12(1):1029–43.

[72] Saxena J, Tanner RS. Effect of trace metals on ethanol production from synthesis gas by the ethanologenic acetogen, *Clostridium ragsdalei*. Journal of Industrial Microbiology and Biotechnology 2011;38(4):513–21.
[73] Gruber K, Puffer B, Kräutler B. Vitamin B_{12}-derivatives - enzyme cofactors and ligands of proteins and nucleic acids. Chemical Society Reviews 2011;40(8):4346–63.
[74] Kundiyana DK, Huhnke RL, Wilkins MR. Effect of nutrient limitation and two-stage continuous fermentor design on productivities during '*Clostridium ragsdalei*' syngas fermentation. Bioresource Technology 2011;102(10): 6058–64.
[75] Abubackar HN, Veiga MC, Kennes C. Biological conversion of carbon monoxide to ethanol: effect of pH, gas pressure, reducing agent and yeast extract. Bioresource Technology 2012;114.
[76] Xu H, Liang C, Yuan Z, Xu J, Hua Q, Guo Y. A study of CO/syngas bioconversion by *Clostridium autoethanogenum* with a flexible gas-cultivation system. Enzyme and Microbial Technology 2017;101:24–9.
[77] Phillips JR, Klasson KT, Clausen EC, Gaddy JL. Biological production of ethanol from coal synthesis gas - medium development studies. Applied Biochemistry and Biotechnology 1993;39–40(1):559–71.
[78] Kundiyana DK, Huhnke RL, Maddipati P, Atiyeh HK, Wilkins MR. Feasibility of incorporating cotton seed extract in *Clostridium* strain P11 fermentation medium during synthesis gas fermentation. Bioresource Technology 2010;101(24):9673–80.
[79] Saxena J, Tanner RS. Optimization of a corn steep medium for production of ethanol from synthesis gas fermentation by *Clostridium ragsdalei*. World Journal of Microbiology and Biotechnology 2012;28(4):1553–61.
[80] Abubackar HN, Veiga MC, Kennes C. Production of acids and alcohols from syngas in a two-stage continuous fermentation process. Bioresource Technology 2018; 253:227–34.
[81] Doll K, Rückel A, Kämpf P, Wende M, Weuster-Botz D. Two stirred-tank bioreactors in series enable continuous production of alcohols from carbon monoxide with *Clostridium carboxidivorans*. Bioprocess and Biosystems Engineering 2018;41(10):1403–16.
[82] Abubackar HN, Fernández-Naveira Á, Veiga MC, Kennes C. Impact of cyclic pH shifts on carbon monoxide fermentation to ethanol by *Clostridium autoethanogenum*. Fuel 2016;178:56–62.
[83] Zhu H, Shanks BH, Heindel TJ. Enhancing CO-water mass transfer by functionalized MCM41 nanoparticles. Industrial & Engineering Chemistry Research 2008; 47(20):7881–7.
[84] Zhu H, Shanks BH, Choi DW, Heindel TJ. Effect of functionalized MCM41 nanoparticles on syngas fermentation. Biomass and Bioenergy 2010;34(11):1624–7.
[85] Ruthiya KC. Mass transfer and hydrodynamics in catalytic slurry reactors technische. Netherlands: Universiteit Eindhoven; 2005. p. 67–98. https://doi.org/10.6100/IR584834.
[86] Kim YK, Park SE, Lee H, Yun JY. Enhancement of bioethanol production in syngas fermentation with *Clostridium ljungdahlii* using nanoparticles. Bioresource Technology 2014;159:446–50.
[87] Kim YK, Lee H. Use of magnetic nanoparticles to enhance bioethanol production in syngas fermentation. Bioresource Technology 2016;204:139–44.
[88] Asimakopoulos K, Gavala HN, Skiadas IV. Reactor systems for syngas fermentation processes: a review. Chemical Engineering Journal 2018;348:732–44.
[89] Devarapalli M, Atiyeh HK. A review of conversion processes for bioethanol production with a focus on syngas fermentation. Biofuel Research Journal 2015;2(3): 268–80.
[90] Bredwell MD, Worden RM. Mass-transfer properties of microbubbles. 1. Experimental studies. Biotechnology Progress 1998;14(1):31–8.
[91] Bredwell MD, Srivastava P, Worden RM. Reactor design issues for synthesis- gas fermentations. Biotechnology Progress 1999;15(5):834–44.
[92] Munasinghe PC, Khanal SK. Syngas fermentation to biofuel: evaluation of carbon monoxide mass transfer coefficient (k_La) in different reactor configurations. Biotechnology Progress 2010;26(6):1616–21.
[93] Shen Y, Brown R, Wen Z. Enhancing mass transfer and ethanol production in syngas fermentation of *Clostridium carboxidivorans* P7 through a monolithic biofilm reactor. Applied Energy 2014;136:68–76.
[94] Devarapalli M, Atiyeh HK, Phillips JR, Lewis RS, Huhnke RL. Ethanol production during semicontinuous syngas fermentation in a trickle bed reactor using *Clostridium ragsdalei*. Bioresource Technology 2016;209:56–65.
[95] Shen Y, Brown R, Wen Z. Syngas fermentation of *Clostridium carboxidivoran* P7 in a hollow fiber membrane biofilm reactor: evaluating the mass transfer coefficient and ethanol production performance. Biochemical Engineering Journal 2014;85:21–9.
[96] Abrini J, Naveau H, Nyns E-J. *Clostridium autoethanogenum*, sp. nov., an anaerobic bacterium that produces ethanol from carbon monoxide. Archives of Microbiology 1994;161(4):345–51.
[97] Shen Y, Brown RC, Wen Z. Syngas fermentation by *Clostridium carboxidivorans* P7 in a horizontal rotating packed bed biofilm reactor with enhanced ethanol production. Applied Energy 2017;187:585–94.
[98] Mohammadi M, Younesi H, Najafpour G, Mohamed AR. Sustainable ethanol fermentation from synthesis gas by *Clostridium ljungdahlii* in a continuous stirred tank bioreactor. Journal of Chemical Technology and Biotechnology 2012;87(6):837–43.
[99] Kundiyana DK, Huhnke RL, Wilkins MR. Syngas fermentation in a 100-L pilot scale fermentor: design and process considerations. Journal of Bioscience and Bioengineering 2010;109(5):492–8.

CHAPTER 16

Recycling of Technologic Metals: Fundamentals and Technologies

STEFAN LUIDOLD, PHD

GENERAL ASPECTS OF METAL RECYCLING

The globally increasing demand for metals during the past century puts permanent pressure on natural resources. Consequently, metals are a favored sector for decoupling growths from resource consumption and environmental damages. The exigency of that will become increasingly relevant in the future because of the ongoing worldwide need for metals. Reasons for this are rapid industrialization in the developing countries and the development of modern, metal-intensive technologies in the developed countries, especially for the transformation toward green technologies. Therefore metal recycling and, consequently, resource efficiency have enormous significance because they minimize the exploitation of natural resources. Furthermore, the utilization of secondary resources diminished not only the environmental impacts caused by mining but also the contamination of the environment by partly toxic wastes. Considering that our modern, highly developed technologies are notably based on technologic metals, which are often labeled as critical owing to their not abundant deposits, their preservation and reuse as far as possible are of vital importance [1].

This group of elements comprises not only the rare earth elements (REEs) (scandium, yttrium, lanthanum, cerium, etc.) as well as the high-melting refractory (tungsten, molybdenum, niobium, tantalum, etc.) and precious metals (silver, gold, platinum, palladium, etc.) but also further metals such as lithium, indium, gallium, cobalt, and antimony. These metallic elements, which partially exhibit very diverging physical and chemical properties, are generally relatively scarce in the earth's crust and their complex extraction requires special methods beyond the conventional metallurgy. In addition, some of these elements did not form their own ores but are only extractable as by-products from the winning of common metals such as aluminum, zinc, or lead.

The recycling of metals started a long time ago, because it was more resource- and cost-efficient than just to get rid of scraps and other secondary resources and starting all over again with mining of ores as primary resources. In the past, the relatively straightforward recycling concentrated on specific metals, as most products were relatively simply built. Therefore it follows the so-called material (and metal)-centric approach. However, due to the technologic progress during the past decades, products became increasingly complex, combining any imaginable metal or other materials. Thus their recycling becomes increasingly challenging and requires a much more product-centric approach [1].

Anyhow, the recovery of metals from secondary resources predominantly concentrates to date both on the large mass flows of common metals (iron, aluminum, copper, magnesium, lead, zinc, etc.) and on the very valuable precious metals (silver, gold, platinum, palladium, etc.). The required technologies are partly very well developed for both groups because of the economic considerations on the one hand and to some extent based on reducing energy consumption, landfill demand, emissions to the environment, and further aspects regarding sustainability on the other hand.

Contrariwise, the reclamation of technologic metals from alternative raw materials instead of ores and concentrates did not attract much attention for a long time, with only some exceptions because of their relatively low production volume and value. Beyond that, the availability of raw materials for these elements was not considered significant because of their comparably low prices, until some years ago, in opposite to the fossil fuels, whose supply is carefully observed since the oil crisis in the 1970s. The efforts for a massive extension of renewable energy sources (wind power, photovoltaic, etc.), which require huge amounts of technologic

Sustainable Resource Recovery and Zero Waste Approaches. https://doi.org/10.1016/B978-0-444-64200-4.00016-5

metals, caused a strongly increased attention of the industry on the existing supply risks and bottlenecks for these elements. This trend was boosted by the interim limited export of rare earth metals by China, resulting in an extreme rise in their prices. Finally, this progress initiated worldwide activities to develop proper technologies to extract these critical raw materials from diverse wastes, residues, or other alternative sources.

Nevertheless, the recycling rates of individual metals, especially for technologic metals, are still comparably low due to the limited technologies that are available on an industrial scale for their reclamation from secondary resources. Furthermore, these are often not sufficiently optimized for the recovery of technologic metals. As a result, the recycling rates of these elements amount to below 1% in many cases and also their recycled contents (RCs) are still low, as illustrated in Figs. 16.1 and 16.2. In this connection, the recycling rate indicates the percentage of an individual element's content in the end-of-life products, which is reintroduced to the scrap market and subsequently used for the production and fabrication of the metal, whereas the RC specifies the proportion of secondary materials from the scrap market in its total production.

The achievement of higher recycling rates and RCs for technologic metals is impeded by several factors, such as the constantly rising complexity of the products and their very low contents of respective elements in most cases. Thereby, it should be mentioned that the recycling of such metals from end-of-life products often comprises an appropriate beneficiation and treatment of the corresponding waste streams prior their introduction to suitable process steps of the primary metallurgy. In doing so, the minor metals are usually by-products of common metals. However, the applied process routes cannot deal with any combinations of elements because of the fundamental limitations regarding the simultaneous recoverability of different metals. This fact is indicated by the so-called metal wheel for end-of-life products in Fig. 16.3, which reflects the physics-based modeling of the destinations of different elements as a function of interlinked metallurgical processes. Here, each slice represents not a single process but the complete infrastructure for base or carrier metal refining. Consequently, the complexity of consumer products requires an industrial network of many metallurgical production infrastructures to maximize the recovery of as many as possible metals from end-of-life products.

For example, the current entire infrastructure for the recycling of tin enables the recovery of silver, which dissolves in tin during its pyrometallurgical treatment. Furthermore, copper, bismuth, lead, and zinc are extractable, whereat they are transferred to dust, slime, speiss, or slag during the smelting of tin. Subsequently, these metals are mainly recovered by hydrometallurgical process steps. Indium, as indium oxide, accumulates in the earlier mentioned streams, from where the element can be separated, or is lost in benign

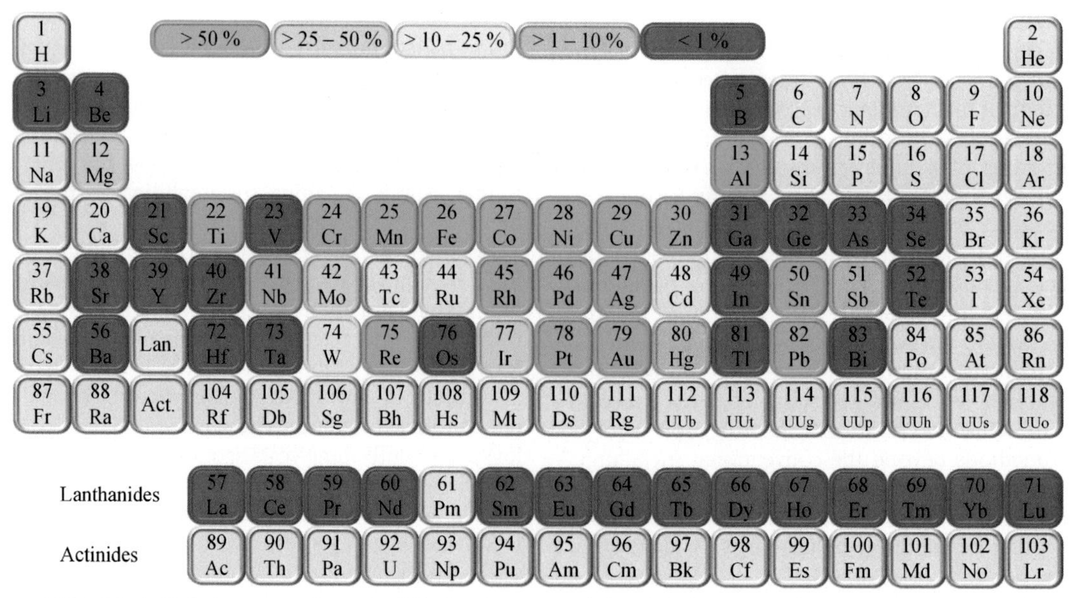

FIG. 16.1 Global average for end-of-life (post-consumer) functional recycling (end-of-life recycling rate [EOL-RR]) [2].

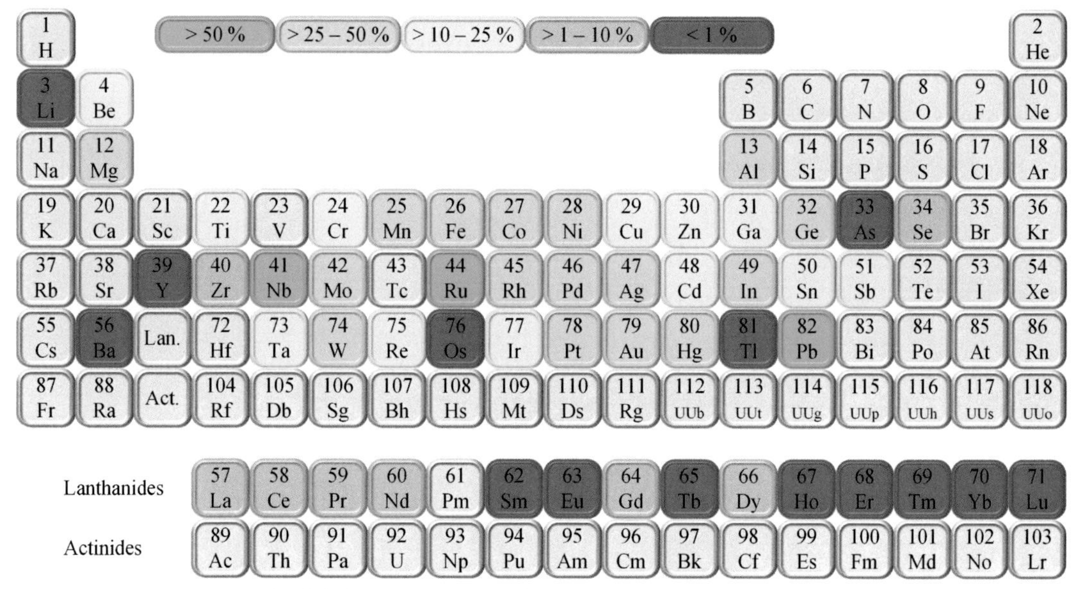

FIG. 16.2 Global average for end-of-life (post-consumer) recycled content [2].

low-value products. However, many other technologic metals (titanium, tantalum, tungsten, zirconium, REEs, etc.) are lost via low-value products. Consequently, a sophisticated beneficiation or processing of scraps and other secondary resources is mandatory before the smelting or reduction step to separate such unfavorable combinations of elements from each other.

Fig. 16.4 represents a generalized flow sheet for the recovery of multiple metals from a range of secondary raw materials. The products are not only various chemicals such as hydrochloric acid, copper sulfate, nickel chloride, and various vanadium chemicals but also several concentrates that are further processed in subsequent process routes to extract the respective metals, e.g., platinum group metals (PGMs) or nickel.

REFRACTORY METALS

The members of this group of elements are characterized by their high melting points in combination with an ignoble behavior. The latter resulted in relatively high corrosion resistances at room temperature due to passivation. However, they easily react with many nonmetals at elevated temperatures making their production and recycling more difficult. This set of metals includes at least zirconium, hafnium, vanadium, niobium, tantalum, chromium, molybdenum, and tungsten. Technetium and rhenium also fulfill this definition, whereat the former has no significant technical application but the latter is partially denominated as a refractory metal (RM). Sometimes even titanium is considered a part of this group of elements despite its lower melting temperature due to its comparable chemical behavior and reduction process that is similar to those of zirconium and hafnium [4].

The major process steps of hydrometallurgy and powder metallurgy, which transforms the ores, scraps, and other resources to (in many cases powdery) RMs, can be summarized as follows:

- alkaline or acidic digestion of concentrates from ores and oxidized scraps, residues, etc.;
- purification of crude solution by precipitation and ion exchange;
- liquid/liquid or solid/liquid extraction of oxidic RM anions;
- precipitation of RM hydroxides or crystallization of RM salts from high-purity aqueous solutions;
- calcination of RM salts and reduction of RM oxides to RM powders.

The subsequent processing of these materials occurs either by a powder metallurgical route or via smelting before some final processing steps according to Fig. 16.5.

The most important RMs are tungsten, molybdenum, tantalum, and niobium, which are presented in the following subsections. Their applications are characterized by two major issues [5]:

- Their utilization mainly rests upon their unique properties.

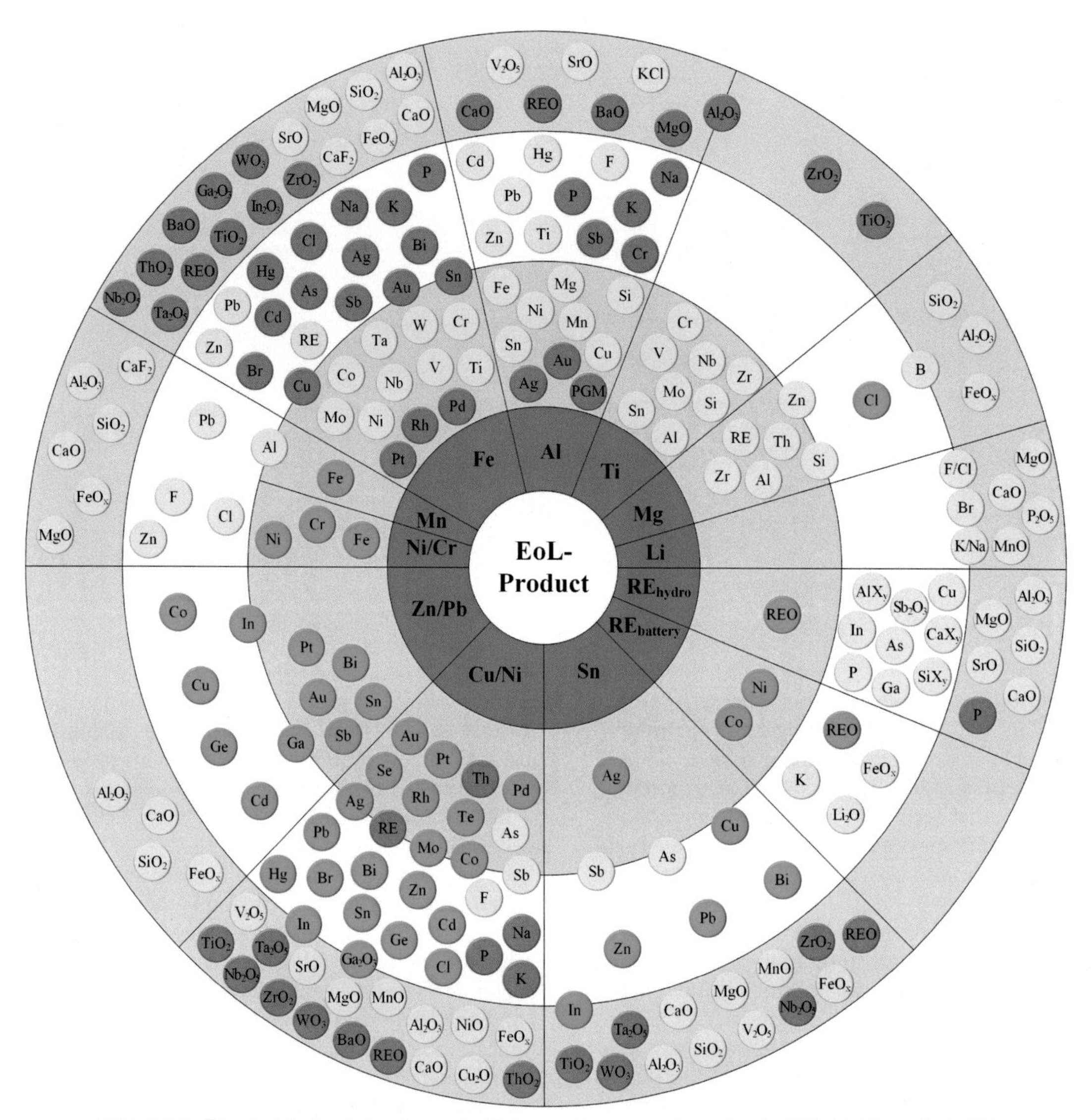

FIG. 16.3 The metal wheel showing metal linkages for processing of end-of-life (EoL) products [1].

- The mass and value percentages of RM are almost always very low. However, they influence the functionality and serviceability quite significantly and disproportionately high.

Consequently, these elements are indispensable and barely substitutable in many modern technologies and products. The RMs are established in almost all sectors of industrial production, and this widespread utilization alone provokes tough challenges for an economic and ecological recycling [5].

Tungsten

The major application areas of this RM include high-temperature and vacuum technologies, electronics, and electric engineering, as well as energy, light, and X-ray technologies. In particular, tungsten serves as a main constituent in the tool industry and in wear protection. In this connection, tungsten carbide (hard phase) together with cobalt (ductile phase, binder) forms cemented carbides, which represent both hard and ductile composites. The latter provides the backbone

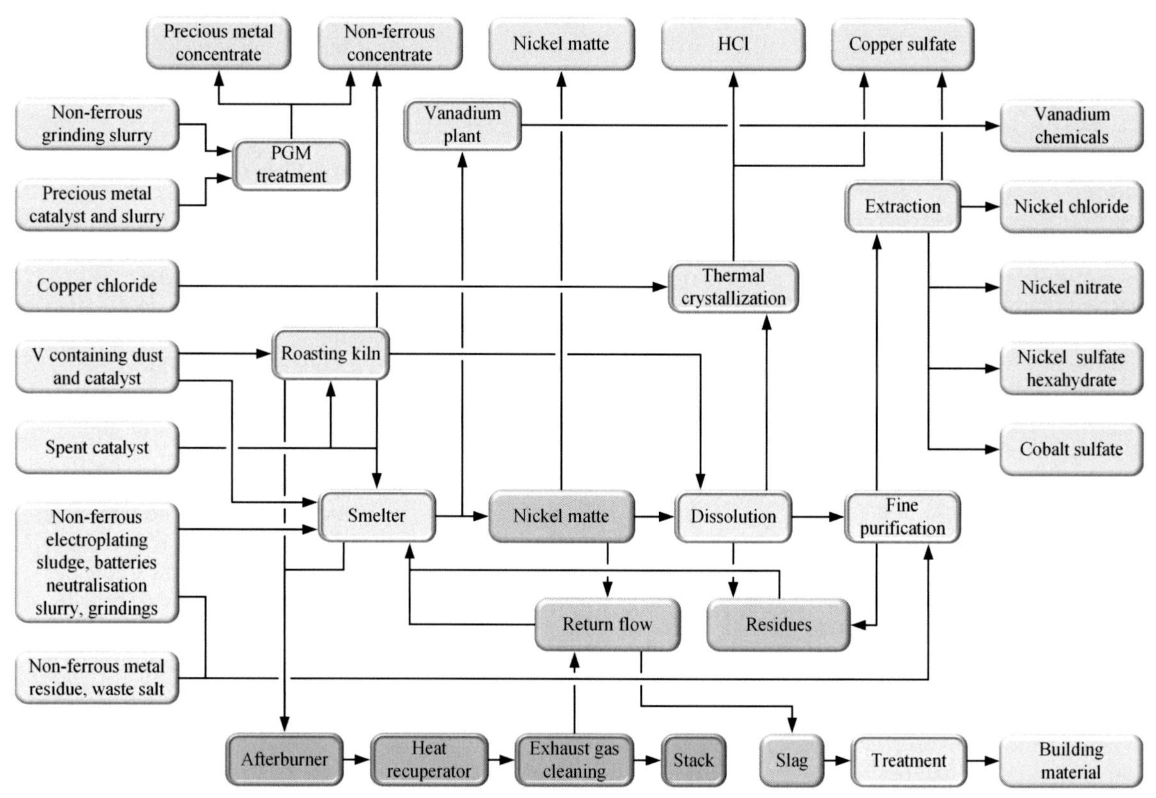

FIG. 16.4 Recycling process at Nickelhütte Aue [3]. *PGM*, platinum group metal.

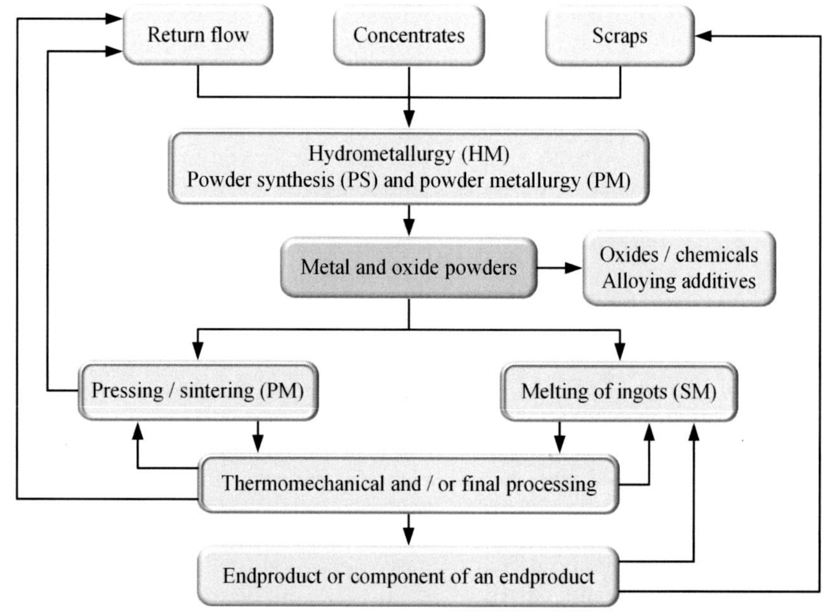

FIG. 16.5 The technology of refractory metals from their educts to final products, including possible return flows [5]. *SM*, smelting.

for all modern chipping and metal-forming processes and is responsible for about two-thirds of the global tungsten consumption [5].

The recycling routes for tungsten are commonly divided into three groups: direct recycling, semidirect recycling, and indirect recycling. Till today, no industrial process exists for the second group, the semidirect approach. However, many concepts are therefore investigated on laboratory scale. Remelting of W-containing scraps is possible only for tungsten-containing steels, stellites, and superalloys, which have W contents in the one-digit percentage range. Smelting of pure tungsten products is in general not practicable because the microstructure of such ingots does not fit to the subsequent thermomechanical processing [5].

The so-called zinc process represents the primary and mostly applied technology for cemented carbide scraps within the direct recycling routes. Thereto, the lumpy material is placed in graphite crucibles, covered with zinc plates, and heated in an inert atmosphere to 800–900°C. The zinc melts and forms intermetallic phases together with the cobalt binder of the composites. The associated significant increase in volume causes bloating and fracturing of the scrap chunks. Heating further to about 1000°C at a low pressure evaporates zinc, which will be condensed and recovered outside the furnace chamber. The remaining solids comprise porous pieces, which can be milled (e.g., by the coldstream process) to a powdered mixture of the same composition as the original fed scrap and finally applied for the production of cemented carbides by pressing and sintering. However, such reclaimed materials contain some process-related contaminants as well as debris from coating materials, which cause tiny defects and therefore negatively influence the characteristics of the cemented carbides. Consequently, only correctly sorted scrap without contaminations can be treated by this process. However, it does not generate solid or liquid residues that require different treatments and/or disposal [5].

Although the process has been used for the recycling of cemented carbides for several decades, no detailed knowledge can be found in literature about the reaction sequence on a microscopic scale. Hence, Ebner et al. [6] focused their research on the phase formation, which takes place during the infiltration of zinc into the binder metal. Finally, the deepened insight should result in a broader applicability of this technology.

Contrariwise, the direct or chemical recycling decomposes all metallic components down to the atomic level to enable their subsequent separation and recovery as pure metals. Because of the existence of tungsten as metal or carbide in most scraps, an oxidation and mechanical conditioning is mandatory at first. In principle, air suffices as an oxidant in roasting plants (multiple hearth, pusher, or rotary furnaces) for the lumpy post-consumer scrap containing tungsten, but its digestion by alkaline salt melts ($NaCO_3$, $NaNO_3$, $NaOH$, Na_2SO_4, and O_2) represents a procedurally interesting and economic alternative. Subsequently, the resulting reaction mass has to pass through practically the entire primary route for the production of tungsten. This includes its dissolution in alkaline solvents and filtration of the remaining insoluble residues. The residues comprise, for instance, cobalt, nickel, tantalum, copper, and silver and can be further treated for the recovery of these metals. The next stages serve for the purification of the solution by two precipitation steps to remove several contaminants (Si, Mo, V, Al, etc.) and a liquid/liquid extraction (solvent extraction [SX]) to eliminate sodium ions by their exchange with ammonium ions. An evaporative crystallization of the purified aqueous phase resulted in ammonium paratungstate powder, which is converted to tungsten oxides by calcination in air (yellow oxide, WO_3) or in an inert atmosphere (blue oxide, WO_{3-x}). Finally, these oxidic powders are reduced to metallic tungsten powder by hydrogen, which can be subsequently converted to tungsten carbide (WC) for the production of cemented carbides. Thus this recycling process, realized on an industrial scale, represents the other extremum of all reclamation routes of tungsten. The products of this chemical conversion results in the same powder properties, completely independent of whether the feed material consists of ore concentrates or scraps. Furthermore, this route features the highest tolerance regarding impurities but requires large amounts of energy and reagents and results in several by-products that are no longer utilizable [5].

The remaining group of semidirect recycling routes is situated between the prior discussed extrema due to their basic principle of selectively dissolving at least one phase, whereas at least one phase remains unaffected. In case of cemented carbides, these are the principal components: metallic binder (mainly cobalt) and various coatings, as well as the carbide phase (WC with additives). Despite the lacking application of such process concepts on an industrial scale, research still takes place in this area. Regarding cemented carbides or similar materials, the investigations aim at the dissolution of the metallic binder and partly also of the coatings by chemical (mostly acidic solutions, e.g., Kücher et al. [7]) or electrochemical methods, preferably without significant attack on the hard phase, the carbides. As a result, the remaining carbide skeleton should be suited for return to the production of cemented carbides after milling and some adjustments (e.g., grain size distribution and carbon balance). The spent leaching solution will be utilized for the reclamation of valuable metals, which originates mainly from the binder and also from different coatings, etc. Finally, the avoidance of wastewater by recirculation of the aqueous phase constitutes the best case in view of sustainability, ecology, and zero waste.

Hence, a successful realization of such a semidirect recycling process will enable the separation of detrimental impurities (coatings, adhesions, etc.) in contrast to the zinc process or other direct routes. Also, a semidirect recycling process does not require the entire complex procedure as in case of the indirect chemical recycling techniques. However, several challenges have to be mastered for this approach to reach an economic as well as ecological alternative to the existing industrial processes:

- sufficient aggressive conditions to dissolve binder, coatings, and other contraries from the carbide skeleton within a reasonable period;
- insignificant attack to the tungsten-containing phase to avoid tungsten losses and other disturbing effects;
- efficient recovery of valuable metals from spent leachates to conserve resources and diminish solid wastes;
- recirculation of the aqueous phase to minimize the generation of wastewater.

So far, most investigations have focused on the first two challenges, which generally contradict each other. Their results indicated that selective leaching requires much more time than the complete oxidation and digestion in the indirect route, even for small lumps of scrap such as cutting inserts. Furthermore, a work by Ebner et al. [8] also addresses the recovery of further metals (e.g., Co, Ta) from spent leaching solutions by their concentration in solid matter, whereas the remaining liquid should fit the prerequisites for its discharge to the sewage system.

The International Tungsten Industry Association (ITIA) estimated in 2012 that recycling covers about 30%–40% of the entire tungsten supply, whereat 10% are process wastes. In addition, the United Nations Environment Programme (2011) rated the share of secondary raw materials in tungsten production (RC) to 25%–50% and the percentage of old scrap (postconsumer scrap) on the total amount of scrap to >50% [9].

Tantalum

In addition to the general characteristics of the RMs, tantalum has some outstanding properties, which enable its major applications. The easy to control electrochemical oxidation of tantalum resulted in thin, amorphous, and well electric insulating (dielectric) oxide layers on the surface. This effect is utilized in electrolytic capacitors with high specific capacities. It serves as a construction material for specialized plants in the chemical industry as well as for high-temperature and vacuum technologies because of its dense, well-adherent oxide layer, which greatly reduces further corrosion, and its excellent plastic deformability at room temperature. Furthermore, several special catalytic and physicoelectric properties of tantalum oxides and other compounds permit their numerous applications in optics, optoelectronics, electronics, and catalysis [5].

Three essential facts have to be considered regarding tantalum scraps and their recycling [5]:

- The application of tantalum capacitors in electronic devices is interconnected with a very strong dilution that disables an appreciable recovery from waste electronic and electric equipment (WEEE) economically.
- Tantalum equipment in plant engineering and construction, such as heat exchangers, reactors, and parts for high-temperature furnaces, as well as spent tantalum sputter targets of the semiconductor industry, can be easily collected. Consequently, their recycling rates reach high values.
- The reutilization of tantalum works quite well in the recycling of used cemented carbides and in the remelting of spent components consisting of Ta-containing alloys (turbine blades, etc.). Contrariwise, chemicals comprising tantalum are virtually not recycled.

Regarding the principal process design, several similarities exist between the chemical recycling of tungsten and tantalum, but in detail, there are some essential differences. For the reclamation of tantalum from scraps and residues, these secondary resources, together with ore concentrates, are either directly leached (in case of sufficiently high Ta contents) in strong acids (HF/H_2SO_4) or processed to a synthetic concentrate by pyrometallurgical techniques. After purification of the crude solution (particularly removal of As, Bi, and Sb) a liquid/liquid extraction separates niobium and tantalum from each other and effectuates further purification. The resulting aqueous phase containing tantalum is further treated either by precipitation and calcination to obtain Ta_2O_5 or by crystallization to synthesize the salt K_2TaF_7. Finally, both compounds are convertible to tantalum powders by reduction with liquid sodium or gaseous magnesium, washing, and drying, which are further processed to final products via, e.g., alloying and doping, pressing and sintering, or smelting and metal forming [5].

Compared with the rate of postconsumer scrap recycling for tungsten, tantalum only amounts to 17.5%. The main sources, therefore, are sputter targets (semiconductor industry) and Ta-containing superalloys (turbine blades, etc.). The prime reason, which prevents from reaching higher recycling rates, is the tantalum capacitors. These components of electronic devices provoke about 40% of tantalum consumption, from there virtually no recovery of tantalum occurs.

Molybdenum

Today, this element exhibits a very broad application range as a metal or an alloy. This includes the following sectors [10]:

- lamp and lighting industries
- electronic and semiconductor industries
- high-temperature and vacuum furnace construction
- glass and ceramic industries
- casting technology and metal working
- coatings
- nuclear technology
- medicine

The first major step for its production comprises the conversion of the raw materials to pure molybdenum oxide, MoO_3. In case of primary sources, such as Mo ores or Mo-containing Cu ores, molybdenum sulfide is enriched and separated from gangue and copper by gravimetric methods and flotation. If necessary, several detrimental contaminants are subsequently removed by leaching prior the roasting of the sulfide to crude molybdenum oxide. Crude molybdenum oxide can be refined to pure molybdenum oxide by either sublimation or hydrometallurgical techniques and further converted to molybdenum powder via reduction with hydrogen. Finally, powder metallurgical methods serve for the manufacture of the final products. The utilization of secondary sources, e.g., spent catalysts, follows the same route after some treatment steps. These include roasting and subsequent alkaline roasting of the catalysts, leaching of the reaction mass, and selective precipitation of molybdenum compounds to separate molybdenum from vanadium and other elements. Regardless of the applied precipitant, the product obtained is either molybdenum sulfide, which can be transferred to the roasting step of the primary route, or calcium molybdate, which must be further treated by hydrometallurgical techniques for the reclamation of pure molybdenum oxide. At least from this stage on, no differences are detectable whether primary or secondary sources were used as feedstock [10].

Regarding the recycling of molybdenum from postconsumer scrap, much lesser opportunities exist in comparison to tungsten. High-purity molybdenum scrap is directly converted to powder for spraying processes, whereas slightly contaminated Mo scrap and TZM (Mo alloy with titanium and zirconium) are reintroduced to the closed loop via oxidation and sublimation, which was investigated for Mo alloys by Ratschbacher and Luidold [11]. Molybdenum scraps with higher impurity content are converted together with other raw materials to ferromolybdenum [12].

Furthermore, steel and superalloy scraps are feedstocks for the reuse of molybdenum. They account for about 30% of the total molybdenum demand. However, no extraction of molybdenum takes place in this case but their significant Mo content finds use in similar products [13].

Niobium

Niobium exhibits similar but slightly lower resistance against chemicals corrosive to tantalum. However, 90% of its total demand arises from the production of high-strength low-alloy steels, whereas rolled niobium—based products account for 4% and superalloys and compounds (oxides and chemicals) account for 3% each [5]. In general, its application areas comprise the iron and steel industry, superalloys, chemical industry, mechanical engineering, electronics, optical industry, aerospace, superconductors, and nuclear technologies [12].

The commonly used raw materials for the extraction of niobium are ores and slags from tin production. In case of the mineral pyrochlore, $Ca_2Nb_2O_7$, pyrometallurgical treatment by sintering and melting removes detrimental impurities such as sulfur, phosphorus, and lead from the physically enriched ore. The obtained synthetic niobium concentrate can be treated together with tantalum sources (see Tantalum section) until their separation via liquid/liquid extraction. Subsequently, niobium oxide hydrate is precipitated from the obtained niobium-containing aqueous solution and calcined to pure niobium pentoxide, Nb_2O_5. An aluminothermic reduction and several melting cycles in an electron beam melting furnace for the removal of excess aluminum, oxygen, and residual impurities converts the latter to high-purity niobium ingots. Alternatively, the reduction of Nb_2O_5 by gaseous magnesium [6] yielded niobium powders of high specific surface area [14].

Reuse of niobium occurs by recycling Nb-containing steels and superalloys; however, the recovery of scraps specifically for niobium content remains negligible. The amount of recycled niobium is not available, but it may be as much as 20% of apparent consumption [15].

RARE EARTH ELEMENTS

This group of elements (REEs) comprises 17 metals, which are often subdivided into the light REEs (Sc and La-Eu) and the heavy REEs (Y and Gd-Lu). However, the assignment of yttrium is controversial. REEs are requisite for many key technologies and not replaceable by other materials or technologies in short or medium term [16].

The REEs have an ever-growing variety of applications in the modern technology, which can be broadly classified as magnets, catalysts, electronics inclusive phosphors, glass, ceramics, alloys, and others. Fig. 16.6 depicts the share of the global consumption among these applications [17].

REEs are actually not rare but exhibit an average level in the earth's crust of 150–200 ppm. In contrast, many mined and industrially used metals have lower average values (e.g., 55 ppm copper and 70 ppm Zn). Anyhow, securing a reliable supply chain for this group of elements constitutes a great challenge because of several reasons. One reason is the significant content of radioactive elements, such as uranium and thorium, in most REE deposits, and another is the so-called balance problem. The latter reason results from deviations between the distribution of individual REEs in the mined ores and that of their global demand, which

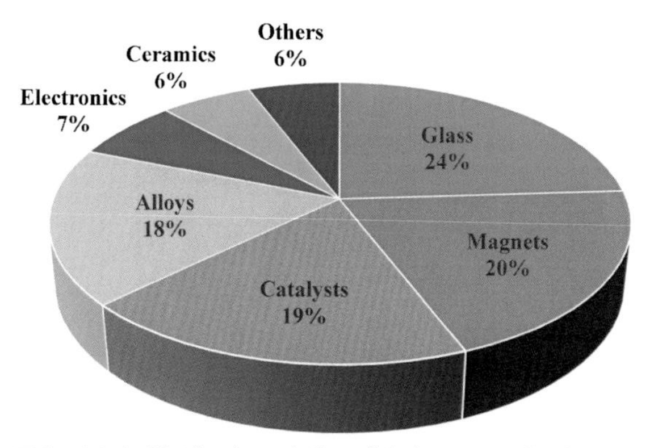

FIG. 16.6 Distribution of the global rare earth element consumption to their application areas [17].

can be provoked by a declined availability of the more important REEs and vice versa. However, an answer to this problem is the recycling of those REEs, which exhibit the lowest ratio between supply and demand.

Therefore, and mainly due to the extreme rise in prices in the years 2010 and 2011, numerous research projects have been started for the recycling of these elements from various end-of-life products. Despite the extremely broad application areas of REEs, almost all projects focused on the processing of nickel metal hydride batteries, phosphors from fluorescent lamps, permanent magnets from motors and generators (e.g., Stuhlpfarrer et al. [18]), and several types of catalysts due to their comparatively high REE contents that provide the best chance to establish an economic recycling route. Many studies on a laboratory scale reported on a successful extraction of rare earth oxide (REO) mixtures with high recovery rates, but no complete recycling route for this group of elements exists so far on the industrial scale [19].

After the reclamation of REO mixtures from secondary resources, further processing to individual pure metals should be possible in the primary route due to the extreme complexity of the subsequent separation of the individual REEs from each other via liquid/liquid extraction (SX) or ion exchange to obtain large enough flow rates for industrial facilities. Another option that seems feasible is the separate processing of individual scrap types, which have a sufficient homogeneous composition (e.g., NdFeB permanent magnets), to remove all impurities and contaminants and to synthesize new feedstock for this type of magnets. However, sufficient blending with virgin materials will be required to adjust the accurate composition.

Several years ago the industrial separation of the individual REEs via SX applied almost exclusively mixer/settler extractors that are arranged in groups of 10–100 separation stages. For example, the plant of the former company Rhodia possesses about 1000 of such units to cut out nearly all components, the individual REEs. Facilities of similar dimension exist in China. For the separation of one REE from a mixture, about 100 stages are sufficient. In parallel, ion exchange procedures were used for special cases to produce small amounts of particular REEs of exceptionally high purity [20].

Although liquid/liquid SX represents a well-established technology, whose application for the separation of REEs on industrial scale started already in 1959 [20], much research is still ongoing concerning further improvements and diverse novel approaches.

An overview of that topic was published by, for example, Kislik [21]. Furthermore, many studies indicated that solvents with ionic liquids can significantly improve the separation efficiency. These materials that are often termed green and designer solvents have been recommended with good prospects. Consequently, further work on investigations and also on tests in small-scale industrial applications are expected [22].

A review about the recycling of REEs provides not only the typical compositions of potential feeds (Table 16.1) but also many simplified process flow sheets regarding the recycling of REEs from various secondary sources. The authors concluded that concerning the recycling of REEs from various scraps, significant technologic progresses are being made and a few commercial recycling plants are already operating. However, further improvement in the recycling rate of REEs requires efforts for the improvement of the logistics (collection, dismantling, and transport) to bring the materials to processing centers, which exist or have to be constructed. Ultimately, to access economical routes, the so-called soft incentives such as strategic independence from foreign sources, sustainability, and stewardship must be considered in many cases [23].

PRECIOUS METALS

Silver, gold, and the PGMs (Pt, Pd, Rh, Ir, Ru, and Os) are summarized as precious metals because of their high oxidation resistance and their vicinity in the periodic table of the elements. By reason of their extremely high value, well-developed technologies already exist

TABLE 16.1
Typical Compositions of Potential Recycling Feeds [23].

Potential Recycling Feeds	Typical Composition (%)
CATALYSTS	
Fluid cracking	3.5 REEs (mostly La+Ce, Pr, and Sm)
Auto converter	< 3.5 REEs (mostly Ce)
Styrene	6–12 CeO_2
PERMANENT MAGNETS	
SmCo	18–30 Sm, 50 Co, < 20 Fe, 8 Cu, 4 Zr, and < 9 Gd
NdFeB	50–60 Fe, 1–2 Dy, 20–30 Nd, 1 B, 4 Co, and < 7 Pr
RECHARGEABLE BATTERIES	
Ni metal hydride	48 Ni, 3.5 Co, 13 Fe, 15 La, 1 Ce, 3 Pr, 10 Nd, and 0.3 Sm
Polishing powders	40–65 Ce, 1–6 La, 0.1–0.7 Nd, 5–10 Si, 1–2 Pb, and 1–2 Fe
ALLOYS	
Mg	2–3 Y or La
Ferroalloys	1–4 Y
Superalloys	0.3–1.3 Y and < 1 La
Misch metal	50–60 Ce, 20–30 La, 2–14 Nd, and 5–8 Pr
GLASS	
Decoloring	0.65 CeO_2
Stabilization	<2.5 CeO_2
Coloring	<5% REE
PHOSPHORS	
Red	60 Y_2O_3 and 3 Eu_2O_3
Green	30 Tb
Blue	10 Eu

for their extraction and winning from both primary and secondary resources leads to high yields in the metallurgical processes. However, essential losses at the collection and processing of WEEE as well as their increasing complexity still boost research and development for the recovery of contained valuable metals.

Regarding the primary metallurgy, silver is mainly a by-product of lead from Pb/Zn concentrate processing, but it can also be extracted from Cu/Ni or Ag/Au ores. In case of primary production of lead, silver continues along with lead in the reduction step and is subsequently separated from it by addition of zinc to the melt to precipitate Ag/Zn mixed crystals (the Parkes process). The crystals are converted to silver by vaporization of zinc and oxidation of lead, followed by refining electrolysis. The extraction route of gold from ores depends on the nature of the raw materials and can be carried out, for example, by the classic cyanidation. Therefore the feedstock is finely grounded and suspended in an aerated sodium cyanide solution to dissolve gold. After subsequent filtration to remove the insoluble residue, the precious metal can be cemented by addition of zinc or adsorption on activated carbon. The obtained solid can be converted to 99.99 wt% Au by treatment with sulfuric acid, roasting in air, smelting with fluxes, and final refining techniques (wet chemical treatment, Miller process, and Wohlwill electrolysis). The wastewater from the leaching constitutes, in general, an ecological problem, but the contained toxic cyanide is decomposed by air in combination with solar radiation (ultraviolet light) via nontoxic cyanate to carbon dioxide and ammonia or systematically destroyed by treatment with hydrogen peroxide [24].

The PGMs occur together with mineral deposits and are often associated with gold, silver, nickel, and copper. The finely ground ores are predominately treated by flotation and subsequent smelting in an electric arc furnace separates the iron-containing silicate slag from the copper matte comprising precious metals. Peirce-Smith converters or top-blown rotary converters are used for the desulfurization of copper matte and further separation of iron as a fayalite slag. A slow, controlled cooling of the ingots induces the formation of two phases, whereat the platinum group elements are concentrated in a metallic Ni/Co/Fe phase with a yield of up to 98%, which is isolated by magnetic separation. A subsequent pressure leaching with sulfuric acid and oxygen removes the base metals, whereas the remaining concentrate of PGMs (40%–90%) requires a complex treatment with various process steps (selective dissolution or precipitation, SX, and ion exchange) for the separation of PGMs from each other. The final stage is the thermal decomposition or reduction of the obtained salts, mostly carried out in hydrogen atmosphere [25].

Regarding the recovery of precious metals from secondary resources, three fundamental influencing factors are taken into account [26]:

- effective metal value of the material (contents and prices of individual precious metals),
- composition and structure of the matter (influences the applicable recycling technology as well as costs and yields),
- application area and life cycle (closed or open precious metal circuit).

Consequently, the selection of a proper recycling process depends on material composition, content of precious metals, characteristics of matrix, and the further included metals and organic contaminants. Two approaches are applied in general: specialized processes for certain feedstocks and the rather universal technologies for varying materials. However, most of the recovery of precious metals takes place via the secondary copper route, and therefore belongs to the second approach. In doing so, the precious metals follow copper until the refining electrolysis, where they are separated as anode slime. The anode slime is treated in a specialized process to separately reclaim the individual precious metals as well as some other valuable elements. Furthermore, modern integrated smelters, e.g., the plant Umicore in Hoboken, Belgium, combine several advantages of both the approaches [26].

The focus of the research concerning the recycling of precious metals lies on WEEE. The recycling rates for this fraction of end-of-life products amounts to 40% in the European Union, to 12% in the United States and Canada, and to 24%–30% in China and Japan. Awasthi and Li [27] give an overview and comparison of different technologies, which are examined and developed by various researchers. They noted that the main part of that type of scrap is still recycled in the informal sector, although numerous sophisticated technologies are available. In addition to chemical methods (Fig. 16.7), many research projects deal with microorganisms for the leaching, which can be classified into three major groups: autotrophic bacteria (e.g., *Thiobacillus* sp.), heterotrophic bacteria (e.g., *Pseudomonas* sp., *Bacillus* sp.), and heterotrophic fungi (e.g., *Aspergillus* sp., *Penicillium* sp.). However, heavy metals maybe provoke cellular effects that can inhibit metabolism. Even low concentrations of these elements are able to not only manipulate the morphologic characteristics of

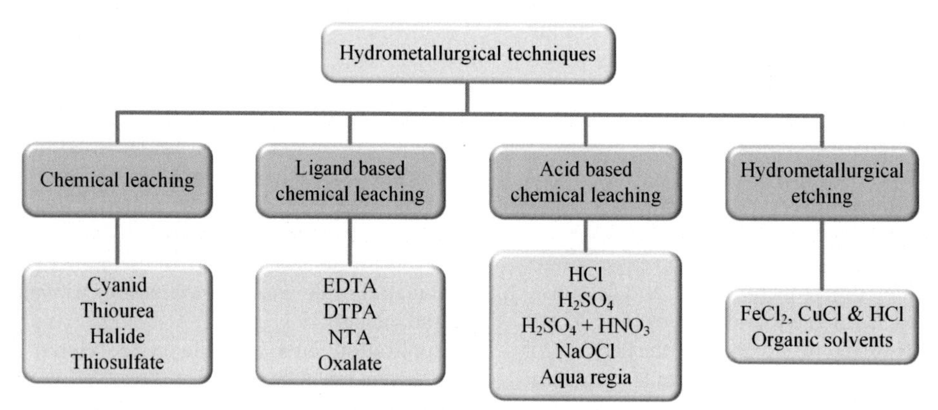

FIG. 16.7 Overview of the different types of leaching methods [27]. *DTPA*, diethylenetriaminepentaacetic acid; *EDTA*, ethylenediaminetetraacetic acid; *NTA*, nitrilotriacetic acid.

fungal strains but also inhibit their growth. Additionally, protoplasmic poisoning and denaturation of proteins and nucleic acid can be caused by too high metal concentrations. Despite these problems, some researchers were able to successfully maintain the growth of the microbes in the laboratory [27].

Li et al. [28] published a review about the hydrometallurgical recovery of metals from waste printed circuit boards (WPCBs). The WPCBs constitute the most complex and valuable components in electric and electronic equipment. They comprise more than 40 metals of wide and varying range of concentrations. Their recovery is of great interest to protect the environment and human beings (e.g., Pb, Cr, As, Cd, and Hg) and due to their economic value (e.g., Cu, Sn, Ag, Au, and PGM). The presented overview about the current status of metal reclamation from WPCBs by hydrometallurgical techniques concluded that a mechanophysical disassembly and size reduction will remain dominant for the pretreatment prior an industrial hydrometallurgical process. It promotes the subsequent leaching step, exhibits low costs, and does not require harmful chemicals. An additional deduction comprises that more research efforts are required on the modern leaching methods to establish environmentally benign and easily recycled leaching agents and reductants. The common or mild leaching methods are not suitable due to environmental, economic, or technical reasons. Furthermore, selective leaching in several steps can facilitate the downstream refining processes. The latter require a range of methods, in which the employment of process combinations and development of new technologies probably depict the trend of this area [28].

FURTHER SPECIAL METALS

In addition to above-discussed groups of elements, several more metals are of particular relevance for highly developed technologies, such as lithium, indium, gallium, germanium, and Co. Of these, lithium received special attention because of its application in high-performance rechargeable batteries (lithium-ion batteries [LIBs]). Today, this type of energy storage is not only utilized in portable electronic devices (smartphones, tablets, laptops, etc.) and tools but also utilized in increasing proportions for electromobility. Electromobility comprises the development of not only electric and hybrid automobiles but also electrically driven bicycles, motorcycles, and other similar equipment (Segway, scooters, boards, etc.). The strongly rising applications of LIBs and their appearance as waste with some time delay have provoked since several years much attention to the availability of lithium and consequently on its recovery from such batteries, which induced many research activities.

Lithium, with a density of 0.534 g/cm^3, is the least dense of all known solids at room temperature and is highly reactive. Its main primary resources are mineral deposits and natural brines; however, ordinary seawater, which contains only 0.17 ppm Li, cannot be considered as a viable feed for lithium production. The extraction of this metal from ores starts by digestion of the ore in acidic or alkaline solutions, which are further processed to recover lithium as a carbonate, hydroxide, or chloride. In case of brines, solar ponds are applied for the evaporation of water, which resulted in the enrichment of lithium and the precipitation of the accompanying, less-soluble salts, because the subsequent processing of the concentrated brine for the reclamation of lithium carbonate requires a Li concentration of at least 5000 ppm. In the end, these semifinished products constitute the starting material for both metallic lithium and a broad range of lithium-containing compounds [29].

In 2016, the global demand for lithium products was approximately distributed to batteries (43%), ceramic and glasses (28%), lubricating greases (7%), polymer production (5%), mold flux powders for continuous casting (4%), air treatment (3%), and other uses (10%). Regarding LIBs, electric vehicles, hybrid electric vehicles, and plug-in hybrid electric vehicles accounted for around 25% of the market by volume; cellular phones and smartphones, 19%; laptop computers and computer tablets, each 16%; electric bicycles, 5%; power tools, 4%; household devices, 2%; and other uses, 13%. However, the worldwide LIB consumption increased by an average of 22% per year from 2010 through 2016, reaching an estimated storage capacity of 70 GWh in 2016 [30].

Concerning the prospects of collection and conditioning of Li-containing end-of-life products as well as the total volume and concentration of lithium in the products of each particular application area, LIBs exhibit the most promising secondary source for the recovery of this metal. Consequently and owing to the enormous increase in their utilization in various

TABLE 16.2
Industrial Battery Recycling Processes All Over the World [31].

Company	Battery Type	Process	Location
Toxco	Ni and Li based	Cryomilling (Li) Pyrometallurgy (Ni)	Trail, BC, Canada Baltimore, OH, USA
Salesco System	All	Pyrometallurgy	Phoenix, AZ, USA
OnTo Technology	Li based	Solvent extraction	Bend, OR, USA
AERC	All	Pyrometallurgy	Allentown, PA, USA Hayward, CA, USA West Melbourne, FL, USA
Dowa	All	Pyrometallurgy	Japan
Japan Recycle	All	Pyrometallurgy	Osaka, Japan
Sony Corporation and Sumitomo Metal Mining Co.	All	Pyrometallurgy	Japan
Xstrata	All	Pyrometallurgy + electrowinning	Glencore Nikkelverk, Sudbury ON, Canada
Accurec	All	Pyrometallurgy	Mülheim, Germany
DK	All	Pyrometallurgy	Duisburg, Germany
AEA Technology	Li based		Sutherland, Scotland
Batrec Industrie AG	Li based, Hg	Pyrometallurgy	Wimmis, Switzerland
AFE Group (Valdi)	All	Pyrometallurgy	Zurich, Switzerland Rogerville, France
Citron	All	Pyrometallurgy	Zurich, Switzerland Rogerville, France
Euro Dieuze/SARP	All	Hydrometallurgy	Lorraine, France
SNAM	Cd, Ni, MH, Li	Pyrometallurgy	Saint-Quentin-Fallavier, France
IPGNA Ent. (Recupyl)	All	Hydrometallurgy	Grenoble, France
Umicore	All	Pyrometallurgy + electrowinning	Hoboken, Belgium

commodities, numerous researchers focus on the development of appropriate recycling processes for several years. Additionally, recycling plants are already operated by various companies over the whole world, which are summarized in Table 16.2 [31].

Today, only up to 3% of LIBs are recycled for the recovery of valuable metals and, moreover, most of them did not focus on the reclamation of lithium. Hence, the recycling rate for lithium amounts to only $<1\%$. Nevertheless, utilization of LIBs as a secondary resource for lithium exhibits a very high importance regarding the environment as well as the lithium market. The reasons are the evolution of the global LIB market and its future perspective, the average life of LIBs and the accumulating end-of-life scrap, and the predicted future scarcity due to limited extraction capacities [31].

Regarding the numerous concepts for the recycling of LIBs, Ordonez et al. [32] systematically reviewed this topic, in which various physical and chemical process steps are principally applied. These include, for instance, mechanical separation, thermal treatments, mechanochemical methods, and dissolution processes, on the one hand, and acid leaching, bioleaching, SX, precipitation, and electrochemical techniques, on the other hand. They not only listed various processes and technologies but also presented the respective objectives and the resulting major findings and limitations.

Furthermore, Yun et al. [33] described the recycling of LIB packs from electric vehicles. At first, the whole battery pack consisting of battery modules and a battery management system is dismantled to extract single modules that are further fragmented into individual cells. Subsequently, various mechanical and chemical process stages convert the cells into several marketable fractions, which are typically arranged as shown in Fig. 16.8. However, most solutions that are proposed in the literature are still at the stage of conception.

Over the whole process, several challenges have to be mastered to close critical gaps within the complete recycling route, which require further research and development [32]:

- automatic and intelligent recovery system (industrial standards for electric vehicles can decrease the number of different designs of battery packs);
- efficient and safety disassembly of battery packs (high energy content of spent LIBs, risk of fire or even explosions by shortcuts, etc.);
- adjustment of chaos in recycling market (different recycling policies, recycling conception of government should be redesigned);

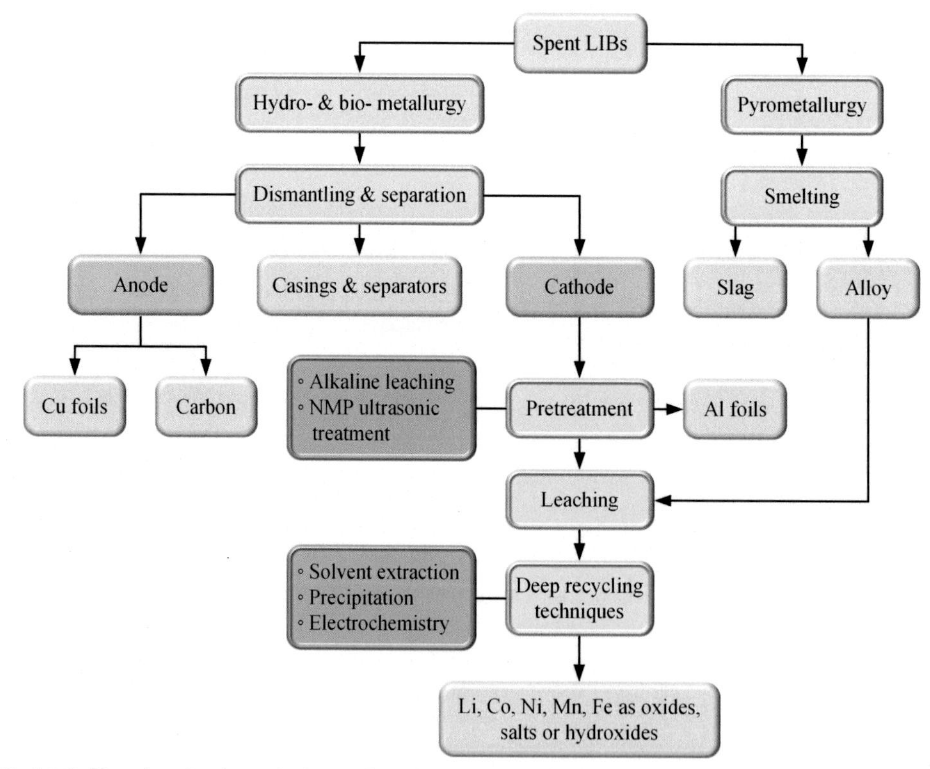

FIG. 16.8 Flowchart for the typical recycling of lithium-ion batteries (LIBs) [33]. *NMP*, *N*-methylpyrrolidone.

- lack of recovery processes for slag, electrolyte, and anode;
- application in industrial scale (most research on hydro- and biometallurgy resulted in very complex processes with high-cost operations);
- development of recycling methods for new batteries with various components.

CONCLUSIONS AND PERSPECTIVES

The summary of the state of the technology and of the science indicates that the recycling of technologic metals is of, with a few exceptions, only marginal relevance for the coverage of the commodity requirements in spite of the strongly increasing demand. Their reclamation from more and more complex end-of-life products as well as other secondary resources still represents several major challenges and therefore needs further research and development. The existence and availability of a comprehensive network of different metallurgical infrastructures constitutes an additional imperative to recover as many valuable basic materials in place of only one or a few metals from individual complex scraps or residues. The increased recycling rates for various metals can reduce the amount and the risk potential of landfilled residues. At the best, such networks involve both primary and secondary facilities, which focus on base metals and technologic metals. Finally, the evaluation of potential recycling routes or feedstocks should incorporate not only financial ratios but also the so-called soft incentives such as strategic independence from foreign sources, sustainability, and product stewardship. However, providing soft incentives will not take place by itself but can be initiated by applying legal regulations and providing diverse incentives.

REFERENCES

[1] Reuter MA, Hudson C, van Schaik A, Heiskanen K, Meskers C, Hagelüken C. Metal recycling: opportunities, limits, infrastructure, a report of the working group on the global metal flows to the international resource panel. UNEP; 2013.

[2] Graedel TE, Allwood J, Birat J-P, Buchert M, Hagelüken C, Reck BK, Sibley SF, Sonnemann G. Recycling rates of metals – a status report, a report of the working group on the global metal flows to the international resource panel. UNEP; 2011.

[3] Recycling, Nickelhütte Aue GmbH. http://www.nickelhuette.com. Accessed 12.09.2018.

[4] Luidold S. Verfahrenstechnik bei der Primärmetallurgie und dem Recycling von Technologiemetallen (Process engineering for primary metallurgy and recycling of technological metals), habilitation thesis, Nonferrous Metallurgy. Austria: Montanuniversitaet Leoben; 2012.

[5] Gille G, Meier A. Recycling von Refraktärmetallen (Recycling of refractory metals). In: Thome-Kozmiensky K-J, Goldmann D, editors. Recycling und Rohstoffe (Recycling and raw materials), vol. 5. Neuruppin ,Germany: TK Verlag; 2012. p. 537–60.

[6] T. Ebner, S. Luidold, H. Antrekowitsch, C. Storf, C. Czettl, Phase formation in Co(Ru)-Zn diffusion couples, in: Proceeding 19th plansee seminar, Reutte, Austria.

[7] Kücher G, Luidold S, Czettl C, Storf C. Lixiviation kinetics of cobalt from cemented carbide. International Journal of Refractory Metals and Hard Materials 2018; 70:239–45.

[8] Ebner T, Luidold S, Honner M, Antrekowitsch H, Czettl C. Conditioning of spent stripping solution for the recovery of metals. Metals 2018. https://doi.org/10.3390/met8100757.

[9] Rohstoffrisikobewertung – wolfram (risk assessment of raw materials – tungsten). 2014. https://www.deutsche-rohstoffagentur.de/DE/Gemeinsames/Produkte/Downloads/DERA_Rohstoffinformationen/rohstoffinformationen-19.pdf.

[10] Sebenik RF, Burkin AR, Dorfler RR, Laferty JM, Leichtfried G, Meyer-Grünow H, Mitchell PCH, Vukasovich MS, Church DA, Van Riper GG, Gilliland JC, Thielke SA. Molybdenum and molybdenum compounds. In: Ullmann's encyclopedia of industrial chemistry. 7th ed., A16. Weinheim, Germany: WILEY-VCH Verlag GmbH & Co. KGaA; 2005.

[11] K. Ratschbacher, S. Luidold, Oxidation kinetics of a molybdenum-tantalum alloy, in: Proceedings of european metallurgical conference (EMC), GDMB, Düsseldorf, Germany, 771–782.

[12] Gille G, Gutknecht W, Haas H, Schnitter C, Olbrich A. Die Refraktärmetalle Niob, Tantal, Wolfram, Molybdän und Rhenium (The refractory metals niobium, tantalum, tungsten, molybdenum and rhenium). In: Chemische Technik: prozesse und Produkte (Chemical techniques: processes and products), vol. 6b. Weinheim, Germany: WILEY-VCH Verlag GmbH; 2006.

[13] Molybdenum. Mineral commodity summaries. U.S. Geological Survey; 2018. https://minerals.usgs.gov/minerals/pubs/commodity/.

[14] Eckert J. Niobium and niobium compounds. In: Ullmann's encyclopedia of industrial chemistry. 7th ed., A17. Weinheim, Germany: WILEY-VCH Verlag GmbH & Co. KGaA; 2005.

[15] Niobium (columbium). Mineral commodity summaries. U.S. Geological Survey; 2018. https://minerals.usgs.gov/minerals/pubs/commodity/.

[16] McGill I. Rare earth metals. In: Handbook of extractive metallurgy. Weinheim, Germany: WILEY-VCH; 1997. p. 1693–741.

[17] Krishnamurthy N, Gupta CK. Extractive metallurgy of rare earths. 2nd ed. Boca Raton, USA: CRC Press, Taylor & Francis Group; 2016.

[18] Stuhlpfarrer P, Luidold S, Schnideritzsch H, Antrekowitsch H. New preparation and recycling procedure to recover rare earth elements from magnets by using a closed loop treatment. Trans Inst Min Metall C Mineral Processing and Extractive Metallurgy 2016;125:204–10.
[19] Dutta T, Kim K-H, Uchimiya M, Kwon EE, Jeon B-H, Deep A, Yun S-T. Global demand for rare earth resources and strategies for green mining. Environmental Research 2016;150:182–90.
[20] Richter H, Schermanz K. Seltene erden (rare earths). In: Chemische Technik – prozesse und Produkte (Chemical techniques: processes and products), vol. 6b. Weinheim, Germany: WILEY-VCH Verlag GmbH; 2006.
[21] Kislik VS. Solvent extraction: classical and novel approaches. Amsterdam, Netherland: Elsevier; 2012.
[22] Wang LY, Guo QJ, Lee MS. Recent advances in metal extraction improvement: mixture systems consisting of ionic liquid and molecular extractant. Separation and Purification Technology 2019;210:292–303.
[23] Ferron CJ, Henry P. A review of the recycling of rare earth metals. Canadian Metallurgical Quarterly 2015;54(4): 388–94.
[24] Beck G, Beyer H-H, Gerhartz W. Edelmetall-Taschenbuch (Precious metal paperback). Heidelberg, Germany: Hüthig; 2000.
[25] Brumby A, Hagelüken C, Lox E, Kleinwächter I. Edelmetalle (precious metals). In: Chemische Technik: prozesse und Produkte (Chemical techniques: processes and products), vol. 6b. Weinheim, Germany: WILEY-VCH Verlag GmbH; 2006. p. 163–78.
[26] Hagelüken C. Edelmetallrecycling – status und Entwicklungen (Recycling of precious metals – status and developments). In: Sondermetalle und Edelmetalle: vorträge beim 44. Metallurgischen Seminar (Special metals and precious metals: speeches at the 44th Metallurgical Seminar). Germany: GDMB, Clausthal-Zellerfeld; 2012. p. 163–78.
[27] Awasthi AK, Li J. An overview of the potential of eco-friendly hybrid strategy for metal recycling from WEEE. Resources, Conservation & Recycling 2017;126:228–39.
[28] Li H, Eksteen J, Oraby E. Hydrometallurgical recovery of metals from waste printed circuit boards (WPCBs): current status and perspectives – a review. Resources, Conservation & Recycling 2018;139:122–39.
[29] Wietelmann U, Bauer RJ. Lithium and lithium compounds. In: Ullmann's encyclopedia of industrial chemistry. 7th ed., A15. Weinheim, Germany: WILEY-VCH Verlag GmbH & Co. KGaA; 2005.
[30] Lithium. Minerals yearbook. U.S. Geological Survey; 2018. https://minerals.usgs.gov/minerals/pubs/commodity/.
[31] Swain B. Recovery and recycling of lithium: a review. Separation and Purification Technology 2017;172: 388–403.
[32] Ordonez J, Gago EJ, Girard A. Process and technologies for the recycling and recovery of spent lithium-ion batteries. Renewable and Sustainable Energy Reviews 2016;60:195–205.
[33] Yun L, Linh D, Peng X, Garg A, Phung LE ML, Asghari S, Sandoval J. Resources, Conservation and Recycling 2018;136:198–208.

CHAPTER 17

Waste Electric and Electronic Equipment: Current Legislations, Waste Management, and Recycling of Energy, Materials, and Feedstocks

PANAGIOTIS EVANGELOPOULOS, PHD • EFTHYMIOS KANTARELIS, PHD • WEIHONG YANG, PHD

INTRODUCTION

As innovation in technology progresses and the innovation cycles become even shorter, the production of electronic equipment increases and its replacement accelerates, which makes electric and electronic equipment (EEE) a rapidly growing source of waste.

The United Nations University has estimated that only for the year 2014, 41.8 million tons of electronic waste (e-waste) was generated, which can become a serious risk for living organisms and the environment [1]. It has also been recorded that the increasing rate has reached 3.0%–5.0% or even higher every year, which is approximately three times faster than the conventional municipal solid waste stream [1,2]. This situation arises from people's constant desire for newer, faster, and more efficient technologic possessions as well as the brainwashing marketing by the producers, which introduce more and more goods to the market. Therefore more and more "old" equipment become obsolete, which sooner or later will be discarded, leading to the continuously growing fraction of e-waste.

The figures and statistics gathered from different organizations globally have already shown the severity of the situation, with the increasing amounts of waste electric and electronic equipment (WEEE). The US Environmental Protection Agency reports that 3.36 million tons of new electronic products were sold in 2014 in the United States, whereas only 1.4 tons of WEEE were recycled the same year [3]. Based on China's government statistics, 25 million TVs, 5.4 million refrigerators, 10 million washing machines, 1 million air conditioners, 12 million computers, 6.0 million printers, and 40 million mobile phones were discarded in 2009 [4]. Furthermore, almost 3.8 million tons of e-waste was collected in the European Union in 2015 as every European citizen discards 14–24 kg of WEEE every year [2].

Despite the increasing amount of WEEE globally, the real problem is that only a low percentage globally is being treated properly in suitable recycling and recovery facilities, even though it is well known that the e-waste fraction contains a variety of hazardous materials [5–7]. WEEE consist of not only several valuable materials such as plastics, conventional metals, ceramic materials, and rare earth metals but also toxic components such as heavy metals and persistent organic pollutants. Therefore the benefits from proper waste treatment of this fraction are numerous, as e-waste can provide industries with materials, minimizing the cost of virgin material extraction as well as the environmental risks from the hazardous components. Hence, adoption of more sustainable strategies for e-waste handling and treatment is essential.

In order to deeply understand the growing problem of WEEE generation and poor handling, it is necessary to trace the different stages of the fraction. First, each electric and electronic device takes a while before it is considered as waste, as consumers keep their obsolete electronics for a certain period before they throw them away. In the United States, more than 4.0 million tons of devices were kept in storage in 2009 for end-of-life management [4]. Some obsolete devices are traded nationally, whereas others are shipped to the

Sustainable Resource Recovery and Zero Waste Approaches. https://doi.org/10.1016/B978-0-444-64200-4.00017-7

developing countries where older technology can be used for a few more years.

The reuse of electronics in the developing countries can extend the life period of the products and contributes to their development in terms of following up the technologic innovations, but usually, these countries lack environmental restrictions as well as proper waste management techniques and treatment facilities. Therefore by exporting huge quantities of old electronics to the developing countries under the umbrella of charity, the potential e-waste stream will not be treated under controlled conditions and this can cause major environmental hazards globally (Fig. 17.1 [8].

CURRENT GLOBAL LEGISLATIONS

The European Union has taken the necessary steps for reducing this growing waste stream by adapting the Directive 2002/96/EC and later this was updated to 2012/19/EU about the management use and recycling of WEEE. Through these derivatives, each member state must adopt national legislations in order to meet the specific objectives, which allow each nation to tailor legislations according to each different circumstances existing in the national level. Specifically, the restriction of the use of certain hazardous substances (RoHS I) in EEE, which was introduced on February 13, 2003, sets strict maximum limits for lead, mercury, cadmium, hexavalent chromium, polybrominated biphenyls, or polybrominated diphenyl ethers used in specified types of EEE. This legislation has prevented high amounts of banned substances from being disposed and potentially released into the environment. Furthermore, the producers have already adopted this directive by implementing changes in product design for the products sold not only in the European Union but also worldwide. The final revision of this directive has been made in 2011, following the technologic improvements and thus setting new goals and restrictions on chemical compounds for the upcoming years [10,11].

The more important elements of the new Directive can be summarized as follows:

- the rules will gradually be applied to all EEE, cables, and spare parts produced until 2019;
- a review of the list of RoHS until July 2014 and periodically thereafter;
- more transparent and cleaner rules for allowing exemptions from the substance ban;
- improving the correlation with REACH (Regulation on the Registration, Evaluation, Authorisation and Restriction of Chemicals);
- photovoltaic panels are excluded from the new Directive so that the European Union can meet its objectives for renewable energy and energy efficiency [12].

These directives aim to reduce the cost of waste handling in the municipality level while also extending the producer's responsibility by requesting to finance the collection, treatment, recovery, and environment-friendly proper disposal of WEEE according to each

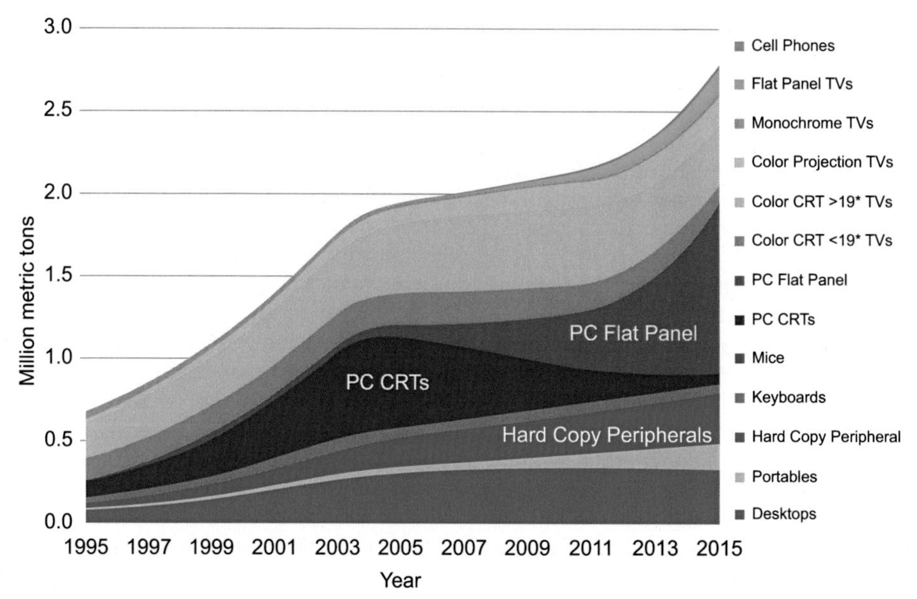

FIG. 17.1 Total e-waste disposition by product type 1995–2015. *CRT*, cathode-ray tube. (Adapted with permission from Ref. [9].)

product's characteristics. The underlying concept of this directive is to indirectly impose producers to develop more environment-friendly products with more recyclable materials and that are simpler to dismantle, which will eventually reduce the cost of the environmental fee for the end-of-life management of their products [8]. Similar legislations about extended producer responsibility (EPR), which makes the producers financially responsible for the end-of-life management of their products, have also been introduced in Japan, South Korea, Taiwan, China, and in several states of the United States [8,13].

On the other hand, the appliance of the EU Directive from each member state varies. Netherland and Sweden have legislated EPR from the early stages of the EU Derivative. In Netherlands, all the e-waste categories, except waste from the information and communications technology, are being compulsorily managed by an EPR organization, whereas in Sweden the management is being performed by a nonprofit organization. This is the reason why the collection rate is higher in these countries than that in other countries that lack national legislation and an organized collection system that could increase the collection efficiency (Fig. 17.2) [14].

According to the European Union, EEE is categorized as shown in Table 17.1.

The statistics available for the e-waste fractions are limited; therefore Fig. 17.3 illustrates the percentages of the total e-waste mass collected in 2015 in each fraction's category. Furthermore, it should be noted that the percentages may have been changed due to the development of new technologies, which can affect the WEEE composition. As expected, the first category "large household appliances" represents more than half of the total waste fraction.

However, studies indicate that it is not practical to have a skip for each of the 10 different categories of WEEE, and it is better to make some simplifications by categorizing WEEE into the following groups:

1. Refrigeration equipment—requires specialist treatment under the ozone depleting substances regulations.
2. Large household appliances (excluding refrigeration equipment), which have a metal-rich content and can be easily reprocessed together.
3. Cathode-ray tube (CRT)-containing equipment—must be handled separately due to broken monitor glass.

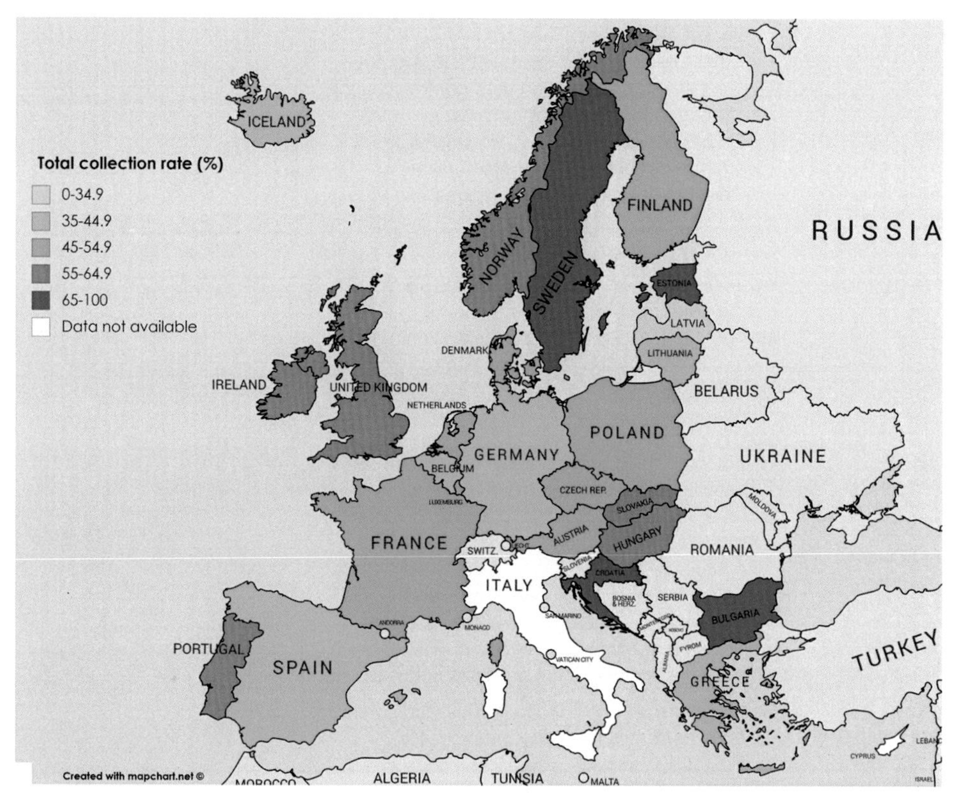

FIG. 17.2 The total collection rate (%) of waste electric and electronic equipment in 2016. (Adapted from Ref. [15].)

TABLE 17.1
Categories of Electric and Electronic Equipment and Indicative Equipment According to the Directive 2012/19/EU and the Targets for Recovery, Recycling, and Reuse [11].

	Category	Indicative Equipment	Targets for Recovery (%)	Targets for Recycling and Reuse (%)
1.	Large household appliances	• Appliances used for refrigeration, conservation, and storage of foods (refrigerators, freezers etc.). • Appliances used for cooking and other processing of food (cookers, electric stoves, electric hot plates, microwaves, etc.). • Appliances used for heating rooms, beds, seating furniture (electric radiators, electric heating appliances, etc.). • Appliances used for exhaust ventilation and conditioning equipment (air conditioners, electric fans, etc.) • Washing machines, cloth dryers, dish washing machines.	>80	>75
2.	Small household appliances	• Appliances used for sewing, knitting, weaving, and other processing for textiles (vacuum cleaners, carpet sweepers, etc.).	>70	>50
3.	Information technology and telecommunications equipment	• Centralized data processing (mainframes, minicomputers, printer units, etc.). • Personal and laptop Computers. • Printing and copying equipment. • Pocket and desk calculators. • Other equipment for the collection, storage, processing, presentation, and communication of information by electronic means. • Equipment of transmitting sound images or other information by telecommunications (cellular phones, telex, telephones, etc.).	>75	>65
4.	Consumer equipment and photovoltaic panels	Products and equipment for the purpose of recording or reproducing sound or images, including signal or other technologies, for the distribution of sound and image than by telecommunications (radio and television sets, video cameras, video recorders, audio amplifiers, musical instruments, etc.).	>75	>65
5.	Lighting equipment		>70	>50
6.	Electric and electronic tools[a]	• Equipment for turning, sanding, grinding, sawing, cutting shearing, drilling, making holes, punching, folding, bending, or similar processing of wood, metal, and other materials. • Equipment for riveting, nailing, screwing, removing rivets, nails, and screws. • Tools for welding and soldering. • Tools for mowing and gardening activities, etc.	>70	>50
7.	Toys, leisure, and sports equipment	Electric trains and cars, video games, and consoles. Computers for biking, diving, running, etc.	>70	>50

TABLE 17.1
Categories of Electric and Electronic Equipment and Indicative Equipment According to the Directive 2012/19/EU and the Targets for Recovery, Recycling, and Reuse [11].—cont'd

	Category	Indicative Equipment	Targets for Recovery (%)	Targets for Recycling and Reuse (%)
8.	Medical devices[b]	Appliances for detecting, preventing, monitoring, treating, and alleviating illness, injury, or disability (nuclear medicine equipment, dialysis equipment, radiotherapy and cardiology equipment, etc.).	Not defined	
9.	Monitoring and control instruments	Monitoring and control instruments used in household and industrial installations (smoke detectors, heating regulators, measuring weight, adjusting appliances, etc.).	>70	>50
10.	Automatic dispensers	All appliances that automatically deliver all kinds of products.	>80	>75

[a] With the exception of large-scale stationary industrial tools.
[b] With the exception of implanted and infected products.

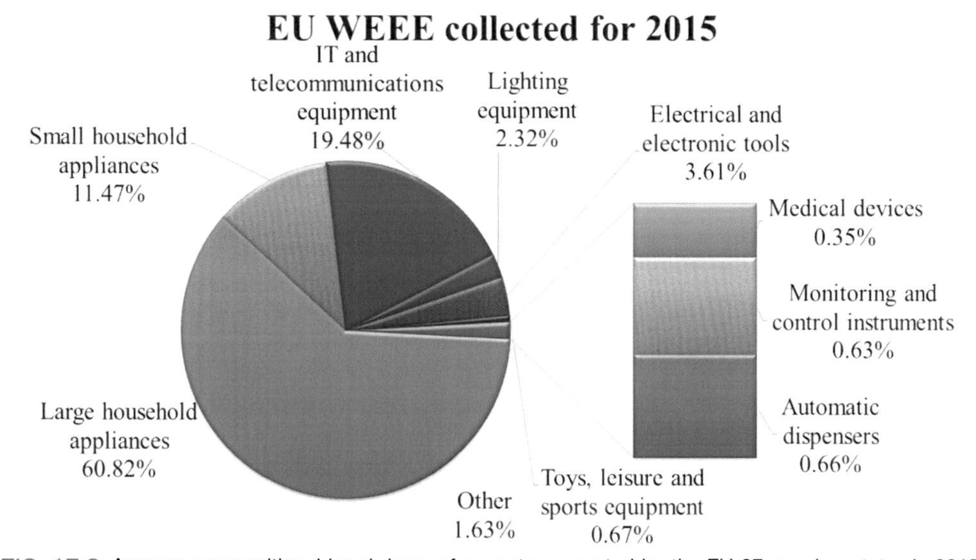

FIG. 17.3 Average compositional breakdown of e-waste generated by the EU 27 member states in 2015, including the 10 different waste electric and electronic equipment categories. (Adapted from Ref. [2])

4. Linear and compact fluorescent tubes—to prevent contamination and enable recycling.
5. All other WEEE.

These categories are based on the management of each waste fraction because they require different handling and treatment methods, whereas the categories set by the European Union are based on the collection and statistical purposes of WEEE [16].

COMPOSITION OF WASTE ELECTRIC AND ELECTRONIC EQUIPMENT

The challenges of WEEE treatment are related not only to the rapidly growing quantity but also mainly to the complexity of WEEE composition. In order to develop a recycling process that could be both environmentally friendly and economically viable, in-depth characterization of the materials streams is needed [17].

Lack of publicly available information of the exact composition as well as the dynamic composition of WEEE poses uncertainties to companies when making the necessary choices in terms of investment in recycling facilities and equipment.

There are plenty of studies in the literature focusing on the chemical composition and the characterization of the e-waste fraction. Widmer et al. [18] summarize findings from the recycling of the e-waste stream by the SWICO/S.EN.S recycling system in Switzerland, which are also presented in Table 17.2. Other studies have been focusing on examining more representative samples from three different regions of Germany according to their land development structure index, resulting in a more objective chemical composition of WEEE. The sampling procedure was also of a large extent because 5 tons of e-waste was examined, but this study is focusing only on the composition of a small e-waste fraction that can be discarded in the municipal's electronic waste bins (Table 17.2) [7].

The composition of WEEE varies depending on the age and type of the discarded item, as the evolution of technology and the new functions of electric equipment can vary the raw materials used in the manufacturing processes. Several methods of characterization of the different fractions have been reported. They include investigation of material composition by means of Fourier transform infrared (FTIR) spectroscopy for identification of the polymer type, energy-dispersive X-ray fluorescence for identification of the heavy metals and halogens, HPLC-UV/MS (high-performance liquid chromatography-ultraviolet/mass spectrometry) for flame retardants, gas chromatography/high-resolution mass spectrometry for PBDD/F (polybrominated dibenzo-*p*-dioxin and dibenzofuran) detection, Raman spectroscopy, etc. [19,20]. The application of combined methods provides a faster analysis of WEEE plastics. Faster and more efficient characterization of the chemical composition of WEEE can improve the recycling techniques and proper management of WEEE in terms of separation and blending different flows for material reuse [20].

TABLE 17.2
Average Material Composition on WEEE (% w/w) [4,7,18].

Material	Ongondo et al.[4]	Dimitrakakis et al.[7]	Widmer et al.[18]
Plastics	33.64%	19%	15.21
Ferrous metals	15.64%	38%	60.20%
Nonferrous metals	3.81%	28%	
Electronic components	23.47%		
"Bonded" materials	8.02%		4.97%
Cables	5.43%		1.97%
Printed wiring boards	2.89%		1.71%
Others	3.89%	10%	4.08%
Batteries	1.53%		
Rubber	0.54%		
LCDs	0.14%		11.87%
Glass		4%	
Wood		1%	

Most types of WEEE contain varying quantities and types of plastics. The need for different polymers in different electronic components makes their recycling even more complicated. The plastics that are commonly encountered in EEE are acrylonitrile butadiene styrene (ABS), polycarbonate (PC), PC/ABS blends, high-impact polystyrene (HIPS), and polyphenylene oxide blends (Table 17.3) [19].

In general, there are two different polymer fractions that are available on the market: a well-defined polymer product, which comes from plastic fractions during dismantling (like during the course of CRT glass recovery), and another kind of the polymer-containing fraction from WEEE processing, which is a by-product of the metal recovery processes and is isolated from the bulk WEEE by means of shredders, magnetic separators, and cyclones [21].

Epoxy resins are widely used materials in electronics manufacture, and they are used in conductive adhesives, flip chip encapsulation, bonding of leads, die coatings, surface mounting adhesives, encapsulation, and conformal coatings.

The largest single flame retardant, cost wise, is tetrabromobisphenol A (TBBPA) that is used predominantly in printed wiring boards (printed circuit boards) [22]. The amount of TBBPA in a finished resin used in electronic equipment may contain about 18%–21% Br [22].

Other flame retardants commonly used in WEEE are shown in Table 17.4.

The National Electrical Manufacturers Association (NEMA) classes of different printed circuit boards are listed in Table 17.5.

TABLE 17.3
Main Polymers Used in the Manufacture of the Most Common WEEE Items Collected [20].

WEEE Item	Polymer Composition
Printers/faxes	PS (80%), HIPS (10%), SAN (5%), ABS, PP
Telecoms	ABS (80%), PC/ABS (13%), HIPS, POM
TVs	PPE/PS (63%), PC/ABS (32%), PET (5%)
Toys	ABS (70%), HIPS (10%), PP (10%), PA (5%), PVC (5%)
Monitors	PC/ABS (90%), ABS (5%), HIPS (5%)
Computer	ABS (50%), PC/ABS (35%), HIPS (15%)
Small households appliances	PP (43%), PA (19%), ABS-SAN (17%), PC (10%), PBT, POM
Refrigeration	PS&EPS (31%), ABS (26%), PU (22%), UP (9%), PVC (6%)
Dishwashers	PP (69%), PS (8%), ABS (7%), PVC (5%)

ABS, acrylonitrile butadiene styrene; *HIPS*, high-impact polystyrene; *PA*, polyamide; *PBT*, polybutylene terephthalate; *PC*, polycarbonate; *PET*, polyethylene terephthalate; *POM*, polyoxymethylene; *PP*, polypropylene; *PPE*, polyphenylene ether; *PS*, polystyrene; *PU*, polyurethane; *PVC*, polyvinyl chloride; *SAN*, styrene acrylonitrile; *UP*, Unsaturated Polyester Resins; *WEEE*, waste electric and electronic equipment.

TABLE 17.4
Characteristics of Commercial Flame Retardants [22].

Commercial Flame Retardants	Br Chemical %	Br %	Sb_2O_3
DBDPO	11–13	9–11	2–5
TBPME	14–16	9–11	2–5
BOE	17–21	9–11	2–5
OCBDPO	18–20	14–16	2–5
TBPE	21–24	15–17	2–5
TBBPA	20–23	12–14	2–5

BOE, brominated oligomer epoxy; *DBDPO*, decabromodiphenyl oxide; *OBDPO*, octabromodiphenyl oxide; *TBBPA*, tetrabromobisphenol A; *TBPE*, tribromophenoxy ethane; *TBPME*, tetrabromophthaldiphenyl ethane.

Additionally, WEEE may contain several metals and inorganic compounds that can be associated with environmental and health hazards (Table 17.6). The most toxic metals contained in this fraction of waste are antimony (Sb), barium (Ba), beryllium (Be), cadmium (Cd), chromium (Cr), and lead (Pb) [24]. All these metals are used in electronics' manufacturing process as impurities for different polymers, for instance, Pb is used as a stabilizer in polyvinyl chloride (PVC) cables; Sb in flame retardant formulation; and some for improving conventional metals' properties, for instance, chromium can improve steel hardness and beryllium can improve copper's strength; and some such as Cd and Ba are used for their electric properties [8]. These heavy metals, as well as their oxides and compounds, are highly toxic for the living organisms, as even in small quantities they are associated with carcinogenesis, skin diseases, and respiratory infections. Therefore health and safety measures and extensive air and flue gas cleaning systems are necessary in the whole process of recycling, reuse, and recovery and to prevent the leakage of these pollutants.

Chemical analysis of metals and other compounds have been carried out by using techniques such as leaching, loss of ignition, and inductively coupled plasma atomic emission spectroscopy (Table 17.7) [25].

Typical analysis of different elements and compounds in WEEE is listed in Table 17.8.

ENERGY, MATERIALS RECOVERY, AND FEEDSTOCK RECYCLING OPTIONS FOR WEEE

The recycling of WEEE has been proven to be a valuable option not only for metal recovery but also for the energy savings compared with the extraction of virgin ore. The energy savings of recycled aluminum over virgin materials can reach up to 95%, as the extraction of aluminum from bauxite can be really energy consuming compared with remelting the metallic form at a low temperature of 660°C. Similarly, copper recycling can save up to 85%; iron or steel, 74%; and zinc, 60%. Other materials can also be recycled from WEEE fractions, such as plastics, which, if they are pure enough, can also be remelted and recycled with 80% less energy consumed compared with the production of new materials [29].

The precious metals extraction consumes significant amounts of energy because of their low concentration in mining ores. Recovering, for example, Au from WEEE can be easier and less energy consuming, as 1 ton of e-waste from personal computers can exceed the amount of gold that can be recovered from 17 tons of gold ore [30].

TABLE 17.5
The NEMA Classes of Copper-Clad Laminate Printed Circuit Boards.

NEMA Grade	Resin, Reinforcement	Description	Typical Uses
XXP, XXXPC	Phenolic, paper	Hot or cold punching	Inexpensive consumer items such as calculators
FR-1 and FR-2	Phenolic, paper	Flame retardant XXP/XXXPC	Where flame retardancy is required
FR-3	Epoxy resin, paper	Flame retardant	High insulation resistance
CEM-1,-3	Epoxy, glass cloth, or glass core	Punchable epoxy, properties between XXXPC and FR-4	Radios, smoke alarms, lower cost than FR-4
G-10	Epoxy, paper, glass	Excellent electric devices, water resistant, not flame retardant	Computers, telecom, costlier than FR-2
FR-4	Epoxy, glass cloth	Like G10 but flame retardant, T_g ~130°C	Where flame retardancy is required
FR-5	Epoxy, glass cloth	Like FR-4, more heat resistant, higher T_g than FR-4	Military, aerospace, where specified
FR-6	Polyester resin, glass mat	Flame resistant	Low capacitance or high impact apps

NEMA, National Electrical Manufacturers Association.
Adapted from Ref. [22].

TABLE 17.6
Environmental Toxicity and Recyclability of Commercial Flame Retardants [23].

	Melting Range (°C)	Environmental Toxicity	Recyclability
DBDPO	300–315	Dioxin/furan	Excellent
TBPME	445–458	No issues	Excellent
BOE	120–140	No issues	Good
OCBDPO	70–150	Dioxin/furan	Excellent
TBPE	223–225	Potential	Fair
TBBPA	197–181	Dioxin/furan	Poor
ATO(Sb_2O_5)	656	Carcinogenic	Excellent

ATO, antimony-doped tin oxide; *BOE*, brominated oligomer epoxy; *DBDPO*, decabromodiphenyl oxide; *OCBDPO*, octabromodiphenyl oxide; *TBBPA*, tetrabromobisphenol A; *TBPE*, tribromophenoxy ethane; *TBPME*, tetrabromophthaldiphenyl ethane.

A range of techniques is currently applied for retrieving components and materials from WEEE. The implementation of one or a combination of more technologies depends on the material characteristics and overall process economics. The overall recycling process from final user includes a collection and transportation step to a suitable treatment facility. The treatment essentially consists of a scheme of: sorting/disassembly (step 1), size reduction (step 2), and separation/refining (step 3) (Fig. 17.4).

Dismantling and Sorting

The first step of WEEE treatment is achieved almost exclusively by manual intervention (primary recycling). The primary recycling includes manual dismantling where whole components of the e-waste fractions are separated into metals, plastics, glass parts, and hazardous materials (Fig. 17.5). The weight of the total fraction is reduced significantly because the metallic and plastic covers can be easily dismantled, sorted, and transferred directly for reuse or recycling. Concerning

TABLE 17.7
Reported Composition of Personal Computers and Printed Circuit Boards.

	Kim et al. [26]	Evangelopoulos et al. [27]
Pb	13,500	49,611
Fe	1400	10,300
Sn	32,400	1530
Cu	156,000	338,690
Ba		1645
Ni	2800	1340
Zn	1600	9410
V		14.8
Au	420	6.61
Ti		1372
Co		3.23
Pd	100	11.6
Mn		78
Ag	1240	398
Sb		40.8
Cr		237
Cd		0.23
Hg		3.65
As		0.264
Pt		0.01
Mo		0.187

TABLE 17.8
Annual Production and Consumption of Metals in Electric and Electronic Equipment Production and Their Ratio [28].

Metal	Production (tons/year)	Demand for Electric and Electronic Equipment (tons/year)	Demand/ Production (%)
Ag	20,000	6000	30
Au	2500	300	12
Pd	230	33	14
Pt	210	13	6
Ru	32	27	84
Cu	15,000,000	4,500,000	30
Tin	275,000	90,000	33
Sb	130,000	65,000	50
Co	58,000	11,000	19
Bi	5600	900	16
Se	1400	240	17
In	480	380	79

the hazardous materials, CRT glass and LCD screens contain Hg and they are treated in special mercury recovery facilities or incinerated in authorized hazardous waste treatment units with extensive flue gas cleaning system. Batteries and conductors are normally sent to other processes for recovery of Cd, Ni, Hg, and Pb [8]. This step remains the most popular because simplicity and low cost is ensured; however, it deals with the recycling of uncontaminated single-type waste [31].

Although manual sorting is efficient and relatively precise, it slows down the next steps of the recovery and recycling process and can become a bottleneck for the whole process. Therefore several new technologies have been introduced for making the dismantling and sorting processes automated.

One solution for automatic sorting is based on infrared (IR) spectroscopy. Lamps with a selected range of IR wavelength (600–2500 nm) are used and the materials reflect the light differently according to their chemical structure. Analysis of the data from the reflection of different materials can distinguish the differences in structure among plastic-coated cardboard, ordinary cardboard, and different kinds of plastics even if their color looks similar. Then the sorting can be performed by a jet of pressurized air, sending the selected material to a different conveyor from the rest e-waste fraction [33].

Two main wavelength regions are used, the near-infrared (NIR) spectrum and the mid-infrared (MIR) spectrum (Fig. 17.6). CMYK (cyan, magenta, yellow, key) sorts paper or carton that has been printed using CMYK.

MIR works on a similar principle to NIR but projects light in the MIR range. The French company Pellenc ST has been piloting this technology since 2008 as a more efficient way to separate paper and cardboard. The new MIR method brings efficiency levels up to 90%, which is an improvement of around 30%.

Other automated techniques involve visual spectrometry that recognizes all colors that are visible and works for both transparent and opaque objects;

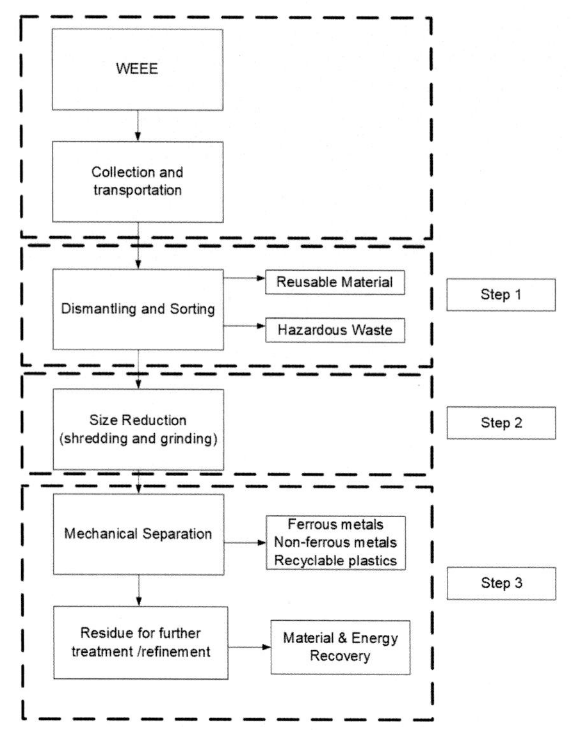

FIG. 17.4 Different steps in waste electric and electronic equipment (WEEE) processing. (Adapted from Ref. [8])

FIG. 17.5 Manual dismantling of personal computers. (Adapted with permission from Ref. [32].)

electromagnetic technique sorts metals with electromagnetic properties, as well as sorts metals from nonmetals and recovers stainless steel or metallic compounds; RGB sorts specifically in the color spectrums of red, green, and blue for specialized applications; and X-ray sorts by recognizing the atomic density of materials.

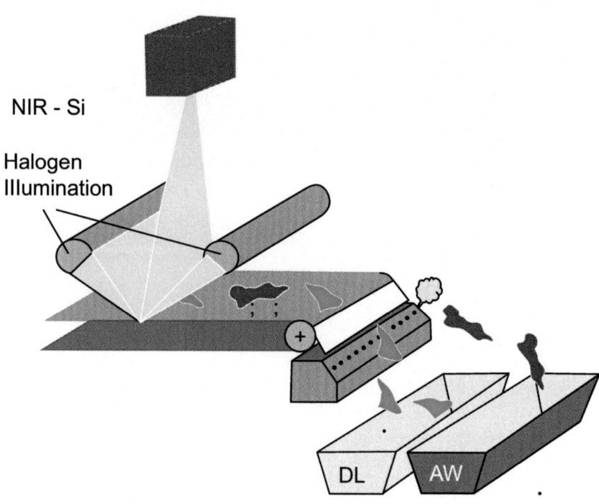

FIG. 17.6 Schematic sorting of different wastes according to their light reflection. *AW*, accepted ware; *DL*, dump load; *NIR*, near infrared. (Adapted from Ref. [33])

According to the Directive 2002/96/EC, hazardous components and materials that must be removed from any separately collected e-waste stream can be categorized as follows [10]:

- capacitors containing polychlorinated biphenyls (PCBs);
- mercury-containing components such as switches or backlighting lamps;
- batteries;
- printed circuit boards of mobile phones and other devices, if the surface area of the circuit board is greater than 10 cm^2;
- toner cartridges;
- plastics containing brominated flame retardants;
- asbestos waste and components that contain asbestos;
- CRTs;
- Freon and hydrocarbons;
- gas discharge lamps;
- LCDs, together with their casing where appropriate, of a surface area greater than 100 cm^2 and all those backlighted with gas discharge lamps;
- external electric cables;
- components containing refractory ceramic fibers;
- components containing radioactive substances above exemption thresholds;
- electrolyte capacitors containing substances of concern.

Sorting and segregation of different fractions of WEEE and hazardous components in the early processing chain pays dividends in the later stages of treatment. The second processing step includes a variety of impaction and shredding methods that are well advanced [34].

Shredding and Grinding

Size reduction can be performed through mechanical shredding, crushing, and grinding processes to achieve size homogeneity of the e-waste mixture. Size homogeneity is essential for all the solid waste sorting processes because the total volume decreases and the efficiency of sorting increases. Furthermore, the proper particle size of the shredded fraction usually is less than 5 mm, which can be achieved through several different shredders in a row [21]. Fig. 17.7 shows different types of industrial mechanical shredders, which can also be used for the WEEE fraction.

Maximum separation of materials from WEEE can be achieved by shredding the wastes to small (or even fine) particles generally below 5 or 10 mm [21]. Studies, both in the laboratory and industrial scales, on personal computers and printed circuit boards have shown that after secondary shredding the main metals present are in the −5 mm fraction and show liberation of about 96.5%–99.5% [21,36].

Mechanical Separation and Sorting

The next step of WEEE processing is the sorting/separation based on the differences in the physical characteristics of the materials, such as weight, size, shape, density, electric conductivity, and magnetic characteristics [21].

The magnetic separation is usually performed with a magnetic drum that attracts the ferrous materials to its surface and deposits all the nonferrous metals and the other materials to a different fraction. In the past decades, more intelligent magnetic separators have been developed, both in design and in operation, mostly because of the use of rare earth alloy permanent magnets, which have the ability of providing very high field strengths and gradients.

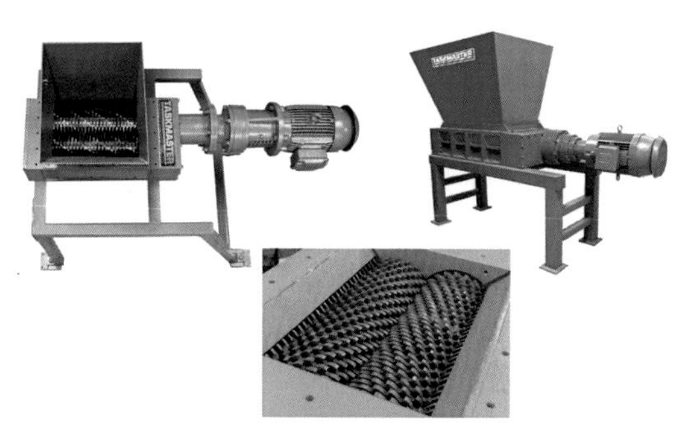

FIG. 17.7 Different industrial shredders. (Adapted from Ref. [35].)

Separation based on the electric conductivity sorts the materials into their different conductivity characteristics. There are three main techniques that can perform this separation, namely, the corona electrostatic separation, the triboelectric separation, and the most commonly used the Eddy current separator. The Eddy current separator, which was developed during the past decade, applies powerful magnetic field to the rest materials in order to repel the nonferrous metals such as Cu and Al, while the rest material fraction continue to the density separation (Fig. 17.8) [21].

Density separation includes a bath with specific density liquid set by the current separation procedure, which separates the low-density recyclable plastics on the top of the bath but the flame retardant–containing high-density plastics sink to the bottom [37].

The techniques in step 2, coupled with the different available separation methods in step 3, can achieve significant recovery of materials (Fig. 17.9). However, the heterogeneity and high complexity of WEEE make it difficult to obtain high recycling rates for some materials and further processing is required.

There are several recycling routes for the treatment of WEEE using pyrometallurgy, hydrometallurgy, pyrohydrometallurgy, electrometallurgy, biohydrometallurgy, or their combinations. The two major routes for processing e-waste are the hydrometallurgical and the pyrometallurgical processes, which usually are being followed by electrorefining or electrowinning for selected metal separation [38].

Pyrometallurgical Treatment

The pyrometallurgical processes are mainly used in the metallurgical industry for ore metal extraction and refining. The mineral processing could provide an alternative route for metals recovery from WEEE, especially this method increases the recovery rates of precious metals in order to upgrade the mechanical separation, which is not so efficient.

The pyrometallurgical process includes a furnace or a molten bath where the shredded scrap of the WEEE fraction is burned to remove the plastic materials, while the refractory oxides and some metal oxides form a slag phase. After this procedure, the recovered materials

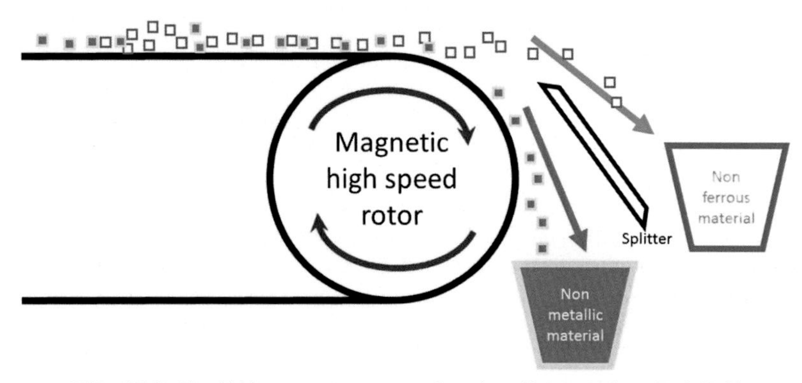

FIG. 17.8 The Eddy current separator function. (Adapted from Ref. [21].)

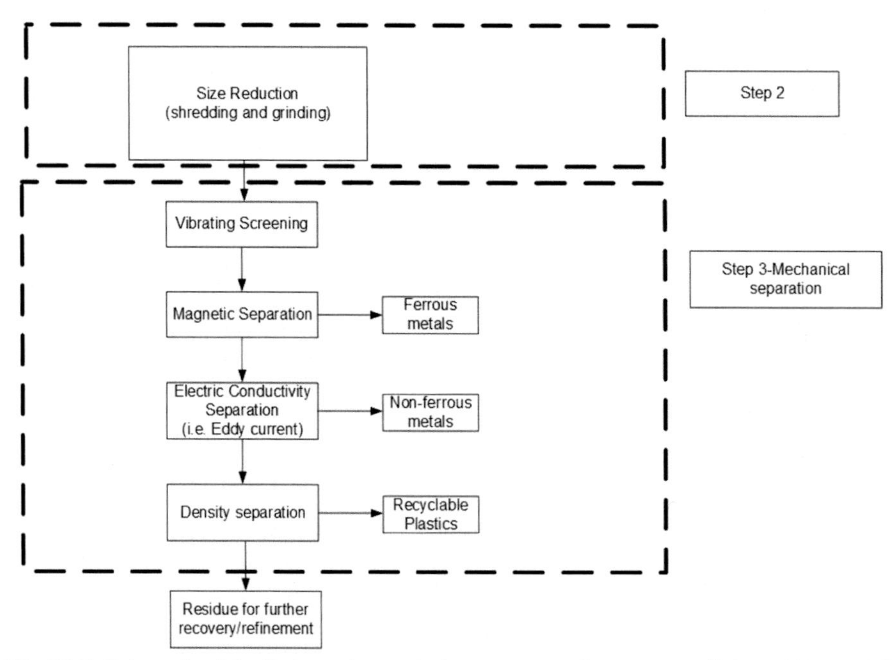

FIG. 17.9 Schematic of the first step in a typical e-waste recycling process. (Adapted from Ref. [8])

are being refined by chemical processing. The plastic fraction is not being recovered as material, but the plastic fraction and other flammable materials are used in the feeding as energy agents to reduce the total cost of the process. It should be stated, however, that as the mineral processing is designed to treat and purify metal ores, by applying this method to the treatment of WEEE, some limitations can occur because of the differences in the size of the particles involved and the material content of those two systems. On the other hand, there are a few industries globally that have extended their fields by implementing "integrated" smelters, which are a combination of the original mineral process with the addition of several extra chemical units for processing e-waste [39].

Globally, the industries that treat e-waste together with metal scrap by pyrometallurgical processes are Noranda process in Quebec, Boliden Rönnskär smelter in Sweden, Outotec in Finland, Dowa in Japan, Aurubis in Germany, and Umicore at Hoboken, Belgium. The Boliden Rönnskär smelter process is presented schematically in Fig. 17.10.

Generally, pyrometallurgical processes are efficient because they can maximize the recovery rates of precious metals, but there some limitations to this process. First, the remaining plastics in the e-waste fraction

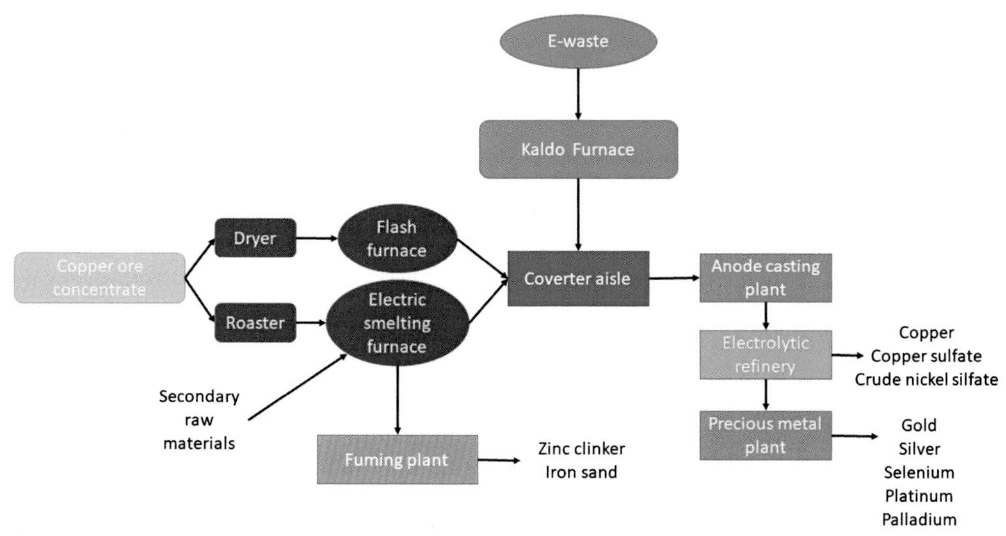

FIG. 17.10 Schematic of the processes used to recover copper and precious metals from ore concentrate, copper scrap, and e-waste at Boliden's Rönnskär smelter in Sweden. (Adapted from Ref. [8])

cannot be recovered in terms of material recovery and can only be used as an energy agent for replacing the necessary coke for the energy requirements of the process. Some metals such as iron and aluminum, together with the remaining glass and ceramic components, which can occur in the input material, cannot be easily recovered, as they end up in the slag phase. Moreover, the environmental hazards of the process are high, as toxic emissions such as dioxins are generated during the smelting process because of the halogenated flame retardants. Finally, extra steps of treatment are necessary because the recovery and purity of precious metals with this method can only be achieved partially [38].

Hydrometallurgical Treatment

The procedures used for metals recovery from WEEE fractions and primary ores are very similar because the same acid and caustic leaching is followed [38]. Generally, leaching is the process of extracting a soluble constituent from a solid by using a solvent. For electronic wastes, first, leaching is removing the base metals and other impurities and then different acids, such as HNO_3-, H_2SO_4-, and HCl-based solutions, are employed for precious metals recovery. Furthermore, other techniques can be used for further purification, such as precipitation of impurities, solvent extraction, adsorption, and ion exchange to isolate and concentrate the metals of interest. Finally, the solutions are treated by electrorefining process, chemical reduction, or crystallization for metal recovery [38,40].

The hydrometallurgical process can be divided into four main steps [41].

1. Mechanical treatment: Before chemical treatment, mechanical treatment is necessary for granulating total the fraction.
2. Leaching: The e-waste fraction passes through a series of acid or caustic leaching procedures, where the soluble component is extracted from a solid by using a solvent. Strong acids are mainly used because of their high efficiency and their ability to leach not only base metals but also precious metals. Mainly for base metals, the most common leaching agent used is nitric acid; for copper recovery, sulfuric acid or aqua regia; for precious metals such as gold and silver, thiourea or cyanide; and for palladium, hydrochloric acid and sodium chlorate.
3. Separation and purification: The solutions with the leached metals pass through a separation and purification process to discard impurities.
4. Precious metals recovery: The final step of the hydrometallurgical process includes electrorefining processes, chemical reduction, or crystallization step, which can separate the precious metals.

PCBs in mobile phones have been subjected to oxidative sulfuric acid leaching that dissolve Cu and Ag partially, whereas oxidative leaching using Cl dissolves Pd and Cu. Cyanidation reported to recover

Au, Ag, Pd, and Cu. Optimized flow sheet permitted the recovery of 93% of Ag, 95% of Au, and 99% of Pd [42]. Moreover, Cu, Pb, and Sn recovery from PCB can be achieved by dissolution in acids and treatment in an electrochemical process to recover the metals separately [43]. Another approach of hydrometallurgical treatment of PCBs involves leaching with NH_3/NH_5CO_3 solution to dissolve Cu. The remaining solid residue is then leached with HCl to recover Sn and Pb. Heating of the produced $CuCO_3$ converts it to CuO [44].

Generally, extraction efficiencies of around 90% for Y, Cu, Au, and Ag; 93% for Li; and 97% for Co, with respective purities of 95%, 99.5%, 80% (A and Ag), 18%, and 43%, have been reported for different hydrothermal treatment of WEEE; however, the relatively low impurities are not suitable for direct commercialization (minimum purity of 99% would be needed), but these products are still marketable to companies for final refinement. Moreover, a final wastewater treatment step should be considered [45]. Life cycle assessment of hydrometallurgical treatment of PCBs identified the equivalent emissions during treatment, which are listed in Table 17.9.

TABLE 17.9
Emissions for the Treatment of 100 tons of WEEE Residues: PCB Granulate From PCBs [45].

Impact Categories	PCB Granulate from PCBs
Global warming (kg CO_2 eq.)	7.8×10^5
Abiotic depletion (kg Sb eq.)	2.2
Acidification (kg SO_2 eq.)	2044
Eutrophication (kg phosphate eq.)	137
Ozone layer depletion (kg R11 eq.)	0.05
Photochemical ozone creation (kg ethane eq.)	154

PCB, polychlorinated biphenyl; *WEEE*, waste electric and electronic equipment.

A summary of the different studies involving hydrometallurgical treatment of WEEE is shown in Table 17.10.

Even though the reported recovery yields of different metals are quite high, limitations to hydrometallurgical treatment have been identified and have slowed down its industrial implementation. The limitations are summarized in the following [38].

1. The timescale of hydrometallurgical processing is long, which can implicate the overall recycling scheme.
2. The fine particle size that is needed for efficient dissolution results in 20% loss of precious metals during the liberation process.
3. Several leachates that are used, such as CN, are extremely dangerous and high safety standards are required. Moreover, extremely corrosive halide leaching requires specialized equipment to avoid corrosion.
4. There is loss of precious metals during dissolution and subsequent treatment, which affects the overall recovery efficiency and process economics.

The industrial processes for recovering metals from WEEE are based on combined technology from pyrometallurgical, hydrometallurgical, and electrometallurgical processes. In most cases, the e-waste fraction is blended with other metallic fraction and is being

TABLE 17.10
Hydrometallurgical Treatment of Waste Electric and Electronic Equipment.

References	e-Waste Component	Leaching agent(s)	Recovered Metals
[42]	Printed Circuit boards (mobile phones)	H_2SO_4, HCl, HCN	Cu, Ag, Pd, Au
[44]	Printed Circuit boards	NH_3/NH_5CO_3, HCl	Sn, Pb, Cu
[46]	Printed circuit boards	*Aqua regia*	Au, Ag, Pd
[47]	Computer circuit boards	HNO_3, epoxy resin, *aqua regia*	Au
[48]	Printed Circuit boards	HNO_3	Pb, Cu, and Sn after electro dissolution in HCl

processed in either Cu or Pb smelters. In the final stage of Cu production, the pure Cu (99.99% purity) is produced by electrorefining, while the precious metals are separated and recovered by hydrometallurgical processes. There are only a few companies globally whose main objective is the extraction of metals from e-waste, namely, Umicore, Noranda, Boliden, Kosaka, Kayser, and Metallo-Chimique [38].

Umicore

Umicore's integrated smelters and refinery is located at Hoboken, Belgium, and can process a high range of different metals such as precious metals (Au, Ag, Pt, Pd, Rh, Ir, Ru), special metals (Se, Te, In), secondary metals (Sb, Sn, Ar, Bi), and, of course, base metals (Cu, Pb, Ni). The average annual capacity of the total unit is 250,000 tons of feed material, and the annual production of over 50 tons of platinum group metals, 100 tons of gold, and 2400 tons of silver constitute to the world's largest recycling facility for the extraction of precious metals [49].

The fraction of WEEE that concludes to this unit is either nonferrous fraction or mostly printed circuit boards, which are fed to the smelting process. The plastics are partially replacing the necessary coke for the process as an energy agent. The copper from the e-waste conclude to the leaching and electronwinning process where it is purified while the precious metals are being removed. The slag from the Cu smelter is fed to the Pb blast furnace that produces impure Pb, which will be purified in different steps. Finally, the precious metals from all the other processes are fed to the precious metals refinery where cupellation occurs.

Boliden

Boliden's Rönnskär smelter is located in Skelleftehamn in the northern part of Sweden and has an annual production of more than 200,000 tons of Cu; 13 tons of Au; 34,000 tons of zinc clinker; 28,000 tons of Pb; and 485 tons of Ag. The total capacity of the unit was increased from 45,000 tons to 120,000 tons per year. During the past years, some modifications of the process were made and the Kaldo furnace was built to process more low-grade WEEE fraction such as printed circuit boards and nonferrous metals [51,52].

The crushed WEEE fraction with high Cu content is fed directly into the electric smelting furnace, whereas the printed circuit boards and nonferrous fraction of WEEE are fed into the Kaldo furnace where they are combusted with a supply of oxygen and air. The plastics contained in the fraction are also burned, contributing to the process' energy requirements. The Kaldo furnace products, which include a mixture of Cu, Ag, Au, Pt, Pd, Ni, Se, and Zn, are further processed to the anode casting plant, the electrorefinery, and the precious metals plant. The flow sheet of the whole process takes place in Boliden's unit, which is presented in Fig. 17.10.

Noranda

Noranda's smelter is named Horne and it is located in Rouyn-Noranda in Quebec, Canada. This commercial pyrometallurgical process is the largest and most advanced recycling plant of its kind in North America and has a unique ability to process complex feeds such as the e-waste. More specifically, the feed material is composed of certain industrial copper scrap and WEEE such as circuit boards either low grade or high grade, cell phones, pins/punchings/integrated circuits/components, lead frames/trims/bare boards, sweeps, insulated consumer wire/degaussing wire, and copper yokes. The total annual capacity reaches 840,000 tons of copper and precious metal–bearing materials [53].

The process includes some pretreatment of the e-waste, and together with the copper concentrate the e-waste is fed into the reactor, where the molten bath reaches 1250°C. Plastics and other combustible materials contained in WEEE are burned to produce energy for the maintenance of high temperatures and for the reduction of energy costs. All the impurities, including Fe, Pb, and Zn, are converted into oxides and end up in a silica-based slag. The melted liquid Cu is further processed in the converter to improve its purity, whereas the residue that contains the precious metals is further processed by electrorefining of anodes [38].

Thermochemical Treatment

As mentioned earlier, WEEE is a mixture of various materials whose effective separation is the key for development of a recovery system.

Combustion

Generally, combustion of WEEE is not a sustainable option anymore, as it contributes only to energy recovery, which is the last means of recycling and it has to be regarded as the last resort.

Unlike other fossil-fueled incineration plants, waste-to-energy (WtE) plants have significantly lower energy efficiencies (13%–24%) because of lower steam temperatures, fouling, and slugging. Apart from these problems, there is emission of acidic gases such as HCl, SO_x, NO_x, HF, and volatile organic compounds (such as polyaromatic hydrocarbons, PCBs, and polychlorinated dibenzo-*p*-dioxins and dibenzofurans [PCDD/Fs]), which are harmful. Moreover, solid residuals of the final process cannot be recovered even up to a point and constitute a serious problem due to their heavy metal content [54].

Among the potential pollutants, dioxins attract the most attention. The formation of PCDD/Fs is catalyzed by Cu present in the fly ash (where Cl_2 is more active than HCl during the formation of PCDD/Fs [55]) in a process also known as de novo synthesis of PCDD/F and maximum formation occurs at about 300°C [56]. The emissions during combustion pose an environmental threat and a well-advanced air pollution control system should be coupled to WtE plants. Emissions from the combustion of WEEE components are listed in Table 17.11.

The main problem is associated with the PCDD/F and PCBs. The destruction of such compounds that might be formed can be achieved by application of high-temperature processing (>1300°C); however, the costs (both capital and operating) associated with such treatment have incentivized researchers for other routes involving catalytic processes at lower temperatures [59].

The two main catalytic routes reported in the literature include total oxidation of chlorinated hydrocarbons at temperatures between 300 and 550°C over supported noble metal catalysts (for example, Pt, Pd, and Au) and catalytic hydrodechlorination, in which chlorinated hydrocarbons are transformed in the presence of hydrogen into alkanes and HCl. Commonly used catalysts are supported Ni, Pd, and Pt. Although hydrodechlorination offers economic and environmental advantages, it is not used often [60–63].

The use of WEEE in cement kilns and power plants faces several limitations because of the chlorine and certain heavy metal content and the particle size of mixed plastic waste from the WEEE [64].

TABLE 17.11
Yields of Gases and Volatiles (mg/Kg) From the Combustion of Different WEEE Components [55,57,58].

	WCB (600°C) [57]	NmfWCB (600°C) [57]	WCB (850°C) [57]	NmfWCB (850°C) [57]	EW (500°C) [58]	EC (500°C) [58]	PCB (1400°C) [55]
HBr	12,700	45,900	11,800	59,600			~1800
Br_2	1000	3700	6700	5800			~400
HCl	400	5300	500	6500			
Cl_2	100	500	100	300			
CO	89,000	288,000	94,800	295,600	39,000	24,000	
CO_2	441,600	595,000	677,400	506,200	210,000	90,000	
Methane	2910	8310	30	30	1687	576	
Ethylene	930	1540	–	–	512	127	
Propylene	310	340	–	–	337	191	
Acetylene	330	830		150	20	24	
Propyne	120	250	–	–	–	–	
Benzene	650	1290	–	–	18	68	
Toluene	–	1070	–	–	–	268	
Bromomethane	200	3740	–	–			
Acetone	–	5490	–	2220			
Cyanogen bromide	–	–	160	750			
Total PCDD/F & PCBS (pg WHO2005 TEQ/g)					6.806	3.057	

–, not detected or <10 mg/kg; *EC*, electronic circuit; *EW*, electronic waste; *Nmf*, no metal fraction; *PCB*, printed circuit board; *WCB*, waste circuit board.

Gasification

Gasification refers to partial oxidation of carbonaceous material (by reaction with a controlled amount of oxygen and/or steam) at elevated temperatures to produce a mixture of syngas (CO and H_2), CO_2, and other light hydrocarbons. It is a way to convert waste into energy carriers, with simultaneous transformation into a less voluminous substance, resulting in a more sustainable and effective waste management.

The produced gas can be used in many applications such as in lime and brick kilns, in metallurgical furnaces, as raw material syngas, as synthetic natural gas, in the Fischer-Tropsch process, in chemical synthesis, or even combusted in gas turbines, providing higher efficiencies and market opportunities [54,65]. The remaining solid material in the form of slug (high-temperature gasification >1000°C) could be used as additive in building materials.

However, owing to the more strict rules and requirements for higher recycling rates both for plastics and metals, this does not represent a viable option for e-waste anymore.

Many processes have been developed through the years that have used plastic-containing waste as feedstock (mainly municipal solid waste). Most of the processes employed high temperatures for maximization of gaseous products and destruction of unwanted halogenated hydrocarbons. Some of the processes are the Texaco process, Thermoselect, Sky Gas, and the Sustec Schwarze Pumpe-Sekunddärohstoff Verwertungs Zentrum (SVZ) process [64,66].

However, the requirements for low metal content and low halogen content (in the feed should not exceed 10% in order to avoid the problems of corrosion) in the feedstock make them inappropriate for WEEE treatment for maximum energy and material recovery. Moreover, the costs of preparation and treatment are high, limiting this method of recovery to mixed plastic wastes that are difficult to separate, even polluted [64]. Nevertheless, gasification (especially steam gasification) at lower temperatures (<900°C) coupled to sophisticated gas treatment could serve for both material and energy recovery.

Pyrolysis

Another method for enabling recycling of both organic and inorganic matter of WEEE is pyrolysis. Pyrolysis aims to provide added value materials (long polymeric chains of the plastic to reusable monomers, hydrocarbon fuels having carbon number distribution in the range of C_1-C_{50}, and valuable aromatic solvents such as benzene, toluene [67]) and to separate the plastic and inorganic fractions, which is in accordance with the principles of sustainable development [68]. In addition, life cycle assessment studies have shown that pyrolysis provides significant resource savings without an impact on climate change or landfill space [69], and therefore it is regarded as the most promising processing route for WEEE plastics [70,71].

Thermal decomposition of the polymers in the absence of oxygen leads to the formation of gases, oil, and char, which can subsequently be used as chemical feedstock or fuels. The degradation of the plastic fraction enables the separation of the organic, metallic, and glass fiber fractions, which makes recycling of each fraction more viable. During the course of pyrolysis, the long polymeric chains break down to smaller fragments (molecules); the quality and quantity of those molecules depend on the process conditions such as temperature, heating rate pressure, and residence time.

For most plastics, pyrolysis starts at around 300°C; however, the actual onset depends on additives and characteristics of the different materials. The decomposition is a series of complex reactions and the heat for reactions that is required can be provided by the combustion of gases produced after removal of some valuable and some hazardous compounds.

Generally, high temperatures (>600°C) favor the production of smaller and simpler gaseous molecules, whereas low temperatures (~400°C) favor the formation of liquid products (oil) [72]. Other authors distinguish those pyrolysis regimes in *thermal cracking* (>700°C), in which gaseous products are aimed, and *thermolysis* (400−500°C), in which liquid products are desired [68]. As understood, those limits are somehow arbitrary and depend on the actual feedstock and process conditions. Buekens has categorized the factors that affect the product distribution, as shown in Table 17.12.

Several researchers have studied the pyrolysis of plastics, providing essential information about the different degradation mechanisms. However, plastic-containing WEEE differ greatly from model compound studies of flame-retarded plastics, which are present in WEEE.

The most common WEEE component that has been studied as a basis for pyrolytic recycling of WEEE is printed circuit boards or epoxy resins that represent the main polymeric fraction of WEEE [27,73−79]. Phenols and substituted phenols are the most abundant compounds found in the liquid products, and brominated compounds (such as mono- and dibrominated phenols, bisphenol A, and mono-, di-, tri-, and tetrabrominated bisphenol A) are also produced (Fig. 17.11) [75−77,80].

TABLE 17.12
Factors Affecting the Product Distribution of Pyrolysis [72].

Factor	Effect
Chemical composition of material	The primary pyrolysis product relates directly to the composition of the material and also to the mechanism of decomposition (thermal or catalytic)
Pyrolysis temperature and heating rate	Higher heating rates and temperatures favor the formation of smaller molecules
Residence time	Longer residence time favors secondary reactions that lead to increased gas yields and coke and tar formation
Reactor type	Determines the heat and mass transfer rates and the residence time of the products
Pressure	Lower pressures minimize the condensation reactions between the reactive vapors and this results in less coke and heavy ends formation

Generally, higher temperatures and longer times make debromination more extensive at the expense of liquid yield [73,75–77].

Contamination of oil by bromine-containing phenols (potentially hazardous compounds) emitted during heating of polymers flame retarded with TBBPA-based flame retardants. So reduction of the amount of brominated phenols in the pyrolysis oil in favor of less toxic substances is a way to add value to the whole recycling process [81].

The size of the circuit board particles has been identified as a crucial parameter for the pyrolysis behavior and the onset of pyrolysis. Heat transfer controls the degradation process of printed circuit boards if the particle size of the material exceeds 1 cm^2 due to heat transfer limitations [79]. It has been indicated that pyrolysis of printed circuit boards improves Cu recovery by leaching as compared with nontreated boards [78].

A combined approach of pyrolysis with simultaneous separation of metals and glass components using molten salts is also under development [74]. This indicates the versatile role of pyrolysis in the overall recycling of WEEE. HIPS and ABS (halogenated or not) are other polymers that have been studied as model compounds that represent the behavior of plastics in WEEE [82,83]. Other studies have used these plastic materials by impregnating several types of flame retardants [84], whereas other studies have focused on real WEEE materials collected from recycling facilities [81].

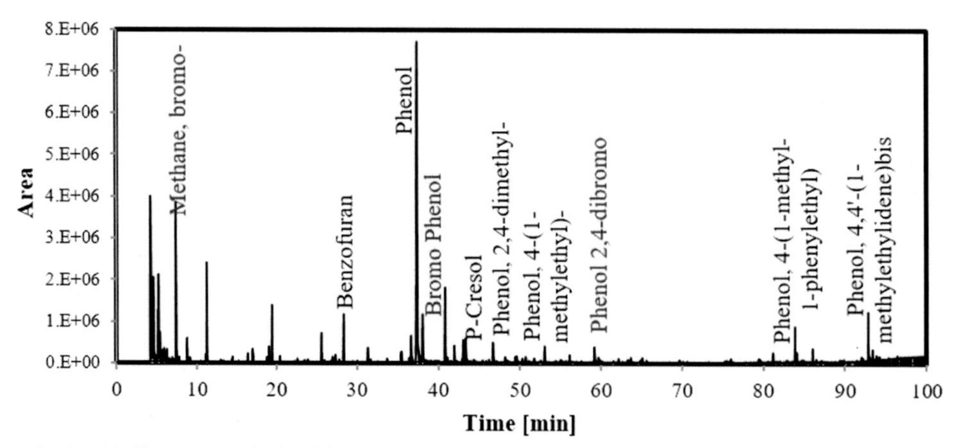

FIG. 17.11 Pyrograms obtained from the pyrolysis of printed circuit boards. (Adapted from Ref. [73].)

Characterization of products indicates that there are many halogenated compounds in the resultant liquid product. Hall and William's study [85] on mixed WEEE fraction from a WEEE recycling site indicates that pyrolysis gases were mainly halogen free, while the liquid fraction contained valuable chemicals such as phenol, benzene, and toluene [27].

As indicated in the above-discussed studies, the resultant oil from pyrolysis, due to the presence of flame retardants, is heavily contaminated with halogenated compounds and must be further processed in order to be considered as a marketable product [86]. Understanding their formation and facilitating their removal from the final product is a major consideration in developing processes for mixed plastic wastes [66]. The necessity of Br and Cl removal from the final oil product requires upgrading of the oil products or taking dehalogenation measures during the pyrolysis process and directly obtain oils that are not contaminated. Dehalogenation is of major importance for the pyrolysis process and different options are discussed in the section Dehalogenation of Pyrolysis Oil: Catalytic Upgrading.

Pyrolysis technologies. Research on pyrolysis processes has been conducted using several reactor types such as fluidized bed reactors, rotating cone reactors, rotary kilns, tubular or fixed bed reactors, and stirred batch reactors. The product distribution of each reactor depends on the process condition as summarized in Table 17.12. Most pyrolysis studies are conducted in batch or semibatch equipment, and therefore it is hard to study the influence of operating parameters on a real continuous process. The main results have been presented earlier, and in this section, pyrolysis technologies that slightly differ from the rest will be discussed.

Vacuum pyrolysis. A method to facilitate thermal decomposition of the plastic matter of WEEE is the application of vacuum pyrolysis. Studies focusing on printed circuit board scraps aimed for solder recovery and separation of glass fibers and metals have been reported [87–89].

During pyrolysis, vacuum application reduces the decomposition temperature and shortens the residence time of the produced vapor. Moreover, it minimizes secondary cracking and condensation reactions among the vapors, thus maximizing the liquid product.

Cu recovery rates of 99.86% were reported with Cu grade of 99.5% [87], while the recovery of solder is possible after heating the solid residue at 400°C and applying centrifugal forces [88].

The obtained oil mainly consisted of phenolic and benzofuranic structures. However, studies showed that pyrolysis oils contained a significant number of brominated compounds, the most abundant of which was 2-bromophenol and 2,6-dibromophenol. Therefore the oils could be used as fuels or chemical material resources after proper treatment [87–89].

Vacuum pyrolysis has been studied as a part of a treatment step for In recovery from LCD panels. Results from pyrolysis indicate that liquids could be used as chemical feedstock or fuels [90].

Microwave pyrolysis. Microwave pyrolysis is another pyrolysis alternative that has been studied for the treatment of WEEE. Microwave irradiation is an efficient way to heat up the material and initiate its thermal decomposition.

WEEE material subjected to microwave irradiation has shown reduction indicating that it is a favorable method to treat residual wastes and to increase the solid metal-to-organics ratio, making the waste more valuable for further recycling treatment [91].

Addition of activated carbon has been found to enhance the heating of the material, by absorption of the microwave irradiation, improving its decomposition products [92]. Liquid products contained significant amount of Br; however, significant debromination can be achieved downstream of the pyrolysis reactor by treatment with $CaCO_3$ [92].

Commercial demonstration projects. There have been several attempts to commercialize pyrolysis technologies through the years. The closest process to commercialization has been reported to be the Haloclean process, which converts WEEE into debrominated pyrolysis oil, noble metal–rich solid residue, and gaseous bromine. Sea Marconi Technologies has built a plant with a capacity of 3000 t/y [93].

Another process for WEEE treatment that has been reported to be in operation in larger scale is the catalytic depolymerization process, in which e-waste is converted to diesel fuel by pyrolysis followed by catalytic treatment. A key part of the process is the heating of the material, which is achieved by means of mechanical friction. This technique is said to give very accurate control of the temperature in the process and prevents overheating, which would otherwise lead to the production of dioxins. However, little information is available about the reliability and wear of the different process components [93].

During this process, high-grade diesel is produced. Part of the produced diesel is directly converted into electricity by means of powerful generators that is then used to supply the entire recycling plant [93].

The Veba Combi Cracking Unit KAB process is another example of a pyrolysis-hydrogenation process able to process WEEE with high halogen content. The process was initially designed for coal liquefaction and it was further modified to treat vacuum distillation crude oil residues into synthetic crude. Since 1988, it has been substituted with chlorine-containing waste, especially PCBs. The plant was able to treat 10/h of waste but it had to close down in 2003 because of process economics [64,66].

Another plant that has been closed down because of economical competiveness is the BASF thermolysis plant. This process had a capacity of 15 ktons/year of plastic waste for energy and materials recovery. The process included a liquefaction step at low temperatures (300–350°C) followed by a pyrolysis step in a tubular cracking reactor at higher temperatures (400–450°C). The main products of the process was naphtha for ethylene and propylene production, olefins, aromatics, and a heavy fraction, which was supplying a gasifier for producing synthesis gas (Fig. 17.12) [64,66,95].

Dehalogenation of pyrolysis oil: catalytic upgrading. As presented by the abovementioned research studies and demonstration attempts, the maximum output of the pyrolysis process can be achieved only when the liquid product is marketable or can be used for fueling a recycling plant. Therefore dehalogenation of the liquid product is required.

In fact, it is a crucial step, for the recycling of WEEE plastics, to efficiently remove organic halogen [96]. Dehalogenation could be carried out before, during, or after the decomposition of WEEE plastics [97].

Dehalogenation before waste electric and electronic equipment decomposition. This method of dehalogenation requires a two-staged pyrolysis process: in the first stage, a low-temperature treatment (~300°C) produces mainly halogenated compounds and in the second stage, pyrolysis at higher temperatures produces halogen free/or oil with reduced halogen content. However, a distinction between Cl-containing and Br-containing compounds should be made. While Cl-containing compounds exhibit a low-temperature release of HCl at the low-temperature region, Br-containing retardants without synergistic agents (such as Sb_2O_3) do not decompose earlier than the rest of the WEEE material and their decomposition occurs at the same temperature range as the rest of the WEEE and thus different methods of dehalogenation must be

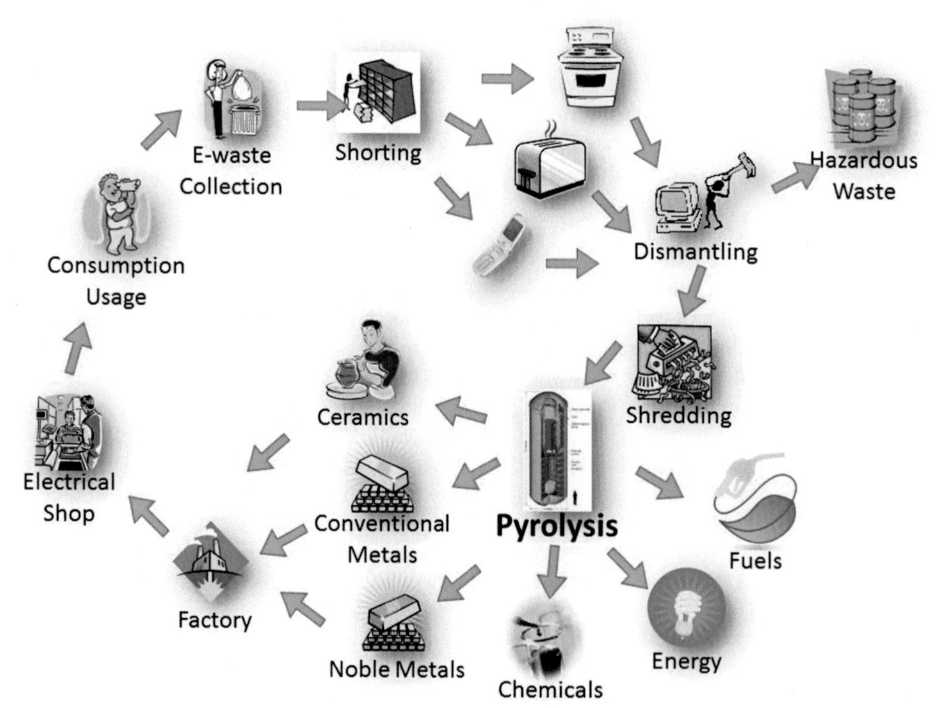

FIG. 17.12 Closing the loop of waste electric and electronic equipment by using pyrolysis technology. (Panagiotis Evangelopoulos (author).)

used. Nevertheless, behavior similar to Cl-containing flame retardants is observed when Sb_2O_3 is present [97]. Efforts have been made to remove the halogens by using solvents or supercritical CO_2 with positive results but scaling up of the process might be challenging [81,98]. The introduction of a two-staged process implies that the capital and operating costs are higher than those of the one-staged pyrolysis process and that the composition of the processed material is well known and no significant fluctuations in the compositions/or processed material are expected.

An alternative combined pyrolysis-gasification approach for the recovery of Br and Cl from WEEE was proposed by the Pyromaat process in which pyrolysis of WEEE takes place at low temperatures to produce metal-rich char, which is subsequently subjected to pyrometallurgical treatment. The Br-containing vapors and gases are injected into a high-temperature gasification process where they are converted to syngas. Halides are then recovered by performing a caustic wash of the produced gas [99].

Dehalogenation during waste electric and electronic equipment decomposition. Dehalogenation during the course of pyrolysis can be achieved by controlling the pyrolysis parameters and/or inclusion of additives/or catalysts. The control of the reaction parameters involves the slow heating of the material and the long residence time of the produced vapors. In that way the organic brominated compounds that are formed in the first stages of the decomposition spend longer time in the hot zone, which allows for their destruction [100]. However, lower yields of oil are to be expected, while the long size of the reactor suggests increased costs compared with those of the typical fixed or fluidized bed reactors.

In a similar approach to prolong the residence time of the produced vapors in a hot environment, Hlain et al. [125] introduced a reflux condenser operating at 200°C at the top of the pyrolysis chamber and managed to reduce the bromine content in the final liquid product by almost 10 times. Similarly, Luda et al. [75] have tried to thermally dehalogenate Br-containing plastics. They pyrolyzed 2,4-dibromophenol with a series of plastics for extended periods and managed to successively convert the harmful brominated compounds into HBr and a highly polyaromatic char.

Generally, the reason for the use of additives is to transform waste-bound halogens to metal halides (M_xX_y) or metal oxohalides ($M_xO_yX_z$).

Dehalogenation by introduction of additives in the pyrolysis reactor was achieved by Blazsó et al. [96] while pyrolyzing PCB scraps in the presence of NaOH or sodium-containing silicates. Introduction of the additives resulted in enhanced bromomethane evolution and suppression of the formation of brominated phenol.

Zhu et al. [101] investigated the decomposition of PVC samples with different Ca-based additives ($CaCO_3$, CaO, $Ca(OH)_2$) by thermogravimetry-FTIR analysis. It was shown that Ca is an effective additive to capture hydrogen halides. Without additives, Cl was completely found as HCl after degradation. Ca-based additives at Ca/Cl molar ratios of 1.5 for $Ca(OH)_2$, 1.5−2.5 for CaO, and 2.6 for $CaCO_3$ were suggested. Ca-based additives during pyrolysis of brominated ABS have also shown good activity in reducing the bromine and antimony contents in liquid products at the expense of the final liquid yield [102]. Similarly, Alston and Arnold [69] have identified that Ca-based fillers (mainly $CaCO_3$) used in PVC can act as sorbents for Cl removal in the form of calcium chloride ($CaCl_2$).

Various combinations of cracking catalysts and sorbents for halogenated compounds ($CaCO_3$ and red mud) decreased the amount of all heteroatoms in the pyrolysis oils of PCBs; after pyrolysis at 300−540°C the oils were passed into a secondary catalytic reactor [103]. ZnO has also been used for determination of debromination efficiency during treatment with TBBPA [104]. It has been shown that 64−70 wt% of ZnO has been converted to $ZnBr_2$. However, the onset of volatilization of $ZnBr_2$ is reported to be around 340°C [105]. The same group have studied the effects of Sb_2O_3 in the debromination of TBBPA and indicated that the onset of vaporization of the formed $SbBr_3$ is at ~340°C, that is, as soon as it is formed [106].

The effects of both antimony oxide (Sb_2O_3) and bismuth subcarbonate $(BiO)_2CO_3$ on the debromination of thermally stable (decomposition >500°C) decabromodiphenyl oxide (DBDPO) have been studied by Bertelli et al. [107]. It has been shown that at a temperature range of 600−700°C, DBDPO undergoes debromination forming $BiBr_3$ and $SbBr_3$, which are volatile. The reaction mechanism is different with BiOBr formation at the initial stages and then to $BiBr_3$. The reducing environment (reduction of the BiOBr to metallic Bi) competes with the debromination reaction. From the two different oxybromide bromides formed, the richer in bromine is less stable.

Debromination using ZnO, Fe_2O_3, La_2O_3, CaO, and CuO of TBBPA and real printed circuit boards has been studied by Terakado et al. [108,109]. The addition of the different metal oxides resulted in reduction of at least 30 wt% in CuO and reached 99 wt% in the case of La_2O_3 in a temperature range of 400−800°C However, the presence of glass fibers could reduce the effectiveness of calcium oxides due to the formation of silicates [109].The effect of ZnO addition during the treatment of real printed circuit board sample in an oxidative environment was also investigated by the same authors. They concluded that the presence of an oxidative atmosphere resulted in the production of more Br_2 and HBrO instead of HBr [110].

The pyrolytic and oxidative decomposition of TBBPA in the presence of Pd or PdO has been studied by Kuzuhara and Sano [111]. Results indicated that Pd did not have any activity in debromination regardless of the temperature or oxygen presence. PdO showed debromination activity with debromination reaction occurring at a temperature range of 280−450°C, whereas treatment at higher temperatures indicated bromide decomposition.

TBBPA and TBBPA-diglycidyl ether mixtures have been pyrolyzed in the presence of Cu, Ag, and Au. It has been shown that Au is insensitive to the presence of brominated compounds, while, in line with the studies mentioned earlier, Cu showed 50% conversion to bromides. Ag showed similar activity to Cu. The same group also assessed the debromination efficiency of PbO present (or intentionally added) during the decomposition of TBBPA, which proceeded with an efficiency of approximately 70% (80% maximum). However, volatilization of $PbBr_2$ started at temperatures as low as 315°C, while complete volatilization could also be achieved during isothermal treatment at a temperature of 650°C in an oxidative environment [112]. Other metal oxides that have been used for dehalogenation of PVC and containing streams include Co_3O_4, Cr_2O_3, MoO_3 [113].

Catalytic pyrolysis of waste electronic and electric equipment. In this dehalogenation method a suitable catalyst is used to carry out the cracking reactions and transform the organic halogenated compounds during pyrolysis (in situ). The presence of catalyst lowers the reaction temperature and time. The additional benefit of catalytic degradation is that the products are of a much narrower distribution of carbon atom number with a peak at lighter hydrocarbons [114].

Y and ZSM-5 zeolite catalysts have been used for the removal of organobromine compounds during the pyrolysis of HIPS and ABS that were flame retarded using decabromodiphenyl ether. Generally, use of zeolite catalysts increases the amount of gaseous hydrocarbons produced during pyrolysis and decreases the amount of pyrolysis oil, with increased amount of coke being formed on the catalyst surface. Amounts of valuable products such as styrene and cumene are reduced; however, other useful compounds such as naphthalene are formed instead.

Both the zeolites (especially the Y zeolite) are very effective in removing volatile organobromine compounds but they are found to be less effective in removing antimony bromide from the volatile pyrolysis products [115]. HIPS and PC (both model compounds and in commercial level from CD fraction) were tested in the presence of ZSM-5 and MgO catalysts. Catalytic treatment seemed to lower the monomeric fraction (compared with thermal pyrolysis) in favor of other desirable chemicals. No information regarding halogen content in the liquid was presented [116].

Waste fluid catalytic cracking (FCC)catalysts were also used by Hall et al. [117] in order to observe the influence on brominated HIPS and ABS plastics. Unlike the Y zeolites investigated by the same researchers, FCC catalyst did manage to reduce bromine content but it did not alter the product distribution significantly.

Blazsó and Czégény [118] have investigated the introduction of other zeolite catalysts during the analytic pyrolysis of TBBPA flame retardant in an attempt to identify debrominating effects. The catalysts used included NaY, 13X, 4A, and Al-MCM-41 zeolites. It was found that zeolites with small pore size, such as 4A, show limited debromination activity, whereas the medium-sized pores of NaY showed enhanced activity toward debrominated products. Surprisingly Al-MCM-41 did not show any debromination activity, but all brominated bisphenols have been cracked to bromophenol and dibromophenol.

Similarly, polystyrene plastics containing brominated flame retardants have been pyrolyzed at 400°C in the presence of 13X and 4A zeolites as well as hydrotalcite (a layered double hydroxide composed of metal complex hydroxide $\left[M^{2+}_{1-x}M^{3+}_{x}(OH)_2\right]^{x+}\left[(A^{n-})_{\frac{x}{n}}\cdot nH_2O\right]^{x-}$, ($x = 0.22 - 0.33$) where M^{2+} and M^{3+} are Mg^{2+} and Al^{3+} metal ions, respectively, and A^{n-} is an anionic species [119]). As in the study by Blazsó and Czégény [118], the zeolite 4A showed limited debromination activity and produced a more viscous product. Hydrotalcite and the zeolite 13X promoted cracking to low-molecular-weight products; however, debromination of the liquid using the 13X zeolite did not exceed 35%. On the other hand, the use of hydrotalcite not only promoted cracking but also resulted in 90% reduction in Br in the liquid products [120].

Sakata et al. [121] have used the iron oxide carbon composites (Fe–C(1)[TR00301], Fe–C(2)[TR97305], and Fe–C(3)[TR99300] containing magnetite, as well as iron oxide catalysts (α-Fe2O3,γ-Fe2O3), for the dehalogenation of mixed plastic/PVC-derived oil (plastic derived oil). It has been concluded that iron oxides act not only as catalysts but also as sorbents for Cl and Br, thus achieving dehalogenation of the oil. In the same study, iron oxide and calcium carbonate carbon composite catalysts were also used during thermal degradation by using the same waste plastics mixture. It was demonstrated that a one-step dehalogenation process can also possibly achieve complete dehalogenation. The same researchers have also reported the use of FeOOH as a dehalogenation catalyst sorbent [122,123].

From an economic perspective, finding the optimum catalysts and thus reducing the cost even further will make this process an even more attractive option. This option can be optimized by reuse of catalysts and the use of effective catalysts in lesser quantities. This method seems to be the most promising to be developed into a cost-effective commercial polymer recycling process to solve the acute environmental problem of plastic waste disposal [114].

Hydrodehalogenation: upgrading of produced pyrolysis oil. The dehalogenation of produced pyrolysis oil has the advantage of decoupling the oil production and upgrading processes and the goal is to remove the halogenated compounds after optimized pyrolysis process [97]. This can be done either by implementation of the catalytic routes mentioned in the catalytic pyrolysis of waste electronic and electric equipment or by the hydrodehalogenation (HDH) process. HDH includes reaction of pyrolysis oil in the presence of hydrogen-donating media and catalysts resulting in a hydrogenated liquid and production of valuable hydrogen halides.

Vasile et al. [103] have investigated the use of commercial hydrogenation DHC-8 catalyst and NiMo/AC catalysts. Results showed that brominated compounds were eliminated after processing with Br being converted to HBr. Wu et al. [124] demonstrated that HDH of chlorobenznene is feasible in the liquid phase in the presence of NaOH and different catalysts (Ni/AC, Ni/γ-Al_2O_3, Ni/SiO_2, Raney Ni) at relatively mild conditions. Several other halogenated hydrocarbons that could be found in the pyrolysis oil from WEEE have been reported to successfully dehalogenate over Rh-, Ru-, Au-, Pt-, Pd-, and Ni-based catalysts (108–112), achieving complete dehalogenation, and demonstrated that HDH is feasible producing a marketable organic product.

CONCLUSIONS AND PERSPECTIVES

The successful recycling route of WEEE should aim for maximum output of different recycled items and for comprehensive processes, which include recovering and recycling of the ceramic and organic fractions. The complexity of WEEE poses a serious challenge for the implementation of an efficient recycling route. A disassembly stage is required to remove hazardous components, with manual dismantling still being the core operation; however, automated dismantling processes have been developed. Crushing and separation are the key points for successfully improving further treatment. Physical recycling is a promising recycling method without environmental pollution and with capital cost, low energy cost, and diversified potential applications of products. However, separation between the metallic and the nonmetallic fraction from the WEEE has to be enhanced.

Metal recovery can be performed by traditional pyrometallurgical approaches on metal-concentrated WEEE fractions. Even though pyrometallurgical processing is attractive and commercially available, in order to be of economic interest for the Cu smelters, the Cu content should be higher than 5% wt, which is not generally the case for WEEE [66]. Compared with pyrometallurgical processing, hydrometallurgical treatment targets specific metals recycling, with high recovery yields of metals. However, environmental concerns, process requirements, and economics slow down its implementation in industrial scale.

High-temperature gasification technologies have been tested for a long time with several limitations due to the halogen and metal content of the WEEE. Although in the short term, it does not seem probable that a WEEE gasification unit could be viable, lower gasification temperatures coupled to sophisticated gas treatment facilities could be proved effective in the future. Pyrolytic approach is attractive because it allows recovery of valuable products from gases, oils, and residues. The main limitation of pyrolysis processes is the halogenated oil that is obtained. Therefore the development in catalytic processing/upgrading of oil (in situ or ex situ) is the most crucial step for maximum exploitation of all product streams and economic viability. Recent developments in catalytic processing indicate that WEEE pyrolysis will represent an economical and sustainable route. Moreover, pyrolysis can be benefitted by new developments and requirements (RoHS) in electronics industry and by the use of more environment-friendly substances.

REFERENCES

[1] Baldé CP, Wang F, Kuehr R, Huisman J. The Global E-Waste Monitor 2014;2015.
[2] Eurostat. Waste electrical and electronic equipment (WEEE) by waste operations. 2015. http://appsso.eurostat.ec.europa.eu/nui/show.do?dataset=env_waselee&lang=en.
[3] Environmental Protection Agency. Electronic products generation and recycling in the United States, 2013 and 2014. 2013. p. 1–14.
[4] Ongondo FO, Williams ID, Cherrett TJ. How are WEEE doing? A global review of the management of electrical and electronic wastes. Wastes Management 2011;31: 714–30.
[5] Wang R, Xu Z. Recycling of non-metallic fractions from waste electrical and electronic equipment (WEEE): a review. Wastes Management 2014;34:1455–69.
[6] Dias P, Javimczik S, Benevit M, Veit H. Recycling WEEE: polymer characterization and pyrolysis study for waste of crystalline silicon photovoltaic modules. Wastes Management 2017;60:716–22.
[7] Dimitrakakis E, Janz A, Bilitewski B, Gidarakos E. Small WEEE: determining recyclables and hazardous substances in plastics. Journal of Hazardous Materials 2009;161:913–9.
[8] Swedish Environmental Protection Agency. Recycling and disposal of electronic waste. Health Hazards & Environmental Impacts; 2011.
[9] United States EPA. Preliminary assessment of the flow of used electronics in selected States . 2016. p. 97. Illinois , Indiana , Michigan , Minnesota , Ohio , and Wisconsin.
[10] European Parliament and Council. Directive 2002/96/EC of the European Parliament and of the Council on waste electrical and electronic equipment (WEEE). Official Journal of European Union 2003;L 37:24–38.
[11] European Parliament and Council. Directive 2012/19/EU of the European Parliament and of the Council on waste electrical and electronic equipment (WEEE). Official Journal of European Union 2012;13:1–24.
[12] European Commission. Fewer risks from hazardous substances in electrical and electronic equipment. PRESS RELEASES - Press release - Environment; 2011.
[13] Atasu A, Özdemir Ö, Van Wassenhove LN. Stakeholder perspectives on E-waste take-back legislation. Production and Operations Management 2013;22:382–96.
[14] Gottberg A, Morris J, Pollard S, Mark-Herbert C, Cook M. Producer responsibility, waste minimisation and the WEEE Directive: case studies in eco-design from the European lighting sector. The Science of the Total Environment 2006;359:38–56.
[15] Eurostat. Waste electrical and electronic equipment (WEEE). 2018.
[16] Dalrymple I, Wright N, Kellner R, Bains N, Geraghty K, Goosey M, Lightfoot L. An integrated approach to electronic waste (WEEE) recycling. Circuit World 2007;33: 52–8.
[17] Huisman J. Waste electrical and electronic equipment (WEEE) review. 2008. p. 1–11.
[18] Widmer R, Oswald-Krapf H, Sinha-Khetriwal D, Schnellmann M, Böni H. Global perspectives on e-waste. Environ Impact Assess Rev 2005;25:436–58.
[19] Schlummer M, Gruber L, Mäurer A, Wolz G, van Eldik R. Characterisation of polymer fractions from waste electrical and electronic equipment (WEEE) and implications for waste management. Chemosphere 2007;67: 1866–76.
[20] Taurino R, Pozzi P, Zanasi T. Facile characterization of polymer fractions from waste electrical and electronic equipment (WEEE) for mechanical recycling. Wastes Management 2010;30:2601–7.
[21] Cui J, Forssberg E. Mechanical recycling of waste electric and electronic equipment: a review. Journal of Hazardous Materials 2003;99:243–63.
[22] Weil ED, Levchik S. A review of current flame retardant systems for epoxy resins. Journal of Fire Sciences 2004; 22:25–40.
[23] Menad N, Björkman B, Allain EG. Combustion of plastics contained in electric and electronic scrap. Resources, Conservation and Recycling 1998;24:65–85.
[24] Tsydenova O, Bengtsson M. Chemical hazards associated with treatment of waste electrical and electronic equipment. Wastes Management 2011;31:45–58.
[25] Yamane LH, de Moraes VT, Espinosa DCR, Tenório JAS. Recycling of WEEE: characterization of spent printed circuit boards from mobile phones and computers. Wastes Management 2011;31:2553–8.
[26] Kim BS, chun Lee J, Seo SP, Park YK, Sohn HY. A process for extracting precious metals from spent printed circuit boards and automobile catalysts. Journal of Occupational Medicine 2004;56:55–8.
[27] Evangelopoulos P, Kantarelis E, Yang W. Investigation of the thermal decomposition of printed circuit boards (PCBs) via thermogravimetric analysis (TGA) and analytical pyrolysis (Py-GC/MS). Journal of Analytical and Applied Pyrolysis 2015;115.
[28] Buekens A, Yang J. Recycling of WEEE plastics: a review. Journal of Material Cycles and Waste Management 2014;16:415–34.
[29] Cui J. Mechanical recycling of consumer electronic scrap. 2005. p. 156.
[30] Rankin WJ. Minerals, metals and sustainability – meeting future material needs. CSIRO Publishing; 2011.
[31] Achilias D, Andriotis L, Koutsidis I, Louka D, Nianias N, Siafaka P, Tsagkalias I, Tsintzou G. Recent advances in the chemical recycling of polymers (PP, PS, LDPE, HDPE, PVC, PC, Nylon, PMMA). Material Recycling – Trends and Perspectives 2012.
[32] Serva, Manual sorting of WEEE, [n.d.].
[33] Menad N, Guignot S, van Houwelingen JA. New characterisation method of electrical and electronic equipment wastes (WEEE). Wastes Management 2013;33: 706–13.

[34] Capel C. Waste sorting - a look at the separation and sorting techniques in today's European market. 2008.
[35] Franklin Miller Inc. Industrial shredders. 2018.
[36] Zhang S, Forssberg E. Mechanical separation-oriented characterization of electronic scrap. Resources, Conservation and Recycling 1997;21:247–69.
[37] Goosey M, Kellner R. Recycling technologies for the treatment of end of life printed circuit boards (PCBs). Circuit World 2003;29:33–7.
[38] Khaliq A, Rhamdhani M, Brooks G, Masood S. Metal extraction processes for electronic waste and existing industrial routes: a review and Australian perspective. Resources 2014;3:152–79.
[39] Bridgwater E, Anderson C. CA site WEEE capacity in the UK an assessment of the capacity of civic amenity sites in the United Kingdom to separately collect waste electrical and electronic equipment. 2003. p. 74.
[40] Luda MP. Recycling of printed circuit boards. In: Kumar S, editor. Integrated Waste Management. Rijeka: IntechOpen; 2011.
[41] Cui J, Zhang L. Metallurgical recovery of metals from electronic waste: a review. Journal of Hazardous Materials 2008;158:228–56.
[42] Quinet P, Proost J, Van Lierde A. Recovery of precious metals from electronic scrap by hydrometallurgical processing routes. Miner Metall Process 2005;22: 17–22.
[43] Veit HM, Diehl TR, Salami AP, Rodrigues JS, Bernardes AM, Tenório JAS. Utilization of magnetic and electrostatic separation in the recycling of printed circuit boards scrap. Wastes Management 2005;25: 67–74.
[44] Liu R, Shieh RS, Yeh RYL, Lin CH. The general utilization of scrapped PC board. Wastes Management 2009;29: 2842–5.
[45] Rocchetti L, Vegliò F, Kopacek B, Beolchini F. Environmental impact assessment of hydrometallurgical processes for metal recovery from WEEE residues using a portable prototype plant. Environmental Science and Technology 2013;47:1581–8.
[46] Park YJ, Fray DJ. Recovery of high purity precious metals from printed circuit boards. Journal of Hazardous Materials 2009;164:1152–8.
[47] Sheng PP, Etsell TH. Recovery of gold from computer circuit board scrap using aqua regia. Waste Management & Research 2007;25:380–3.
[48] Mecucci A, Scott K. Leaching and electrochemical recovery of copper, lead and tin from scrap printed circuit boards. Journal of Chemical Technology and Biotechnology 2002;77:449–57.
[49] Rombach E, Friedrich B. Recycling of rare metals. Handbook of Recycling. State-of-the-Art Practical Analytical Science 2014:125–50.
[50] Umicore. Exploring Umicore precious metals refining. 2005. p. 1–12.
[51] Boliden Group. One of the largest recycler of electronic material. 2018.
[52] Boliden Group, Rönnskär B. A world leader in recycling electronics. 2018 [n.d.].
[53] G. Recycling, Horne Smelter, [n.d.].
[54] Malkow T. Novel and innovative pyrolysis and gasification technologies for energy efficient and environmentally sound MSW disposal. Wastes Management 2004; 24:53–79.
[55] Ni M, Xiao H, Chi Y, Yan J, Buekens A, Jin Y, Lu S. Combustion and inorganic bromine emission of waste printed circuit boards in a high temperature furnace. Wastes Management 2012;32:568–74.
[56] Kikuchi R, Sato H, Matsukura Y, Yamamoto T. Semi-pilot scale test for production of hydrogen-rich fuel gas from different wastes by means of a gasification and smelting process with oxygen multi-blowing. Fuel Processing Technology 2005;86:1279–96.
[57] Ortuño N, Conesa JA, Moltó J, Font R. Pollutant emissions during pyrolysis and combustion of waste printed circuit boards, before and after metal removal. The Science of the Total Environment 2014; 499:27–35.
[58] Moltó J, Font R, Gálvez A, Conesa JA. Pyrolysis and combustion of electronic wastes. Journal of Analytical and Applied Pyrolysis 2009;84:68–78.
[59] Van Der Avert P, Weckhuysen BM. Low-temperature destruction of chlorinated hydrocarbons over lanthanide oxide based catalysts. Angewandte Chemie International Edition 2002;41:4730–2.
[60] Bonarowska M, Burda B, Juszczyk W, Pielaszek J, Kowalczyk Z, Karpinski Z. Hydrodechlorination of CCl2F2 (CFC-12) over Pd-Au/C catalysts. Applied Catalysis B: Environmental 2001;35:13–20.
[61] Feijen-Jeurissen MMR, Jorna JJ, Nieuwenhuys BE, Sinquin G, Petit C, Hindermann JP. Mechanism of catalytic destruction of 1,2-dichloroethane and trichloroethylene over γ-Al2O3and γ-Al2O3supported chromium and palladium catalysts. Catalysis Today 1999;54: 65–79.
[62] Nutt MO, Heck KN, Alvarez P, Wong MS. Improved Pd-on-Au bimetallic nanoparticle catalysts for aqueous-phase trichloroethene hydrodechlorination. Applied Catalysis B: Environmental 2006;69:115–25.
[63] Coute N, Ortego JD, Richardson JT, Twigg MV. Catalytic steam reforming of chlorocarbons: trichloroethane, trichloroethylene and perchloroethylene. Applied Catalysis B: Environmental 1998;19:175–87.
[64] Huisman J, Magalini F, Kuehr R, Maurer C, Ogilvie S, Poll J, Delgado C, Artim E, Szlezak J, Stevels A. 2008 review of directive 2002/96 on waste electrical and electronic equipment (WEEE) - final report. 2007. p. 1–347. Comm. by Eur. Comm. Contract No 07010401/2006/442493/ETU/G4.
[65] Yamawaki T. The gasification recycling technology of plastics WEEE containing brominated flame retardants. Fire and Materials 2003;27:315–9.
[66] Goodship V, Stevels ALN. Waste electrical and electronic equipment (WEEE) handbook. 2012.

[67] Ray R, Thorpe R. A comparison of gasification with pyrolysis for the recycling of plastic containing wastes. International Journal of Chemical Reactor Engineering 2007;5:1–14.
[68] Achilias DS, Roupakias C, Megalokonomos P, Lappas AA, Antonakou V. Chemical recycling of plastic wastes made from polyethylene (LDPE and HDPE) and polypropylene (PP). Journal of Hazardous Materials 2007;149:536–42.
[69] Alston SM, Arnold JC. Environmental impact of pyrolysis of mixed WEEE plastics part 2: life cycle assessment. Environmental Science and Technology 2011;45: 9386–92.
[70] Luda MP, Euringer N, Moratti U, Zanetti M. WEEE recycling: pyrolysis of fire retardant model polymers. Wastes Management 2005:203–8.
[71] Brebu M, Sakata Y. Novel debromination method for flame-retardant high impact polystyrene (HIPS-Br) by ammonia treatment. Green Chemistry 2006;8:984–7.
[72] Buekens A. Introduction to feedstock recycling of plastics. In: Scheirs J, Kaminsky W, editors. Feedstock recycling and pyrolysis of waste plastics: converting waste plastics into diesel and other fuels; 2006.
[73] Evangelopoulos P. Pyrolysis of waste electrical and electric equipment (WEEE) for energy production and material recovery. The Royal Institute of Technology; 2014.
[74] Riedewald F, Sousa-Gallagher M. Novel waste printed circuit board recycling process with molten salt. Methods (Orlando) 2015;2:100–6.
[75] Luda MP, Balabanovich AI, Zanetti M, Guaratto D. Thermal decomposition of fire retardant brominated epoxy resins cured with different nitrogen containing hardeners. Polymer Degradation and Stability 2007;92: 1088–100.
[76] Luda MP, Balabanovich AI, Zanetti M. Pyrolysis of fire retardant anhydride-cured epoxy resins. Journal of Analytical and Applied Pyrolysis 2010;88:39–52.
[77] Lin KH, Chiang HL. Liquid oil and residual characteristics of printed circuit board recycle by pyrolysis. Journal of Hazardous Materials 2014;271:258–65.
[78] Mankhand TR, Singh KK, Kumar Gupta S, Das S. Pyrolysis of printed circuit boards. International Journal of Metallurgical & Materials Engineering 2013; 1:102–7.
[79] Quan C, Li A, Gao N. Thermogravimetric analysis and kinetic study on large particles of printed circuit board wastes. Wastes Management 2009;29:2353–60.
[80] Evangelopoulos P, Kantarelis E, Yang W. Experimental investigation of the influence of reaction atmosphere on the pyrolysis of printed circuit boards. Applied Energy 2017;204.
[81] Evangelopoulos P, Arato S, Persson H, Kantarelis E, Yang W. Reduction of brominated flame retardants (BFRs) in plastics from waste electrical and electronic equipment (WEEE) by solvent extraction and the influence on their thermal decomposition. Wastes Management 18 June 2018 (in press, corrected proof).
[82] Hall WJ, Williams PT. Pyrolysis of brominated feedstock plastic in a fluidised bed reactor. Journal of Analytical and Applied Pyrolysis 2006;77:75–82.
[83] Hall WJ, Williams PT. Fast pyrolysis of halogenated plastics recovered from waste computers. Energy & Fuels 2006;20:1536–49.
[84] Bhaskar T, Matsui T, Uddin MA, Azhar M, Kaneko J, Muto A, Sakata Y. Effect of Sb2O3in brominated heating impact polystyrene (HIPS-Br) on thermal degradation and debromination by iron oxide carbon composite catalyst (Fe-C). Applied Catalysis B: Environmental 2003;43:229–41.
[85] Hall WJ, Williams PT. Analysis of products from the pyrolysis of plastics recovered from the commercial scale recycling of waste electrical and electronic equipment. Journal of Analytical and Applied Pyrolysis 2007;79: 375–86.
[86] Jie G, Ying-Shun L, Mai-Xi L. Product characterization of waste printed circuit board by pyrolysis. Journal of Analytical and Applied Pyrolysis 2008;83:185–9.
[87] Long L, Sun S, Zhong S, Dai W, Liu J, Song W. Using vacuum pyrolysis and mechanical processing for recycling waste printed circuit boards. Journal of Hazardous Materials 2010;177:626–32.
[88] Zhou Y, Wu W, Qiu K. Recovery of materials from waste printed circuit boards by vacuum pyrolysis and vacuum centrifugal separation. Wastes Management 2010;30: 2299–304.
[89] Zhu P, Chen Y, you Wang L, Zhou M. A new technology for recycling solder from waste printed circuit boards using ionic liquid. Waste Managment Research 2012;30: 1222–6.
[90] Ma E, Xu Z. Technological process and optimum design of organic materials vacuum pyrolysis and indium chlorinated separation from waste liquid crystal display panels. Journal of Hazardous Materials 2013;263:610–7.
[91] Andersson M, Knutson Wedel M, Forsgren C, Christéen J. Microwave assisted pyrolysis of residual fractions of waste electrical and electronics equipment. Minerals Engineering 2012;29:105–11.
[92] Sun J, Wang W, Liu Z, Ma C. Study of the transference rules for bromine in waste printed circuit boards during microwave-induced pyrolysis. Journal of the Air and Waste Management Association 2011;61:535–42.
[93] Global Watch Service, Great Britain Department of Trade and Industry. WEEE recovery: the european Story : report of a DTI global watch mission. Department of Trade and Industry; 2006.
[94] H. Hornung, A.; Koch, W.; Seifert, Haloclean and pydra - a dual staged pyrolysis plant for the recycling waste electronic and electrical equipment (WEEE), [n.d].
[95] Azapagic A, Emsley A, Hamerton I. Polymers: the environment and sustainable development. Wiley; 2003.

[96] Blazsó M, Czégény Z, Csoma C. Pyrolysis and debromination of flame retarded polymers of electronic scrap studied by analytical pyrolysis. Journal of Analytical and Applied Pyrolysis 2002;64:249–61.
[97] Yang X, Sun L, Xiang J, Hu S, Su S. Pyrolysis and dehalogenation of plastics from waste electrical and electronic equipment (WEEE): a review. Wastes Management 2013;33:462–73.
[98] mnim Altwaiq A, Wolf M, Van Eldik R. Extraction of brominated flame retardants from polymeric waste material using different solvents and supercritical carbon dioxide. Analytica Chimica Acta 2003;491:111–23.
[99] Boerriger H. Bromine recovery from the plastics fraction of waste of electrical and electronic equipment (WEEE) with stages gasification phase 2: production of bromine salt in staged gasification to determine technical feasibility of bromine recovery. 2001.
[100] Miskolczi N, Hall WJ, Angyal A, Bartha L, Williams PT. Production of oil with low organobromine content from the pyrolysis of flame retarded HIPS and ABS plastics. Journal of Analytical and Applied Pyrolysis 2008;83:115–23.
[101] Zhu HM, Jiang XG, Yan JH, Chi Y, Cen KF. TG-FTIR analysis of PVC thermal degradation and HCl removal. Journal of Analytical and Applied Pyrolysis 2008;82:1–9.
[102] Jung SH, Kim SJ, Kim JS. Thermal degradation of acrylonitrile-butadiene-styrene (ABS) containing flame retardants using a fluidized bed reactor: the effects of Ca-based additives on halogen removal. Fuel Process Technology 2012;96:265–70.
[103] Vasile C, Brebu MA, Karayildirim T, Yanik J, Darie H. Feedstock recycling from plastics and thermosets fractions of used computers. II. Pyrolysis oil upgrading. Fuel 2007;86:477–85.
[104] Grabda M, Oleszek-Kudlak S, Shibata E, Nakamura T. Influence of temperature and heating time on bromination of zinc oxide during thermal treatment with tetrabromobisphenol A. Environmental Science and Technology 2009;43:8936–41.
[105] Grabda M, Oleszek-Kudlak S, Shibata E, Nakamura T. Vaporization of zinc during thermal treatment of ZnO with tetrabromobisphenol A (TBBPA). Journal of Hazardous Materials 2011;187:473–9.
[106] Rzyman M, Grabda M, Oleszek-Kudlak S, Shibata E, Nakamura T. Studies on bromination and evaporation of antimony oxide during thermal treatment of tetrabromobisphenol A (TBBPA). Journal of Analytical and Applied Pyrolysis 2010;88:14–21.
[107] Bertelli G, Costa L, Fenza S, Marchetti E, Camino G, Locatelli R. Thermal behaviour of bromine-metal fire retardant systems. Polymer Degradation and Stability 1988;20:295–314.
[108] Terakado O, Ohhashi R, Hirasawa M. Thermal degradation study of tetrabromobisphenol A under the presence metal oxide: comparison of bromine fixation ability. Journal of Analytical and Applied Pyrolysis 2011;91:303–9.
[109] Terakado O, Ohhashi R, Hirasawa M. Bromine fixation by metal oxide in pyrolysis of printed circuit board containing brominated flame retardant. Journal of Analytical and Applied Pyrolysis 2013;103:216–21.
[110] Terakado O, Kuzuhara S, Takagi H, Hirasawa M. Thermal decomposition of printed circuit board in the presence of zinc oxide under inert and oxidative atmosphere: emission behavior of inorganic brominated compounds. Engineering 2018;10:606–15.
[111] Kuzuhara S, Sano A. Bromination of Pd compounds during thermal decomposition of tetrabromobisphenol A. Engineering 2018;10:187–201.
[112] Oleszek S, Grabda M, Shibata E, Nakamura T. Fate of lead oxide during thermal treatment with tetrabromobisphenol A. Journal of Hazardous Materials 2013;261:163–71.
[113] Sivalingam G, Karthik R, Madras G. Effect of metal oxides on thermal degradation of Poly(vinyl acetate) and Poly(vinyl chloride) and their blends. Industrial & Engineering Chemistry Research 2003;42:3647–53.
[114] Panda AK, Singh RK, Mishra DK. Thermolysis of waste plastics to liquid fuel. A suitable method for plastic waste management and manufacture of value added products-A world prospective. Renewable & Sustainable Energy Reviews 2010;14:233–48.
[115] Hall WJ, Williams PT. Removal of organobromine compounds from the pyrolysis oils of flame retarded plastics using zeolite catalysts. Journal of Analytical and Applied Pyrolysis 2008;81:139–47.
[116] Antonakou EV, Kalogiannis KG, Stephanidis SD, Triantafyllidis KS, Lappas AA, Achilias DS. Pyrolysis and catalytic pyrolysis as a recycling method of waste CDs originating from polycarbonate and HIPS. Wastes Management 2014;34:2487–93.
[117] Hall WJ, Miskolczi N, Onwudili J, Williams PT. Thermal processing of toxic flame-retarded polymers using a waste fluidized catalytic cracker (FCC) catalyst. Energy & Fuels 2008;22:1691–7.
[118] Blazsó M, Czégény Z. Catalytic destruction of brominated aromatic compounds studied in a catalyst microbed coupled to gas chromatography/mass spectrometry. Journal of Chromatography A 2006;1130:91–6.
[119] Kwon T, Tsigdinos GA, Pinnavaia TJ. Pillaring of layered double hydroxides (LDH's) by polyoxometalate anions. Journal of the American Chemical Society 1988;110:3653–4.
[120] Morita N, Wajima T, Nakagome H. Reduction in content of bromine compounds in the product oil of pyrolysis using synthetic hydrotalcite. International Journal of Chemical Engineering 2015;6:262–6.
[121] Sakata Y, Bhaskar T, Uddin MA, Muto A, Matsui T. Development of a catalytic dehalogenation (Cl, Br) process for municipal waste plastic-derived oil. Journal of Material Cycles and Waste Management 2003;5:113–24.

[122] Brebu M, Bhaskar T, Murai K, Muto A, Sakata Y, Uddin MA. Thermal degradation of PE and PS mixed with ABS-Br and debromination of pyrolysis oil by Fe- and Ca-based catalysts. Polymer Degradation and Stability 2004;84:459–67.

[123] Brebu M, Bhaskar T, Murai K, Muto A, Sakata Y, Uddin MA. Removal of nitrogen, bromine, and chlorine from PP/PE/PS/PVC/ABS-Br pyrolysis liquid products using Fe- and Ca-based catalysts. Polymer Degradation and Stability 2005;87:225–30.

[124] Wu W, Xu J, Ohnishi R. Complete hydrodechlorination of chlorobenzene and its derivatives over supported nickel catalysts under liquid phase conditions. Applied Catalysis B: Environmental 2005;60:129–37.

[125] Hlaing ZZ, Wajima T, Uchiyama S, Nakagome H. "Reduction of Bromine Compounds in Oil Produced from Brominated Flame Retardant Plastics via Pyrolysis Using a Reflux Condenser,". *International Journal of Environmental Science and Development* 2014;5: 207–11.

CHAPTER 18

What do Recent Assessments Tell Us About the Potential and Challenges of Landfill Mining?

JOAKIM KROOK, PHD • NICLAS SVENSSON, PHD • STEVEN VAN PASSEL, PHD • KAREL VAN ACKER, PHD

INTRODUCTION

Landfill mining has been proclaimed as an alternative strategy to address the unwanted impacts of waste deposits [1,2]. In real-life projects, such excavation and processing of deposited waste have mainly been used to facilitate traditional objectives such as remediation, land reclamation, or creation of landfill airspace [3]. A key target of the recent landfill mining research, however, is to go beyond these local motives and enhance the recovery of materials and energy resources by employing more advanced processing technologies [4,5]. Although such an ambitious approach clearly displays a wider societal potential, it also adds complexity to the implementation and assessment of the pros and cons of landfill mining.

Recent reviews display multiple challenges for sound implementation of landfill mining in terms of a general need for further development of the know-how and the technology, as well as a better understanding of influencing policy and market conditions [3,5]. In essence, the field suffers from a deficit in knowledge about how resource recovery from landfills could be executed in a cost-efficient manner together with clear societal benefits. An important barrier for the development of the area is the so far often speculative nature of research using models with a wide range of real-life assumptions. Besides a few trials, often small-scale pilot trials, there is a general lack of empirical knowledge regarding real-life implications and consequences of landfill mining implementation [3]. However, the mere complexity of such resource recovery initiatives also contributes to this situation. For instance, the economic and environmental impacts of landfill mining are influenced by a large number of site-specific, technical, organizational, policy, and market factors and conditions [6]. Understanding how all these influencing elements interact and jointly contribute to the outcome is indeed a challenging endeavor.

For any emerging concept, systems analysis methods could be useful for dimensioning the potential, assessing the feasibility of specific solutions, and not the least for guiding knowledge and technology development toward critical factors for implementation [7,8]. When it comes to landfill mining, quite a few economic and environmental assessments have been conducted during the recent years, but there is not yet any systematic synthesis of this body of knowledge. In order to facilitate the further development of the area, it is thus useful to review the main findings of these studies, the consistency of their applied methodologies, and what specific challenges they display for the realization of such unconventional projects.

Aim and Scope of the Study

This study involves a review of recent economic and environmental assessments of landfill mining. By analyzing their main contributions and weaknesses, our aim is to guide and specify the needs for future knowledge and technology development within the area. The review is structured according to the following research questions:

- *What can we learn from the assessments about critical factors for the performance of landfill mining?*
- *How have the assessments been realized in terms of methodological choices and principles and what are the implications on their validity and usefulness?*
- *Which needs for knowledge and technology development can be deduced from the assessments and how could these challenges be addressed in future research?*

Sustainable Resource Recovery and Zero Waste Approaches. https://doi.org/10.1016/B978-0-444-64200-4.00018-9

The selected literature was extracted from the scientific databases Scopus and Web of Science and targets published, peer-reviewed journal articles in which the full value-chain of landfill mining has been quantitatively assessed in specific case studies. To avoid overlap with previous reviews within the field [3,5], only assessments published after 2013 were included. This restriction on the date of publication was thus applied to emphasize recent developments within the landfill mining area.

In the first step of the review, we focused on the empirical findings of the different case studies by simply summarizing their main results, with special emphasis on identifying reoccurring critical factors for economic and environmental performance. The second part of the review is more analytical. Here, we used state-of-the-art knowledge from environmental systems analysis theory to assess the strengths and weaknesses of the methodologies applied in the assessments. This benchmarking exercise was conducted to specify the validity and usefulness of the obtained results and to identify reoccurring empirical constraints and knowledge gaps throughout the landfill mining value chain. The methodological review was structured according to the key elements of any economic and environmental assessment, i.e., scenario development, procedures for handling parameter data, and selected modeling principles. Finally, our findings from both the empirical and methodological review of the assessments were used to identify key areas for future research on landfill mining.

CHARACTERISTICS AND MAIN RESULTS OF THE REVIEWED ASSESSMENTS

In total, 12 articles published during the period 2013–17 were selected for the review. Five of the articles focus on economic assessments [9–13], two focus on environmental impacts [14,15], and the rest involve studies addressing both topics [16–20], as seen in Table 18.1. Most of the studied landfills are municipal solid waste (MSW) deposits, which could be regarded as small- to medium-sized in a European context, i.e., containing 0.1–2 million metric tons of deposited mass. It is worth noting that one of the large-sized deposits, the REMO landfill in Belgium with about 16 million metric tons of waste, has been the case for several of the reviewed assessments [16,18,20]. Virtually all the studies include somewhat of a reference case displaying "business as usual" if the landfill in question is not to be mined. This reference serves as a benchmark to which the environmental impacts and economics of landfill mining are contrasted. Doing-nothing (implying no further actions or costs) or final closure and aftercare represent the two most common reference cases.

The lack of knowledge and experience from full-scale projects is apparent, meaning that most of the studies involve ex ante assessments of landfill mining scenarios developed from expert judgments or, less commonly, small-scale pilot trials. Only one of the studies is an ex post assessment of a realized, full-scale landfill mining project [10]. Regarding the material processing schemes, the assessments are quite evenly distributed between scenarios based on mobile separation units [9,10,13,15,19] and more advanced sorting plants [12,14,16–18,20]. The subsequent treatment and recovery of separated energy carriers and materials often follow state-of-the-art waste management practices in different countries. However, for waste-to-energy, some main differences can be found among the studies in terms of employed technologies (i.e., waste incineration and plasma gasification) and the organization of related costs and benefits.

Important Processes for Economic and Environmental Performance

The results from the economic assessments imply that the choice between project internal or external thermal treatment of extracted refuse-derived fuel (RDF) has overarching implications on cost and benefit profiles. In projects where such fuel is sent to external waste-to-energy plants (e.g., Refs. [9,13]), process-related (e.g., labor, energy, auxiliary materials, and maintenance of employed processing lines) and material-flow-related (e.g., gate fees, disposal costs and transports of extracted materials) cash flows are reported as the main expenditures. Also, in projects involving internal thermal treatment (e.g., Ref. [18]), these main processes remain significant but capital investments and operational expenditures related to the (new) waste-to-energy plant typically dominate the cost profile.

Regardless of the employed waste-to-energy approaches, the total project cost of the different cases is virtually always significantly higher than the anticipated revenues for separated materials and energy resources [12,13,17–20]. Only one of the cases, involving metal recovery from a waste incinerator ash deposit, concludes cost-effectiveness [10]. In addition, there are a few case studies indicating that landfill mining could be potentially profitable (e.g., [9,11,16]). However, such conclusions rely on assumptions about changed

TABLE 18.1
Main Characteristics and Results of the Reviewed Articles.

Publication & Topic	Type of Landfill, Size, & Location	Type of Assessment	Reference Case	LFM Processing Technology	Treatment of Energy Resources and Materials & Reclaimed Land/ Landfill Airspace	Main Costs for LFM Scenarios in Descending Order of Magnitude	Main Benefits for LFM Scenarios in Descending Order of Magnitude	Net Outcome
Frändegård et al. [14] *Environment*	MSW landfill, 1 Mt, Sweden	Ex ante	Solely remediation of landfill, no existing LFG collection	Different scenarios for mobile & advanced separation units	WtE: RDF recovery by incineration WtM: metals, aggregates, plastics (advanced separation) Relandfilling: fines, hazardous & mixed residue Land/airspace: N/A	Incineration of plastics	*Avoided:* heat, plastics, & metals	LFM better than remediation
Van Passel et al. [16] *Economic*	REMO MSW/IW landfill, 16 Mt, BEL	Ex ante (extrapolated from pilot trial)	Do nothing, existing LFG collection & utilization	Advanced separation plant	WtE: internal RDF recovery by gas plasma WtM: fines, aggregates, metals Relandfilling: N/A Land/airspace: land (but value not accounted for)	CAPEX/OPEX of WtE Material processing & separation Contingency (mixed) Excavation	Electricity sales Green energy certificates Mixed material sales	ELFM potentially profitable, given the nonconditions
Jain et al. [15] *Environment*	Typical MSW landfill, 1 Mt, USA	Ex ante (extrapolated from other full-scale LFM operations)	Do nothing, no existing LFG collection Relocate, LFG	Mobile separation unit	WtE: RDF recovery by incineration WtM: metals, soils Relandfilling: ash from incineration Land/airspace: N/A	Incineration	*Avoided:* Metals production, coal for electricity	LFM better than alternatives
Danthurebandara et al. [4] *Environment/ Economic*	MSW open dump site, 1 Mt, Sri Lanka	Ex ante	Do nothing, no existing LFG collection	Advanced separation plant	WtE: RDF to cement industry or to incineration WtM: fines, metals, aggregates, glass Relandfilling: hazardous & mixed residues	Env: (1) transportation, (2) transportation & incineration Eco: transportation	Env: *Avoided:* (1) coal combustion, (2) electricity, metal production Eco: (1) RDF selling price, (2) electricity revenues	Env: ELFM beneficial Eco: ELFM nonprofitable

Continued

TABLE 18.1
Main Characteristics and Results of the Reviewed Articles.—cont'd

Publication & Topic	Type of Landfill, Size, & Location	Type of Assessment	Reference Case	LFM Processing Technology	Treatment of Energy Resources and Materials & Reclaimed Land/ Landfill Airspace	Main Costs for LFM Scenarios in Descending Order of Magnitude	Main Benefits for LFM Scenarios in Descending Order of Magnitude	Net Outcome
Danthurebandara et al. [17] *Environment/ Economic*	REMO MSW/IW landfill, 16 Mt, Belgium	Ex ante (extrapolated from pilot trial)	Do nothing, existing LFG collection & utilization	Advanced separation plant	WtE: internal RDF recovery by gas plasma WtM: fines, aggregates, metals, glass Relandfilling: solid waste Land/airspace: land	Env: thermal treatment Eco: operation costs WtE, investment WtE	Env: Eco: electricity revenues, ELFM support, green energy fraction	Env: ELFM nonbeneficial Eco: not explicitly stated
Frändegård et al. [9] *Economic*	Typical MSW landfill, 0.1 Mt, Sweden	Ex ante	Solely remediation of landfill, no existing LFG collection	Mobile separation unit	WtE: external or internal RDF recovery by incineration WtM: metals, aggregates, fines Relandfilling: fines & mixed residues Land/airspace: land (small amount reclaimed)	Landfill tax for relandfill (if applicable) OPEX WtE (for internal RDF recovery) Material processing & separation	Heat/electricity sales (internal RDF recovery) Material sales (metals) Land reclamation	LFM cost-effective for solely remediation under certain rare conditions
Winterstetter et al. [20] *Environment/ Economic*	REMO MSW/IW landfill, 16 Mt, Belgium	Ex ante (extrapolated from pilot trial)	Landfill closure & aftercare, existing LFG collection & utilization	Advanced separation plant	WtE: internal RDF recovery by gas plasma/incineration WtM: metals, aggregates, fines Relandfilling: fines and mixed residues	Env: WtE emissions Eco: CAPEX/ OPEX of WtE, material processing and separation, excavation and storage	Env: *Avoided*: electricity generation, metals production, LFG emissions Eco: electricity sales, avoided	Env: net addition of climate emissions Eco: ELFM not profitable
Wagnor and Raymond, [10] *Economic*	Ecomaine ash landfill, 0.7 Mt, USA	Ex post (assessment of realized project)	Do nothing	Mobile separation unit	WtM: metals Relandfilling: processed ash Land/airspace: landfill airspace (small amount reclaimed)	Excavation & processing equipment Fuel, labor, & maintenance	Metals recovery New landfill airspace (due to metal recovery)	LFM profitable

Zhou et al. [11] *Economic*	Yingchun MSW landfill, 1.5 Mt, China	Ex ante	Only landfill aftercare	N/A (assumes 100% separation efficiency)	WtE: internal/external RDF recovery by incineration WtM: metals, soil-type fertilizer, aggregates, glass Relandfilling: mixed residues Land/airspace: land or landfill airspace	Excavation & hauling equipment Material processing & separation Transport	Land reclamation Heat/electricity sales (internal RDF recovery) Material sales (fertilizer)	LFM potentially profitable under nonconventional conditions
Wolfsberger et al. [16] *Economic*	Typical MSW landfill, 0.7 Mt, Austria	Ex ante	Do nothing	Mobile separation unit	WtE: external RDF recovery by incineration WtM: metals, aggregates Relandfilling: fines & mixed residues Land/airspace: not accounted for	Gate fees & disposal (including transport) Material processing and separation	Material sales (metals)	LFM not profitable
Hermann et al. [19] *Economic*	Styria MSW landfill, 0.7 Mt, Austria	Ex ante	Landfill closure & aftercare	Mobile separation unit	WtE: external RDF recovery by incineration WtM: metals Relandfilling: fines & mixed residues Land/airspace: land	Gate fees & disposal Material processing & separation	Land reclamation Material sales (metals)	LFM not profitable
Kieckhafer et al. [12] *Economic*	Pohlsche Heide MSW landfill, 2.6 Mt, Germany	Ex ante (extrapolated from pilot trial)	Landfill closure & aftercare	Different scenarios for mobile & advanced separation units	WtE: external RDF recovery by incineration WtM: metals, aggregates, plastic (advanced separation) Relandfilling: fines & mixed residues Land/airspace: land or landfill airspace	Gate fees & disposal (including transport) Material processing & separation Capital investments	Land reclamation New landfill airspace Material sales	LFM not profitable

For the net results, only climate impacts are presented for the environmental assessments.
CAPEX, capital expenditure; *ELFM*, enhanced landfill mining; *IW*, industrial waste; *LFG*, landfill gas; *LFM*, landfill mining; *MSW*, municipal solid waste; *OPEX*, operational expenditure; *RDF*, refuse-derived fuel; *WtE*, waste-to-energy; *WtM*, waste-to-material.

policy and market environments going beyond current conditions (e.g., increasing raw material prices and governmental support).

In most of the case studies, a fundamental problem is that in current markets, only a small share of the processed resources will generate any significant income (i.e., metals) while the remaining will involve low revenues (e.g., aggregates) or, even worse, disposal costs (e.g., gate fees for RDF, disposal of fines, and mixed residues). This market situation is also the reason why some studies conclude that the employment of more advanced and thereby expensive material separation plants does not pay off financially (e.g., Ref. [12]). Indeed, approaches that involve internal thermal treatment of RDF extend the material-flow-related revenues beyond the sales of metals to also include income from generated heat or electricity. However, such increased revenue streams are typically followed and surpassed by a massive need for financial investments or increased operational costs. Several studies, therefore, conclude that the cost-effectiveness of internal thermal treatment approaches relies on governmental support in terms of, for instance, green energy certificates or investment subsidies [11,16–18,21].

Even though previous reviews of landfill mining stress the importance of indirect benefits in terms of reclaimed land, landfill airspace, or avoided closure and aftercare [3], the conditions for such benchmarking costs and revenues are often poorly described in the assessments or not even accounted for. When properly included, their importance range from insignificant to being the main economic drivers dependent on case-specific settings. Given the above-described markets for secondary resources, it is, however, clear that identifying landfills related to (extraordinarily) high indirect benefits (e.g., land or landfill void space values) or alternative costs (extensive landfill aftercare needs) is an essential step in developing cost-efficient landfill mining operations [11,16,18].

When it comes to environmental assessments, the only impact that has been comprehensively studied in the reviewed literature is climate impact. The reported results are somewhat opposing, where some studies conclude that landfill mining would lead to reduced climate impacts [14,15,17], whereas others have found that such projects would instead result in net contributions to global warming [18,20]. In essence, these reported variations of the climate impact rely on the extent to which landfill mining contributes to avoided emissions from long-term landfill gas generation and replaced energy generation and material production. This is in turn influenced by a wide range of interrelated parameters ranging from landfill-specific factors (e.g., material composition, landfill gas potential and treatment), over project settings (e.g., efficiency of selected material processing technologies), and system conditions (e.g., assumptions about avoided primary material and energy production) to modeling choices (e.g., applied landfill gas and carbon footprint models and procedures for accounting of biogenic emissions from thermal treatment processes). Many of these potentially important parameters and conditions are, however, not clearly described or elaborated on in the reviewed studies, making comparisons and valuations of the validity of the results difficult. Nevertheless, reduced landfill gas impacts and metal recovery often seem to constitute the key processes for avoided climate emissions, whereas waste-to-energy processes could be significant in terms of both added and avoided emissions, largely depending on the structure of the background energy system.

Applied Procedures for Assessing Critical Factors for Performance

A general feature of the reviewed cases is a lack of systematic and fine-grained assessments of what actually builds up the economic and environmental performance of landfill mining. Instead, the emphasis is on assessing the overall potential and feasibility of engaging in such projects, as seen in Fig. 18.1. This means that when it comes to critical factors for performance, the results merely highlight main process steps (e.g., material processing and separation, waste-to-material, and waste-to-energy), while contributions of underlying factors in terms of the numerous model parameters that build up each of these processes largely remain unknown, or at least implicit. The identification of critical factors is further complicated by a low level of transparency regarding project-specific conditions and different manners for how the subsequent processes of landfill mining and their constituents have been aggregated.

Admittedly, several studies include some sort of sensitivity analysis of a few, selected single model parameters. This is especially so when it comes to economics, and some of the most common types of parameters assessed are market prices for metals [9,19], generated electricity [18], and reclaimed land or landfill airspace [12]. These parameters are, however, not necessarily identified as critical in the studies, but the sensitivity assessments rather elaborate on at what realistic or unrealistic (future) revenue streams landfill mining could approach breakeven. Another type of even less frequently occurring sensitivity analysis aims to identify important performance drivers. Such

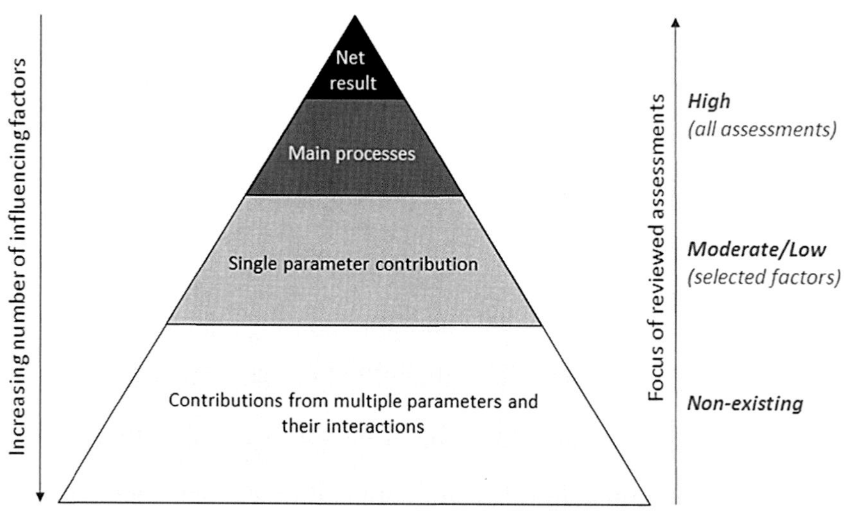

FIG. 18.1 Critical factors for the performance of landfill mining could be assessed on different levels, ranging from just identifying the main process steps as important to specifying the contributions of multiple and interrelated model parameters.

assessments often focus on the contribution from selected, single model parameters related to the efficiency of, e.g., material separation [9] and waste-to-energy processes [16–18].

Although such variations of one selected parameter (i.e., the so-called one-at-a-time sensitivity analysis) at a time within a certain value range could provide useful insights, they are also insufficient for developing a sound understanding of the economic and environmental principles of landfill mining [22]. Apart from being arbitrarily determined, the applied value ranges for the parameters are often very narrow—something that directly influences the result of the sensitivity analysis, i.e., the importance of a certain parameter. However, what is more important is that the performance of landfill mining is virtually always influenced by several hundred such model parameters. Many of these parameters are strongly interrelated throughout the landfill mining value chain, meaning that their real criticality depends on downstream and upstream parameter values and is thus a matter of such combination effects rather than the possible variation of a single model parameter (cf. [6]).

APPLIED METHODOLOGICAL CHOICES AND PRINCIPLES

Virtually all the reviewed studies constitute ex ante assessments, meaning that they aim to forecast economic or environmental outcomes of engaging in future landfill mining projects. Although forecasts are inherently uncertain, they are now common practice, not the least within the financial world. However, the fact that landfill mining is yet an unconventional practice makes these ex ante assessments particularly challenging [7,8]. In contrast to established practices, such emerging concepts suffer from lack of knowledge, practical experience, and records of accomplishment. This epistemological deficit has several potential methodological implications on the design, execution, analysis, and presentation of the obtained results. Here, we will discuss some examples of such challenges and how they have been dealt with in the reviewed assessments.

Procedures for Scenario Development

From systems analysis theory, we know that the scenario(s) under consideration constitute the foundation of any assessment. Apart from determining spatial and temporal boundaries, such "models" specify the included processes and how their interactions, together with system boundary conditions (e.g., policy and markets), affect relevant material, energy, economic, and environmental flows. In this section, we will focus on how the reviewed assessments have dealt with three fundamental parts of such scenarios along the landfill mining value chain: selection of landfills and reference cases (i.e., business as usual if landfill mining is not realized), choice of material processing and treatment schemes, and assumptions about resource quality and accessible markets for

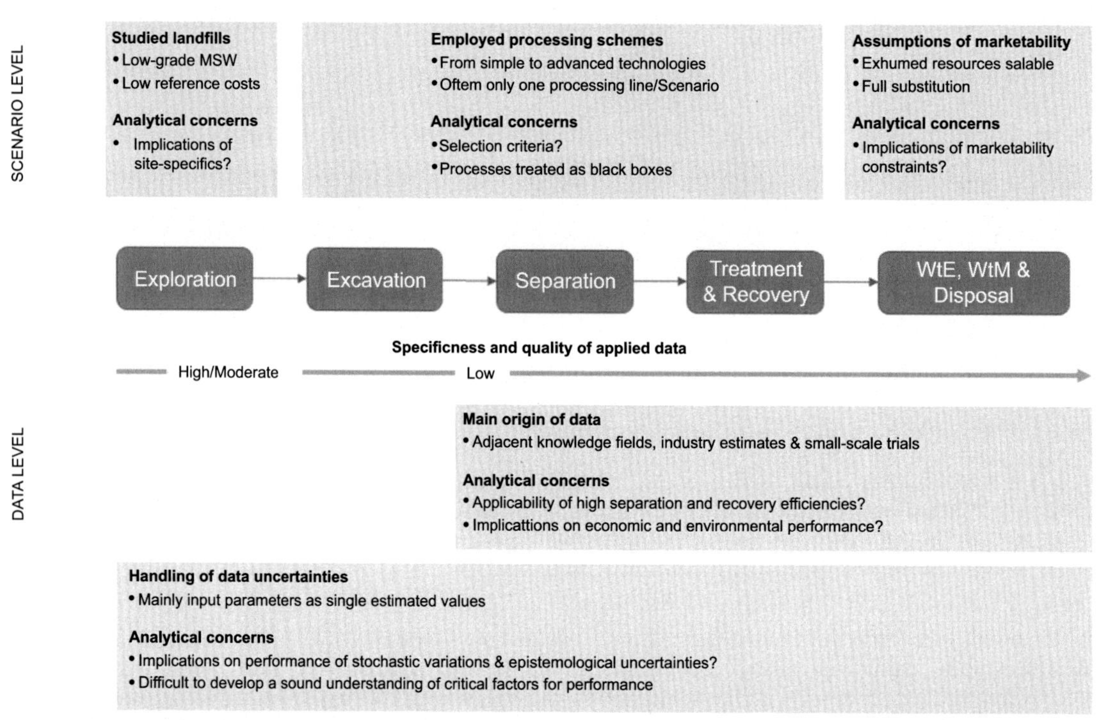

FIG. 18.2 Overview of methodological characteristics and implications of the reviewed assessments throughout the landfill mining value chain, divided into the scenario and data levels. *MSW*, municipal solid waste; *WtE*, waste-to-energy; *WtM*, waste-to-material.

extracted materials and energy resources, as shown in Fig. 18.2.

When it comes to the studied landfills, they seem to have been more or less "randomly picked" at least in terms of their resource potential and suitability for landfill mining. Such an understanding is further strengthened by the fact that none of the reviewed studies elaborate on if the deposit in question is to be considered as a particularly good, bad, or moderate case for landfill mining. Typically, the studied landfills seem to involve heterogenic and low-grade MSW deposits, and the related reference cases imply low-to-moderate landfill management costs and impacts (see also Table 18.1). The absence of explicit discussions about the importance of such site-specific settings makes it hard to fully understand the implications of the results from the different case studies. This is especially so given that landfill material compositions, costs and environmental impacts of landfill management options, values of reclaimed land or landfill airspace, and so on could vary widely between different cases and regions.

Although most landfill mining scenarios seem to involve similar main process steps, significant differences can be found in the type and advancement of employed technologies for material separation. Here, the cases display a quite wide spectrum of equipment ranging from simple or standard mobile units to highly advanced processing plants. There are, however, several issues related to how this very central part of landfill mining scenarios is dealt with in the conducted assessments. To start with, most of the studies only assess one landfill mining scenario involving a specific material separation scheme, and they often do so without providing any clear criteria for why this particular processing line was selected. How was the separation scheme developed, based on what knowledge was the scheme developed, what is the empirical evidence and specifications of requirements, and are there any alternatives that potentially could be more suitable or cost-effective in this particular case? Instead, selected separation schemes are more or less taken for granted and only dealt with on a very general level, often by referring to industry expertise/estimates or some previous, small-scale trials. This lack of description of the constituents, capabilities, and limitations of selected processing lines makes it difficult to evaluate their actual validity and suitability for landfill mining.

A similar situation exists for the selection of thermal treatment methods, where especially the possibilities and limitations of using emerging technologies, such as plasma gasification for the treatment of RDF retrieved from landfills, are poorly described. In some studies, this general lack of transparency regarding the selected material separation and treatment schemes is explained by confidentiality reasons, but equally important is probably the low level of knowledge and practical experience of processing previously deposited materials. The employed separation and treatment schemes could therefore merely be considered as conceptual and are largely reliant on experiences from adjacent fields of knowledge.

It is a well-known fact that the beneficial implementation of landfill mining relies on the possibility that the excavated masses can be separated out and then recovered, or at least disposed of, in some affordable and environmentally just way. Material outputs and their anticipated marketability, therefore, constitute a key issue of any landfill mining scenario. Still, only a few studies discuss the important issue of resource quality, whereas the rest seem to assume that the exhumed resources such as waste fuel, aggregates, and various metals will be readily accepted by the existing material and energy markets. Such an assumption is by no means straightforward and if not true, this will have significant implications on the economic (e.g., revenues, disposal costs, amount of reclaimed land or landfill airspace) and environmental (e.g., impacts of disposal, type and extent of avoided primary production) outcomes.

Empirical Support and Handling of Parameter Data

In order to specify the material and energy flows and related economic and environmental implications of the developed landfill mining scenarios, different types of data need to be collected and introduced into the assessment model in terms of numerous input parameters. In the reviewed assessments, there is a general tendency that both the data quality and its relation to the specific project gradually decreases along the landfill mining value chain. The used landfill material composition, for instance, often originates from specific data obtained from logbooks or waste sampling and characterization of the deposit in question. However, for all subsequent processes in terms of excavation, separation, and treatment and recovery, the applied data primarily originates from adjacent knowledge fields. Typically, experiences from the processing, treatment, disposal, and recovery of other conventional types of wastes, such as fresh MSW, serve as a foundation for specifying material and energy flows as well as related economic costs and revenues. When it comes to environmental assessments, for instance, the environmental impact parameters are to a high degree based on generic processes from Life Cycle Inventory (LCI) databases—processes that often have little to do with the specific landfill mining project. For any emerging concept, such a practice is more or less unavoidable. However, the main concern is the lack of discussions about to what extent such empirical data or industry estimates from adjacent fields are applicable to the specific processes of landfill mining and how this affects the validity of the study results.

To take another example, several studies use extremely high efficiencies of employed separation technologies, which in practice means that 80%–90% of the different deposited materials and energy resources in the landfills are assumed extractable. Such assumptions are often combined with quite optimistic estimates regarding the obtained resource quality, economic value, and potential for environmental savings due to the substitution of primary production. Although they may be applicable to the processing and recovery of some conventional wastes, there is an obvious risk that this transfer of (unconditioned) knowledge from adjacent fields might result in overestimating both the economic and environmental potential of landfill mining. Admittedly, there are also a few exceptional studies using more conservative data regarding these aspects, thereby acknowledging the often inevitable degradation of materials situated in landfills and the practical difficulties of processing previously deposited masses. These studies are commonly related to specific pilot trials and material characterization efforts.

The concern of using data from adjacent knowledge fields is further strengthened by the fact that the input data parameters are typically represented as one single, estimated value. This practice of neglecting the likely event of parameter variation constitutes an important weakness of virtually all the reviewed studies, which could be used as an argument to undermine the validity of their results. Even in well-established practices, stochastic variations of such parameters often occur due to, for instance, the changing conditions in various processes and markets. Just this kind of "naturally" occurring variations could have significant impacts on plausible net outcomes of a process, investment, or industrial project. However, when it comes to landfill mining, epistemological uncertainties also come into play, meaning that the possible ranges for parameter variation are much larger because of an incomplete or inaccurate scientific understanding of the studied processes [23]. If fully taken into account, such stochastic and epistemological parameter distributions would, in many cases, have major implications on what actually could be said about the net economic and environmental outcomes of landfill mining, as shown in Fig. 18.3. Typically, the result would then not constitute a single (and implicitly highly uncertain) value but would rather display a wide range of plausible outcomes—ranges that for a specific scenario even could include both positive and negative net results.

Apart from influencing the validity of the obtained results, this ignorance of both scenario and parameter uncertainties has several implications for the further development of the landfill mining area as well as of specific projects (cf. [7]). Constraining an assessment to only one scenario means that alternatives and potentially more promising paths for realizing landfill mining might be overlooked. Given the early stage of development, there are presumably plenty of options for the development of such projects, ranging from different landfills and site-specific factors, processing lines, and technologies to policy and market environments [3]. Furthermore, the current practice of using only single input parameter values in the assessments disqualifies the use of more advanced approaches to model uncertainties [22]. This is the main reason why most of the conducted assessments cannot provide any systematic and detailed information on the critical factors for performance.

Modeling Principles for Economic and Environmental Impacts

Although some of the economic assessments focus on direct costs and benefits [9,10], the most common metric used to study the profitability of landfill mining is the net present value (NPV) [11–13,16–20]. Such

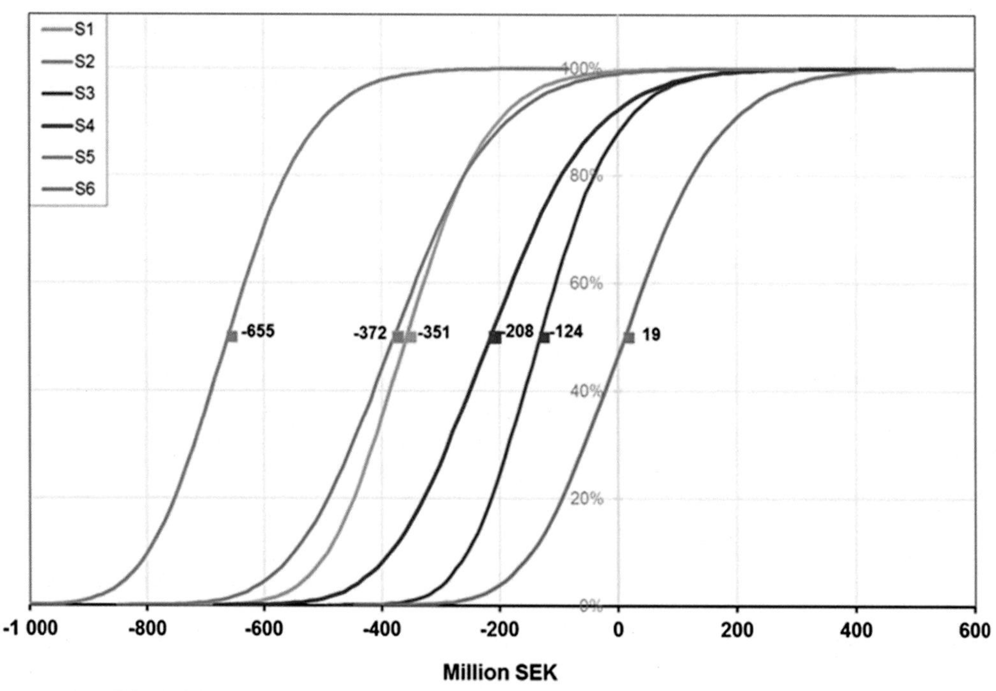

FIG. 18.3 Principal illustration of the cumulative probability distribution (*Y*-axis) of the net economic outcome (*X*-axis) for different landfill mining scenarios (S1–S6) when stochastic and epistemological parameter uncertainties are accounted for. The squares represent the expected net result for each scenario, comparable to the generated single value result obtained in most of the reviewed assessments, whereas the lines display the full range of plausible outcomes due to variations in input parameters. *SEK*, Swedish Kronor. (Derived from Refs. [23,25].)

capital budgeting is frequently used in many lines of business to evaluate whether or not it is worthwhile to invest in a project. In principle, the NPV indicates the potential profit of a landfill mining project by subtracting the investment costs from the total sum of cash flows during a predetermined time. In contrast to direct costs and benefits, the NPV is sensitive to when different cash flows occur and the time value of money is accounted for by a discount rate incorporating the inflation, investment risks, and rates of return into the assessment [24]. From a practical point of view, this means that the present value of money is lower the farther we go in the future and the higher the selected discount rate is. For long-term projects such as landfill mining, this way of budgeting is somewhat challenging, given that large up-front investments often need to be done while many of the revenue cash flows will come in small portions distributed over the years (e.g., electricity and material sales), or only materialize in a distant future (e.g., avoided landfill aftercare and reclaimed land). There are, however, different private and public views on to what extent the value of money should be downplayed by time, and among the reviewed assessments, the applied discount rates range within the interval of 0%–15%, something that obviously could have significant effects on the obtained results.

Virtually all the economic assessments are performed from the perspective of one single (private) actor in terms of the landfill owner or project manager. However, several studies also underline the key challenge that many of the proclaimed benefits of landfill mining currently only materialize on the societal level. A few assessments, therefore, include some environmental economics principles, for instance, by inserting a monetary value on avoided climate impacts (inverted carbon tax) and renewable energy (green energy certificates) [11,16–18,21]. However, when it comes to broader socioeconomic values such as nature restoration, employment, local health, and material autonomy issues, our knowledge is still largely limited. Admittedly, some studies discuss or even semiquantify a few such potential societal benefits, but real-life landfill mining projects would presumably generate several socioeconomic consequences, in terms of both benefits and costs.

Just as for the economic modeling, the environmental evaluations rely on assumptions and simplifications in order to be able to assess the complex systems at hand. In an ideal life cycle assessment, all relevant environmental impacts arising from the emissions and resource extractions from the processes within the studied system should be evaluated. Several of the articles study multiple environmental impact categories, but there is very little discussion to clarify the selection of categories and to explain the consequences of these choices [17,18,25]. Furthermore, for systems that are mainly based on ex ante modeling, it is hard to get specific data that covers a multitude of impacts. This means that for the studies that present highly local environmental impacts such as toxicity (e.g., Ref. [18]), generic data from commercial LCI databases is used with little to no connection to the actual site-specific toxicity problems. Most of the time, a discussion on how the results relate to the studied scenarios is lacking.

All the studies show the impact on climate change for the studied scenarios. These impacts can mainly be attributed to various energy processes throughout the whole chain of landfill mining activities. Normally, these emissions are divided into climate gases of fossil or biogenic origin. The biogenic emissions have historically often been regarded as carbon neutral when biomass has been used as some kind of biofuel. This neutrality has been put into question on the account that added focus should be on changes in carbon stocks than just on sequestration and release of carbon dioxide [26]. In the case of enhanced landfill mining, this means that biomass-based waste in the landfills could be regarded as a carbon sink, and when excavated and then used in a waste-to-energy facility, the ensuing emissions should be taken into account and be assessed alongside the waste of fossil origin. Only one of the environmental assessments actually includes these emissions, although one other article mentions the concept but does not include it [14,20]. If the modeling principle of counting biogenic emissions from waste-to-energy practices is included, the climate change impact will be substantially influenced.

CONCLUDING DISCUSSION ON KEY AREAS FOR FUTURE RESEARCH ON LANDFILL MINING

This review of recent landfill mining assessments reveals different types of challenges for the further development of the area. Here, we will first focus on some methodological challenges related to the emerging character of landfill mining and, in particular, the role of systems analysis in such early-stage research. We will then discuss the landfill mining value chain to specify key research questions and topics for addressing identified limitations and gaps in knowledge. Finally, we will put the contributions of conducted assessment into a societal perspective and discuss their implications on policy and market conditions and interventions.

Systems Analysis as a Learning Tool for Guiding Future Research

Although conducted assessments have provided valuable knowledge on the potential of landfill mining, the usefulness of applying systems analysis in such an early phase of conceptual development has only been partially explored. In essence, there is a need for more explorative studies focusing on learning and guiding research toward key challenges and potentials rather than assessments aiming to obtain an accurate result for a specific scenario [8]. Given the current deficits in knowledge and real-life experiences, such deterministic attempts to forecast the net outcome of a planned project will most likely be wrong anyway, displaying a high risk to either overestimating or discriminating the potential of different solutions in a too early stage of development [7]. In addition, these deterministic assessments offer limited guidance on how landfill mining projects could be further developed toward cost-effectiveness and improved environmental performance.

Such an understanding has several implications for the design and execution of future assessments on landfill mining. To start with, the usefulness of explorative approaches relies on the full recognition of empirical constraints and knowledge gaps occurring on the scenario and parameter levels [8]. From a learning perspective, such uncertainties need to be made explicit, brought to the forefront of the analysis, and allowed to expand in order to facilitate identification of critical challenges and alternative paths of development. When it comes to landfill mining, this means that assessment of several scenarios is encouraged to better understand implications of different site-specific settings, choices of processing lines and technologies, and policy and market conditions [6,20,23]. Even in specific projects, such an openness to different alternatives and conditions could be useful, given the often early stage of development. The fact that many landfill mining projects will be implemented first in the future also implies a need to consider temporal dynamics, e.g., effects of technology learning, project upscaling, and changes in policy and market environments [7].

Another important feature of explorative approaches is the use of advanced approaches to model uncertainties such as global sensitivity analysis and regression analysis [6]. From a validity perspective, this is fundamental because it makes it possible to accurately display and internalize the level of uncertainty into the results [20,22,23]. The use of advanced sensitivity methods also enables systematic and fine-grained assessments of what actually builds up the performance of landfill mining, specifying contributions of multiple model parameters as well as their combined effects [6]. Such detailed knowledge on the interrelations of factors occurring on the site, project, and system levels is a necessity for identifying which specific conditions and settings are most critical for performance and thus where to focus to learn more and decrease the epistemological deficits, which is something that previous assessments have failed to do.

Apart from guiding future landfill mining research toward the most critical challenges for beneficial implementation, explorative assessments can also contribute to several hands-on applications. Establishing a systemic and detailed understanding of what settings that build up profitability and environmental performance in different situations is, for instance, essential for developing landfill mining prospecting methods, project setup blueprints, and specifications of requirements for material processing and recovery schemes [6]. However, the employment of such explorative studies relies on the provision and collection of knowledge and data displaying different options for realizing landfill mining as well as possible ranges of variation in the efficiency and performance of corresponding processes [7,8]. This leads us to the next section, in which we discuss research needs for learning more about the constituents of the landfill mining value chain.

Applied Research to Open Up the Black Boxes of the Landfill Mining Value Chain

Our review shows a massive need for research addressing various limitations and gaps in knowledge throughout the landfill mining value chain. In the very beginning of this chain, the fundamental question of which site and local settings that constitute landfills suitable for mining yet remains unresolved. Conducted assessments more or less miss this issue owing to their case-specific nature, where only the feasibility of mining one single deposit is considered (see Table 18.1). Often, these landfills do not appear as particularly suitable for mining too because of their low grades in valuable resources such as metals and low to moderate alternative cost for management and aftercare of the deposits. The presumably high importance of such local settings [3,6,12] calls for prospecting methods that enable a more strategic selection of landfills for mining. In many regions, such applied research could depart from already existing surveys of landfills containing some general characteristics of the deposits, e.g., age, size, types of landfilled waste, aftercare or remediation needs, and landfill infrastructure in place. A few previous studies on this topic (e.g., Ref. [27]) indicate that such information is accessible but that it needs to be adjusted as well as complemented with other local conditions of relevance for landfill mining, e.g., the actual and future use of the location, proximity to other landfills and treatment and recovery plants, surrounding land values, and needs for landfill void space in the region.

However, in order to transform such landfill mapping efforts into functional prospecting methods, several knowledge-related challenges still need to be addressed. To start with, there is a general lack of in-depth information about the material composition and characteristics of landfills in most regions [27,28]. Typically, such detailed knowledge only exists for a few particularly studied landfills, whereas for most other deposits, it has to be estimated based on logbooks or extrapolations from records of historical waste streams. Understanding the implications of all these site and local settings also relies on the development of a systemic and detailed knowledge on how they jointly contribute to the performance of landfill mining in different situations and settings [6].

In the processing line of landfill mining, the research challenges circulate around the question of which material and energy resources that actually can be separated out and at what quality levels [3]. Despite the fact that quite a few studies have dealt with this topic over the years (e.g., [29–31]), our knowledge about the specific challenges of processing deposited waste and the efficiency and capacity of the different technological schemes for doing so are still limited. The fact that most of the reviewed feasibility assessments instead employ experiences and data from the separation and recovery of fresh waste further demonstrates this lack of scientific understanding.

In order to develop a sound understanding about what is technically and economically feasible to separate out and recover in a landfill mining context, there are no real alternatives but to go from the often-seen laboratory studies and small-scale trials to practice [3,31]. More specifically, and in contrast to previous studies on this topic, there is a need for well-planned pilot projects in which the resource use, separation efficiencies, and material flows of different processing lines are assessed and carefully documented in a scale comparable to real-life projects. This is needed to facilitate an in-depth learning about how different process parameters and related uncertainties influence the efficiency and performance of technologies in different situations and settings. Furthermore, we also suggest that the realization of such research should be guided by a drive for the continual development of the processing schemes where repeated trials, modifications, and evaluations are used to gradually improve their efficiency. This is required to facilitate continuous learning processes and overcome limitations of ad hoc trials involving once-in-a-lifetime tests of some readily available equipment not adjusted for the purpose [3,28].

When it comes to the development of suitable processing lines for landfill mining, the area would also benefit from a strategic approach in terms of clear goals on what materials and energy resources to actually produce, and which resource quality levels to aim for. In some situations and settings, it might be justified to use simple mobile separation equipment, whereas in others, more advanced processing schemes could be beneficial [3]. In order to account for such implications, developers of processing schemes cannot act in isolation but need to consider various interrelations with both upstream (e.g., material composition and characteristics of the deposited waste) and downstream processes (e.g., accessible markets for extracted commodities). Here, systems analysis methods could play an important role by contributing with such specific guidance facilitating the selection and further development of the processing schemes and technologies [6].

When approaching the very end of the landfill mining value chain, the issue of resource quality and marketability of extracted materials and energy resources deserves some special attention. Apart from a few sporadic studies contrasting the properties of some extracted materials with available legislative and end-user requirements (e.g., Ref. [32]), this topic has not yet been thoroughly dealt with, but exhumed resources are rather just assumed to be of high quality and salable on existing markets. There are, however, some ongoing initiatives, e.g., the EU-funded NEW-MINE project, in which process development including resource quality and upgrading technologies is in focus. Still, the fundamental importance of resource quality in process development and the need for detailed characterization of the generated outputs from such schemes should be more strongly emphasized in the research. The reason for this is its presumably large impact on both the economic feasibility and environmental motives for realizing such projects. Worth noting here is also that the marketability of the extracted resources is not just a matter of their intrinsic quality [3]. Knowledge of the existing market structures, competition, policy measures, and supply and demand dynamics is equally important for facilitating sound estimates on viable recovery routes. For instance, it is largely unclear if recyclers and incinerators are willing or even have the capacity to accept supplementary resources extracted from landfills, which by the way also will compete with more high-quality materials obtained from source separation programs.

Common Knowledge and Broader Scope to Support Policy and Market Interventions

The review also points to some inconsistencies when it comes to applied modeling principles in the economic and environmental assessments. Although somewhat theoretic, such methodological aspects are of the utmost importance for the further development of the area, given that certain modeling choices (e.g., selected discount rates, investment models, procedures for accounting for biogenic climate emissions) could have a significant impact on the conditions and motives for such projects [14,20]. In order to facilitate comparisons, synthesis, and common knowledge building within the area, it might, therefore, be fruitful to discuss and streamline what models, boundaries, and metrics to use when assessing the performance of landfill mining. This also raises the policy-relevant and essential question of how to evaluate the consequences of this type of unconventional practices. Several of the reviewed studies, for instance, stress the need for new or adjusted policy measures to increase economic interests to engage in landfill mining [11,16–18,21]. However, such claims about policy support yet seem a bit far-fetched, given that our knowledge about the environmental and societal impacts are still largely limited. In order to better support policy-making, future research must, therefore, address the challenges related to the application of broader assessment frameworks, including global to local environmental impacts as well as socioeconomic consequences [16,33]. Again, such a research endeavor should initially be guided by a will to learn and better understand the various cause-effect relationships rather than attempt to obtain an accurate result, simply because we are not yet in the position of being able to provide such findings and claims.

REFERENCES

[1] Dickinson W. Landfill mining comes of age, vol. 9; 1995.
[2] Hogland W. Remediation of an old landsfill site: soil analysis, leachate quality and gas production. Environmental Science and Pollution Research International 2002;(1):49–54.
[3] Krook J, Svensson N, Eklund M. Landfill mining: a critical review of two decades of research. Waste Management 2012;32:513–20.
[4] Danthurebandara M, Van Passel S, Machiels L, Van Acker K. Valorization of thermal treatment residues in enhanced landfill mining: environmental and economic evaluation. Journal of Cleaner Production 2015;99: 275–85.
[5] Tom Jones P, Geysen D, Tielemans Y, van Passel S, Pontikes Y, Blanpain B, Quaghebeur M, Hoekstra N. Enhanced landfill mining in view of multiple resource recovery: a critical review. Journal of Cleaner Production 2013;55:45–55.
[6] Laner D, Cencic O, Svensson N, Krook J. Quantitative analysis of critical factors for the climate impact of landfill mining. Environmental Science and Technology 2016;50: 6882–91.
[7] Fleischer T, Decker M, Fiedeler U. Assessing emerging technologies—methodological challenges and the case of nanotechnologies. Technological Forecasting and Social Change 2005;72:1112–21.
[8] Villares M, Işıldar A, van der Giesen C, Guinée J. Does ex ante application enhance the usefulness of LCA? A case study on an emerging technology for metal recovery from e-waste. International Journal of Life Cycle Assessment 2017;22:1618–33.
[9] Frändegård P, Krook J, Svensson N. Integrating remediation and resource recovery: on the economic conditions of landfill mining. Waste Management 2015;42:137–47.
[10] Wagner TP, Raymond T. Landfill mining: case study of a successful metals recovery project. Waste Management 2015;45:448–57.
[11] Zhou C, Gong Z, Hu J, Cao A, Liang H. A cost-benefit analysis of landfill mining and material recycling in China. Waste Management 2015;35:191–8.
[12] Kieckhäfer K, Breitenstein A, Spengler TS. Material flow-based economic assessment of landfill mining processes. Waste Management 2017;60:748–64.
[13] Wolfsberger T, Pinkel M, Polansek S, Sarc R, Hermann R, Pomberger R. Landfill mining: development of a cost simulation model. Waste Management and Research 2016;34:356–67.
[14] Frändegård P, Krook J, Svensson N, Eklund M. Resource and climate implications of landfill mining. Journal of Industrial Ecology 2013;17:742–55.
[15] Jain P, Powell JT, Smith JL, Townsend TG, Tolaymat T. Life-cycle inventory and impact evaluation of mining municipal solid waste landfills. Environmental Science and Technology 2014;48:2920–7.
[16] Van Passel S, Dubois M, Eyckmans J, de Gheldere S, Ang F, Tom Jones P, van Acker K. The economics of enhanced landfill mining: private and societal performance drivers. Journal of Cleaner Production 2013;55: 92–102.
[17] Danthurebandara M, Van Passel S, Van Acker K. Environmental and economic assessment of 'open waste dump'

mining in Sri Lanka. Resources, Conservation and Recycling 2015;102:67–79.
[18] Danthurebandara M, Van Passel S, Vanderreydt I, Van Acker K. Assessment of environmental and economic feasibility of enhanced landfill mining. Waste Management 2015;45:434–47.
[19] Hermann R, Baumgartner R, Vorbach S, Wolfsberger T, Ragossnig A, Pomberger R. Holistic assessment of a landfill mining pilot project in Austria: methodology and application. Waste Management and Research 2016;34: 646–57.
[20] Winterstetter A, Laner D, Rechberger H, Fellner J. Framework for the evaluation of anthropogenic resources: a landfill mining case study – resource or reserve? Resources, Conservation and Recycling 2015;96:19–30.
[21] Winterstetter A, Laner D, Rechberger H, Fellner J. Evaluation and classification of different types of anthropogenic resources: the cases of old landfills, obsolete computers and in-use wind turbines. 2016. https://doi.org/10.1016/j.jclepro.2016.05.083.
[22] Saltelli A, Annoni P, D'Hombres B. How to avoid a perfunctory sensitivity analysis. Procedia – Social and Behavioral Sciences 2010;2:7592–4.
[23] Frändegård P, Krook J, Svensson N, Eklund M. A novel approach for environmental evaluation of landfill mining. Journal of Cleaner Production 2013;55:24–34.
[24] Brealey R, Myers S, Allen F. Principles of corporate finance. Mcgraw-hill Education; 2016.
[25] Esguerra JL, Svensson N, Krook J, Van Passel S, Van Acker K. The economic and environmental performance of a landfill mining project from the viewpoint of an industrial landfill owner. In: Conference proceedings of the 4th international symposium on enhanced landfill mining, February 5–6, Mechelen, Belgium; 2018. p. 289–396.
[26] Johnson E. Goodbye to carbon neutral: getting biomass footprints right. Environmental Impact Assessment Review 2009;29:165–8.
[27] E. Wille, T. Behets & L. Umans. Mining the anthropocene in flanders: Part 1 – landfill mining. Conference proceedings of the 2nd International symposium on enhanced landfill mining, 14–16 October, Houthalen-Helchteren, Belgium, pp. 55-69.
[28] Johansson N, Krook J, Eklund M. Transforming dumps into gold mines – experiences from Swedish case studies. Environmental Innovations and Societal Transitions 2012;5:33–48.
[29] Chang S, Cramer R. The potential for reduction of landfill waste by recycling and mining of construction and demolition waste at the White Street Landfill, Greensboro, North Carolina. Journal of Solid Waste Technology and Management 2003;29:42–55.
[30] Rettenberger G. Results from a landfill mining demonstration project. In: Proceedings sardinia, Fifth International landfill symposium, Cagliari, Italy; 1995. p. 827–40.
[31] Kaartinen T, Sormunen K, Rintala J. Case study on sampling, processing and characterization of landfilled municipal solid waste in the view of landfill mining. Journal of Cleaner Production 2013;55:55–66.
[32] Johansson N, Krook J, Frändegård P. A new dawn for buried garbage? An investigation of the marketability of previously disposed shredder waste. Waste Management 2017;60:417–27.
[33] Burlakovs J, Kriipsalu M, Klavins M, Bhatnagar A, Vincevica-Gaile Z, Stenis J, Jani Y, Mykhaylenko V, Defanas G, Turkadze T, Hogland M, Rudovica V, Kaczala F, Rosendal R, Hogland W. Paradigms on landfill mining: from dump site scavenging to ecosystem services revitalization. Resources, Conservation and Recycling 2017;123:73–84.

Index

Note: Page numbers followed by "f" indicate figures, "t" indicate tables.

P

R

S

X

Y

Z